Water Science and Technology Library

Volume 114

The aim of the *Water Science and Technology Library* is to provide a forum for dissemination of the state-of-the-art of topics of current interest in the area of water science and technology. This is accomplished through publication of reference books and monographs, authored or edited. Occasionally also proceedings volumes are accepted for publication in the series. *Water Science and Technology Library* encompasses a wide range of topics dealing with science as well as socio-economic aspects of water, environment, and ecology. Both the water quantity and quality issues are relevant and are embraced by *Water Science and Technology Library*. The emphasis may be on either the scientific content, or techniques of solution, or both. There is increasing emphasis these days on processes and *Water Science and Technology Library* is committed to promoting this emphasis by publishing books emphasizing scientific discussions of physical, chemical, and/or biological aspects of water resources. Likewise, current or emerging solution techniques receive high priority. Interdisciplinary coverage is encouraged. Case studies contributing to our knowledge of water science and technology are also embraced by the series. Innovative ideas and novel techniques are of particular interest.

Comments or suggestions for future volumes are welcomed.

Vijay P. Singh, Department of Biological and Agricultural Engineering & Zachry Department of Civil and Environment Engineering, Texas A&M University, USA
Email: vsingh@tamu.edu

All contributions to an edited volume should undergo standard peer review to ensure high scientific quality, while monographs should also be reviewed by at least two experts in the field.

Manuscripts that have undergone successful review should then be prepared according to the Publisher's guidelines manuscripts: https://www.springer.com/gp/authors-editors/book-authors-editors/book-manuscript-guidelines

Ahmed Sefelnasr · Mohsen Sherif · Vijay P. Singh
Editors

Water Resources Management and Sustainability

Solutions for Arid Regions

Springer

Editors
Ahmed Sefelnasr
National Water and Energy Center
United Arab Emirates University
Al Ain, United Arab Emirates

Mohsen Sherif
National Water and Energy Center
United Arab Emirates University
Al Ain, United Arab Emirates

Vijay P. Singh
Department of Biological and Agricultural Engineering
Texas A&M University
College Station, TX, USA

ISSN 0921-092X ISSN 1872-4663 (electronic)
Water Science and Technology Library
ISBN 978-3-031-80519-6 ISBN 978-3-031-80520-2 (eBook)
https://doi.org/10.1007/978-3-031-80520-2

This Springer imprint is published by the registered company Springer Nature Switzerland AG
The registered company address is: Gewerbestrasse 11, 6330 Cham, Switzerland

Preface

Water is vital for all types of life. Without water, life cannot be sustained. The lack of freshwater resources in arid and semi-arid regions constitutes the major deterrent to sustainable development. The rapid population growth, rising standards of living and per capita water consumption, expansion in urbanization, industrial and agricultural activities, and climate change pose considerable challenges on the use and management of water resources. Those challenges define how government, industry, and the community engage with the use and management of water resources in the future. The issue of water resources scarcity and management is at the top of the research agenda of many countries around the globe and this is typically true for countries located in arid regions including UAE.

The problem of water scarcity is increasingly prevalent, and numerous nations have begun to endure protracted periods of drought. Concurrently with the substantial decline in the per capita share of freshwater, the quality of the water that is accessible in areas experiencing stress has deteriorated. However, water continues to be misapplied, wasted, and occasionally contaminated. In order to address the water scarcity issue in arid and semi-arid regions, enhanced water resource management and utilization are required. Additionally, non-traditional water resources and water treatment technologies should be given consideration. The development of low-energy desalination systems that have minimal negative environmental effects is required. In the interim, purified effluent must be utilized to its maximum capacity in order to alleviate the current strain on conventional resources, particularly groundwater.

Selected papers that were accepted, presented, and discussed during the International Conference on Water Resources Management and Sustainability: Solution for Arid Regions, held in Dubai, United Arab Emirates, from 26–28 February 2024, are included in this issue of the Water Science and Technology Library. The conference was organized by the National Water and Energy Center of UAE University.

The conference is designed to facilitate and encourage new perspectives on how science and innovative technologies can transform water management and sustainability in arid and semi-arid regions around the world. It addressed current challenges and priorities in water management and provided a forum to share knowledge, experiences, research, and new discoveries. It brought together international experts working in industry, academia, and government to share expertise, form partnerships, and discover solutions in the field of water resource management.

With the participation of large national, regional, and international water scientists, engineers, experts, academics, policy makers, and industry members, from more than 70 countries, this version of the conference served as one of the largest gatherings of water specialists in 2024. Covering both technical and policy-focused topics, the three-day conference program comprised of keynote presentations, oral presentations, and poster sessions showcasing recent innovations and research findings.

The papers comprising this book *Water Resources Management and Sustainability: Solutions for Arid Regions* are organized in a logical technical sequence and are divided into five parts that cover key aspects of water resources in arid regions.

Part I deals with rainfall and floods and their related attributes in the arid regions including flood mitigation and risk assessment. It also includes papers addressing hydrological topics, problems, and principles. **Part II** addresses the droughts and their analysis, assessment, and impacts. In **Part III**, groundwater as a significant constituent of the water budget in arid regions is elaborated. Threats, recharges, quality, and management options of this resource were also discussed in this section. The papers discuss rational decision making, challenges, the use of information technology, integrated management, modeling for groundwater resources management, and begin with a perspective on the management of this resource. **Part IV** deals with water quality. The articles in this part explore contemporary methods for evaluating water quality and implementing treatment protocols. **Part V** covers the subjects pertaining to the management of water resources. This subject has garnered considerable interest in several arid regions worldwide, particularly in the Gulf area. The articles under this area address laws pertaining to the preservation, recycling, and protection of resources. It is essential to optimize the development, management, and use of the scarce freshwater resources. The studies included in this section have also examined decision support systems for water resources planning, management, and water rights assessments and control.

The book will be of interest to researchers and practitioners in the field of hydrology, environmental engineering, agricultural engineering, earth sciences, and watershed and range sciences, as well as to those engaged in water resources planning, development, and management in arid and semi-arid areas. Graduate students and those wishing to conduct research in hydrology, environmental science, water engineering, as well as water resources planning, development, and management will find the book to be of value.

This book will appeal to researchers and professionals in the fields of hydrology, environmental engineering, agricultural engineering, earth sciences, and related fields. It will also be valuable to individuals involved in water resources planning, development, and management in arid and semi-arid regions. Moreover, this book will be valuable for graduate students and those interested in doing research in the fields of hydrology, environmental science and engineering, and water resources planning, development, and management.

Al Ain, United Arab Emirates — Ahmed Sefelnasr
Al Ain, United Arab Emirates — Mohsen Sherif
College Station, USA — Vijay P. Singh

Acknowledgements

The organizing committee would like to express its gratitude to the national and international academic, industry, and government experts, as well as individuals who participated in the conference. Sincere thanks are extended to the members of the scientific committee for their support in the review process of the conference program. In addition, appreciation is due to the UAE University and the personnel of the National Water and Energy Center. The conference attracted a large number of national and international well-known experts, from more than 70 countries, covering the five continents, who have been at the forefront of water resources management and sustainability in arid and semi-arid regions.

Contents

About the Editors

Prof. Ahmed Sefelnasr In addition to a Ph.D. in hydrogeology, Prof. Sefelnasr also possesses a B.Sc. and M.Sc. in geology and hydrogeology, respectively. He holds a degree in water resources management and groundwater modeling as his expertise in this field is acknowledged internationally. He is equipped with over 25 years of experience in research pertaining to groundwater systems. Prof. Sefelnasr authored numerous notable publications in the field of water resources. At Assiut University in Egypt, Prof. Sefelnasr holds the position of full professor of hydrogeology and water resources management. Presently employed as a senior researcher and expert on projects pertaining to water resources policies and management, transboundary aquifers, protection and development of groundwater resources, water cooperation and governance, artificial storage and recovery, numerical simulations, coastal aquifer, saltwater intrusion, water security, hydrology, and climate change, he is appointed in the National Water and Energy Center, United Arab Emirates University, UAE. Regarding groundwater research, consulting, and advisory work, Prof. Sefelnasr possesses vast international experience encompassing a vast array of academic interests and practical applications. He was engaged in professional work, research, and instruction in numerous countries and regions, primarily in Egypt, Germany, Japan, the UK, and the United Arab Emirates. Prof. Sefelnasr, who previously focused on coupled flow and solute transport processes, groundwater exploration, and water resources management, has developed a strong interest in groundwater resources at present. This includes groundwater governance, advocacy strategies, and sustainable management, with a particular emphasis on arid and semiarid regions.

Prof. Mohsen Sherif Fulbrighter and Ph.D., C.Eng., D.WRE, M. ASCE, Prof. Mohsen Sherif is presently Director of Research and Sponsored Projects at UAE University in the United Arab Emirates, where he also serves as Professor of Water Resources. In the areas of hydrology, hydraulics, fluid mechanics, numerical simulation, climate change, water resources management, artificial recharge and management of groundwater systems, and groundwater flow and contamination, he has accumulated over thirty-five years of teaching and research experience. His work has been featured in over 150 journals, conferences, and book chapters, and he has successfully concluded several extensive initiatives. Professor Sherif was honored with several national and international distinctions, including the Sheikh Khalifa Award for Education, Higher Education/Research, First Round, in March 2008, and two Fulbright Scholarships in 1993 and 1997, respectively. Additionally, the American Academy of Water Resources Engineers bestowed upon him the Designation of Diplomate, Water Resources Engineer. He is Associate Editor of both the *Journal of Hydrology* and the *Journal of Hydrologic Engineering* at present.

Prof. Vijay P. Singh, Ph.D., D.Sc., D.Eng. (Hon.), Ph.D. (Hon.), D.Sc. (Hon.), P.E., P.H., Honor. D.WRE, Dist. M.ASCE, Dist. Hon. M.IWRA, Dist. F.AGGS, Honor. Member AWRA, holds the inaugural Caroline and William N. Lehrer Distinguished Chair in Water Engineering at Texas A&M University's Department of Agricultural Biological Engineering and Department of Civil and Environmental Engineering. Professor Singh has received acclaim for his forty years of distinguished service, teaching, and research in the field of hydrologic and water resources engineering. In addition to hydrologic science and engineering, hydraulic engineering, water resources engineering, environmental engineering, irrigation science, soil and water conservation engineering, entropy-based modeling, copula-based modeling, and mathematical modeling, Singh has made substantial contributions to the state of the art in numerous other specialized fields. In addition to editing 85 books, he has authored 37 books, 1475 refereed journal articles, 122 book chapters, and 330 conference proceedings and papers. One hundred and ten national and international honors have been bestowed upon him by professional organizations. He has been honored with several prestigious awards from ASCE, including the Arid Land Hydraulic Engineering Award (2002), the Ven Te Chow Award (2005), the Torrens Award (2009), the Norma Medal (2010), the Royce Tipton Award (2019), the OPAL Award (2021), and the EWRI Lifetime Achievement Award (2015); the Ray K. Linsley Award, Wetzen Award, and the Founders Award, all bestowed by AIH; the Merriam Improved Irrigation Award, presented by USCID; the Chow Memorial Award, and the Crystal Drop Award Three honorary doctorates have been bestowed upon him: one each from the University of Waterloo, Canada; the University of Basilicata, Italy; and the University of Guelph, Canada. He is a member of AGU, IAHR, WASER, and AWRA, as well as fellow of ASCE, EWRI, AWRA, IE, ISAE, IWRS, IASWC, and IAH. In addition to the National Academy of Engineering, he holds fellowship status at twelve additional engineering and scientific academies.

Part I
Rainfall and Floods

1.1 Introduction

Predicting the trend and frequency of reservoir inflow aids in effective water resource planning and management. The variations in the inflow patterns critically manifest alternations in the hydrological regime of the unregulated catchment under varying climatic and anthropogenic changes (Gautam & Acharya, 2012). Hydrologists and climate researchers have delved into the relationship between the climatic and hydrologic extremes and the factors affecting their transformations (Sharma et al., 2019). Therefore, assessment of the hydrologic response of the catchment at regulating structures, particularly large dams, is indispensable (Salas, 1992). Under the varying hydroclimatic regimes, particularly the extremes, examining inflow patterns before and after the construction of the hydraulic structure provides insights into the adequacy of the designed or existing spillway to dispose of excess water.

As per the International Commission on Large Dams (ICOLD), there are more than 57,000 large dams exists on the globe which serve for irrigation, water supply and flood control. In India, there are about 5701 large dams. Most of the dams were constructed in the last century and hence, became older than 40–60 years. Such structures become vulnerable due to their insufficient capacity under changing climate across the globe. Further, the maximum inflow into such a structure is altered over the temporal scale while their carrying capacity is fixed. The failure of such large structures can lead to flood disaster leading to substantial economic, social, environmental losses and fatalities.

Temporal trends in inflow time series can be ascertained using non-parametric tests, such as Mann–Kendall (MK), Modified Mann–Kendall (MMK), and Spearman's Rho (SR) tests, which are suitable for non-normally distributed time series data (Yue et al., 2002). However, the hydrological inferences of these non-parametric tests statistically depend on the significance level of the test. Additionally, the magnitude of the trend, sample size and variation in the hydroclimatic time series duly influence their hydrological implications (Timbadiya et al., 2013). Innovative trend analysis (ITA; Şen, 2012) is another non-parametric trend detection test, which is simple yet more advantageous. This approach is independent of the time series trend and other parameters influencing conventional non-parametric tests (Şen, 2012). It provides varying patterns in different regimes of the hydroclimatic time series, sub-sectioned into two, i.e., the first half and the second half. In addition to the temporal variability, frequency analysis plays a pivotal role in critically designing hydraulic structures. Among the available statistical distributions, the generalized extreme value type-I distribution best describes the characteristics of extreme inflows, making it widely used for flood frequency analysis (Gehlot et al., 2021; Jibhakate et al., 2023; Timbadiya et al., 2013).

The Ukai reservoir, constructed on the Tapi River in India in 1972, has a storage capacity of 7414 MCM and a spillway discharge capacity of 45,300 m^3/s. Situated approximately 100 km downstream of the reservoir, Surat City, one of the fastest-growing cities in India, lies on the banks of the Tapi River. Given that the Ukai reservoir has been in operation for nearly 50 years, assessing the characteristics of extreme flows into the reservoir before and after its construction is crucial. Recognizing the significance of managing such a large reservoir, this study aims to ascertain the trend of extreme inflows, i.e., annual maximum discharge (AMD) into the Ukai reservoir before (Pre-) and after (Post-) dam construction using the MMK and ITA. Additionally, the generalized extreme value type-I distribution is employed to determine the values of extreme inflows for higher return periods under both scenarios. The findings of this study will be valuable to water resource engineers in making informed decisions regarding the adequacy of the functioning of large hydraulic structures.

1.2 Methods

1.2.1 Modified Mann–Kendall Test

The non-parametric, Modified Mann–Kendall test (MMK), based on the data rank (Hamed & Rao, 1998), is used in the present study to examine the temporal trends in AMD. This test is a modification of the Mann–Kendall test (Kendall, 1975; Mann, 1945), capable of dealing with the correlated data structure, either positive or negative. The equations for the MMK statistical test are given in Eqs. 1.1–1.4.

$$S = \sum_{(i=2)}^{n} \sum_{(j=1)}^{(i-1)} sign(x_i - x_j) \tag{1.1}$$

$$Z = \begin{cases} \frac{S-1}{\sqrt{Var(S)*}} & if\ S > 0 \\ 0 & if\ S = 0 \\ \frac{S+1}{\sqrt{Var(S)*}} & if\ S < 0 \end{cases} \tag{1.2}$$

$$Var(S)^* = CF \times Var(S) \tag{1.3}$$

$$CF = 1 + \frac{2}{n(n-1)(n-2)} \sum_{k=1}^{n-1} (n-k)(n-k-1)(n-k-2) r_K^R \tag{1.4}$$

Here, n is the data series length, and x_i and x_j are the sequential values in the data series. CF is the correction factor taking care of serial correlation in the data series and r_K^R is autocorrelation coefficient. The test statistic Z measures trends about a particular significance level. In the present study, the significance of trends is detected at 5% significance level. The results of the MMK test is dependent on pre-assigned significant level, magnitude of trend, sample size and amount of variation within time series.

1.2.2 Innovative Trend Analysis

According to Şen (2012), the annual peak flow series was divided into two, i.e., the first and second half, concerning time. Each of these two sub-halves is arranged in ascending order. A scatter plot of the annual peak inflow considering the first (second) half as ordinate (abscissa) is useful to conclude the trend in the AMD, as per Şen (2012). In such plot, increasing and decreasing trend occupy upper and lower triangular area respectively, of the square area define by the the domain of variable while trend free time series appears long 45° straight line. This non-parametric test to detect thc trend is independent to pre-assigned significant level, magnitude of trend, sample size and amount of variation within time series Şen (2012).

1.2.3 Generalized Extreme Value Type-I Distribution

The stationary Generalized Extreme Value distribution uses constant location, scale and shape parameters. The Generalized Extreme Value (GEV) type-I, i.e. Gumbel's distribution is fitted to the AMD using maximum likelihood method (Chow et al., 1988) Many researchers propounded that GEV distribution optimally simulated the extreme rainfall series in various hydrologic analyses (Ghosh et al., 2012; Martins & Stedinger, 2000). The cumulative distribution function (F) of GEV is given in Eq. 1.5 (Coles, 2001).

$$F(x;\mu,\sigma,k) = exp\left\{-\left[1+k\frac{(x-\mu)}{\sigma}\right]^{\frac{-1}{k}}\right\},\left(1+k\frac{(x-\mu)}{\sigma}\right) > 0 \quad (1.5)$$

Here μ, σ and k represent the location, scale and shape parameters of the distribution (Coles, 2001). The GEV distribution is of three types, which depend on the values of the shape parameter k. . It represents the Gumbel (type I) or light-tailed if $k = 0$, Fréchet (type II) or heavy-tailed distribution if $k > 0$ and Weibull (type III) or bounded-tailed distribution if $k < 0$ (Ghate & Timbadiya, 2023) The $(1-p)$ th quantile of the GEV (for the return period $T = 1/p$) can be estimated using Eq. 1.6.

$$X_T = \mu + \frac{\sigma}{k}\left\{[-\ln(1-p)]^{-k} - 1\right\} \quad (1.6)$$

The Gringorten plotting positions were obtained and superimposed on the fitted distribution to check suitability of the GEV distribution. The Gringorten plotting position have been obtained by sorting the annual maxima' in ascending order and determining the corresponding return period from (Haung, 2009):

$$T(q_i) = \left(\frac{n+0.12}{i-0.44}\right) \quad (1.7)$$

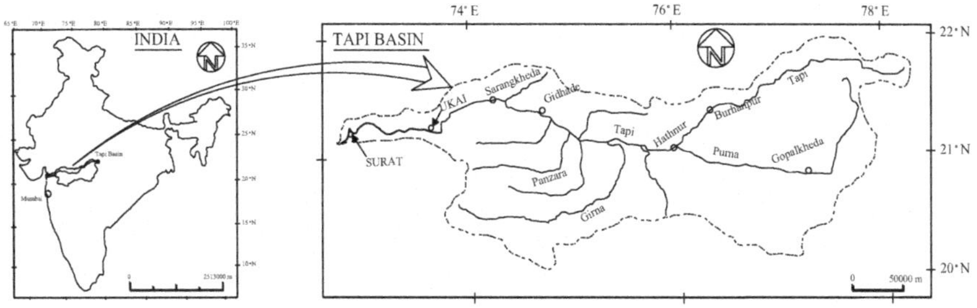

Fig. 1.1 Map of Tapi River basin (Timbadiya et al., 2011)

where i is the rank of year-maximum discharge qi, and n is the total number of year-maxima.

1.2.4 Study Area and Data Collection.

The Tapi basin is the second largest west-flowing basin of India, spreading over an area of 65,145 km$^{2.}$ The basin stretches 587 km from east to west and 210 km from north to south, giving it an elongated shape (Timbadiya et al., 2011). The Ukai project, the largest multipurpose project in the region, has its dam site situated across the Tapi River near Ukai village in Fort-Songadh Taluka of Gujarat, India. The catchment area of the Tapi River upstream of Ukai is 62,225 km^2. Figure 1.1 below depicts the Tapi River basin.

The Ukai reservoir has a live storage capacity of 0.71 M ha-m at FRL 105.156 m. It is equipped with 22 radial gates, each measuring 15.545 m × 14.783 m, and the spillway has a total length of 425.195 m. With a designed maximum outflow of 45,300 m^3/s, the radial gates of spillways play a crucial role in maintaining the live storage of the reservoir, accounting for 78% of the total. The operation of these gates is critical not only for the structural integrity of the dam but also for effective flood control. The water stored in the dam is also used for generation of power up to 300 MW for 8–10 h a day to meet peak demands alongwith the irrigation. According to Central Water Commission (CWC), Govt. of India, the utilizable of water at Ukai dam is 14,500 million m^3.

The annual peak inflow (m^3/s) data were collected from the Surat Irrigation Circle, Narmada, Water Resources, Water Supply and Kalpsar Department, Government of Gujarat, India, from 1939 to 2022. The data details are tabulated in Table 1.1.

The total number of AMD data for Pre- and Post- dam scenario is 34 and 50 which is sufficient to compute the trend and fit with the distribution (Timbadiya & Ghate, 2020). The value of standard deviation of AMD for the Post-dam scenario is less than its corresponding value for the Pre-dam scenario. The value of coefficient of kurtosis indicated that the said data does not follow the normal distribution in both the scenarios. The annual peak flow is shown in Fig. 1.2. It can be seen from the

Table 1.1 Characteristics of AMD in Ukai reservoir on the Tapi River

Sr. no	Scenario	Duration of data	Number of data	Standard deviation (m^3/ s)	Coefficient of skewness	Coefficient of kurtosis
1	Pre-dam	1939–1972	34	10,304.62	1.17	1.08
2	Post-dam	1973–2022	50	7178.88	1.14	1.44

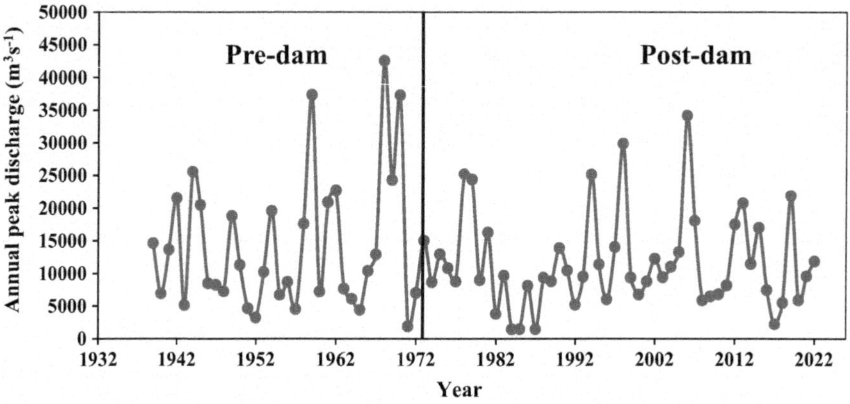

Fig. 1.2 Annual extreme inflow into Ukai reservoir on the Tapi River

figure that the AMD of year 1972 is the highest amongst all the values and falls in the Pre-dam scenario.

1.3 Results and Discussion

The serial correlation, affecting the hydrological inferences of the trend, in the AMD of the Ukai reservoir is accounted for by the MMK test by modifying variance. Before trend detection, the AMD time series were also evaluated for the low and high outliers (Chow et al., 1988). The high and low outliers for the annual extreme inflow data into Ukai dam for the period 1939–1972 (1973–2022) are 76,490.16 m^3/s (69,776.1 m^3/s) and 1553.94 m^3/s (1276.06 m^3/s) respectively. It is observed that all the extreme inflow for Pre-dam and Post-dam fell within the computed outlier values.

The null hypothesis of no-trend is tested through the MMK test. The MMK test statistics ('Z'-Value) for the Pre-dam and Post-dam extreme inflow are found to be 0.0148 and 0.125 at 5% significance level. This indicates no trend in either scenario of extreme inflows in the Ukai dam. The closed look at the innovative trend analysis results shown in Fig. 1.3a, b indicates that the high inflow values increased during the Pre-dam scenario. In contrast, no trend is seen during the Post-dam scenario.

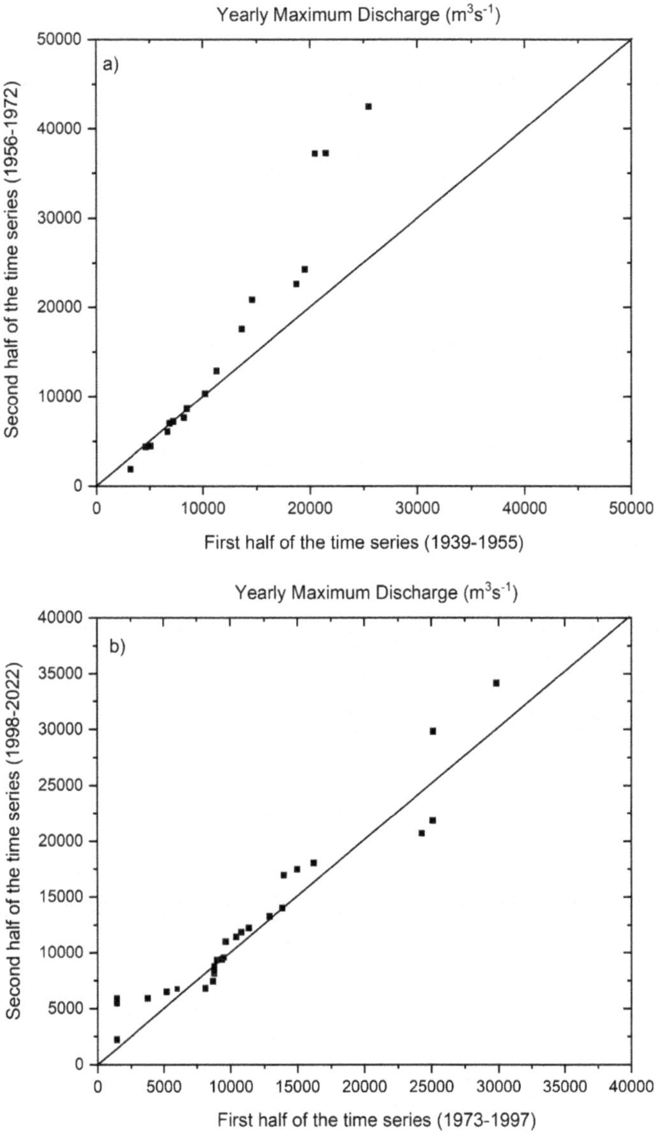

Fig. 1.3 Innovative Trend Analysis on extreme inflow into Ukai dam **a** Pre-dam **b** Post-dam

The lack of trend in the annual extreme inflow in the post-dam scenario using the MMK test and innovative trend analysis methodology contradicts the analysis conducted by Timbadiya et al. (2013). It is noted that the analysis carried out by Timbadiya et al. (2013) uses annual extreme inflow data for 38 years (1972–2009). As the present study is carried out with the most recent corresponding data of 50 years

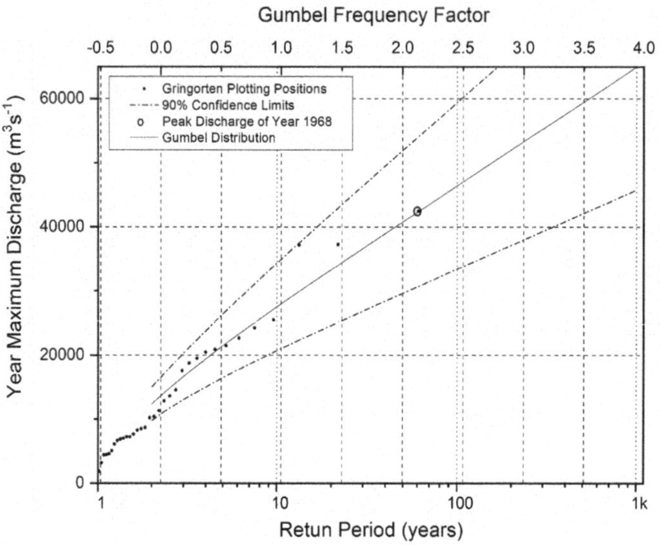

Fig. 1.4 Gumbel extreme value distribution based on the yearly peak discharges for Pre-dam Scenario

(1972–2009) after construction of the Ukai dam, the study can be treated as the most recent work in this area.

The GEV type-I distribution (Huang, 2009) is fitted to the Pre-dam and Post-dam annual extreme inflows into the Ukai reservoir. The fitted distribution, its 90% confidence limits and corresponding Gringorten plotting positions are shown in Figs 1.4 and 1.5 for Pre- and Post- dam scenarios at Ukai gauging stations respectively. From Figs. 1.4 and 1.5, it is evident that, Gringorten plotting positions are within 90% confidence limit of GEV distribution. Hence, the aforesaid fitted distribution is suitable for computation of floods of different return period.

The flood event with higher return period can be extrapolated from the fitted GEV type-I distribution (Huang, 2009). The results of the extrapolation of the Gumbel distribution is shown in Figs. 1.4 and 1.5 respectively for Pre- and Post-dam data. The results of the fitting are also tabulated in Table 1.2. The 50-year return period values of the Pre-dam and Post-dam AMD is 27,522.32 m^3/s and 20,988.26 m^3/s respectively. The 100-year return period values of the Pre-dam and Post-dam AMD is 46,411.17 m^3/s and 34,147.49 m^3/s respectively. It can be inferred that, for all the return period values, the post-dam extreme inflow values are lower than the corresponding values of Pre-dam scenarios. It is interesting to note that the most recent flood into Ukai dam in year 2006 (having magnitude of 34,110 m^3/s) is fitted to about 100-year return period AMD for Post-dam scenario while the same flood is fitted to about 25-year return period for Pre-dam scenario.

This observation shows that the peak discharges are reduced during Post-dam scenario. To a certain this observation a hydrological model for the Ukai dam is

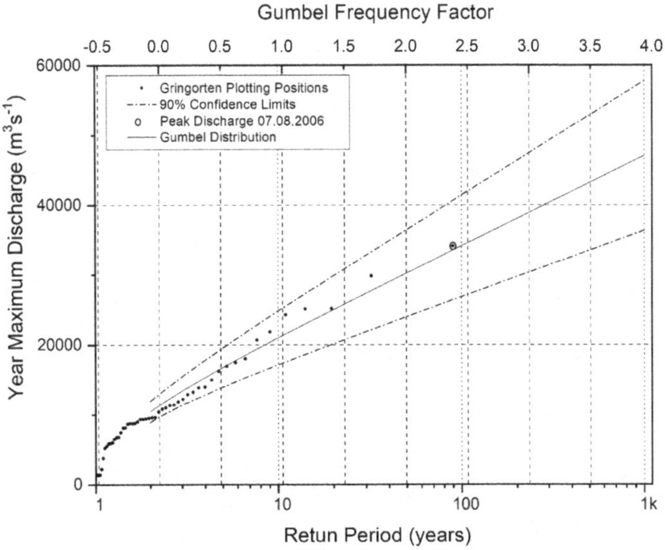

Fig. 1.5 Gumbel extreme value distribution based on the yearly peak discharges for Post-dam Scenario

Table 1.2 Pre-dam and Post-dam inflows corresponding to various return periods

Return period	Inflows (m^3/s)	
	Pre-dam	Post-dam
1 in 5	21,489.97	16,785.72
1 in 10	27,522.32	20,988.26
1 in 50	40,789.57	30,237.38
1 in 100	46,411.17	34,147.49
1 in 500	59,381.04	43,183.17
1 in 1000	64,956.97	47,067.73

needed which is not included in the present study and can be treated as future scope of the work.

The extreme annual inflow is lower for post-dam scenario than the Pre-dam scenario for the corresponding return period indicating the decrease in the magnitude of the extreme inflows into the Ukai reservoirs. Further, the existing design capacity of the spillway gates of the Ukai dam (45,300 m^3/s) corresponds to a 730-year return period.

1.4 Conclusions

The analysis of annual maximum discharge (inflow) into the largest reservoirs of the Tapi River basin, i.e., Ukai, was carried out the using MMK, ITA and GEV type-I distribution using Pre-dam (1939–1972) and Post-dam (1973–2022) data (total 84 years) No significant trend in the annual extreme inflow is indicated by MMK test for the pre- and post-dam scenarios. The generalized extreme value I distribution is well suit to the AMD at Ukai dam. The 100-year return periods for pre-dam and post-dam scenarios extreme inflow are 16,411.17 m^3/s and 34,147.49 m^3/s, respectively. Overall, the extreme inflow values for the given return period decrease in the post-dam scenarios compared to pre-dam scenarios. The existing design capacity of the spillway gates of the Ukai dam (45,300 m^3/s) corresponds to a 730-year return period annual maximum discharge.

The present study is useful to the reservoir operator in deciding on the sufficiency of the existing spillway while comparing its design values at the time of construction. It is also suggested that this type of study be repeated at decadal intervals.

Acknowledgements The author is thankful to the Centre of Excellence on "Water Resources and Flood Management" at SVNIT-Surat for providing computing facilities. The data support from Surat Irrigation Circle, Narmada Water Resources, Water Supply and Kalpsar Department, Govt. of Gujarat, is duly acknowledged. The author is also thankful to the Reviewers for their comments to improve the readability of this paper.

References

Chow, V. T., Maidment, D. R., & Mays, L. W. (1988). *Applied hydrology* (p. 572). MacGraw-Hill. Inc.

Coles, S. (2001). *An introduction to statistical modelling of extreme values* (pp. 208). Springer-Verlag. https://doi.org/10.1007/978-1-4471-3675-0

Gautam, M. R., & Acharya, K. (2012). Streamflow trends in Nepal. *Hydrological Sciences Journal, 57*(2), 344–357. https://doi.org/10.1080/02626667.2011.637042

Gehlot, L. K., Jibhakate, S. M., Sharma, P. J., Patel, P. L., & Timbadiya, P. V. (2021). Spatio-temporal variability of rainfall indices and their teleconnections with El Niño-Southern oscillation for Tapi Basin India. *Asia-Pacific Journal of Atmospheric Sciences, 57*(1), 99–118. https://doi.org/10.1007/s13143-020-00179-1

Ghate, A., & Timbadiya, P. V. (2023). True interval non-stationary intensity-duration-frequency curves under changing climate and effect of temporal discretisation on rainfall extremes. *Journal of Hydrology, 618*, 1–17. https://doi.org/10.1016/j.jhydrol.2023.129136

Ghosh, S., Das, D., Kao, S. C., & Ganguly, A. R. (2012). Lack of uniform trends but increasing spatial variability in observed Indian rainfall extremes. *Nature Climate Change, 2*(2), 86. https://doi.org/10.1038/nclimate1327

Hamed, K. H., & Rao, A. R. (1998). A modified Mann-Kendall trend test for autocorrelated data. *Journal of Hydrology, 204*(1–4), 182–196. https://doi.org/10.1016/S0022-1694(97)00125-X

Huang, Y. (2009). Rapid flood risk assessment using GIS technology. *International Journal of River Basin Management, 7*(1), 3–14. https://doi.org/10.1080/15715124.2009.9635365

Jibhakate, S. M., Timbadiya, P. V., & Patel, P. L. (2023). Multiparameter flood hazard, socioeconomic vulnerability and flood risk assessment for densely populated coastal city. *Journal of Environmental Management, 344*, 118405. https://doi.org/10.1016/j.jenvman.2023.118405

Kendall, M. G. (1975). Rank and product-moment correlation. *Biometrika, 36*, 177–193.

Mann, H. B. (1945). Non-parametric tests against trend. *Econometrica: Journal of the Econometric Society*, 245–259. https://doi.org/10.2307/1907187.

Martins, E. S., & Stedinger, J. R. (2000). Generalized maximum-likelihood generalized extreme-value quantile estimators for hydrologic data. *Water Resources Research, 36*(3), 737–744. https://doi.org/10.1029/1999WR900330

Salas, J. D. (1992). Analysis and modeling of hydrologic time series. In Maidment, D.R. (Ed.), *Handbook of Hydrology* (pp. 19.1–19.72). McGraw Hill, ISBN 0–07–039732–5.

Şen, Z. (2012). Innovative trend analysis methodology. *Journal of Hydrologic Engineering, 17*(9), 1042–1046. https://doi.org/10.1061/(ASCE)HE.1943-5584.0000556

Sharma, P. J., Patel, P. L., & Jothiprakash, V. (2019). Impact of rainfall variability and anthropogenic activities on streamflow changes and water stress conditions across Tapi Basin in India. *Science of the Total Environment, 687*, 885–897.

Timbadiya, P.V., & Ghate, A. (2020). Discussion of Clock hour correction effect on extreme value analysis of rainfall on Western Coast of India by Dauji (2019). *ISH Journal of Hydraulic Engineering*, *28* (sup1), 570–571. https://doi.org/10.1080/09715010.2020.1792362.

Timbadiya, P. V., Mirajkar, A. B., Patel, P. L., & Porey, P. D. (2013). Identification of trend and probability distribution for time series of annual peak flow in Tapi Basin, India. *ISH Journal of Hydraulic Engineering, 19*(1), 11–20. https://doi.org/10.1080/09715010.2012.739354

Timbadiya, P. V., Patel, P. L., & Porey, P. D. (2011). HEC-RAS Based Hydrodynamic Model in Prediction of Stages of Lower Tapi River. *ISH Journal of Hydraulic Engineering, 17*(2), 110–117. https://doi.org/10.1080/09715010.2011.10515050

Yue, S., Pilon, P., & Cavadias, G. (2002). Power of the Mann-Kendall and Spearman's rho tests for detecting monotonic trends in hydrological series. *Journal of Hydrology, 259*(1–4), 254–271. https://doi.org/10.1016/S0022-1694(01)00594-7

Chapter 2
Distributed Manning's Roughness Approach in Calibration and Validation of 1-D Hydrodynamic Model of Lower Tapi River, India

Sandipkumar Rajput and Prem Lal Patel

Abstract Channel roughness plays a critical role in the advancement of hydraulic models used for predicting floods and flood inundation maps. This study presents application of a hydrodynamic model aimed at calibrating the Manning's roughness coefficient (n) for the lower Tapi river in India. The outflows from the Ukai dam and the tidal levels in the Arabian Sea are considered as the upstream and downstream boundary conditions, respectively. The calibration of Manning's n has been performed by taking global as well as distributed parameters, i.e., Ukai dam to Kakrapar and Kakrapar to the Arabian sea, for the selected reaches of river. After several trials, the Manning's value of 0.035 from Ukai dam to Kakrapar and 0.02 from Kakrapar to the Arabian sea are found better with the lowest weighted RMSE. In most of the gauging stations, it is found that simulated water level is higher than the observed water levels. The calibrated Manning's n values have been found to perform well in validation of independent flow event, and the same can be used for design of levees as flood protection measures for the lower Tapi river.

Keywords HEC-RAS · Hydrodynamic model · Lower Tapi River · Manning's roughness · Surat city

2.1 Introduction

The development of a hydraulic model for flood prediction and floodplain mapping is influences by the sensitivity of the channel bed roughness parameter (Parhi et al., 2012). Determining the channel roughness parameter is crucial when examining the open-channel flow, especially in the context of hydraulic modeling. Channel roughness is not a constant parameter; rather, it is influenced by changes in channel

S. Rajput · P. L. Patel (✉)
Department of Civil Engineering, Sardar Vallabhbhai National Institute of Technology, Surat 395007, India
e-mail: plpatel@ced.svnit.ac.in

A. Sefelnasr et al. (eds.), *Water Resources Management and Sustainability*, Water Science and Technology Library 114, https://doi.org/10.1007/978-3-031-80520-2_2

features along the river (Timbadiya et al., 2011b). Discharge, river stage, and other hydraulic characteristics are interconnected and rely on the properties of channel roughness. Channel roughness plays a crucial role in the analysis of open channel flow, particularly when examining and simulating natural river systems. Hydrodynamic models, which simulate the flow dynamics of river channels, have proven to be useful tools in floodplain management (Horritt & Bates, 2002). For development of reliable hydraulic model of a river, the channel roughness values need to be calibrated in accordance with the physical characteristics of the channel and floodplains (Timbadiya et al., 2012). The Manning's coefficient 'n' depends on the resistance to flow caused by the type and amount of vegetation, channel configuration, and so on (Vijay et al., 2007). The roughness of a channel is a factor that is influenced by factors such as surface texture, vegetation, irregularities in the channel, and channel alignment (Ramesh et al., 2000). Further, forecasting river stages is essential for both structural and non-structural flood management techniques (Timbadiya et al., 2014).

Different hydraulic models like HEC-RAS, MIKE11, TUFLOW, etc., are available for hydraulic routing. The River Analysis System (HEC-RAS), created by the Hydrologic Engineering Center (HEC) of the U.S. Army Corps of Engineers, is presently the most favourable in flood modeling for hydrodynamic simulation. Calibration of the roughness parameter in a one-dimensional hydrodynamic model using HEC-RAS is a crucial step for improving the accuracy of flood prediction. The HEC-RAS model utilizes a series of cross-sections of the terrain and conducts flow simulations to calculate the average velocity and water depth for each of these cross-sections. A hydrodynamic model that can simulate river stages and flow hydrographs at various locations along the river is required for flood forecasting, management, and future flood protection work (Timbadiya et al., 2011a).

The present investigation is focused with following objectives: -

(a) Develop a hydrodynamic model of lower Tapi river using distributed roughness parameters, as the Manning's roughness coefficient.
(b) Compare the results from the hydrodynamic model for distributed roughness vis-à-vis global roughness.
(c) Perform the assessment of calibrated model for an independent flood event.

2.2 Materials and Methods

2.2.1 Study Area and Data Sources

The Tapi River, the second largest westward-flowing river in India following the Narmada River. The Tapi river starts from Multai in the Betul district of Madhya Pradesh, situated at a height of 752 m. It spans 724 km in length and eventually flows into the Arabian Sea, a short distance past Surat city. The Tapi has a total drainage area of 65,145 km^2. The basin is an elongated shape with a maximum length of 687 km from east to west and a maximum width of 210 km from north to south

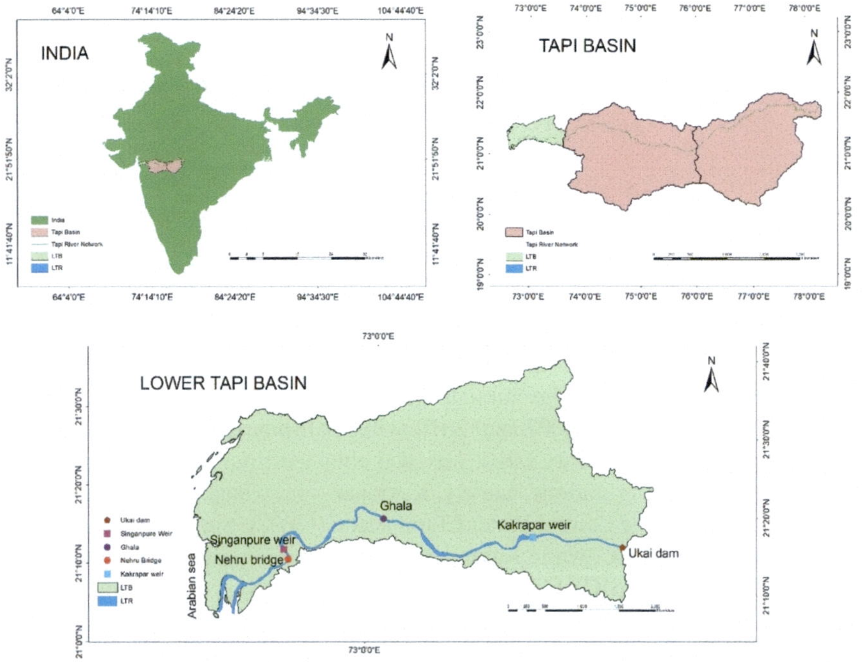

Fig. 2.1 Index map of lower Tapi river

(Jain et al., 2007). Upper Tapi (up to Hatnur Dam), middle Tapi (Hatnur to Ukai Dam), and lower Tapi (Ukai Dam to the Arabian Sea) are the three subbasins of the Tapi basin. The study focuses on a segment of the river that extends from the Ukai dam to the Arabian Sea, covering a distance of 125.68 km. The lower Tapi basin covers an area of 2790 km^2. The Kakrapar and Singanpur weirs are two regulating structures situated along the study stretch, at 22.2 km and 103 km from the Ukai dam, respectively. The Kakrapar weir is an ogee-shaped weir. It has a crest level of 48.78 m above mean sea level (msl) and a discharge coefficient of 1.88 (Subramanya, 1998). The Singanpur weir has a coefficient of discharge of 1.67 and a crest level of 6 m above mean sea level (Timbadiya et al., 2011a). The study area comprises three gauging stations, namely, Kakrapar, Ghala, and Nehru Bridge. The water levels from these stations are used for both calibration and validation of the model. The map of the study area and the alignment of the lower Tapi river are given in Fig. 2.1.

2.2.2 *Data Collection*

To simulate flow in HEC-RAS, essential data includes the channel geometry, along with boundary conditions at both the upstream and downstream ends. Surat Municipal Corporation (SMC) provided geometric data for the lower Tapi river in the form

Table 2.1 Data used for simulation of flow in lower Tapi River

Sr. No.	Flow year	Simulation duration	Output storage interval	Gauge station used for Validation
1	2003	Aug-25, 01:00–Aug-28, 24:00	1 h	Kakrapar and Ghala
2	2007	Sep-20, 01:00–Sep-23, 24:00	1 h	Kakrapar, Ghala and Nehru bridge
3	2010	Aug-25, 01:00–Aug-28, 24:00	1 h	Kakrapar, Ghala and Nehru bridge

of an AutoCAD (.dwg file). The map was used to extract cross-sectional data for the relevant river segment. The Surat Irrigation Circle (SIC) provided the hourly discharge records for 2003, 2007, and 2010, while half-hourly tidal level data, from 6 December 2009 to 5 January 2010, was also obtained from the SIC. Water level data from three gauging stations, namely, Kakrapar weir, Ghala, and Nehru Bridge, were collected from the Central Water Commission (CWC). Singanpur weir details were obtained from SMC, while Kakrapar weir information was obtained from the Surat Irrigation Circle (SIC), Government of Gujarat. The Kakrapar weir has a crest width of 2.4 m, while the Singanpur weir has a 10 m. They have ogee and broad-crested shapes, with estimated discharge coefficients of 1.88 and 1.67, respectively. The description of data used for simulation of flow in lower Tapi river is tabulated in Table 2.1.

2.2.3 *Methodology*

The use of HEC-RAS enables the analysis and simulation of hydraulic conditions in the river systems, assisting in the assessment of water surface profiles along the river. The methodology used in current study for hydrodynamic simulation using HEC-RAS is shown in the form of flow chart in Fig. 2.2. The HEC-RAS model utilizes the Saint-Venant equations, primarily designed for studying unsteady, gradually varied flow in natural channels.

$$\frac{\partial A}{\partial t} + \frac{\partial Q}{\partial x} = 0 \tag{2.1}$$

$$\frac{\partial Q}{\partial t} + \frac{\partial (Q^2/A)}{\partial x} + gA\frac{\partial H}{\partial x} + gA\left(S_0 - S_f\right) = 0 \tag{2.2}$$

where,

Q discharge
A cross-sectional area normal to the flow
H elevation of the water surface above a specified datum

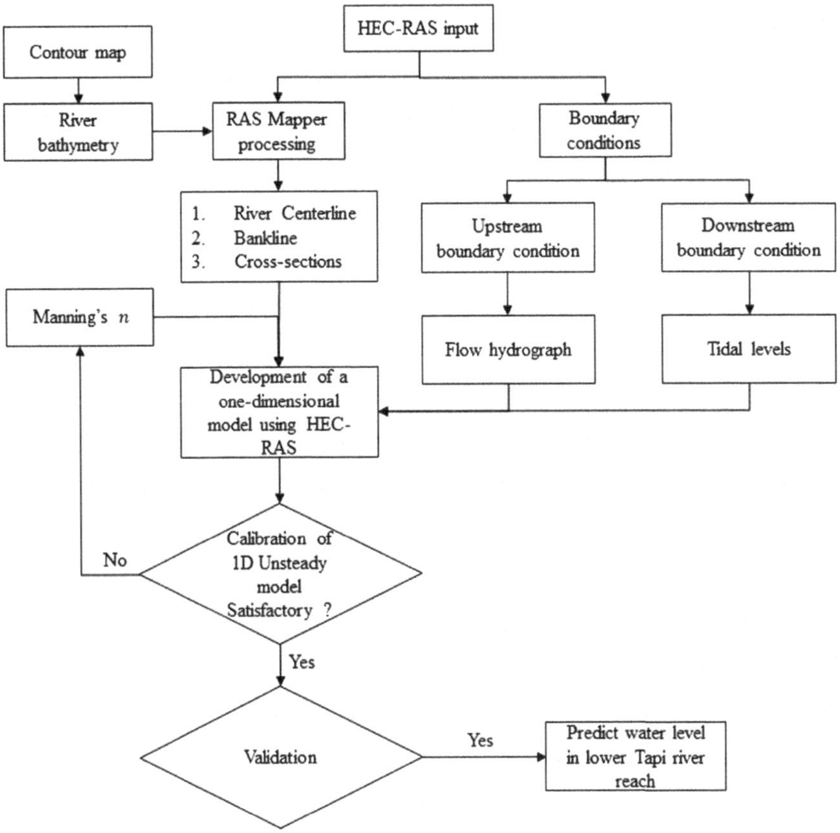

Fig. 2.2 Flow chart of detailed methodology adopted in the present study

g acceleration due to gravity
S_0 bed slope
S_f energy slope
t temporal coordinate
x longitudinal coordinate

These equations (Eqs. 2.1 and 2.2) are solved using widely recognized four-point implicit box finite difference scheme (Chow et al., 1988).

The contour data from the survey was utilized to create the river bathymetry, which, in turn, served as the basis for generating geometric features of the river, including the centerline of the river, banklines, and cross-sections. The model undergoes a two-step process involving calibration and validation. Calibration is carried out using flow data from the years 2003, 2007, and 2010. Manning's Roughness coefficient (n) was used as the calibration parameter in this study. The global and distributed 'n' value is applied along the length of the river, with one value (n_1) used for the segment from Ukai dam to Kakrapar and another value (n_2) for the segment

Table 2.3 RMSE values during calibration of HEC-RAS model using distributed Manning's *n* for the year 2003

Manning's *n*		RMSE (m)		Weighted RMSE (m)
Ukai to Kakrapar	Kakrapar to Arabian sea	Kakrapar	Ghala	
0.02	0.035	0.203	2.142	1.173
0.02	0.03	0.203	1.927	1.065
0.02	0.025	0.203	1.705	0.954
0.035	0.03	0.201	1.921	1.061
0.035	0.025	0.201	1.697	0.949
0.035	**0.02**	**0.201**	**1.463**	**0.832**
0.028	0.022	0.202	1.561	0.882
0.025	0.02	0.202	1.468	0.835
0.03	0.02	0.202	1.466	0.834
0.022	0.028	0.203	1.839	1.021
0.025	0.035	0.202	2.140	1.171

Table 2.4 RMSE values during calibration of HEC-RAS model using global Manning's *n* for the year 2007

Manning's *n*	RMSE (m)			Weighted RMSE (m)
	Kakrapar	Ghala	Nehru bridge	
0.02	**1.878**	**1.445**	**1.498**	**1.607**
0.025	1.819	1.610	1.427	1.619
0.03	1.870	1.763	1.340	1.658
0.035	1.873	1.905	1.341	1.706

2.4 Conclusions

A one-dimensional hydrodynamic model for the lower Tapi river has been developed. This model has been developed and calibrated using flow data of years 2003, 2007, and 2010. The Manning's *'n'* has been calibrated using two approaches, viz. global and distributed approaches. The independent flood event of the year 2004 has been used for assessing the performance of the aforesaid calibrated models. The following are the key conclusions of this study: -

(a) The calibrated value of Manning's '*n*' using the global approach has been found to be 0.02 for entire lower Tapi river.
(b) The calibrated values of Manning's '*n*' using distributed approach have been found to be 0.035 and 0.02 respectively for stretches Ukai to Kakrapar and Kakrapar to the Arabian sea respectively.

Table 2.5 RMSE values during calibration of HEC-RAS model using distributed Manning's *n* for the year 2007

Manning's *n*		RMSE (m)			Weighted RMSE (m)
Ukai to Kakrapar	Kakrapar to Arabian sea	Kakrapar	Ghala	Nehru bridge	
0.02	0.035	1.866	1.911	1.342	1.706
0.02	0.03	1.867	1.767	1.389	1.674
0.02	0.025	1.867	1.612	1.436	1.638
0.035	0.03	1.873	1.761	1.380	1.671
0.035	0.025	1.874	1.607	1.437	1.639
0.035	**0.02**	**1.874**	**1.441**	**1.496**	**1.604**
0.028	0.022	1.869	1.510	1.457	1.612
0.025	0.02	1.871	1.444	1.499	1.605
0.03	0.02	1.872	1.452	1.492	1.605
0.022	0.028	1.868	1.707	1.410	1.662
0.025	0.035	1.868	1.909	1.340	1.706

Table 2.6 RMSE values during calibration of HEC-RAS model using global Manning's *n* for the year 2010

Manning's *n*	RMSE (m)			Weighted RMSE (m)
	Kakrapar	Ghala	Nehru bridge	
0.02	**1.260**	**1.279**	**1.440**	**1.326**
0.025	1.311	1.590	2.362	1.754
0.03	1.264	1.546	1.425	1.412
0.035	1.268	1.667	1.431	1.455

(c) The independent flood event of year 2004 has been validated for global and distributed Manning's '*n*', and their RMSE values have been found to be 1.205 m and 1.161 m respectively. Thus, distributed calibrated model performs slightly better than global calibrated model.

Table 2.7 RMSE values during calibration of HEC-RAS model using distributed Manning's *n* for the year 2010

Manning's *n*		RMSE (m)			Weighted RMSE (m)
Ukai to Kakrapar	Kakrapar to Arabian sea	Kakrapar	Ghala	Nehru bridge	
0.02	0.035	1.258	1.678	1.433	1.456
0.02	0.03	1.258	1.551	1.425	1.411
0.02	0.025	1.260	1.419	1.426	1.368
0.035	0.03	1.266	1.539	1.421	1.409
0.035	0.025	1.266	1.407	1.422	1.365
0.035	**0.02**	**1.266**	**1.270**	**1.437**	**1.324**
0.028	0.022	1.264	1.276	1.441	1.327
0.025	0.02	1.263	1.278	1.441	1.327
0.03	0.02	1.265	1.275	1.441	1.327
0.022	0.028	1.260	1.499	1.425	1.395
0.025	0.035	1.262	1.676	1.434	1.457

Acknowledgements The authors express their gratitude to the Surat Irrigation Circle, Surat Municipal Corporation, and the Central Water Commission (CWC) for providing the data. The authors also acknowledge the infrastructure support from the **Centre of Excellence on 'Water Resources and Flood Management'**, TEQIP-II, Ministry of Education, Government of India.

References

Chow, V. T., Maidment, D. R., & Mays, L. W. (1988). *Applied hydrology*. McGraw-Hill.

Horritt, M. S., & Bates, P. D. (2002). Evaluation of 1D and 2D numerical models for predicting river flood inundation. *Journal of Hydrology, 268*(1–4), 87–99.

Jain, S. K., Agarwal, P. K., & Singh, V. P. (2007). *Hydrology and water resources of India*. Springer.

Parhi, P. K., Sankhua, R. N., & Roy, G. P. (2012). Calibration of channel roughness for Mahanadi River, (India) using HEC-RAS model. *Journal of Water Resource and Protection, 4*(10), 847–850.

Ramesh, R., Datta, B., Bhallamudi, S. M., & Narayana, A. (2000). Optimal estimation of roughness in open-channel flows. *Journal of Hydraulic Engineering, 126*(4), 299–303.

Subramanya, K. (1998). *Flow in open channels*. Tata Mc-Graw-Hill.

Timbadiya, P. V., Patel, P. L., & Porey, P. D. (2011a). HEC-RAS based hydrodynamic model in prediction of stages of lower Tapi River. *ISH Journal of Hydraulic Engineering, 17*(2), 110–117.

Timbadiya, P. V., Patel, P. L., & Porey, P. D. (2011b). Calibration of HEC-RAS model on prediction of flood for lower Tapi River, India. *Journal of Water Resource and Protection.*

Timbadiya, P. V., Patel, P. L., & Porey, P. D. (2012). Calibration of channel and flood plain roughness using 1d/2d integrated hydrodynamic model: A study of Surat city on lower tapi river. In *Proceedings of the 10th international conference on hydroscience and engineering, Orlando, FL, USA* (pp. 4–7).

Timbadiya, P. V., Patel, P. L., & Porey, P. D. (2014). One-dimensional hydrodynamic modelling of flooding and stage hydrographs in the lower Tapi River in India. *Current Science*, 708–716.

Vijay, R., Sargoankar, A., & Gupta, A. (2007). Hydrodynamic simulation of river Yamuna for riverbed assessment: A case study of Delhi region. *Environmental Monitoring and Assessment, 130*, 381–387.

Chapter 3
Flood Risk Management Evaluation and Climate Change Adaptation in Wadi Derna NE Libya

Nawal Alfarrah, Van de Moortel Ivo, and Kristine Walraevens

Abstract Flash floods are a particularly dangerous phenomenon that attacks and troubles people in many areas' foothill regions. Despite water scarcity, continuous prolonged droughts and irregular rainfall in arid and semi-arid areas, recent disasters demonstrate that flooding is more severe and life threatening than ever. For example, catastrophic floods struck eastern Libya in September 2023 following the Daniel storm, resulting in more than 10,000 deaths, at least 90,000 displaced inhabitants, and millions of euros in economic damage to homes, businesses, and infrastructures. There are several factors that make arid and semi-arid regions more vulnerable to severe flooding and unexpected flood damage. On the one hand, the high variety of climate, geography, hydrogeology, and land use poses difficulties for local flood managers. On the other hand, human and system-related challenges such as lack of early warning systems, drought-based management and thorough institutional frameworks have adverse effects on Flood Risk Management (FRM). All these facts emphasize the need for a comprehensive understanding of FRM challenges in arid and semi-arid regions in order to mitigate the negative impacts of floodings on the environment, economy and society. To better understand these challenges, a case study was selected: Wadi Derna in Northeastern Libya. The current flood assessment in the basins was analyzed, and the hazards associated with the flood phenomenon were assessed for various calculated rainfall return periods. To this end, a flash flood index was calculated as a basis to understand the impact of flash floods (FF). A coupling of this index with the FF histories was included to provide a comprehensive overview of the FF vulnerability of the basin. A comprehensive method for mapping of climate and geohazards, as well as their respective impacts, has been developed by Antea Group Belgium. This includes integrating bottom-up assessments of climate vulnerability in targeted communities, models for calculating climate hazards and their impact, spatiotemporal variability of rainfall

N. Alfarrah (✉) · V. de Moortel Ivo
Antea Belgium nv, Buchtenstraat, P.O. Box 9, 9051 Sint-Denijs-Westrem, Belgium
e-mail: nawal.alfarrah@anteagroup.be

N. Alfarrah · K. Walraevens
Laboratory for Applied Geology and Hydrogeology, Ghent University, Krijgslaan 281-S8, 9000 Ghent, Belgium

A. Sefelnasr et al. (eds.), *Water Resources Management and Sustainability*, Water Science and Technology Library 114, https://doi.org/10.1007/978-3-031-80520-2_3

occurrence, climate change projections, and data on planned upstream dam developments. This approach enhances the accuracy and reliability of the final risk maps. Dams' failures would be among the severe indicators of climate change; they are mirrors to our collective understanding of natural hazard risk. Dams, nowadays, cope with events that the designers never conceived, mainly in dry regions, where drought mitigation measures, including storage dams, recharge dams, artificial lakes and embankments, are implemented, while precautionary mitigation measures such as early warning systems are rare or completely missing. The analysis demonstrates that a single management strategy under climate change impacts is insufficient to reduce FF risks. To reduce vulnerability to the impacts of climate change in Libya an integrated strategy is required.

Keywords Climate change · Climate vulnerability · Flash floods · Flood risk management · Water scarcity

3.1 Introduction

Under the impact of the climate change, extreme rainfall is expected to intensify with global warming over large parts of the globe as the concentration of atmospheric moister increases in proportion to the saturation concentrations at a rate of about 6–7% per degree Celsius rise in temperature (Allen & Ingram, 2022; Ingram, 2016; Trenberth et al., 2003). Climate change impacts are one of the main water management challenges in the arid and semi-arid regions.

Storm Daniel, the result of a very low-pressure system which developed into a tropical-like cyclone known as a Medicane, battered Libya with torrential rain on the 10th September 2023 unleashing catastrophic floods in the northeastern coastal cities, the most devastating is Wadi Derna flood resulting in more than 10,000 deaths in the city of Derna.

'Wadi' is an Arabic word that is used to refer to dry stream channels and their valleys, typically in arid regions, such as in the Middle East and North Africa (Jackson & Bates, 1997; Sen, 2008). In the past (before 20 years), rainfall in the wadis can be described as episodic, varying widely on spatial and temporal scales, with many years receiving no or very low precipitation. In recent decades, the frequency of devastating floods and extreme rainfall events has substantially increased due to the impact of climate change (Ali & Mishra, 2018; Kundzewicz, 2013; Milly et al., 2002).

Compared with riverine floods, FF often occur in mountainous areas or in the foothills (WMO, 2007). A spatiotemporal analysis of rainfall variability over the Middle East and North Africa showed high spatiotemporal variability over the region in terms of frequency and intensity of FF events (Lin, 1999; Saber et al., 2017, 2020). Additionally, the maximum rainfall in storms and the number of extreme rainfall events is expected to increase in the future as a consequence of climate change (Tanaka et al., 2020), leading to higher risks of wadi-related FFs in such regions, which requires adaptation of sustainable management plans of hazard risk and calls for a global early warning system. Changes in flooding intensity, duration and frequency, driven by climate change, hydraulic structures or changes in river regimes, can affect human lives and ecology, drive to displace people and can change the characteristics of floodplains/agricultural land. Many countries around the world such as USA, Canada, UK, Pakistan, India, Brazil, the Philippines, Bangladesh, China, Indonesia, Egypt, Jordan, Kuwait, Sudan, Oman, Qatar and Australia have recently experienced severe flooding with disastrous consequences (Dilley et al., 2005; Al-Qudah, 2011; Guha-Sapir et al., 2012; Moawad et al., 2016; Mahmood et al., 2017; Saharia et al., 2017; Tomohiro, 2020; Abdel-Fattah et al., 2018; Al-Mamari et al., 2019; Gaur et al., 2015, 2019; Rubinato et al., 2019; Yang, 2019).

In arid and semi-arid regions globally, FF have been responsible for a high number of mortality and considerable economic losses. Flooding is common in these regions due to the interaction of climate and topography. One of these catastrophic floods hit Riyadh in Saudi Arabia during 2009–2016, resulting in 168 deaths and millions of USD of economic damage (Rahman, 2016). In Jordan in the past five decades, floods have claimed 345 human lives and influenced 24,321 people (Alhasanat, 2017). A flood that occurred in Afghanistan in 2014 led to about 400 fatalities (Alhasanat, 2017). In Iran, flooding raised losses from 17.2 to 132 million USD in the period between 1950 and 2000 (Vatanfada et al., 2009). In 2019, Iran flooding accounted for the death of 78 people around the country and caused over 1.1 billion USD in damage (PDNA, 2019; United Nations in Iran, 2019). In North Africa, floods in Egypt in 2015 were responsible for a huge economic damage and loss of life (Alhasanat, 2017). The latest FF in Derna of 11 September 2023 is considered to be the most devastating massive flood in the arid and semi-arid areas in history and caused over 45.0 billion USD in damage and thousands of losses of life according to official local reports. Despite this, previous studies addressing FF in these regions are still limited.

Unlike European countries, in which the Floods Directive has defined legalization and framework with determined goals and timeframe (European Commission, 2007), most of water scarce regions in Asia and Africa do not have a defined framework and despite being a serious disaster in these regions, flooding has received little attention among responsible authorities. Other major operational challenges to water managers in arid and semi-arid lands result from the water management of dam reservoirs in these areas. Water managers in arid and semi-arid regions aim to ensure a sufficient water supply by storing water during wet periods and then releasing it in times of scarcity (Farrokhzadeh et al., 2020; Liebe et al., 2007). Operational rules of most existing dam reservoirs are only conditioned for long regular drought periods. Hence, drought management plays a main role in risk-based decision processes of reservoir operations (Taraky et al., 2021; López-Moreno et al., 2009; Hannoun & Tietjen, 2023; Mohammed & Scholz, 2017). Another main technical challenge in arid and semi-arid regions of developing countries lies in the lack of field measurements and data availability, as these areas are mostly poorly gauged catchments with insufficient hydro-meteorological data (Alhasanat, 2017; Omidvar & Khodaei, 2008; Ahmed et al., 2023; Hamad, 2020; Al-Qudah, 2011; Fiorillo et al., 2018; Shakeel et al., 2010).

Flood forecasting and early warning play an important role in flood risk management (FRM). Hydrologists and meteorologists claim that with the present progress in measurement and forecasting technologies, the exact place and time of the flood occurrence can be predicted only one hour before it comes out (Stockdon, 2002; Vos et al., 2019; SHMI, 2006). This means that the current forecasting techniques may not cover all types of FFs, especially those from man-made or natural infrastructure failures like Derna case. Therefore, it is difficult to provide accurate and timely early warnings for all situations. Planning for the mitigation of loss of life and damage by choosing the proper methods to reduce losses incurred by FFs, requires that an evaluation of the level of the flood risk in a given area be carried out. Flood risk can be characterized in many ways, three elements are crucial: hazard, exposure, and vulnerability (WMO, 2006).

The main aim of the study is to address and provide an assessment of the unusual nature of the rainfall event due to Daniel storm and the role that the Wadi Derna dams failure played in the massive flooding. Various data, including historical and recent remote rainfall data sensing images were used to achieve the purpose.

The residents of Derna, NE Libya are "extremely vulnerable to flood risk", according to Ashoor (2022), a hydraulic engineer at Omar Al-Mukhtar University in Libya, who warned for the threat of repeated flooding of the seasonal riverbed to Derna, calling the situation "dangerous". The paper highlighted that five floods have been recorded since 1942 and called for immediate action to ensure regular maintenance of the dams to prevent a catastrophic flood in case of a massive flood in the future. One year later, in 2023, the flooding in Derna has damaged buildings, roads and bridges and swept a great part of the city into the Mediterranean Sea. Libya has experienced years of turmoil since the overthrow of Moammar Gadhafi's regime in 2011. On top of this misery, the disaster has hit Libyans hard. After this FF disaster, the city is called "Sad Derna." The massive floods also lead to infrastructure damage, water contamination and landslides, which made the situation even worse and consequences are still continuing. The impact of Derna FF in the city and its location is showing in Figs. 3.1 and 3.2.

3.1.1 Introduction to Wadi Derna and Its Geomorphological Characteristics

Wadi Derna is a part of El-Jabal Al-Akhadar hydrological system. El-Jabal Al-Akhadar is a highland region located south of the coastal belt in the northeast of Libya. It extends along the coast for about 300 km and rises to some 880 m above sea level, cut by several wadis and receiving annual precipitation of 250–600 mm (Hidroprojekat, 1972; Al-Zwam, 1995). The flooded city of Derna is located on the banks of the large valley of Wadi Derna, which exceeds 60 km length, about 10 km width and has an area of 555 km^2. Wadi Derna is one of the most important wadis in El-Jabal El-Akhadar and is one the few wadis in Libya with perennial watercourse

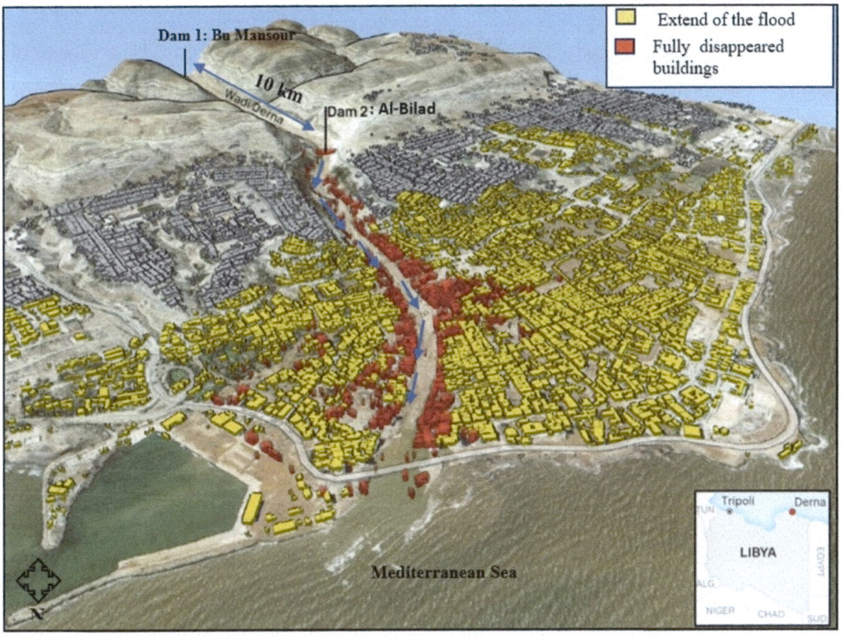

Fig. 3.1 CG Satellite imagery shows the city of Derna, Libya, after the devastating floods that struck parts of the country. This imagery was captured on Sep 14, 2023 with a 3D model of fully disappeared buildings and the extent of the flood. *Source* CG Satellites

throughout the year (Trih, 1971). The wadi is located between the latitudes of 32° 30° to 32° 50 north and the longitudes of 22° to 22° 45 east, and experiences a Mediterranean climate. The elevation of the wadi is up to 300 m above mean sea level (m amsl) in the upstream direction (Fig. 3.3a). The average annual rainfall is 260 mm and mean temperature is about 20 °C (long term average for the years 2000–2023). The rainy season is from October to March with January and December as the wettest months (rainy days are 35–50). The rain is of torrential character, causing surface runoff, originating alluvial and erosion phenomena. Every 3 or 5 years there are particularly heavy rains causing considerable floods.

The geology is formed by hard rock formations and geological structures that have great influence on the basin features. Derna Formation and Al-Abrak Formation have caused the formation of cliffs. The wadi is cut into marly limestones with hornstones and is backfilled with an asymmetrical filling of terrace deposits, sandy coarse gravel with cobbles and boulders, clayey sand and partly cemented conglomerate with carbonate substances. Wadi Derna is controlled by tectonics. The shape of the basin is semi rectangular, its relief ratio is 12.9 m/km and the sinuosity ratio is 1.38. The main activity in the basin is agriculture. Wadi Derna drainage networks with sub-basins extracted from the DEM of 30 m × 30 m resolution is given in Fig. 3.3b.

The basin of Wadi Derna can be divided according to its height above sea level into three parts as follows:

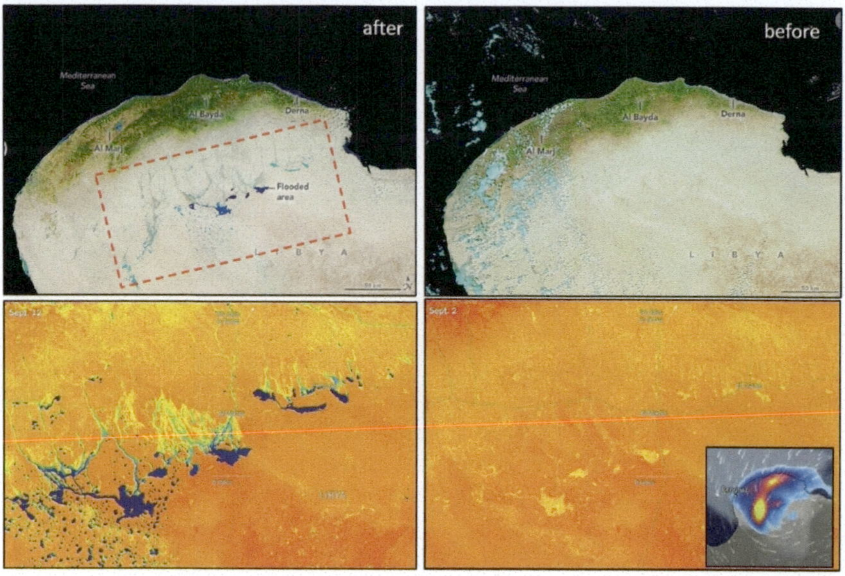

Fig. 3.2 Satellite imagery from September before and after the FF showing that wide lakes had emerged in the desert south of Derna in basins that are generally completely dry at this time of year. *Source* Modified Copernicus Sentinel data, 2023)

The upper part: This part constitutes about 60% of the area of the basin up to approximately 330km^2 and starts from the western border of the basin. It has a height of more than 650 m above mean sea level. Runoff in this part is very rare, Due to low slope, high porosity, and the prevalence of karst craters and dolines. Flooding is of low risk index in this part of the catchment.

The middle part: This part of the basin starts from the heights of Al-Washka (657 m masl) to the dam of Bu Mansour. Its surface area is about 140 km^2 with a length of up to 30 km and an average width of about 4.5 km. It is up to 435 m amsl. It is considered the narrowest part of the basin and comprises many dolines.

The lower part: This part covers the area between the Bu Mansour Dam to the center of Derna, its area is about 83 km^2 and its height is up to 214 m amsl where it begins to decline down to sea level. The valley flanks slope steeply at the Bu Mansour Dam, leading to high runoff intensity, which can turn into dangerous flooding according to Ashoor (2022). Due to the steepness of the valley, the dams were established in this part of the catchment.

The land cover/land use (LC/LU) in Wadi Derna can be classified into six main classes: forest, shrubs, cropland, grassland (>70%), barren land and built up area (Fig. 3.4). The analysis of the soil map shows that the soil in the catchment is mainly composed of clay soil, sandy to clayey alluvial deposits with low salinity.

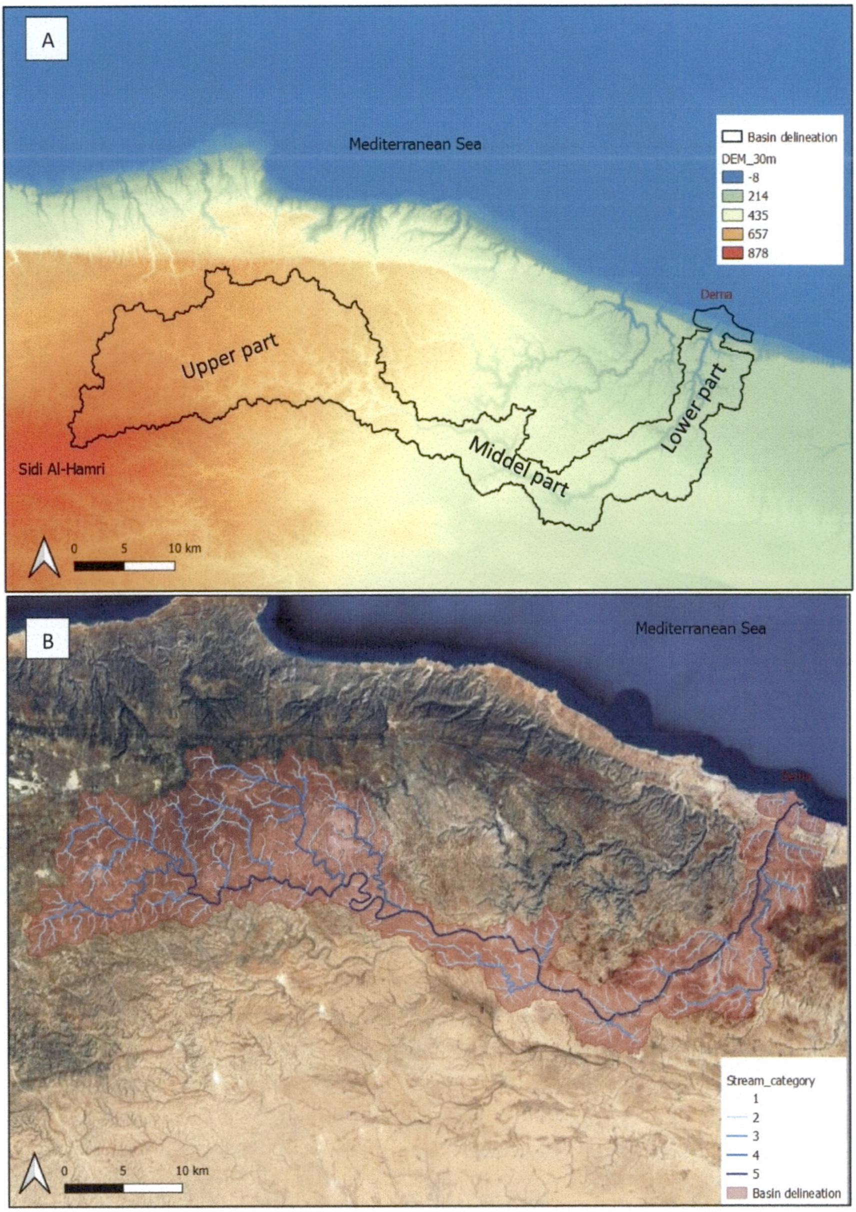

Fig. 3.3 DEM (**a**) and drainage networks with sub-basins extracted from the DEM 30 m × 30 m resolution (**b**)

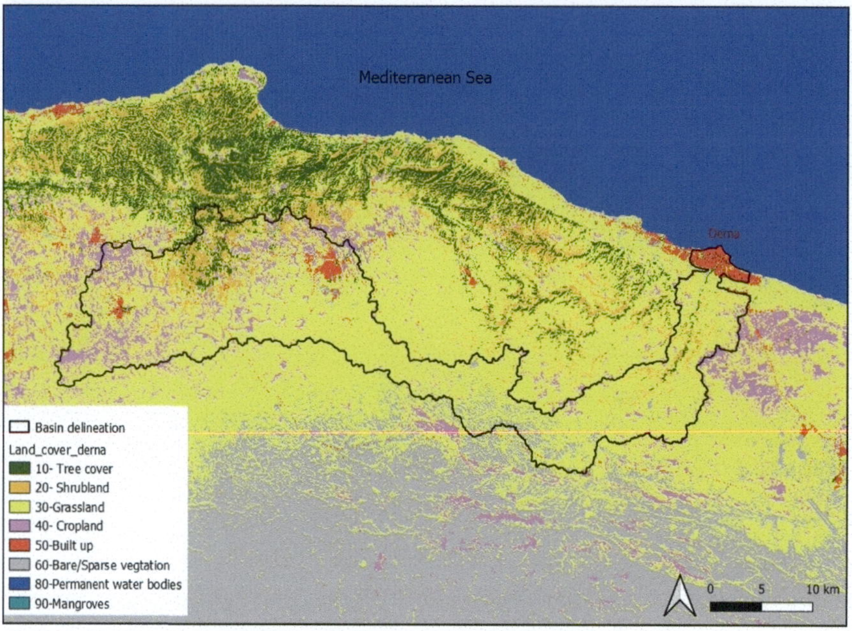

Fig. 3.4 Land use units of Wadi Derna

3.1.2 Flooding History and Characteristics of Wadi Derna Dams

Wadi Derna has been subjected to intensive anthropogenic pressure mainly in the form of agricultural activities and grazing. Two dams were built in the wadi in order to control soil erosion, to be used for irrigation purposes and to protect the city from flooding. Derna's vulnerability to flooding was well known. In fact, the city was flooded several times during the years before the construction of the dams. The historical floodings were: the 1941 flood, which caused great losses to the German army, the flood of 1956, the catastrophic flood of 1959, the 1968 flood, and the 1986 flood for which, despite its size, the dams played their role in preventing damage to the city. The flood of 1959 remains the most painful in the city's memory due to the number of victims and material losses before the September-2023 flood. The flood volume has been estimated by Ashoor (2022), based on the flood of October 1945 and late November 1986. Those events called for average precipitation of 145 and 64.14 mm respectively. The rainfall of 1945 produced a volume flood of 53.36 Mm^3, which represents 40% of annual runoff volume, while the flood of November 1986 was 14.8 Mm^3.

At the beginning of the seventies, Hydraproject company built two cumulus dams (the heart consisting of mud, and the sides of stones and rocks) on the course of the valley, the Al-Bilad Dam (downstream), which is about 10 km south of the heart of the city with a storage capacity of about 1.5 million cubic meters (Mm^3), and the Abu Mansour Dam (upstream) called also Wadi Derna Dam, about 10 km south of the first dam, which is a large dam with a capacity of about 22.5 Mm^3. The reservoir basin is in marly-limestone formations with comparatively low permeability, locally there are more permeable areas and karsts and hence some 'local' leakage points. The catchment area above the downstream dam has been evaluated at 555 km^2. The catchment area of the upstream dam is 476 km^2. The reservoir is for flood control and storage for irrigation. The downstream dam crest is up to 45 m and the crest length is 100 m. The reported upstream dam crest is about 75 m according to Stocky (2003).

The upstream dam is of rock fill with a quite slender section clay score. The upstream slope of the rockfill is hand-placed to form a smooth rip-rap surface at a slope of 1:1.5 The downstream slope has two intermediate berm es at 25 and 35 m and the rock fill is similarly finished to slopes of 1:1.4. The clay core is formed of "clay" vertically bellow the crest roadway with side slopes of 10:1. The dam reservoirs are normally filled by the runoff from springs in the Wadi and its tributaries. These keep the dam reservoir filled at a more continuous level than usual in Libyan dams. The overflow of the upstream dam is designed to carry a flood of 350 m^3s^{-1} (over a bell mouth-type spillway shaft with a diameter of 18m set at 41m level according to Stocky (2003). It seems that the upstream dam has suffered considerable settlement since it was commissioned (>1.5% of its height) in 1978 according to Stocky (2003). The construction of these dams ended the problem of flooding in the city since their construction in 1972–1973 till both dams failed in 2023 September the 11th, which was causing great loss of life and properties.

3.2 Methodology

To model flooding hazards, several 1-D, 2-D, 3-D, simplified conceptual, and data-driven models are available, which include HEC-RAS, TUFLOW, FASTER, SOBEK, DELFT3D, MIKE series, TELEMAC, InfoWorks, HYDRO_AS-2D, STREAM, LISEM (Ata, 2007; Rahmati et al., 2020; Ateeq-Ur-Rehman et al., 2015, 2020; Teng et al., 2017). Some of the models, such as TELEMAC, MIKE, HYDRO_AS and DELFT3D, are suitable for modelling projects with a limited study area. They are very detailed models and require precise boundary conditions/data with good computational resources. Although many flooding models have parallel processing options, their computation cost is often too high to simulate a large region or country. Therefore, for such cases we need a model/tool which should require less computational cost, is easy to use, and can accept remotely sensed data as input, as it is the case in Derna, where very limited information is available. In recent years, a new approach has been gaining popularity for modelling extensive floodplains (such as national-scale flood risk assessment) and data-sparse regions. These models are easy to implement using freely available remotely sensed data. Some examples of these models are STREAM (Aerts et al., 1999) and LISEM models (University of Twente,

2018). Although the STREAM model (Spatial Tools for River Basin and Environment and Analysis of Management Options), developed by the University of Amsterdam, is a GIS-based rainfall-runoff model which simulates flooding, droughts, water availability, and to some extent sediment transport, it has limited functionality regarding some hazards such as slope failure. However, these hazards can be modelled in the LISEM model.

Limburg Soil Erosion Model (LISEM), an open-source model developed by the University of Twente (Netherlands), is selected for this study. It was introduced by De Roo et al. (1996) as a physical and hydrological process-based model to predict runoff and soil erosion. LISEM model can simulate runoff flow, channel flow, channel flooding, erosion and sediment transport processes. In this study, the FF hazard analysis was conducted using Open LISEM applied in the method developed by Antea Group (Glas et al., 2019). The basic processes incorporated in the model are rainfall, interception, surface storage in micro-depressions, infiltration, vertical movement of water in the soil, overland flow, channel flow, detachment by rainfall and throughfall, transport capacity and detachment by overland flow. Schematic diagram of the modelling concept is given in Fig. 3.5.

The driver behind flow in the LISEM model is rainfall intensity. The rainfall is added to each grid cell's current water height, taking slope angle into consideration. The interception by vegetation is simulated using the canopy as simple storage. To produce an analysis of flood hazard, the data as presented in Table 3.1 are required. For the distribution of overland and channel flow routing, a four-point finite difference

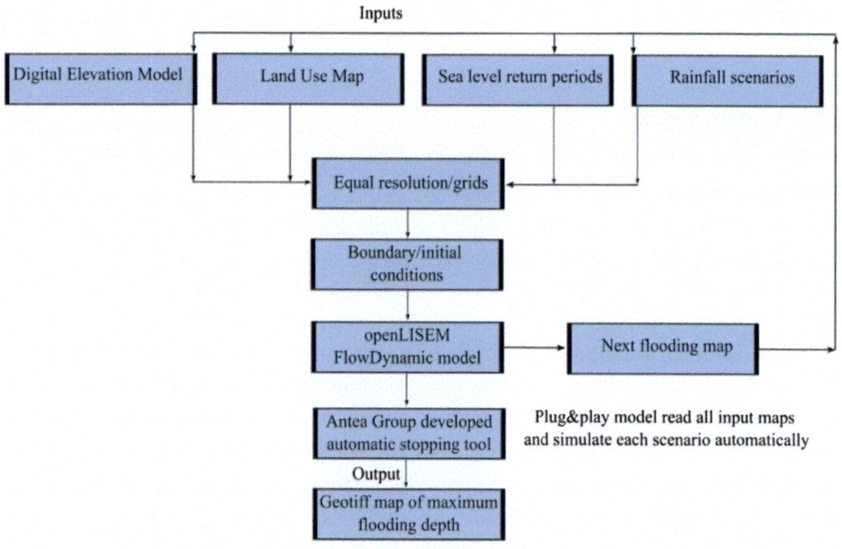

Fig. 3.5 Schematic overview of the general process that was applied to simulate the inland flood heights

Table 3.1 Overview on the data sources used to perform the draft Inland hazard mapping

Dataset (+ infolink)	Data provider	Parameter	Spatial resolution (m)
ESA WorldCover	ESA	Land use/Land cover	10 m × 10 m
SoilGrids	ISRIC World Soil Information	Soil properties: sand, silt, and clay content, as well as soil classes	Undefined
TAMSAT	www.tamsat.org.uk	Meteorological data	0.25° × 0.25° (27.5 km × 27.5 km)
GLO-30	Copernicus	DEM	30 m × 30 m

solution of the kinematic wave is used with Manning's roughness equation. The transport and water flow can be simulated using differential equations that numerically describe the flow. Mass and momentum conservation equations are used in hydrodynamic simulations. For model stability, we use Courant number (CFL) criteria where CFL remain less than 1.0 by changing time steps in simulations. Detailed procedure of the numerical solution can be found in the University of Twente (2018).

All input maps listed in table were preprocessed using the PCRaster program developed by Utrecht University. Vegetation maps were derived from the land use maps. Classification of the different land use classes, index of the roughness used are given in Table 3.2.

The digital elevation model will be used to create slope gradient, channels (width and depth) and drainage direction. Using land cover maps, we will prepare Manning roughness maps.

Manning roughness is the most important calibration parameter in calibrating flow characteristics in shallow waters; it has been adjusted within its physical ranges in the calibration process. These properties were derived from the soil map, as described by Table 3.3.

Utilizing a rainfall map, we will classify the rainfall zones. The TAMSAT precipitation dataset contains long-term records (2000–present) on an hourly basis with a global coverage. The dataset is used because of its long-term records, which is a necessary input variable in the Open LISEM model.

3.3 Results and Discussion

3.3.1 Flood Depth and Impact

One of the main objectives of this work was to identify the causes of the catastrophic flooding and ways to prevent and mitigate the potential consequences of a future flood occurrence in the study area. The simulated model of the 100 years' return period showed a high-risk of the whole floodplain area of Wadi Derna and the city downstream, where utter devastation could occur (Figs. 3.6 and 3.7). Analysis of the

Table 3.2 Classification of the different land use classes, index of the roughness

Land use/cover	Nr	Random roughness (cm)	Manning's (n)	Plant height (m)	Cover
	0	1	2	3	4
Densely Vegetated Farming	1	1.0	0.03	1.0	0.8
Eroded Agricultural Land	2	0.7	0.03	0.5	0.5
Flatland Intensive Farming	3	1.0	0.03	1.0	0.8
Grasslands	4	0.5	0.10	0.2	1.0
Grasslands and Open Wood	5	0.5	0.10	0.2	1.0
Intensive Farming	6	1.0	0.05	1.0	0.8
Mangrove	7	1.0	0.10	5.0	0.9
Mixed Farming	8	1.0	0.10	5.0	0.9
Natural Tropical Forest	9	1.0	0.10	5.0	0.9
Plantation Forest	10	1.0	0.10	5.0	8.0
Rock and Exposed Soil	11	0.5	0.01	0.0	0.0
Rural Settlement	12	0.5	0.05	5.0	0.1
Scrub Forest	13	1.0	0.10	5.0	0.9
Urban Settlement	14	0.5	0.05	5.0	0.1
Water	15	0.1	0.05	5.0	0.9

simulated flood for a 200 years' return period in Derna city (Fig. 3.8), showed a good correlation with the flood discharge value and extent of the event on early September 11, 2023 that was extracted from the Sentinel-2 image. This model was validated using remote sensing interpretation of the real flood event of September 2023. It is clear that the rainfall event in September 2023 was greater than the dam's reservoir capacity. The area where the calculated depth of the flood is > 7m in the model, is corresponding to the most impacted area, including the areas at the city center and at the coast downstream which were totally swept into the Mediterranean (Figs. 3.9 and 3.10).

During the FF there was a sudden rise in the water level in wadis (max. depth 46 m in the upstream dam) and connected small streams, and the flow velocity became very high (up to 15 ms^{-1}) due to steep slopes, limited spaces for retention reservoirs compared to the cumulative collected surface water runoff and due to impermeable surface soil layers (high clay content on the top of marly limestone layers, and already saturated by rainfall prior to the storm). The flash flood in Derna occurred in a matter of a few hours or even in minutes from the triggering event (cascading rainfall and

Table 3.3 Soil properties derived from soil map

Soil texture	Residual water content (–)	Wilting point (–)	Field capacity (–)	Total porosity (–)	Pore-size distribution index (–)	Saturated Hydraulic conductivity (cm/hr)	Wetting front suction (cm)
Sand	0.02	0.033	0.048	0.437	0.694	21.0–23.56	9.62–10.6
Loamy sand	0.035	0.055	0.084	0.437	0.553	5.98–6.11	11.96–14.2
Sandy loam	0.041	0.095	0.155	0.453	0.378	2.18–2.59	21.53–22.2
Loam	0.027	0.117	0.20	0.463	0.252	1.32	17.50–31.5
Silt loam	0.015	0.133	0.261	0.501	0.234	0.68	32.96–40.4
Sandy clay loam	0.068	0.148	0.187	0.398	0.319	0.30–0.43	44.9–53.83
Clay loam	0.075	0.197	0.245	0.464	0.242	0.20–0.23	40.89–44.6
Silty clay loam	0.040	0.208	0.30	0.471	0.177	0.15–0.0	53.83–58.1
Sandy clay	0.109	0.109	0.232	0.430	0.223	0.12	46.65–63.6
Silty Clay	0.056	0.250	0.317	0.479	0.150	0.09–0.10	57.77–64.7
Clay	0.09	0.272	0.296	0.475	0.165	0.06	62.25–71.4

failure of hydraulic infrastructure) that occurred in the upstream, subsequent overtopping, or water breaking through the earth dam, resulting in a landslide dam outburst flood. This event was followed by a large-scale devastating flood that inundated the area upstream of Bu Mansour dam, damaging infrastructure and inducing huge wave flows towards the downstream dam, leading to the second dam failure. This appears to have been a cascade failure, one dam causing the failure of the next dam downstream. The flood caused overtopping of the first dam, leading to erosion and failure, which then released its contents on the next dam downstream. Embankment dams, unlike watertight concrete arch or buttress dams, have been criticized as being vulnerable to extreme weather conditions, especially when the reservoir behind the dam fills up and overtops the structure. Figures 3.9, 3.10 and 3.11 show satellite images and photo captures of the impact of the devastating flood in the city.

The outburst occurred on 11 September 2023 and created a huge FF downstream in Derna City, washing out all infrastructures in its way to the sea. The maximum depth of the flood was more than 45 m near the upstream dam and 4–7 m in vulnerable areas of Derna city. The maximum velocity was 15 ms^{-1}. The peak flood was more than 40 times greater than the average normal flood depth. Due to continuous heavy

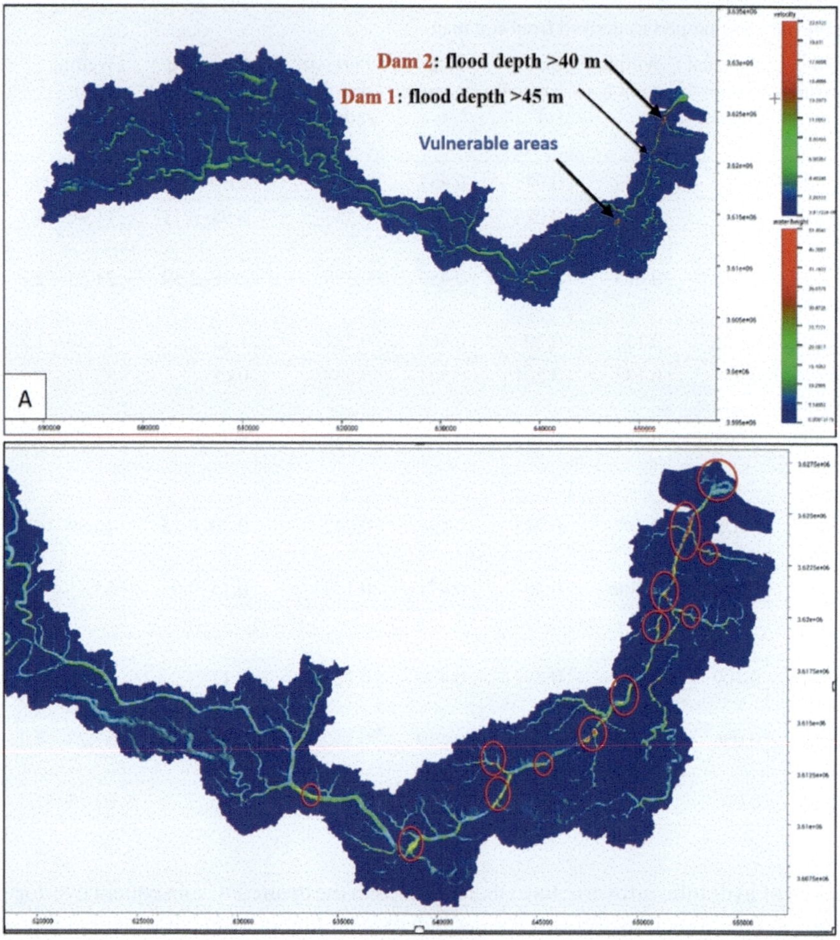

Fig. 3.6 Simulated flood of 100 years' return period 48h (**a**) with arrow indication to the two dams and (**b**) areas with highest depth flood map of Derna catchment (velocity in ms^{-1} and water depth in meters)

rainfall, the water body within the wadi grew 100–500 m wide downstream, 13,000 m long, average of 3 m flood depth and up to 45 m deep in the upstream dam. The total volume of the lake formed downstream near the second dam was about 41 Mm^3.

The force of the flow was so great as to tear away boulders, uproot trees, and destroy bridges and buildings that stood in its path (see Figs. 3.9 and 3.10). The wadi worked as a funnel, forcing the rushing water into the center of the city, leading to catastrophic damage. An analysis of satellite imagery and google foot print showed that about 800 buildings were completely demolished in the city center as a result of the flood and the subsequent landslide to some extent (Fig. 3.11).

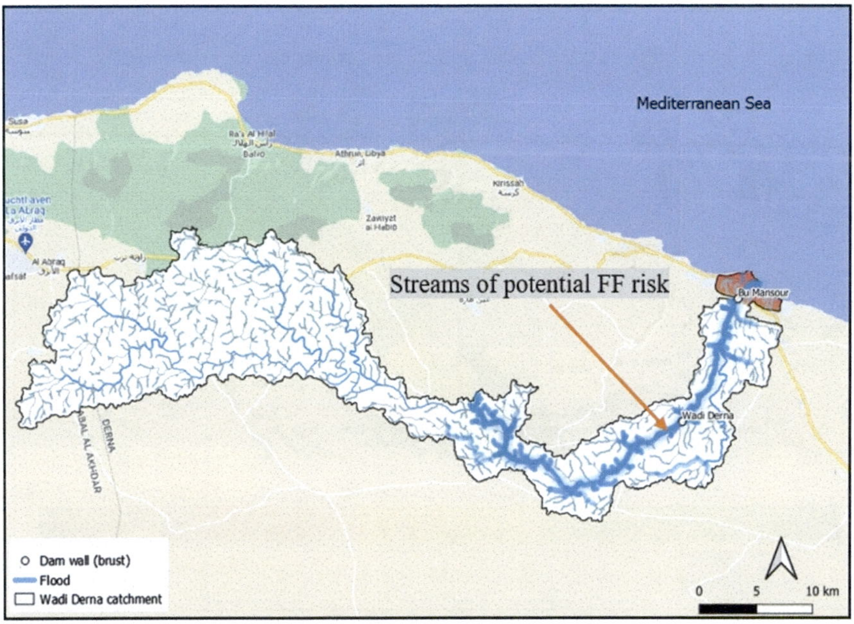

Fig. 3.7 Vulnerable areas to high FF potential risk along Wadi Derna

The FF brought also large amounts of debris and mud. The mud can be very clearly observed in the satellite image where it extended to more than 2 km in the sea (Fig. 3.12). Hydraulic structures and infrastructures in arid and semi-arid regions are subjected to severe sedimentation (Malmon et al., 2004; Hooke, 2019; Abbaszadeh et al., 2019; Tadda, 2023). During storm events, these structures suffer from debris flows and high concentrations of sediments that casus disruption of drainage network, culverts, and bridges like the case in Daniel storm flood event, where this disruption influences the downstream flow regime of Wadi Derna causing increasing the water level at bridges and surrounding inundated areas. Another reported issue in Wadi Derna is the lack of regular maintenance of wadi system to remove debris and sands. Dams should have a regular inspection program that takes into consideration all parts of the dam. Along with the impacts of global climate change, critical issues arise concerning the dam's safety and the adequacy of its spillways against extreme rainfall and floods in Libya.

The investigations unit of the Sanad agency on Derna flood analyzed satellite images obtained from the American company "Maxar" and counted the number of buildings that have been demolished. The data indicate that there were more than 4355 buildings before the flood, and after Storm Daniel, Sanad determined the survival of only 800 buildings in Derna city. By analyzing images by artificial intelligence techniques, it was found that at least 800 buildings had completely collapsed after the floods in the central area of Derna in particular. According to Sanad Technology's test and analysis results, the accuracy of the data and the information extracted reaches

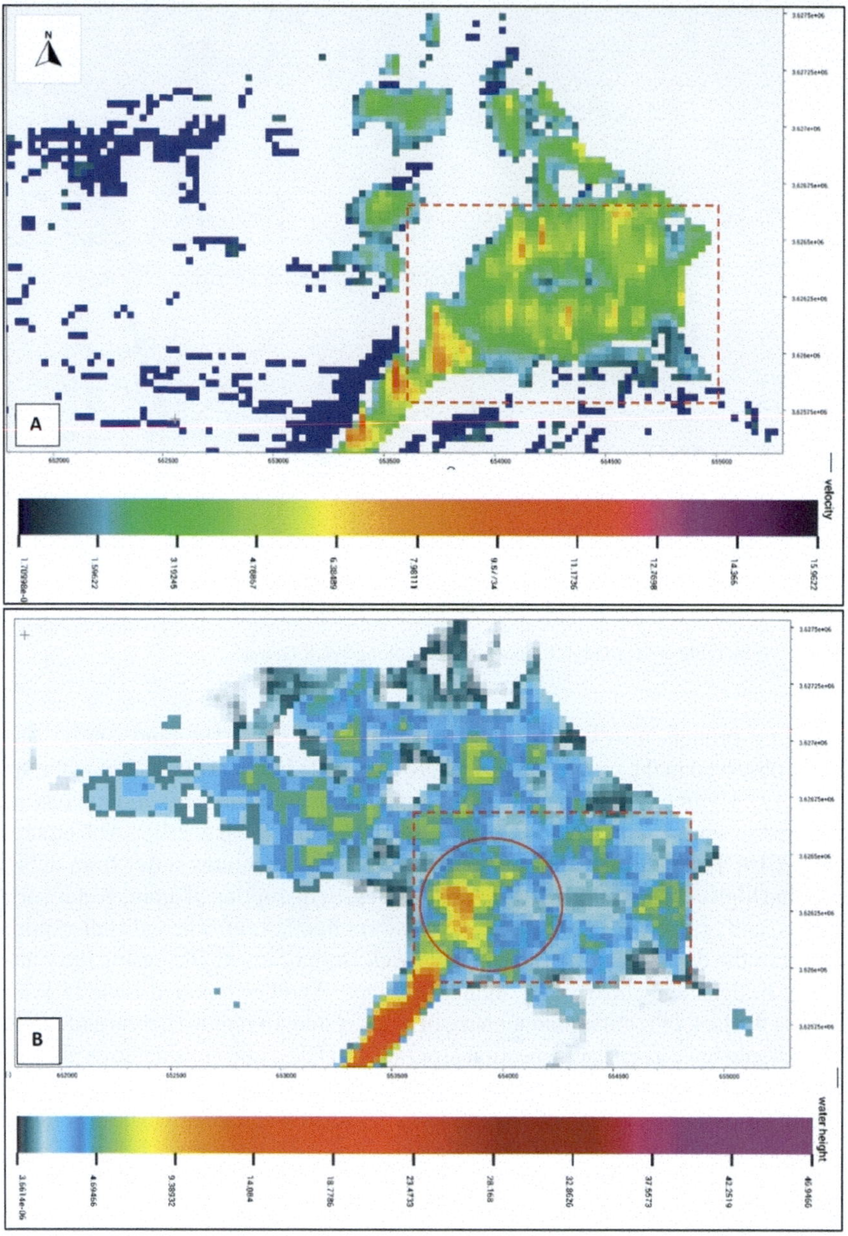

Fig. 3.8 Simulated flood of 200 years' return period (48 h) with simulated flood velocity in ms^{-1} (**a**) and flood depth in meters (**b**) in Derna city with indication of the highly impacted part in the southeast

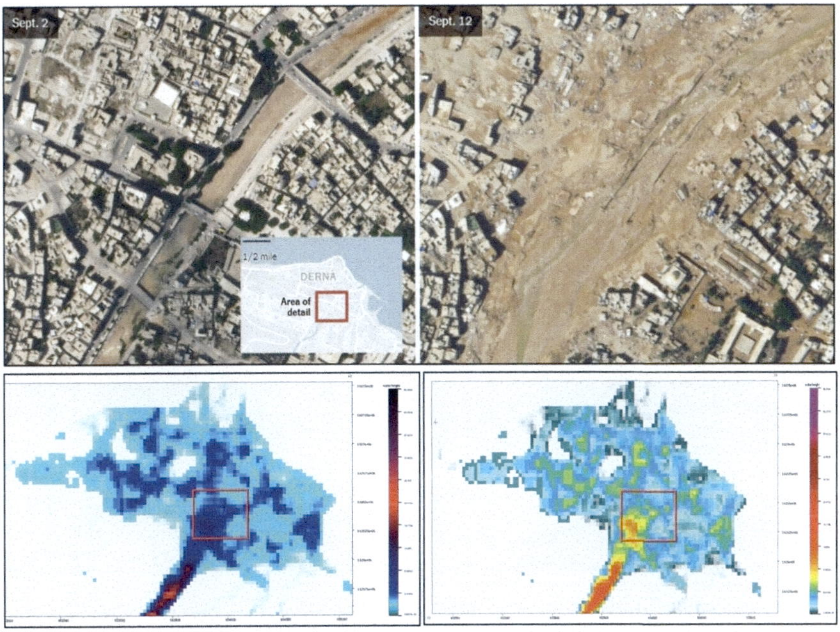

Fig. 3.9 Highly impacted areas by flooding along Wadi Derna flow path

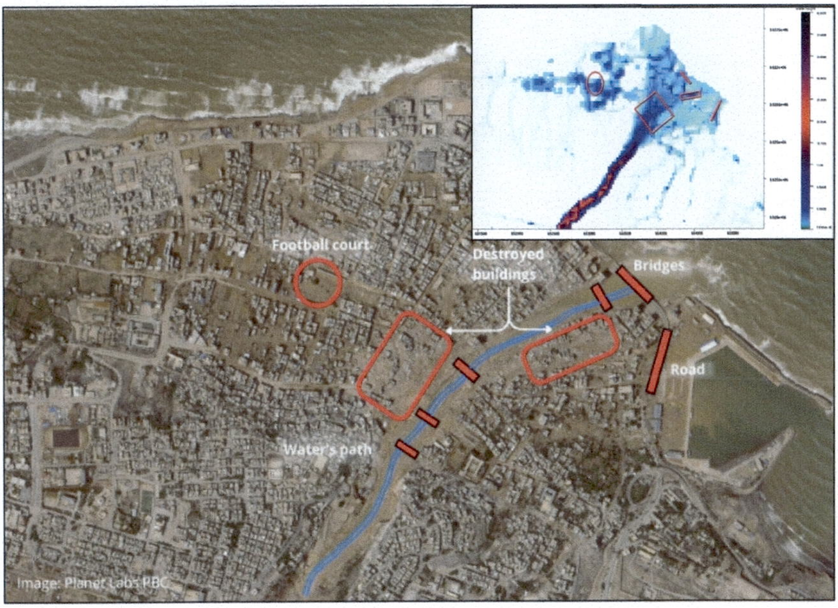

Fig. 3.10 Satellite images show the impact of the FF on the most vulnerable locations in Derna city where flood depth was > 7 m (CNES, 2023)

Fig. 3.11 Satellite images show the view of Derna dam before and after the two dams collapsed during the flood in Derna (right, CNES, 2023) and photos of the debris and landslides (left)

Fig. 3.12 Satellite imagery showing the impact of FF in Derna, Libya, with huge pools of mud-colored water that extends up to 2 km in the sea (CNES, 2023)

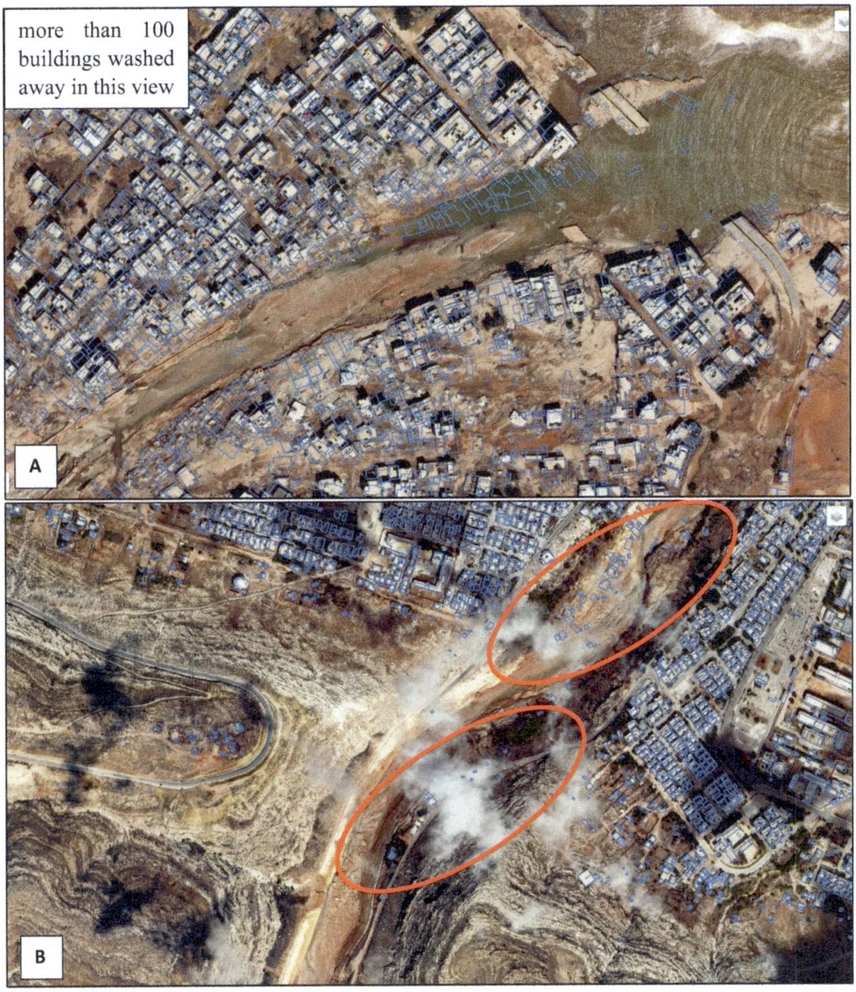

Fig. 3.13 Satellite image with google footprint showing impacted buildings from the flood downstream (**a**) and the informal settlement at the side of the wadi (**b**)

90%. Figure 3.13 represents a satellite image with google footprint showing impacted buildings from the flood downstream and the informal settlement at the side of the wadi which lead to rise the fatality loses from the FF event.

3.3.2 Capacity of the Reservoirs

Another factor making the flood event so devastating was the catchment ground being saturated by water resulting from rainfall prior to the storm. Figure 3.14 shows long-term cumulative rainfall in 24 h (a) and 48 h (b) for a period from 2000 to 2023. The intense rainfall of Daniel storm starts after about 20 h from the onset of the precipitation in the catchment.

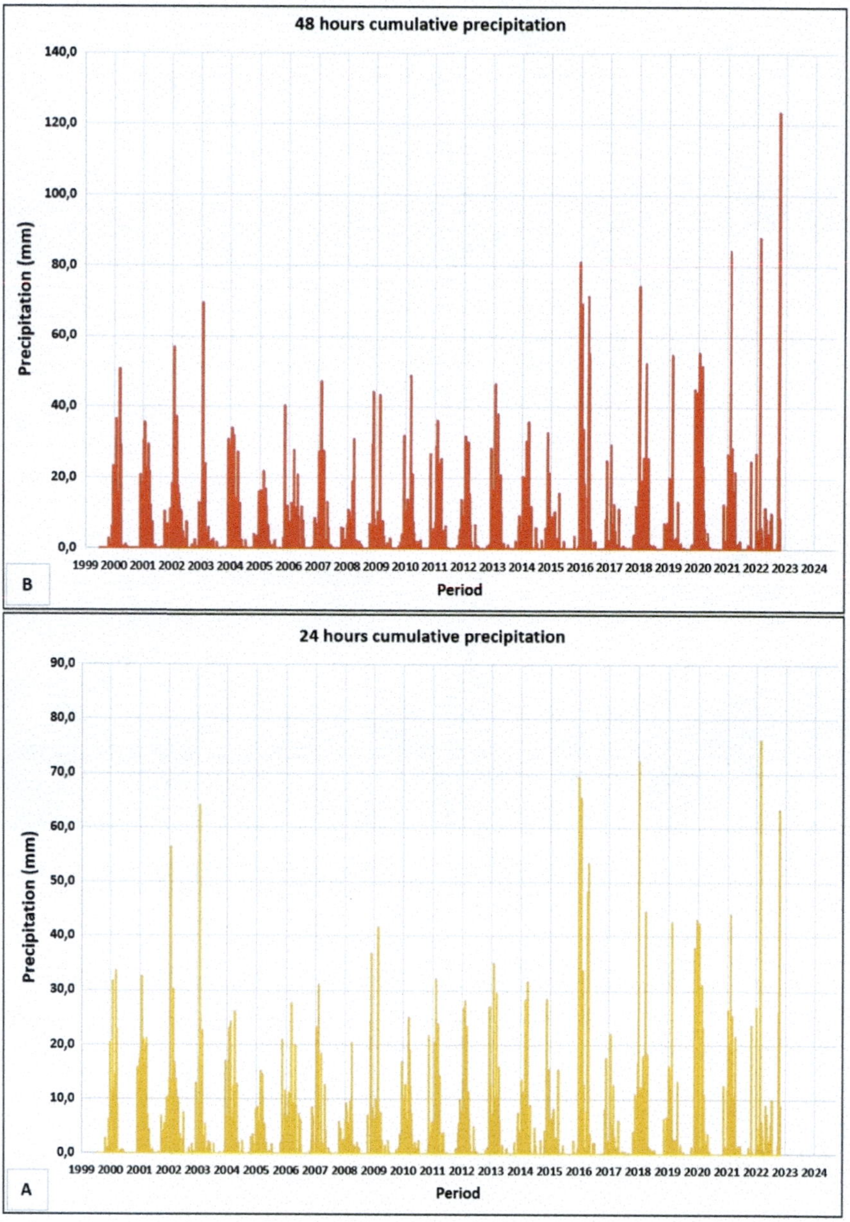

Fig. 3.14 Long-term (2000–2023) 24h (**a**) and 48h (**b**) cumulative rainfall (rainfall data TAMSAT)

The upstream dam of Wadi Derna is located about 13 km south of Derna City at an elevation of about 200 m amsl and the downstream dam "Al Bilad Dam" was located at 1.0 km from the city's southern boundary, at 45 m amsl. The reported designed storage for Wadi Derna Dam was about 22.5 Mm^3 with a height of 75 m, while for Al Bilad Dam downstream, it was only 1.5 Mm^3 with 45 m height (Stocky, 2003). However, the reported values of dam height and storage volume do not align with the topography of the valley which is also reported by Petley (2023). An analysis of the DTM indicates that the elevation of Wadi Derna Dam was only 45 m, with a corresponding storage of 22.5 Mm^3. This estimate is consistent with the findings in Ashoor (2024). The model simulation also shows a maximum flood depth of 46 m water at overtopping of the bank of Wadi Derna Dam.

A analytical analysis of storm runoff volume and the volume from all return periods given in Fig. 3.15 shows that the estimated runoff volume for the watershed was 41.13 Mm^3 for Daniel Storm rain event (average of 145 mm) which corresponds to the 200 years' return period (average 149.89 mm) accumulation of 48 h. Just prior to the dam failure, at 1:30 a.m. on September 11, 2023, the peak discharge reached more than the double volume of the designed peak discharge of 2000 m^3s^{-1} of Wadi Derna Dam. The total runoff generated by Storm Daniel in the Wadi Derna watershed (41.13 Mm^3) is clearly much greater than the capacity of the two dams of 23.5 Mm^3 combined.

Wadi Derna Dam flood discharge is sized to carry 350 m^3s^{-1}. The flood analysis performed, based on extreme rain event return periods and the flood routing by

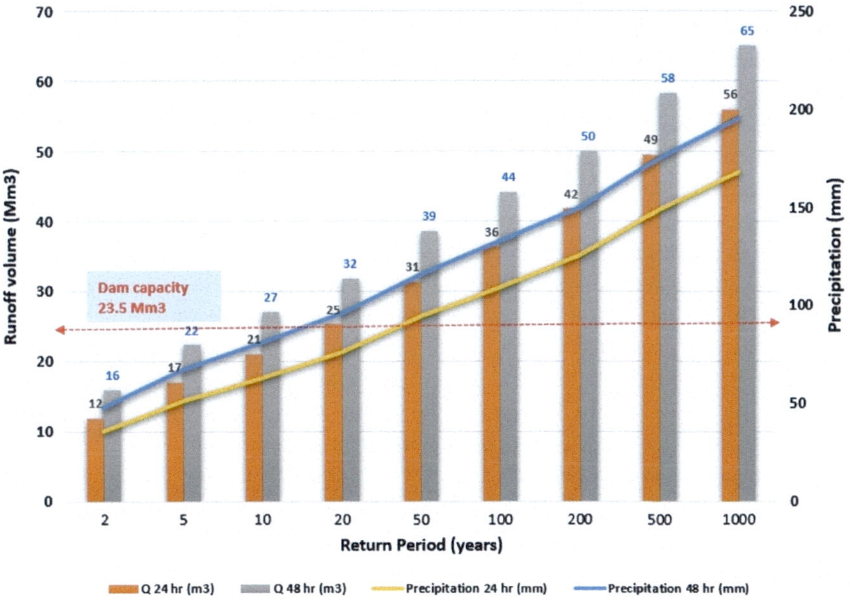

Fig. 3.15 Long-term cumulative rainfall for 24 h and 48 h with the runoff volume for all return periods (rainfall data TAMSAT)

the upstream and downstream dams in their present configuration, shows that the 1000 years' flood at Wadi Derna Dam is 56 Mm^3 (cumulative rainfall of 24 h) and could be 65 Mm^3 (cumulative rainfall after 48 h). Any type of future dam proposed should consider the geometry of the valley, the need to allow the passage of the 1000 years' flood estimated at 65 Mm^3, plus a safety factor. This result is comparable to Stucky Company's proposal in their design flood offer, based on the average rainfall for the design storm of a 1000 years' return period, producing a flood with a volume of 67 Mm^3 meters with a maximum discharge of 2500 $m^3\ s^{-1}$.

Flood protection dams are important to the downstream areas, and their safety is crucial. Approximately 90% of the floodwater in Wadi Derna area came from the Bu Mansour wadi upstream of the dam site (flood depth > 45 m, overtopping the dam), causing a devastating flood in the downstream sector. The constructed tunnel was too small to deviate the reservoir water away from the existing dam site. Additionally, the rainfall event in September 2023 was greater than the dams' reservoir capacity. Another factor making the flood event so devastating was the catchment ground being saturated by water resulting from rainfall prior to the storm.

Finally, the results showed that a new dam proposal is urgently required to prevent the area from future flood events. The failed dams should be replaced with other dams based on a sustainable FRM plan and design. In this case, the reservoir capacity should be based on the Storm Daniel experience, and be at least 41.13 Mm^3, which is equivalent to the runoff volume for a 200 years' return period (see Fig. 3.15) warning that, for comparable storm, even this capacity will not be sufficient. The nature of Derna FF and the rapid response of the catchment (mountainous areas), demonstrates an extremely short time for warning and evacuation in these regions. There is a lack of flood forecasting models for these regions at one hand. On the other hand, no clear operational policies exist in Libya on the amount and time of water release for such a sudden flood potential. The lack of emergency operation rules has exacerbated the problem further. The lack of a digital central database in the country and documentation of past flood events, represents also a significant obstruction to any further flood management. Very few surveys and information are available about prior floods especially, which increases the difficulties for calibration and validation of the model, as well as prediction of future flood characteristics. Another emerging challenge in the country is cross-collaboration within different levels of governmental agencies with diverse sectors. A lack of coordination between various actors with different interests provokes conflicts, thereby posing difficulties in managing floods.

Rapid uncontrolled urbanization in Derna at both sides of the valley has largely increased the losses in this devastating flood. On one hand, the shift in residence of the population to urban areas raises flood exposure. In addition to infrastructures, the design, typology, and construction materials of buildings and houses have been adapted to dry and arid climates without further concerns about extreme rainfall and potential flooding. Nevertheless, the challenge of urbanization is strongly connected to the lack of appropriate flood management plans, which presents issues for sustainable city development.

3.3.3 Dam Failure Scenario

Based on the analysis of local people observations and field observations reported by Ashoor and Ahmed (2024), the overtopping time occurred just after the storm's peak rainfall value at about 1:30 a.m. (see Fig. 3.16), when water spilled over wadi Derna banks lead to the dam failure creating a huge wave that hit the downstream dam at about 2:15 a.m. The flood reached the city after around 50 min from the overtopping. According to eyewitness and field observations, there was also a second wave hitting the city at 3:30 a.m. after the second peak of rainfall at about 3:10 a.m. One of the evidences reported by Ashoor and Ahmed (2024), that many clocks had stopped working after the flood wave arrival time of 2:20 and 03:30 a.m., confirmed that the first major flood inundated the city by 2:20 a.m. and the second wave, that led to rising the water height, arrived at 3:30 a.m. The Bu Mansour dam failure caused an enormous flood wave, overwhelming a great part of the city leading to huge fatality losses and damage of properties. The second wave, which raised the flood depth up to the fourth floor for many buildings (simulated ~ 7 m depth) in the most vulnerable part of the city, lead to reducing the survival rate and increasing the number of missing people.

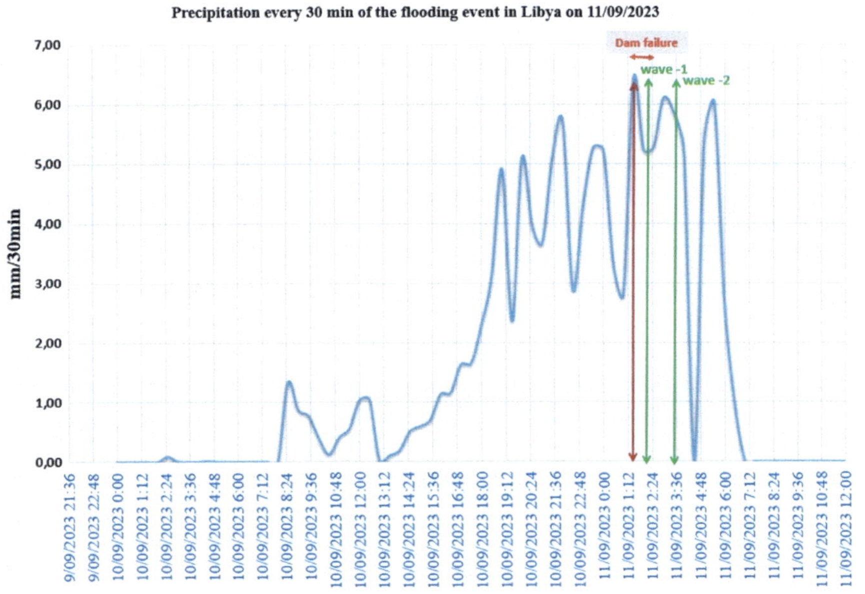

Fig. 3.16 Precipitation every 30 min of the flooding event in Libya and on early 11/09/2023 (rainfall data TAMSAT)

3.4 Conclusion

Libya is typically characterized by water scarcity and regular long-lasting drought due to low irregular annual rainfall. However, storm events in recent decades show that the country is extremely exposed to flooding. After a survival period of 50 years, heavy rainfall caused Wadi Derna Dam to fail and cause a major failure. The major failure of the dam occurred after a peak of intense rainfall. The process of the failure suggests that a combination of downstream slope failure and overtopping was the cause. The failure caused an enormous flood downstream. After about 50 min from the major dam failure, the flood arrived at the city center, where the peak discharge was > 350 m^3s^1. Due to the 50 years' gap between the building and failure of the dam, without reconsidering the catchment required storge this resulted in a disaster with enormous number of human casualties, large scale damage to infrastructure, loss of agricultural land, direct and indirect impact economic loss and harm to the ecosystem, while Derna city was completely destroyed over a surface area of more 70,000m^2 and swept into the sea.

This study shows that the main causes of high fatality losses in Derna include; poor design plan of dams and the presence of informal settlements at the sides of the wadi. Problem did arise when heavy rainfalls from Daniel storm occurred after about 19 h of cumulative precipitation, causing a rapid rise of water level. On one hand, the available capacity of constructed dams reservoirs were not sufficient for such flooding event. On the other hand, the rapid progress of floods leave a very short time by administrations to release the water downstream in due time. Therefore, in Derna's flood, poor design, bad management and planning combined appears to have been the causes of the dam's collapse. Due to instable political situation, dams in Libya have no regular inspection program that takes into consideration all parts of the dam. Along with the impacts of global climate change, critical issues raised concerning the dam's safety and the adequacy of its spillways against extreme rainfall and floods in Libya. The other safety measures relate to settlement in areas downstream at both sides of wadi. In Derna's flood case, there was clearly poor urban planning, thereby increasing exposure. The areas downstream from the dams should not have been constructed without safe and managed urban design. Concerning the high potential of sedimentation and debris flow in the region, there is a significant need for more accurate hydraulic models to investigate culvert and bridge blockages and assess the amount of sediments deposited at those hydraulic structures. The current situation of Libyan dams requires officials to take immediate measures to carry out periodic maintenance of existing dams and check the catchment required storage volume.

Planning based on extreme weather events is mandatory at the design stage of a dam. Simulations should be carried out to investigate several scenarios under current and future climate. Moreover, the effect of cascading events should not be forgotten. A multi-hazard assessment, thereby incorporating the combined impact of inland flooding, coastal flooding due to storm surges, erosion and landslides, would be the way to go.

References

Abbaszadeh, M., Mahdavi, R., & Rezai, M. (2019). Assessment of sedimentation process in flood water spreading system using IRS (P5) and supervised classification algorithms (case study: Dahandar plain, Minab city, south of Iran). *Remote Sensing Application: Society and Environment, 16*, 100269.

Abdel-Fattah, M., & Kantoush, S. A. (2018). Rainfall-runoff modeling for extreme flash floods in Wadi Samail, Oman. *Journal of Japan Society of Civil engineers B1 (Hydraulic Engineering), 74*(5).

Aerts, J. C. J. H., Kriek, M., & Schepel, M. (1999). STREAM (Spatial tools for river basins and environment and analysis of management options): Set up and requirements. *Physics and Chemistry of the Earth, Part B: Hydrology, Oceans and Atmosphere, 24*, 591–595.

Ahmed, A., Al Maliki, A., Hashim, B., Alshamsi, D., Arman, H., & Gad, A. (2023). Flood susceptibility mapping utilizing the integration of geospatial and multivariate statistical analysis, Erbil area in Northern Iraq as a case study. *Science and Reports, 13*, 11919.

Alhasanat, H. A. (2017). *Early warning system guidance to mitigate flash flood impacts in Petra Region, Jordan*. University of Salford.

Ali, H., & Mishra, V. (2018). Increase in subdaily precipitation extremes in India under 1.5 and 2.0 °C warming worlds. *Geophysical Research Letters, 45*, 6972–6982. https://doi.org/10.1029/2018GL078689.

Allen, M. R., & Ingram, W. J. (2002). Constraints on future changes in climate and the hydrologic cycle. *Nature, 419*, 224–232.

Al-Mamari, M. M., & Kantoush, S. A. (2019). Real-Time measurement of flash-flood in a Wadi area by LSPIV and STIV. *Hydrology, 6*(1), 27.

Al-Qudah, K. A. (2011). Floods as water resource and as a hazard in arid regions: a case study in southern Jordan. *Jordan Journal of Civil Engineering, 5*, 148–161.

Al-Zwam, S. M. (1995). *Study in geography natural Al-Jabal Al-Akhdar*. Benghazi University.

Ashoor, A., & Ahmed, E. (2024). *Estimating the depth of surface runoff in the Wadi Derna Basin by integrating GIS techniques and the SCS-CN model*. Preprint report by Research Square. https://doi.org/10.21203/rs.3.rs-3858769/v1.

Ashoor, A. A. R. (2022). Estimating the depth of surface runoff in the Wadi Derna Basin by integrating GIS techniques and the SCS-CN model. *Sebha University Journal of Pure & Applied Sciences, 21*, 22022. https://doi.org/10.51984/JOPAS.V21I2.2137.

Ata, R. (2007). *TELEMAC v7.2 user manual* (Open TELEMAC-MASCARET: Chatou, France).

Ateeq-Ur-Rehman, S., Riaz, Z., Bui, M. D., Rutschmann, P. (2015). Application of a 1-D numerical model for sediment management. In D. F. Lekkas (ed.), Dasu Hydropower Project. *Proceedings of the 14th International Conference on Environmental Science and Technology*. Global CEST.

Ateeq-Ur-Rehman, S., Jutta M., Ward S., Nils B., & Gert ,V. (2020). Numerical modelling of the effects of change in river channel morphology on flooding frequency in the Dijle valley, Belgium, using TELEMAC-2D modelling system. In W. A. Breugem, L. Frederickx, T. Koutrouveli, R. Kulkarni, K. Chu & B. Decrop (Eds.), *Online proceedings of the papers submitted to the 2020 TELEMAC-MASCARET User Conference*. International Marine & Dredging Consultants.

CNES. (2023). *Contains modified Copernicus sentinel data. Includes Pleiades material*. Distribution Airbus DS. Map produced by UNITAR/UNOSAT.

De Roo, A. P. J., Wesseling, C. G., & Ritsema, C. J. (1996). LISEM: A single-event physically based hydrological and soil erosion model for drainage basins. i: Theory, input and output. *Hydrological Processes, 10*, 1107–1117.

Dilley, M., Chen, R. S., Deichmann, U., Lerner-Lam, A., Arnold, M., Agwe, J., Buys, P., Kjekstad, O., Lyon, B., & Yetman, G. (2005). *Natural disaster hotspots. A global risk analysis*. The World Bank.

European Commission, (2007). Directive 2007/60/EC of the european parliament and of the council of 23 October 2007 on the assessment and management of flood risks. Off. J. Eur. Union 2007, 288, 27–34.

Farrokhzadeh, S., Hashemi Monfared, S. A., Azizyan, G., Sardar Shahraki, A., Ertsen, M. W., & Abraham, E. (2020). Sustainable water resources management in an arid area using a coupled optimization-simulation modeling. *Water, 2020*(12), 885.

Fiorillo, E., Crisci, A., Issa, H., Maracchi, G., Morabito, M., & Tarchiani, V. (2018). Recent changes of floods and related impacts in Niger based on the ANADIA Niger flood database. *Climate, 6*, 59.

Gaur, A., Gaur, A., Yamazaki, D., & Simonovic, S.P. (2019). Flooding related consequences of climate change on Canadian cities and flow regulation infrastructure. *Water 11*, 63. https://doi.org/10.3390/w11010063.

Gaur, A., & Simonovic, S. P. (2015). Projected changes in the dynamics of flood hazard in the Grand River Basin Canada. *British Journal of Environment & Climate Change, 5*, 37–51.

Glas, H., Rocabado, I., Huysentruyt, S., Maroy, E., Salazar Cortez, D., Coorevits, K., De Maeyer, P., & Deruyter, G. (2019). Flood risk mapping worldwide: A flexible methodology and toolbox. *Water,* (11), 2371.

Guha-Sapir, D., Vos, F., & Below, R. (2012). Annual disaster statistical review 2011. *Disasters, 52.*

Hamad, S. (2020). Surface runoff estimation of Wadi Ba Al-Arid watershed NE Libya using SCS-CN GIS and RS data. *Iranian Journal of Earth Sciences, 12*(3), 168–175.

Hannoun, D., & Tietjen, T. (2023). Lake management under severe drought: Lake Mead Nevada/Arizona. *Journal of the AMERICAN Water Resources Association, 59*, 416–428.

Hidroprojekat. (1972). *Wadi Derna Projekt: Flood protection of Derna Town.* Ministry of Agriculture, Libya.

Hooke, J. M. (2019). Extreme sediment fluxes in a dryland flash flood. Scientific Reports *9*, 1686. Extreme sediment fluxes in a dryland flash flood|Scientific Reports (nature.com).

Ingram, W. (2016). Extreme precipitation: Increases all round. *Nature Climate Change, 5*, 443–444.

Jackson, J. A., & Bates, R. L. (1997). *Glossary of geology* (4th Edn, 769 p). American Geological Institute.

Kundzewicz, Z. W. (2013). Floods: Lessons about early warning systems. In EEA (European Environment Agency), *Late lessons from early warnings: Science, precaution, innovation* (pp. 347–368). EEA Report no. 1/2013. European Environment Agency.

Liebe, J. R., Andreini, M., van de Giesen, N. C., & Steenhuis, T. (2007). The small reservoirs project: Research to improve water availability and economic development in rural semi-arid areas. In M. Kitissou, M. Ndulo, M. Nagel, M. Grieco (Eds.), *The hydropolitics of Africa* (pp. 325–332). Cambridge Scholars Publishing.

Lin, X. (1999). Flash floods in arid and semi-arid zones. In *International hydrological programme.* UNESCO.

López-Moreno, J. I., Vicente-Serrano, S. M., Beguería, S., García-Ruiz, J. M., Portela, M. M., & Almeida, A. B. (2009). Dam effects on droughts magnitude and duration in a transboundary basin: The Lower River Tagus Spain and Portugal. *Water Resources Research, 45*, W02405.

Mahmood, M. I., Elagib, N. A., Horn, F., & Saad, S. A. G. (2017). Lessons learned from Khartoum flash flood impacts: An integrated assessment. *Science of the Total Environment, 601–602*, 1031–1045.

Malmon, D., Reneau, S., Dunne, T. (2004). Sediment sorting and transport by flash floods. *Journal of Geophysical Research: Earth Surface, 109*, F02005 (Wiley Online Library).

Milly, P. C. D., Wetherald, R. T., Dunne, K. A., & Delworth, T. L. (2002). Increasing risk of great floods in a changing climate. *Nature, 415*, 514–517.

Moawad, B., Moawad, A., Omar, A., & Buhalqem, M. (2016). Flash floods in the Sahara: A case study for the 28 January 2013 flood in Qena Egypt. *Geomatics, Natural Hazards and Risk, 7*(1), 215–236.

Mohammed, R., & Scholz, M. (2017). Adaptation strategy to mitigate the impact of climate change on water resources in arid and semi-arid regions: A case study. *Water Resources Management, 31*, 3557–3573.

Omidvar, B., & Khodaei, H. (2008). Using value engineering to optimize flood forecasting and flood warning systems: Golestan and Golabdare watersheds in Iran as case studies. *Natural Hazards 47,* 281–296. https://doi.org/10.1007/s11069-008-9233-7.

Petley, D. (2023). *Further information about the Wadi Derna dams.* New information is emerging about the dams on Wadi Derna (eos.org). Accessed 02 Dec 2023.

Rahman, M. T., Aldosary, A. S., Nahiduzzaman, K. M., & Reza, I. (2016). Vulnerability of flash flooding in Riyadh, Saudi Arabia. *Natural Hazards, 2016*(84), 1807–1830.

Rahmati, O., Darabi, H., Panahi, M., Kalantari, Z., Naghibi, A., Ferreira, C., Kornejady, A., Karimidastenaei, Z., Mohammadi, F., Stefanidis, S., Bui, D., & Haghighi, A. (2020). Development of novel hybridized models for urban flood susceptibility mapping. *Science and Reports, 10,* 12937.

Rubinato, M., Nichols, A., Peng, Y., Zhang, J., Lashford, C., Cai, Y., Lin, P., & Tait, S. (2019). Urban and river flooding: Comparison of flood risk management approaches in the UK and China and an assessment of future knowledge needs. *Water Science and Engineering 12,* 274–283. https://doi.org/10.1016/j.wse.2019.12.004.

Saber, M., Abdrabo, K. I., Habiba, O. M., Kantosh, S. A., & Sumi, T. (2020). Impacts of triple factors on flash flood vulnerability in Egypt: Urban growth, extreme climate, and mismanagement. *Geosciences, 10*(1), 24.

Saber, M., Kantoush, S., Abdel-fattah, M., & Sumi, T. (2017). Assessing flash floods prone regions at wadi basins in Aswan, Egypt. *Disaster Prevention Research Institute Annuals B, 60*(B), 853–863.

Saharia, M., Kirsteetter, P., Vergara, H., Gourley, J. J., Hong, Y., & Giroud. M.: Mapping flash flood severity in the United States. *Journal of Hydrometeorology, 18,* 397–411.

Sen, Z. (2008). Hydrograph and unit hydrograph derivation in arid regions. *Hydrological Processes, 21,* 1006–1014.

Shakeel, M., Khan, A., & Ullah, S. (2010). Assessment of flash flood causes and associated damages in Dir Valley, Khyber Pakhtunkhwa Pakistan. *International Journal Disaster Risk Reduction, 2016*(16), 216–223.

SHMI (Slovak Hydrometeorological Institute). (2006). *Forward integration of flood warning in areas prone to flash floods.* Country Report on the Third pilot area is located in the North. West of Slovakia in the Senica District around the town of Vrbovce, submitted to the Associated Programme on Flood Management. www.apfm.info/pdf/pilot_projects/Flash_Floods_Slovakia_Pilot.pdf

Stockdon, H. F., Asbury, H., Jr., List, J. H., & Holman, R. A. (2002). Estimation of shoreline position and change using airborne topographic lidar data. *Journal of Coastal Research, 18*(3), 502–513.

Stocky. (2003). *Determination of the 1000-year flood of Derna and BU Mansur Reservoirs.* Ministry of Agriculture, Libya.

Tadda, M., Ahsan, A., Imteaz, M., Shitu, A., Danhassan, U., Muhammad, A. (2020). Operation and maintenance of hydraulic structures. In A. Ahsan (Ed.), *Hydraulic* structures. InTech. Available online: https://www.researchgate.net/publication/343189588_Operation_and_Maintenance_ofHydraulic_Structures. Accessed on 28 July 2023.

Taraky, Y. M., McBean, E., Liu, Y., Daggupati, P., Shrestha, N. K., Jiang, A., & Gharabaghi, B. (2021). The role of large dams in a transboundary drought management co-operation framework—Case study of the Kabul River Basin. *Water, 13,* 2628.

Teng, J., Jakeman, A. J., Vaze, J., Croke, B. F. W., Dutta, D., & Kim, S. (2017). Flood inundation modelling: a review of methods, recent advances and uncertainty analysis. *Environmental Modelling & Software, 90,* 201–216.

The Washington Post. Available online. https://www.washingtonpost.com/news/morningmix/wp/2017/07/17/az-flashflood-sweeps-away-family-relaxing-in-swimming-hole-9-perish

Tomohiro, T., Keiko, K., & Yasuto, T. (2020). Comparison of fluvial and pluvial flood risk curves in urban cities derived from a large ensemble climate simulation dataset: A case study in Nagoya Japan. *Journal of Hydrology, 584,* 124706.

Trenberth, K. E., Dai, A., Rasmussen, R. M., & Parsons, D. B. (2003). Te changing character of precipitation. *Bulletin of the American Meteorological Society, 84*, 1205–1217.

Trih, A. (1971). *Libya geography. Knowledge facility, Alexandria, Egypt.*

United Nations in Iran. (2019). Post Disaster Needs Assessment (PDNA) for Iran floods, Iran. United Nations in Iran, Tehran, Iran.

University of Twente. (2018). *OpenLISEM multi-hazard land surface process model documentation & user manual*. Faculty of Geo-Information Science and Earth Observation.

Vatanfada, J., Avarideh, F., & Tabatabaeei, M. R. (2009). Lessons learned from past floods towards integrated flood management in Iran. In *Proceedings of the Training Workshop of Integrated Flood Management, Esteghlal Hotel. Tehran, Iran*, 11–14 May 2009.

Vos, K., Harley, M. D., Splinter, K. D., Simmons, J. A., & Turner, I. L. (2019). Sub-annual to multi-decadal shoreline variability from publicly available satellite imagery. *Coastal Engineering, 150*, 160–174.

WMO. (2006). *Social aspects and stakeholder involvement in integrated flood management*. APFM Technical Document No. 4, Flood Management Policy Series, Geneva, Switzerland: World Meteorological Organization. www.apfm.info/pdf/ifm_social_aspects.pdf

WMO. (2007a). *Guidance on flash flood management—Recent experience from central and eastern Europe*. APFM, Geneva, Switzerland. World Meteorological Organization. www.apfm.info/pdf/pilot_projects/APFM.CEE.Synthesis_web.pdf.

Yang, L., Smith, J., Baeck, M. L., & Morin, E. (2019). Flash flooding in arid/semiarid regions: climatological analyses of flood-producing storms in central Arizona during the North American Monsoon. *Journal of Hydrometeorology, 20*, 1449–1471.

Chapter 4
Optimization and Simulation Model of the Expert Control System for Navigation Operation of Gabčíkovo Water Structure

Ľudovít Možiešik, Peter Šulek, and Martin Orfánus

Abstract Gabčíkovo Water Structure is a crucial water structure in terms of hydropower and navigation on the Slovak part of the Danube River. Navigation and hydropower are coupled and therefore solved together on the Gabčíkovo Water Structure and Danube river under the Expert Control System (ECS). ECS has two models: Optimization Model and Simulation Model. Optimization Model controls filling and emptying process of the navigation ship lock at the Gabčíkovo Water Structure. In order to perform this process, the software had to be based on theoretical analysis of the relationships of individual physical quantities and parameters expressed by mathematical apparatus. Simulation Model controls discharges and water level regime with respect to navigation safety requirements on the entire section of the Danube river affected by the operation of the Gabčíkovo Water Structure. The software tool for such a complex hydraulic process that would be able to evaluate the impact of the operation had to be developed. The paper describes the principles on which both software work. ECS is a part of the Gabčíkovo Water Structure upgrade and modernization project.

Keywords Gabčíkovo water structure · Navigation operation · Optimization · Simulation

4.1 Introduction

During the model research of the hydraulic filling and emptying system (HFaES) of the ship lock of the Gabčíkovo Water Structure in the 70s and 80s in the Hydrotechnical Laboratory of the Department of Hydraulic Engineering of the Faculty of Civil Engineering of the STU in Bratislava (prof. Kamenský et al.), the geometric shape

Ľ. Možiešik · P. Šulek · M. Orfánus (✉)
Department of Hydraulic Engineering, Faculty of Civil Engineering, STU, Radlinského 11, 810 05 Bratislava, Slovakia
e-mail: martin.orfanus@stuba.sk

A. Sefelnasr et al. (eds.), *Water Resources Management and Sustainability*, Water Science and Technology Library 114, https://doi.org/10.1007/978-3-031-80520-2_4

of the HFAES components (especially the channels in the bottom supplying and discharging water to and from the ship lock) and the course of opening and closing of the caps at the inlets to and outlets from the ship lock were optimized. The aim was to achieve a safe and sufficiently rapid filling or draining of the Gabčíkovo Water Structure ship lock, and strict criteria were set for the limits of this process. The water flow velocity in the canals was not to exceed 10 ms^{-1} (considering the stability and safety of the HFAES structural components), the filling or emptying time of the ship lock was to be 13 min (consideration of sufficient transport capacity of the ship lock). At the same time, the aspect of permissible forces in the mooring ropes of the vessels by which the vessels were moored to the floating battles in the walls of the ship lock and the aspect of exclusion of the phenomenon of cavitation from the whole HFAES were guarded.

The research also recommended the method of opening of control valves (CV) at the inlets (IN) and outlets (OUT) to and from the ship lock—opening and closing at a constant speed of movement of the servo valves, while the time of opening should be 10 min. when filling, 8 min. when emptying. A view of the original hydraulic model is shown in the following photo (Fig. 4.1).

However, due to various technological impacts and additional design changes, the recommendations of the research were not fully followed and some parameters

Fig. 4.1 Model of ship lock of the Gabčíkovo Water Structure in the Hydrotechnical Laboratory of the Department of Hydraulic Engineering of the Faculty of Civil Engineering of STU in the 1970s. *Source* Department of Hydraulic Engineering of the Faculty of Civil Engineering of STU in Bratislava

of the HFaES were changed—the routing of the canals on the outfall portion of the HFaES, the elevation of the canals, and the shape and method of bypassing both the CV IN and the CV OUT were also changed.

The above changes had a negative impact on the course of filling and emptying of the ship lock in terms of flow parameters—cavitation fields were formed in the HFAES channel area (especially at the outlets) and the flow in the post-flood section of the channels at the outlets acquired an intense turbulent character. In addition, during the emptying process, the effect of undercutting of the level in the ship lock below the level of the lower water in the final phase of the ship lock emptying occurred due to the large inertia value of the flowing water (up to 80 cm), which resulted in an unfavorable load on the controls of the lower gates. The following figure shows the HFAES channels with inlets and outlets (Fig. 4.2).

As a result of the above operating parameters, a need was triggered to change the CV movement pattern at the inlets and outlets so that the HFAES control system is able to maintain the established water flow parameters in the HFAES in order to minimize the negative effects of the flow in the HFAES and at the same time optimize

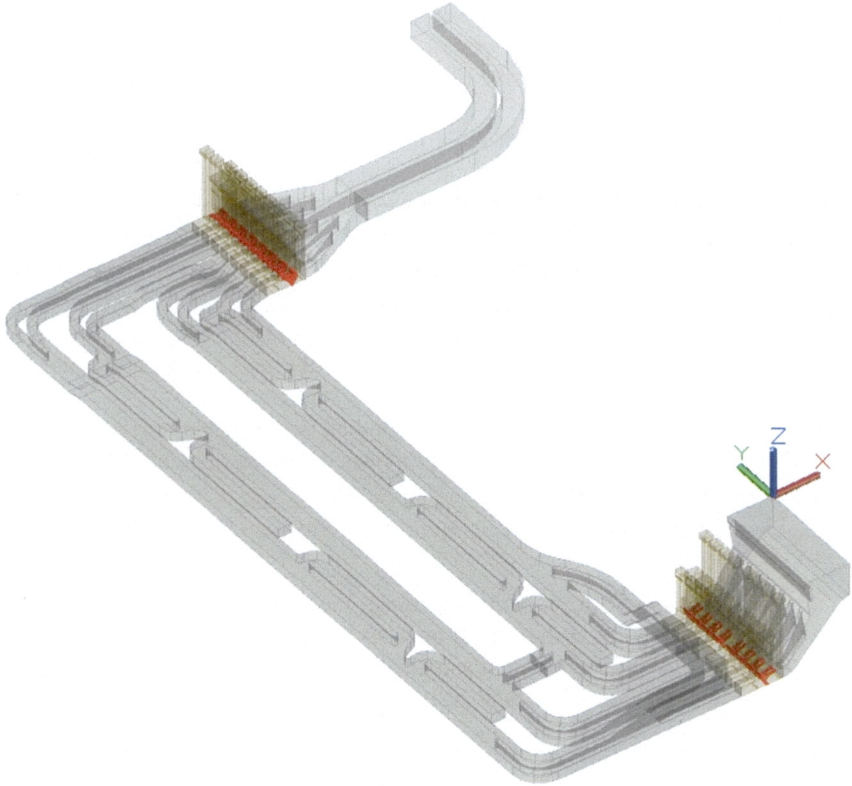

Fig. 4.2 HFAES channels with inlets and outlets

the filling and emptying pattern so that the filling and emptying time is as short as possible.

The ECS-OM is a system that controls the progress of CV movement so that specific criteria are met.

4.2 Methods

In order to ensure that the filling and emptying of the ship lock of the Gabčíkovo Water Structure is safe both in terms of the vessels being navigated and the construction and technological components of the ship lock (concrete structures, CV) and at the same time to make the filling and emptying of the ship lock as short as possible, the following conditions must be ensured by the ECS-OM:

- The ECS shall optimize the filling/emptying behavior when using all four CVs and when using all realistic combinations of CVs used (Bugarskia et al., 2013).
- The average profile velocity in the HFAES channels shall have a set value, allowing 2 different flow velocities to be set during one filling/emptying, with a change between them according to a set criterion (gradient at emptying, level at filling).
- The CV opening speed shall be selectable for two different speeds within a specified time of operation.
- The CVs must eliminate the effect of the inertia of the flowing water by their movement so that, when emptying, the undercutting of the level in the ship lock below the bottom water level does not exceed a specified value, which must be selectable on the basis of a calculation of the relationship between the permissible force in the servo pistons and the position of the bottom water level (Hermans, 2008).
- A manual mode setting shall also be provided for closing the CV in the process of flow braking—2 CV closing speeds shall be adjustable and 2 criteria for changing these speeds shall be provided—% CV opening and gradient.
- The rate of surface movement in the ship lock shall be adjustable and the ECS-OM shall operate in both winter and summer modes of operation at different rates of surface movement.
- The ECS-OM shall operate over the full realistic range of upper and lower water level positions.

In order for ECS-OM to be able to perform these functions, the work on the software was preceded by a theoretical analysis of the relationships of individual physical quantities expressed by mathematical apparatus. This theoretical analysis is part of the ECS-OM (Kanović et al., 2014a, 2014b).

Since the equations expressing the relationship between the time history of CV motion and the inflow/outflow of water to/from the ship lock also include terms expressing hydraulic losses and inertia parameters (filling/emptying coefficient m and the virtual length of the channels influencing the inertia λ) (Kašpar et al., 2021), the simulation of the ship lock filling/emptying process had to be calibrated, and the values of m and λ were calibrated for all the settings of the value of the water flow velocities in the ship lock channels used (Kosmas & Vlachos., 2012).

For the safety of the navigation operation of the VDG it is essential to know the impact of the operation of the VDG on the flow and level regime on the entire stretch of the Danube affected by the operation of the VDG. Due to the complexities of the hydraulic system of the VDG and the complexity of the management of the Gabcikovo Power Hydropower plant (hereinafter VEG) related to the new trends in electricity trading in a liberalized market environment, it is very difficult (if not impossible) to operate the VDG based on the experience of the operator alone. The operation and management of the VDG is conditional on the use of a consistent system that organizes/controls the release of flows through the VDG facilities in such a way that the safety of navigation is not compromised by restricting the fairway parameters (Liu et al., 2014). The current operator of the water structure (Vodohospodárska výstavba, š.p. - VVB) used software tools taken over from the previous operator (Slovak power plants, a.s.) for the preparation and operational management of the operation of the water structure. However, these tools were not suitable because they worked in the so-called hydro-thermal coordination mode and monitored flows and levels only up to the Medvedov Bridge (at that time the limiting ford section of the international waterway was close to this profile). The previous operator of the hydropower plant also had a sufficient amount of comparable energy resources creating suitable conditions for the possibility of operating the hydropower plant in the so-called flow regime and in this way de facto ensured the conditions for safe navigation on the section of the Danube concerned (Možiešik et al., 2017). The current operator's desire to maximize revenues from the sale of electricity makes it necessary to make more intensive use of the regulatory possibilities of the hydropower plant, which, however, has the effect of increasing the risk of endangering the safety of international navigation. Another disadvantage of the software tools used was that they could not simulate the operation of the hydropower plant and the other Gabčíkovo Water Structure facilities or the other associated power plants (Čunovo Hydro power plant, Moshon Hydro power plant, S VII Hydro power plant) on the basis of the specific power output values entered. As a consequence of the above, the need to develop and implement in the hydropower plant operation management system such a software tool that would be able to evaluate the impact of the operation of the Gabčíkovo Water Structure on the water level and flow regime of the Danube and thus ensure the safety of navigation operation on the entire stretch of the Danube affected by the operation of the hydropower plant was triggered (Mundy & Campbell, 2015).

4.2.1 Results and Discussion

The ECS-OM is part of the ship lock control computer system in the ship lock control tower. When the vessels in the ship lock have assumed the final ordered trim and are trimmed for battle, the ECS-OM system's designated rigger issues an instruction to simulate the CV's course of movement, at the same time an instruction is issued to close the appropriate gates. The simulation of the CV movement process takes on the order of up to 3 min, which is shorter than the movement time of the gates. If the simulation runs smoothly (the software reports the opposite and does not start the filling/emptying process), the filling/emptying phase occurs, i.e. the simulated CV movement process (Fig. 4.3).

Very strict conditions are imposed on the accuracy of the simulation compared to the following reality, especially in the final emptying phase, because the ECS-OM should not be mistaken in the calculation of the surface movement in the ship lock by more than an order of magnitude of cm due to the sensitive relationship between the position of the surface in the ship lock and the force acting on the piston of the servo valve controlling the lower gate (Šulek & Kinczer., 2019).

The following figures document the correspondence between the real ship lock emptying parameters and the ECS-OM simulation (level, flow rate) for the opening process shown in the first graph of the following three (Figs. 4.4, 4.5 and 4.6).

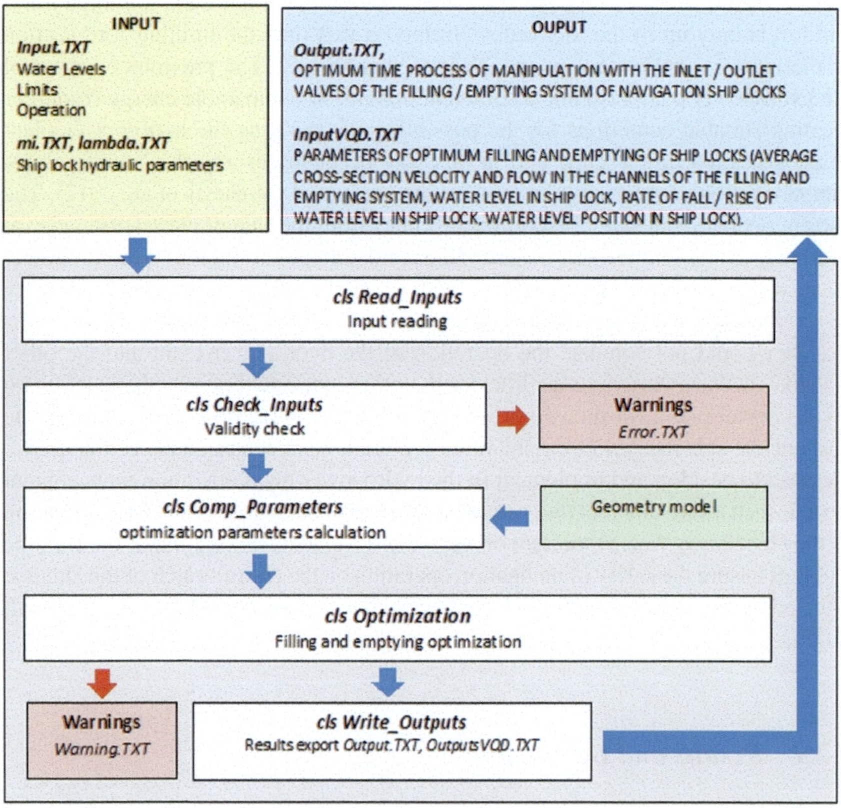

Fig. 4.3 Basic block diagram of ECS-OM

Fig. 4.4 Flow of CV movement at the outlets of the right SHIP LOCK at emptying designed by ECS-OM simulation

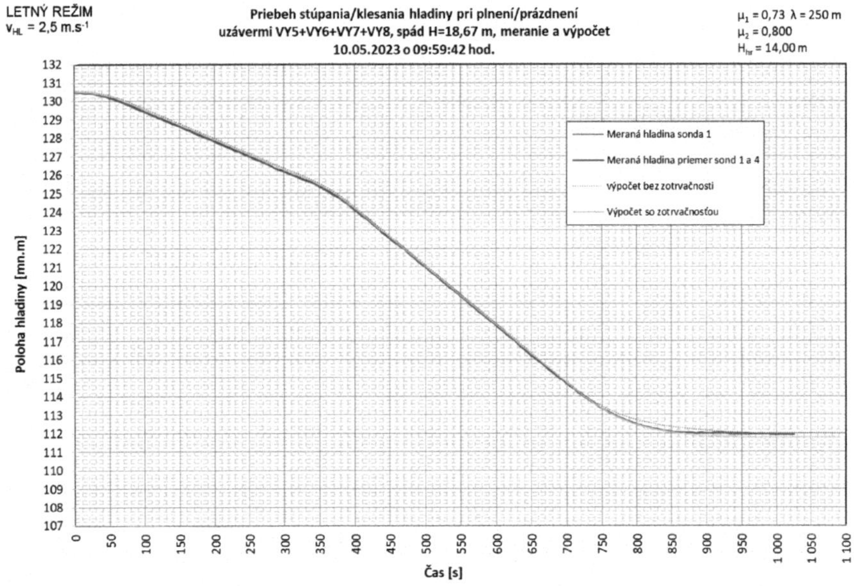

Fig. 4.5 Flow of the SHIP LOCK level movement during emptying—simulated and measured flow

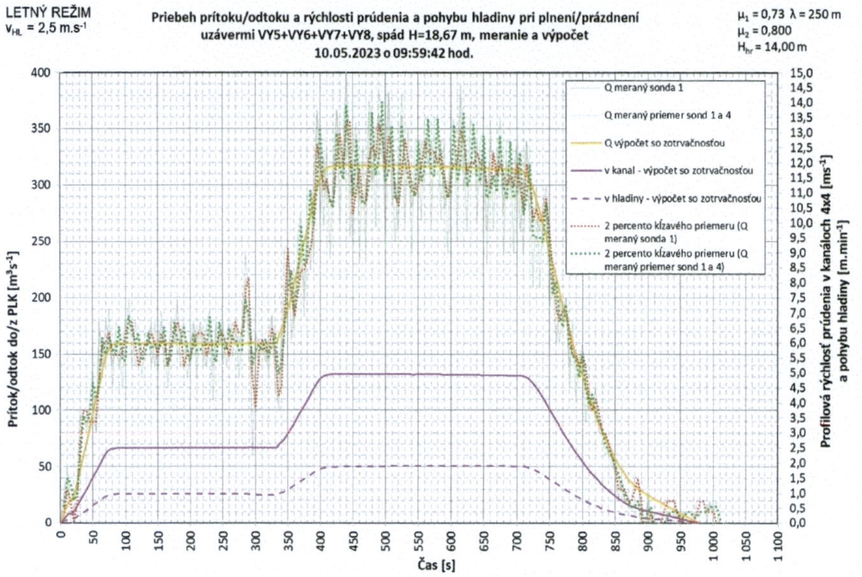

Fig. 4.6 The flow pattern when emptying the right SHIP LOCK—simulated and measured pattern, simulated flow velocities in channels and surface movement

ECS-SIM is a system that can simulate and evaluate the level and flow regime of the Danube to meet the above requirement (Ting & Schonfeld, 1998). In order not to jeopardize and limit the safety of navigation (e.g. by limiting the fairway parameters by underrunning the minimum navigation depths on the ford sections caused by the change of flows on the VEG and the consequent impact of vessels on the bottom of the Danube riverbed), the software solution used to control the water structure must provide the following functions:

- to process hydrological inputs and to model hydraulic conditions and hydraulic connections on the water structure or on the whole stretch of the Danube from Devín to Komárno, including the old Danube riverbed (Zhang et al., 2007),
- to provide a comprehensive picture of the level and flow regime of the Danube in the section Devín - Komárno with the possibility of direct integration of the outputs into the already existing complex information and control system for trading and production of electricity water structure,
- evaluate the proposed manipulation at the water structure for a specified time period (max. 48 h) in terms of the impact on the flow and level regime of the water structure in relation to the limits and restrictions given by the water structure manipulation regulations (e.g. max. and min. operating levels at the water structure objects),
- evaluate the proposed handling on the VDG in relation to the risk of fording or worsening of the conditions on the reported fords on the ford sections between Devín and Komárno,

- to work with a high degree of accuracy in the simulations (especially of the Danube surface regime), as the safety margin of vessels at the fords is only 20 cm. Max. The deviation of the simulation of the water level regime must be -10 cm (i.e. the real level may in reality be only 10 cm lower than the simulated level) or + 20 cm (i.e. the real level may in reality be only 20 cm higher than the simulated level),
- to work with the actual condition of the Danube riverbed and the Hrušov dam and with the actual location of the ford sections and their parameters, as this changes over time due to erosion and sediment transport. A continuous update of the channel morphology should be carried out at least once a year or after each passage of the channel-forming flow.

ECS-SIM forms an integral part of the existing ROVE SELT system implemented at the end customer (VVB). The basic inputs, outputs and interface of the ECS-SIM can be seen from the basic block diagram below (Fig. 4.7).

4.3 Conclusions

A study is currently being carried out to investigate the possibilities of adapting the ECS-OM on the port lock (ship lock) to the changed operating conditions. These changes are due to the need to rehabilitate the defects in outfall channels 2 and 3; the installation of stiffening steel pipes has changed the shape and size of the cross-section of the channels in the outfall section of the HFAES ship lock. As a result, the flow parameters in the most sensitive part of the HFAES just upstream of the CV at the outlets, where the velocity and pressure parameters determine the conditions for the generation and extent of cavitation behind the CV and the degree of flow turbulence, will change. The ECS-OM can respond to this by being adapted to control the movement of each CV and thus the water flow parameters in each channel separately.

ECS-SM is a model that requires a number of inputs that are continuously fed into the model during 24/7 operation—the inputs are mainly related to flow and level data and ford and strait parameters. The model is demanding on the accuracy of the inputs in terms of simulation accuracy requirements. Therefore, in addition to extensive and rigorous calibration of the model, it is important to continuously refine these inputs. This concerns the consumption lines at the gauging stations, the capacity curves of the water release objects and the accuracy of the hydrological forecasts of the SHMÚ. It is true that the more accurate the forecast, the more accurate the simulation of the flow-level regime.

Another factor affecting the accuracy of the ECS-SM is the updating of the model geometry based on channel shape measurements. The Danube is a river with an intensive sediment regime, so care must be taken to ensure that the channel shape in the hydraulic mathematical model replicates reality as closely as possible throughout the lifetime of the model.

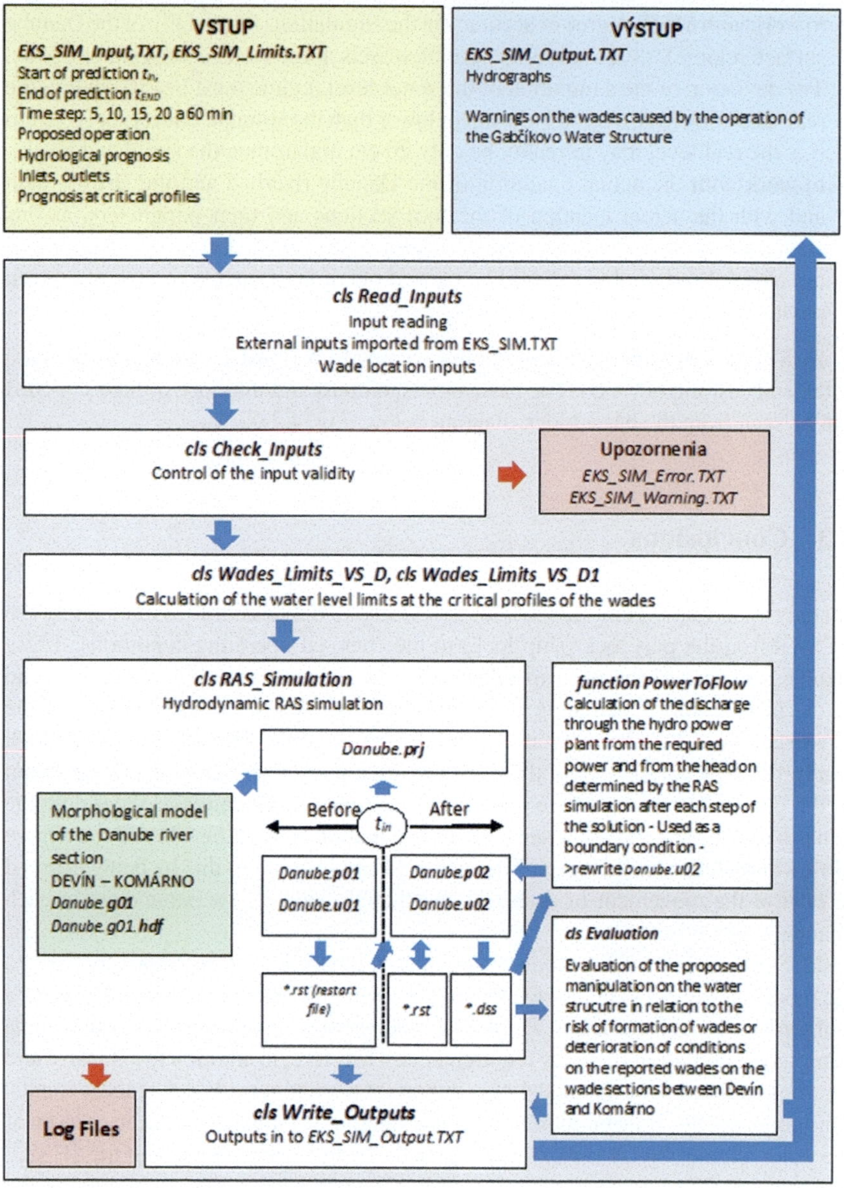

Fig. 4.7 Basic block diagram of ECS-SIM

Acknowledgements The preparation of this article was supported by the Research and Development Agency under contract No. APVV-20-0023

References

Bugarskia, V., Bačkalićb, T., & Kuzmanova, U. (2013). Fuzzy decision support system for ship lock control. *Expert Systems with Applications, 40*(10), 3953–3960.

Hermans, J. (2008). *Optimalisatie van binnenscheepvaart*. Master's thesis, KU Leuven.

Kanović, Ž., Bugarski, V., Bačkalić, T., Jeličić, Z., Petković, M., Matić, D. (2014). Optimization of ship lock control system using swarm-based techniques. *Journal on Processing and Energy in Agriculture, 18*(1), 30–35. ISSN: 1821-4487.

Kanović, Ž., Bugarski, V., Bačkalić, T. (2014). Ship lock control system optimization using GA, PSO and ABC: A comparative review. *Promet—Traffic & Transportation, 26*(1), 23–31. ISSN: 1848-4069.

Kašpar, T., Fošumpaur, P., Králík, M., Zukal, M. (2021). Parametric study of direct filling system of medium-head navigation locks. *Pollack Periodica, 6*(2), 43–49. ISSN: 17881994.

Kosmas, O. T., & Vlachos, D. S. (2012). Simulated annealing for optimal ship routing. *Computers & Operations Research, 39*, 576–581.

Liu, R. J., He, D. D., Wang, L. J., & Xu, H. Y. (2014). Ant colony optimization applied to the three gorges ship lock arrangement optimization. *Applied Mechanics and Materials, 543–547*, 1663–1666.

Možiešik, Ľ., Šulek, P., Orfánus, M., Hruštinec, Ľ. (2017). *Upgrade of Gabčíkovo locks. Feasibility study*, Bratislava.

Mundy, R., & Campbell, J. (2015). *Management systems for inland waterway traffic control*. Technical report. Retrieved from http://www.ctre.iastate.edu/mtc/reports/inland_waterway/volume1.htm.

Šulek, P., & Kinczer, T. (2019). Impact of differential evolution parameters on optimization of hydro-thermal coordination. *Pollack Periodica, 14*(3), 121–130.

Ting, C., & Schonfeld, P. (1998). Integrated control for series of waterway locks. *Journal of Waterway, Port, Coastal and Ocean Engineering, 124*(4), 199–207.

Zhang, X., Qi, H., Fu, X., & Yuan, X. (2007). Hybrid algorithm to minimize total weighted wait-time of ships for navigation co-scheduling in the three Gorges project. In *International conference on transportation, engineering* (pp. 2759–2764).

Chapter 5
Water Resources, Geology and Urbanization Are in a Direct Discipline: The Case Study of Jordan

Taleb Odeh, Alsharifa Hind Mohammad, Faten Alslaty, and Tino Rödiger

Abstract This study investigates the interconnections between water resources, geology, and urbanization in Jordan, a country grappling with severe water scarcity exacerbated by climate change and rapid urban growth. Jordan's diverse geological formations and varied climatic zones influence its topography, water resources, and land cover. Using Remote Sensing (RS) and Geographic Information Systems (GIS), the study assesses land cover changes over four decades, revealing significant urban expansion, especially in areas with higher rainfall, which has strained water resources and depleted aquifers. The decline in rainfall has further impacted agricultural productivity and water availability. To address the water crisis, the study advocates for a holistic approach that combines advanced geospatial technologies, scientific research, policy reforms, and community engagement, emphasizing water conservation, efficient irrigation, wastewater recycling, desalination, and robust water governance to ensure long-term sustainability.

Keywords Geology · GIS · Land cover · Remote sensing · Topography · Water resources

T. Odeh (✉)
Department of Water Management and Environment, Prince Al Hassan bin Talal College for Natural Resources and the Environment, The Hashemite University, P.O. Box 330127, Zarqa 13133, Jordan
e-mail: talebs@hu.edu.jo

A. H. Mohammad
WEEC, Water, Energy and Environment Center, University of Jordan, Amman 11942, Jordan
e-mail: s.jasem@ju.edu.jo

F. Alslaty
Department of Earth Sciences and Environment, Prince Al Hassan bin Talal College for Natural Resources and the Environment, The Hashemite University, P.O. Box 330127, Zarqa 13133, Jordan
e-mail: fatenm@hu.edu.jo

T. Rödiger
Department Catchment Hydrology, UFZ-Helmholtz Centre for Environmental Research, Theodor-Lieser-Str. 4, 06120 Halle, Germany
e-mail: tino.roediger@ufz.de

A. Sefelnasr et al. (eds.), *Water Resources Management and Sustainability*, Water Science and Technology Library 114, https://doi.org/10.1007/978-3-031-80520-2_5

5.1 Introduction

Jordan is situated in the western region of the Middle East, spanning approximately 89,342 km^2 (Fig. 5.1). Its geological composition is intricate, characterized by diverse rock units and geological structures (Bender, 1974; Salameh & Bannayan, 1993). The northern region predominantly comprises limestone, while basalt dominates the eastern sector, granite is prevalent in the southern part, and sandstone prevails in the western and central areas. Additionally, two significant regional geological structures, namely the Syrian arc systems and the Dead Sea Transform (DST) fault system, contribute to the formation of local faults and folds.

The Syrian arc systems manifest as discontinuous folds spanning from Sinai in Egypt, through the middle of Jordan and Palestine, to the southern part of Syria, originating during the Late Cretaceous due to the closure of the Neotethys Ocean. On the other hand, the DST fault system, a geomorphological rift extending over approximately 1000 km from the Maras Triple Junction to the northern section of the Red Sea Rift (Bender, 1974).

Both lithological composition and structural geology exert significant influence over Jordan's landforms, shaping its topography, and consequently impacting climate, water resources, and soil availability (Odeh et al., 2017). Consequently,

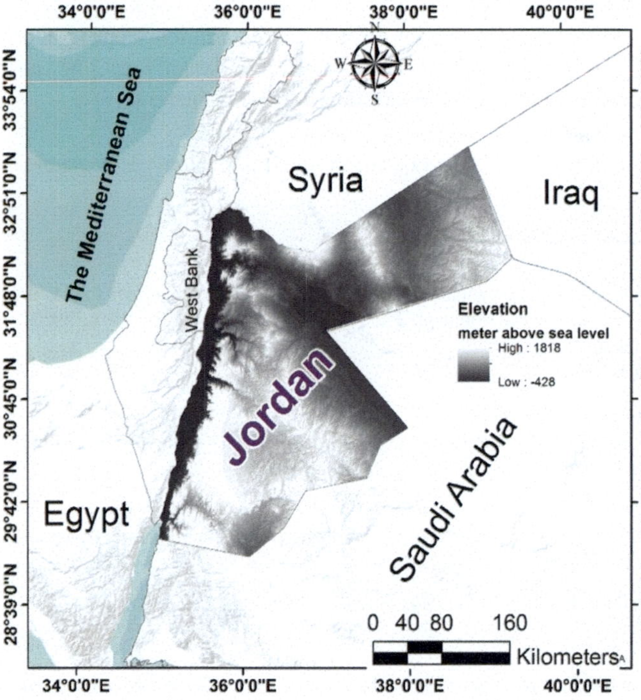

Fig. 5.1 Study area location

Jordan exhibits considerable variation in land cover from a spatial perspective. Anthropogeographically, Jordan has experienced multiple waves of migration from Palestine, Iraq, Kuwait, and Syria over several decades, resulting in substantial alterations to its land cover through urbanization and agricultural activities (Odeh et al., 2017; Salameh & Bannayan, 1993).

The water crisis in Jordan is characterized by a complex interplay of factors stemming from geological, hydrological, climatological, and anthropogenic influences. At its core, Jordan grapples with severe water scarcity, exacerbated by a rapidly expanding population and dwindling water resources (Rödiger et al., 2023). Hydrologically, Jordan faces significant challenges due to its limited precipitation and high rates of evaporation, resulting in meager surface water availability. The country relies heavily on groundwater extraction to meet its water demands, leading to unsustainable depletion of aquifers and declining groundwater levels. Furthermore, the quality of available water resources is compromised by pollution from industrial, agricultural, and urban activities, exacerbating the water crisis (Al-Hasani et al., 2023).

Climatologically, Jordan experiences harsh desert conditions characterized by hot, dry summers and sporadic rainfall, further exacerbating water scarcity issues. Climate change projections indicate potential alterations in precipitation patterns, exacerbating the vulnerability of water resources in the region (Al-Hasani et al., 2023). However, rapid population growth and urbanization place immense pressure on limited water supplies, driving increased demand for domestic, industrial, and agricultural use. Thus, the aim of this research is to assess the impacts of geology and population growth on the spatial distribution of land cover in Jordan and then evaluate the impact of the land cover changes (urbanization namely) on the water resources. However, the effect of climate changes on the water resources would evaluated accordingly.

Inefficient water management practices, inadequate infrastructure, and socioeconomic disparities further compound the water crisis, disproportionately affecting vulnerable populations and exacerbating social and economic inequalities (Tsakiris & Loucks, 2023). Addressing the water crisis in Jordan requires holistic and sustainable approaches that integrate scientific research, technological innovation, policy reforms, and community engagement (Christoforidou et al., 2023; Qiu et al., 2023). Strategies may include enhancing water conservation measures, promoting efficient irrigation techniques, implementing wastewater recycling and desalination projects, strengthening water governance frameworks, and fostering international cooperation for transboundary water management. By adopting a multidisciplinary and collaborative approach, Jordan can mitigate the impacts of the water crisis and ensure the long-term sustainability of its water resources for future generations (Rödiger et al., 2023).

5.2 Methods

The integrated approach of Remote Sensing (RS) and Geographic Information Systems (GIS) has become increasingly prevalent as a methodology for detecting land cover changes (Mohammady, 2021). This synergistic combination leverages the capabilities of RS in capturing detailed spatial data and GIS in analyzing and visualizing these data layers, enabling comprehensive assessments of land cover dynamics over time. Our utilization of this approach spans over twelve years across various case studies in regions like Jordan and Saudi Arabia (Odeh et al., 2017). By integrating RS-derived imagery with GIS analytical tools, we've been able to conduct detailed assessments of land cover transformations, including urban expansion, deforestation, agricultural encroachment, and changes in natural ecosystems (Mohammady, 2021). This approach has provided valuable insights into the drivers and impacts of land cover changes, informing decision-making processes related to land use planning, conservation efforts, and sustainable development initiatives in these regions. Through ongoing research and application, we continue to refine and adapt this integrated approach to address emerging challenges and contribute to our understanding of land cover dynamics at both local and regional scales. However, Fig. 5.2 represents the integrated approach of RS and GIS in order to generate or conceptual model for the influence urbanization on the water resources in Jordan.

Landsat 8 images as a remote sensing data offers distinct advantages for land cover classification due to its advanced imaging capabilities. Firstly, its Operational Land Imager (OLI) sensor captures data across nine spectral bands, enhancing spectral resolution and enabling finer discrimination between different land cover types. Additionally, Landsat 8's spatial resolution of 30 m facilitates detailed imagery, crucial for distinguishing various features such as urban areas, forests, and water bodies (Kazemi Garajeh et al. 2024). With a revisit time of 16 days, Landsat 8 allows for frequent monitoring of land cover changes, including seasonal variations and long-term trends. Therefore, Jordan Land Sat 8 images with the visible bands for four time periods were used as follows: 1984, 1994, 2004, 2014 and 2023 (Fig. 5.3).

Classifying the images into four distinct land cover classes was carried out using the maximum likelihood method within ArcGIS 10.3. The essence of maximum likelihood classification lies in determining the likelihood that a given pixel belongs to each class, a process rooted in analyzing its spectral signature. This likelihood is derived by evaluating the probability density function (PDF) of the pixel's spectral values for each class. By employing a multivariate normal distribution, often utilized for spectral data, this PDF is mathematically expressed. Mathematically, it could be presented as follows:

- $L(\theta \mid X_1, X_2, \ldots, Xx)$: Likelihood function, which quantifies the probability of observing the given satellite image data (features) $X_1, X_2, \ldots, X_n$ given the parameters θ of the probability distribution.
- θ: Parameters of the probability distribution assumed to generate the satellite image data. For instance, in a Gaussian Mixture Model, θ would include mean and variance parameters for each class.

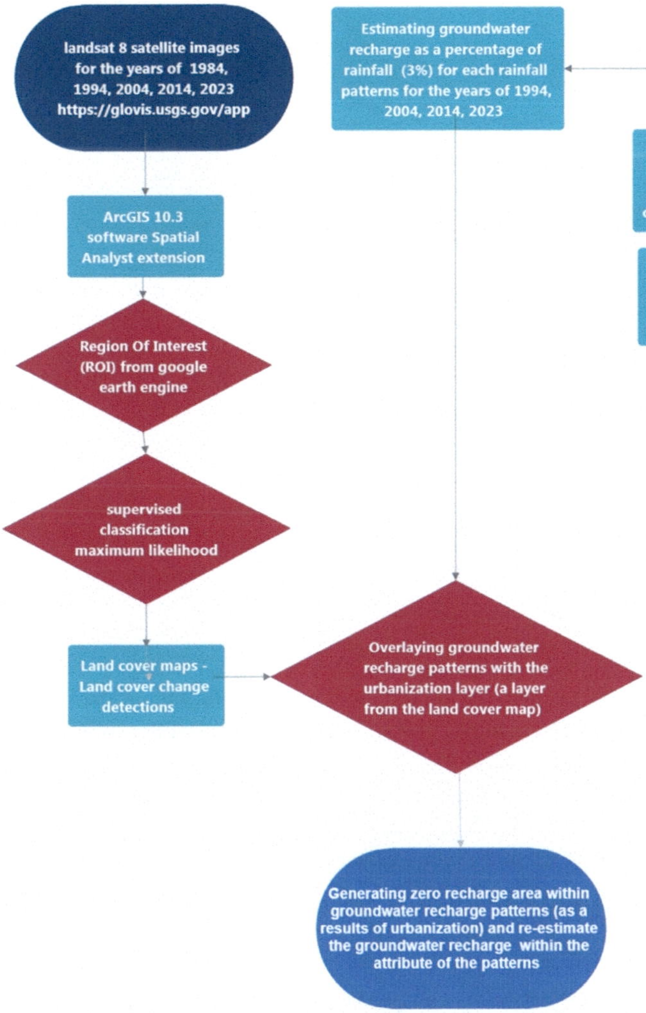

Fig. 5.2 A flowchart illustrating an integrated approach of remote sensing and GIS to estimate the influence of urbanization groundwater recharge across varying time intervals: 1984, 1994, 2004, 2014 and 2023. The interpolation method (IDW) was extensively explained in detail by Odeh et al. in 2019

- $X_1, X_2, \ldots, Xx$: Feature vectors representing observations (pixels) in the satellite image. Each X_i is a vector containing attributes or measurements associated with a particular observation.
- $P(X_i \mid \theta)$: Probability density function (PDF) of each observation X_i given the parameters θ. It describes the likelihood of observing features X_i under the assumed probability distribution characterized by θ.

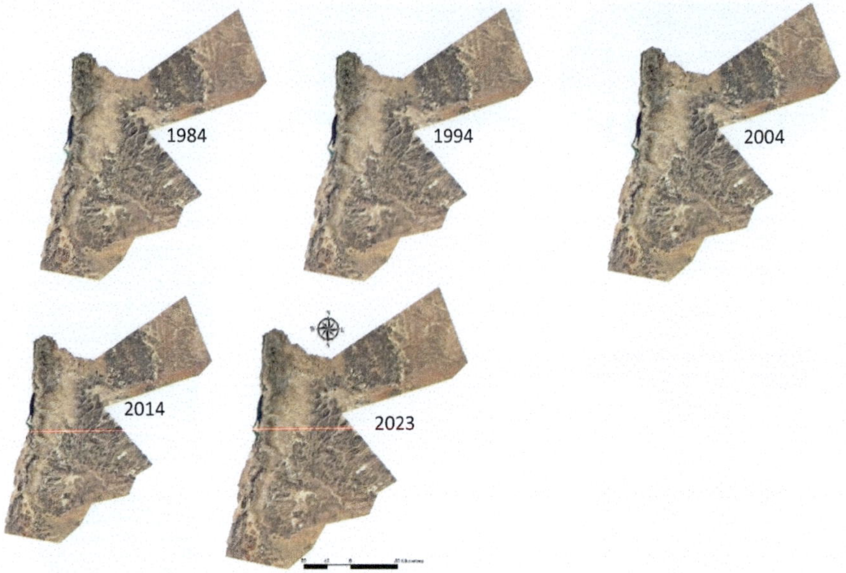

Fig. 5.3 Satellite images of different time period for Jordan

- $\ell(\theta \mid X_1, X_2, \ldots, Xx)$: Log-likelihood function, obtained by taking the natural logarithm of the likelihood function $L(\theta \mid X_1, X_2, \ldots, Xx)$. It simplifies calculations and ensures numerical stability.
- $\hat{}\theta$: Maximum likelihood estimate of parameters θ, obtained by maximizing the log-likelihood function $\ell(\theta \mid X_1, X_2, \ldots, Xx)$. It represents the set of parameters that best explain the observed data according to the maximum likelihood criterion.

we undertook a rigorous validation process. Georeferenced points were carefully selected and cross-referenced with high-resolution imagery from Google Earth to validate the accuracy of our classification results (Zurqani, 2024). This meticulous validation step ensured the reliability and precision of our land cover classification scheme. Additionally, our study benefitted from the abundance of geospatial data available for Jordan from various open sources. Notably, the Ministry of Water and Irrigation provided comprehensive datasets, which were seamlessly integrated into ArcGIS 10.3 for further analysis and interpretation. These geospatial datasets enriched our study by providing valuable contextual information, such as hydrological features, land use patterns, and infrastructure developments, enhancing the depth and comprehensiveness of our findings. By leveraging these diverse geodata sources within a robust GIS platform, we were able to conduct a thorough and multifaceted analysis of land cover dynamics in Jordan, yielding insights that are instrumental for informed decision-making and sustainable land management strategies.

5.3 Results and Discussion

Jordan is a fascinating country with diverse climatic zones despite its relatively small size. Its geographical location, ranging from the shores of the Dead Sea to the heights of the Jordanian Highlands, contributes to this diversity. Accordingly, there are the five major climatic zones as follows:

(1) Mediterranean Climate: Along the western border with Israel and the West Bank, you'll find a Mediterranean climate characterized by hot, dry summers and cool, wet winters.
(2) Arid Desert Climate: Much of Jordan, particularly the eastern and southern regions, falls into this category. These areas experience extremely hot summers and very little rainfall throughout the year.
(3) Semi-Arid Climate: Parts of central Jordan, including the capital city of Amman, have a semi-arid climate. This means they receive slightly more rainfall than the desert regions but still have hot, dry summers.
(4) Subtropical Highland Climate: In the highlands of Jordan, such as the Ajloun and Jerash regions, you'll find a climate with cooler temperatures and more precipitation than the lower-lying areas.
(5) Steppe Climate: This climate type is found in the eastern part of Jordan, characterized by hot summers and cool winters with more rainfall than the desert regions but less than areas with a Mediterranean climate.

However, Jordan has a large variation in the values of topographical elevations that range from − 420 m below within the Dead Sea area to 1854 m above sea level within Jebel Um Ad Dami in the south of Jordan (Odeh et al., 2017). This high relief topography is generated by the underline geological structures that part of regional structures. Geographically, the northwest of Jordan is allocated within Mediterranean area while the southeast with desert area. Both, the high relief topography and the complicated geographical location and high relief topography generated different climatic zones with different rainfall and temperature patterns (Salameh & Bannayan, 1993). Within these broader classifications, there are further nuances and microclimates based on factors like elevation, proximity to bodies of water, and topographical features. The combination of these factors contributes to the existence of approximately 24 climatic zones in Jordan, each with its own unique characteristics and environmental conditions (Fig. 5.4).

Rainfall averages were quantified across various climatic zones during four distinct time intervals: 1984–1994, 1994–2004, 2004–2014, and 2014–2023. These findings are delineated in Table 5.1.

However, Table 5.2 shows a comprehensive overview of rainfall characteristics across various climatic zones. Each climatic zone is identified by a unique number, with corresponding average rainfall values (measured in millimeters per year) over the period from 1974 to 2023. Additionally, the table provides insight into the variability of rainfall within each zone, showcasing maximum and minimum rainfall values experienced during the same period. Standard deviation (SD) values are

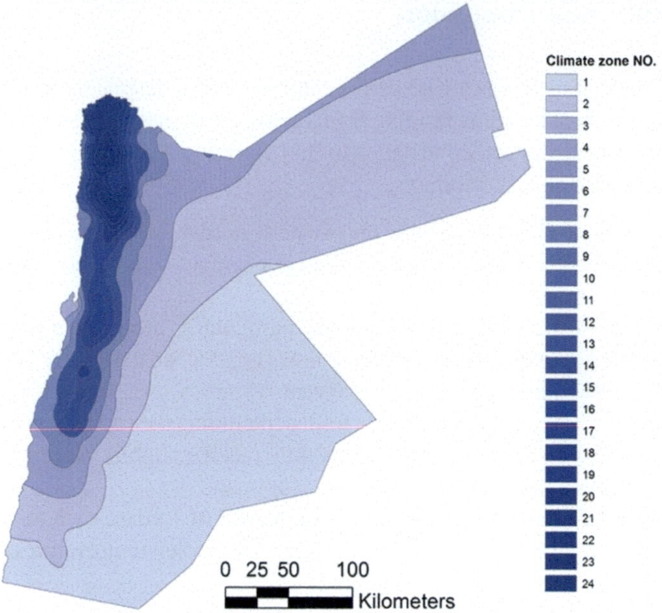

Fig. 5.4 Climatic zones in Jordan

included to indicate the degree of dispersion or spread in rainfall data, reflecting the consistency or variability in annual precipitation patterns within each climatic zone. Overall, this descriptive analysis offers valuable insights into the diverse rainfall dynamics observed across different climatic zones over the specified timeframe, highlighting.

both the average precipitation levels and the range of variability experienced within each zone.

Statistical analysis helps identify long-term trends in climatic variables such as temperature, precipitation, and humidity (Stephenson, 2003). By examining patterns over time, we could determine whether there are significant increases, decreases, or fluctuations in these variables, which can be indicative of climate change or natural variability. Climatic data often exhibit natural variability, including seasonal, annual, and decadal fluctuations. Statistical techniques such as variance analysis and time series analysis allow researchers to quantify this variability, distinguishing between short-term fluctuations and long-term trends. However, Fig. 5.5 shows Jordan has experienced a notable decline in rainfall quantity across its 24 climatic zones, largely attributed to the impacts of climate change. This trend underscores a concerning pattern where once reliable precipitation patterns are now showing signs of disruption (Odeh et al., 2017; Salameh & Bannayan, 1993). As temperatures rise and weather patterns shift, the delicate balance of Jordan's ecosystems is being challenged. From the arid desert landscapes to the semi-arid regions, each climatic zone is witnessing a decrease in the average rainfall quantities. This decline not only affects agricultural

Table 5.1 Average rainfall quantities across climatic zones (1974–2023)

Climatic zone number	Rainfall average (mm/year) 1984–1994	Rainfall average (mm/year) 1994–2004	Rainfall average (mm/year) 2004–2014	Rainfall average (mm/year) 2014–2023
1	13	11	11	10
2	50	44	43	42
3	50	44	43	42
4	100	98	96	97
5	100	95	96	97
6	150	145	146	142
7	150	142	139	135
8	150	142	139	135
9	150	145	146	142
10	200	194	193	192
11	200	192	193	186
12	200	192	193	186
13	200	194	193	192
14	250	243	242	241
15	250	241	242	243
16	300	293	291	290
17	350	346	345	342
18	350	345	343	341
19	350	345	343	341
20	400	391	390	387
21	450	441	440	438
22	500	452	451	485
23	500	452	451	485
24	600	590	589	580

productivity and water availability but also poses significant challenges for local communities dependent on rainfall for their livelihoods (Rödiger et al., 2023).

The water resources, weather they are surface or underground, are controlled directly by climatic zones. The highest amount of rainfall are the highest groundwater recharge and availability and the highest surface water bodies and river availably too (Stephenson, 2003). In Jordan, the highest elevations are mostly found in the northwest where most of the water resources are available therefore most of the urban expansion and agricultural activity growth are allocated as a result of population increasing. However, Fig. 5.6 illustrates the alterations in land cover over the past four decades. It demonstrates a significant increase in urbanization, primarily attributed to the rapid influx of migrants from Syria and Iraq into Jordan (Odeh et al., 2022). The expansion of urbanization into areas with higher rainfall is driven by

Table 5.2 Descriptive analysis of rainfall averages across climatic zones: 1974–2023

Climatic zone number	Average rainfall value (mm/year)	Max. rainfall value (mm/year)	Min. rainfall value (mm/year)	SD value
1	11	13	10	5
2	45	50	42	19
3	45	50	42	19
4	98	100	96	42
5	97	100	95	41
6	146	150	142	63
7	142	150	135	60
8	142	150	135	60
9	146	150	142	61
10	195	200	192	83
11	193	200	186	81
12	193	200	186	81
13	195	200	192	81
14	244	250	241	103
15	244	250	241	102
16	294	300	290	124
17	346	350	342	147
18	345	350	341	146
19	345	350	341	146
20	392	400	387	166
21	442	450	438	188
22	472	500	451	202
23	472	500	451	202
24	590	600	580	253

several factors. Firstly, regions with abundant rainfall often offer better access to water resources, which are crucial for sustaining urban populations and economic activities (Stephenson, 2003). Additionally, these areas tend to boast fertile soils and greater agricultural potential, making them attractive for urban development as agricultural land is converted to accommodate growing populations. Historical settlement patterns may also play a role, as cities often initially develop near water sources or fertile land, leading to the extension of urbanization into adjacent regions with similar environmental characteristics (Mohammady, 2021). Furthermore, existing infrastructure and services in areas with higher rainfall can facilitate urban expansion, providing necessary support systems such as water supply networks and transportation infrastructure (Stephenson, 2003). However, Fig. 5.7 shows that urbanization was expanding toward the climatic zone that have the highest amount of rainfall in

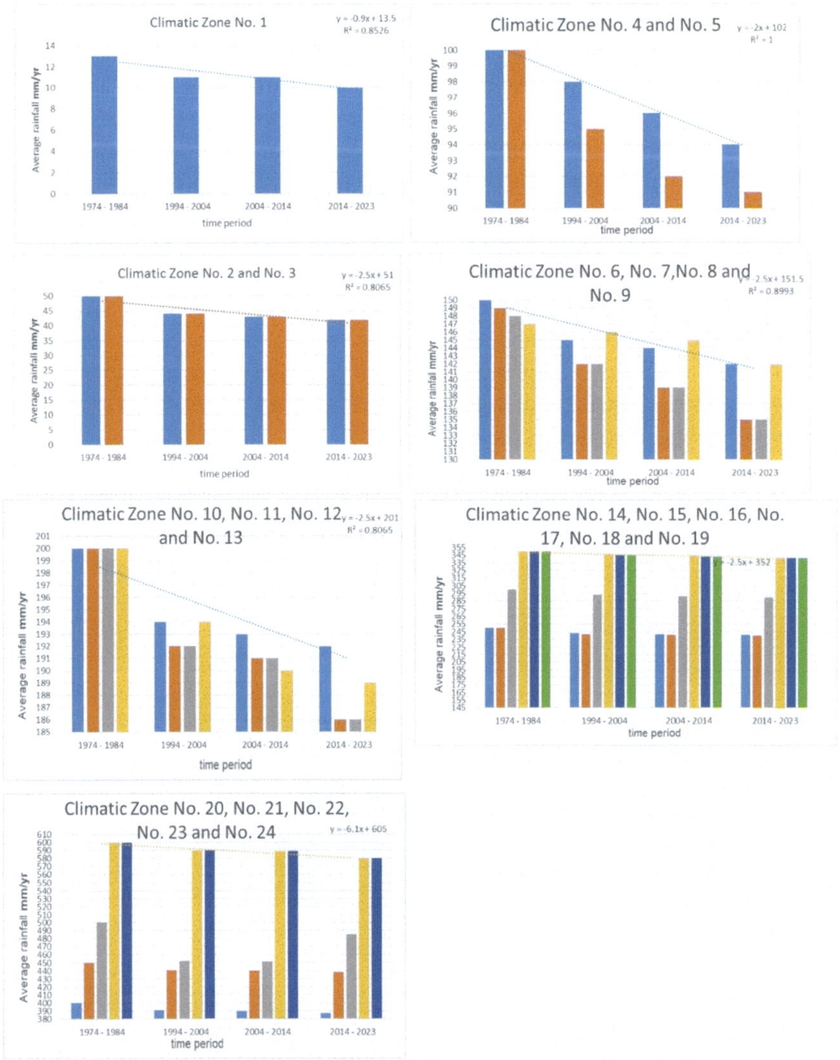

Fig. 5.5 Sequential average rainfall quantity across Jordan's climatic zones (respectively): depicting decreasing trends

Jordan northwest namely. The expansion of urban areas profoundly affects groundwater recharge processes. As cities develop, the construction of impervious surfaces like roads and buildings impedes the natural infiltration of rainwater into the soil, diminishing the amount available for groundwater replenishment (Odeh et al., 2017, 2022). Moreover, urbanization often introduces drainage systems that swiftly channel stormwater away from urban areas, bypassing natural infiltration routes and further reducing groundwater recharge. Human activities associated with urban living, such as industrial operations and increased water usage, exacerbate the strain on groundwater resources. Consequently, as urban populations grow and demand for water rises, groundwater levels can decline, posing significant challenges for both urban and rural communities reliant on groundwater for their water supply. Urban expansion and decreasing rainfall significantly impact groundwater recharge by reducing the amount of water that permeates into aquifers. Urbanization increases impervious surfaces and drainage systems, leading to higher runoff and less infiltration and hence less groundwater recharge (Jyrkama et al., 2005). Figures 5.8 and 5.9 show how the urban expansion moved toward the highest amount of groundwater recharge and hence generates zero spots of groundwater recharge in the highest capability area of recharge. However, the groundwater recharge values are decreasing within the climatic zones as a result of climatic change that decreases the rainfall quantity.

However, there are a direct discipline between geology, water resources population growth and urbanization as it is shown in Fig. 5.10. Each element in the model has a reciprocal relationship with the others. Geological structures shape topography, which in turn influences the visibility of geological units. Topography affects climate by determining elevations, and climate influences topography through geomorphological processes. Climate controls water resources by regulating rainfall, while water

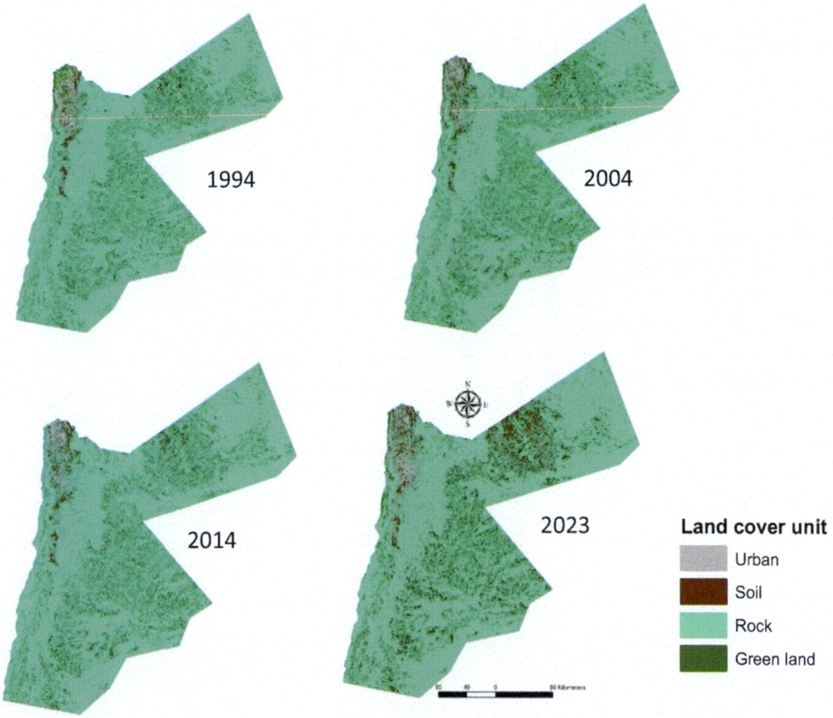

Fig. 5.6 Land cover changes in Jordan during the last four decades

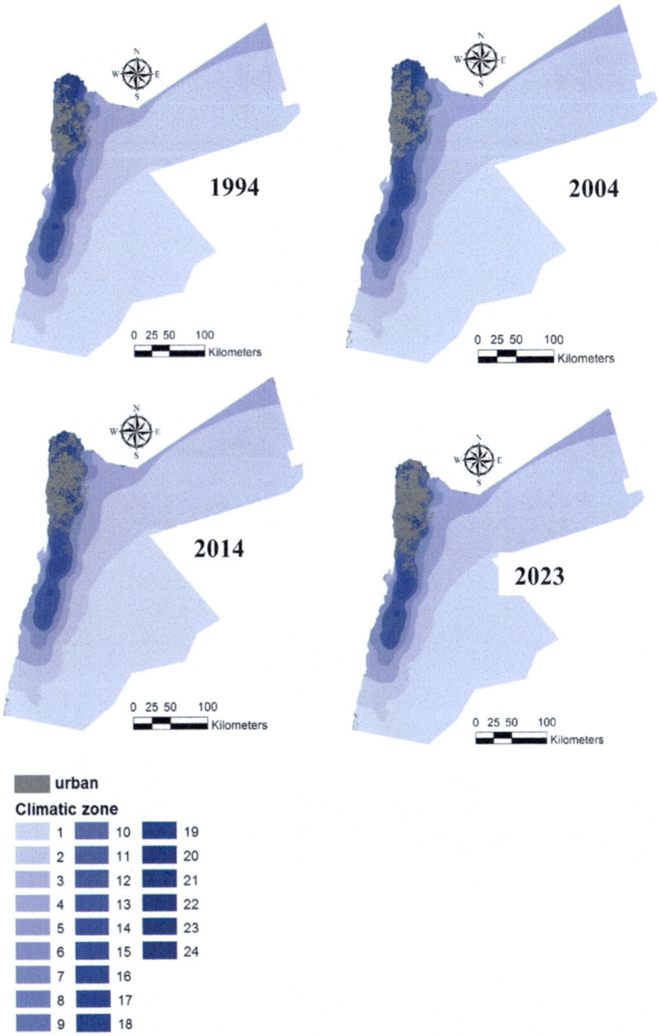

Fig. 5.7 Urban growth across climate zones of Jordan during the last four decades

resources impact climate via the hydrological cycle (Odeh et al., 2017; Stephenson, 2003). Water resources determine population growth by providing drinking water, while populations create new water sources. Population growth drives urbanization, which in turn attracts more people, fostering further urban development. Urbanization reduces groundwater recharge, leading to depletion, while groundwater supports urbanization as a vital resource (Appleyard, 1995; Chou, 1997; Odeh et al., 2022). However, groundwater depletion can have profound effects on geology, leading to land subsidence, sinkhole formation, disrupted hydrological cycles, changes in soil composition, increased geological stress, and impacts on mineral deposits. These

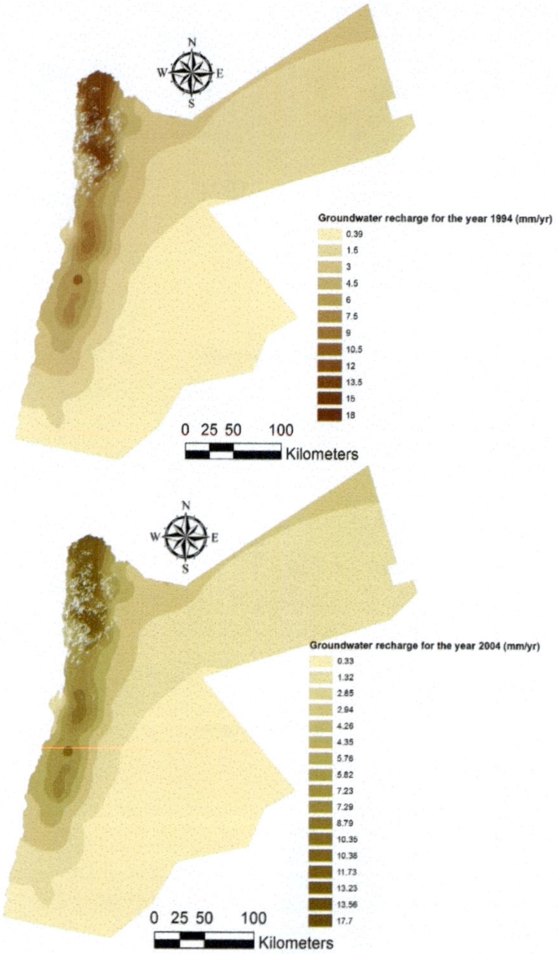

Fig. 5.8 Impact of urban expansion and climate change on decreasing groundwater recharge zones in Jordan during 1994 and 2004. The white spots on the map represents the urbanization area where there is zero groundwater recharge

phenomena are found In Jordan in several places such Dead Sea area and Irbid governorate (Odeh et al., 2019; Salameh & Bannayan, 1993).

5.4 Conclusions

Jordan faces a significant water crisis driven by limited resources, compounded by several interrelated factors. The country's diverse geological formations and complex topography significantly influence its water resources and land cover. Jordan's varied climatic zones, ranging from Mediterranean to arid desert climates, play a crucial role in rainfall distribution and water availability. These climatic variations contribute

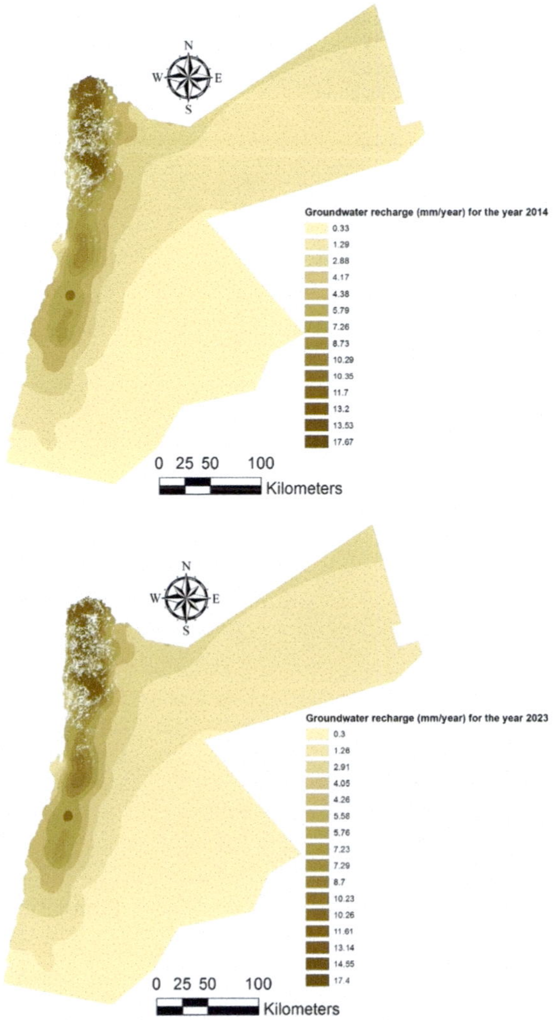

Fig. 5.9 Impact of urban expansion and climate change on decreasing groundwater recharge zones in Jordan during 2014 and 2023. The white spots on the map represents the urbanization area where there is zero groundwater recharge

to the stark differences in water resources across the country, affecting both surface water and groundwater supplies.

Over the past decades, Jordan has experienced a noticeable decline in rainfall due to climate change, which has severely impacted agricultural productivity and further strained water resources. This decline has exacerbated the water crisis, creating challenges for both rural and urban areas. The reduction in rainfall has also led to decreased recharge rates for aquifers, which are a critical source of water for the country. Urban expansion, particularly in regions with higher rainfall, has increased significantly due to population growth and migration. This expansion into fertile and water-rich areas has placed additional pressure on water resources, reducing groundwater recharge and leading to the unsustainable depletion of aquifers. As cities and

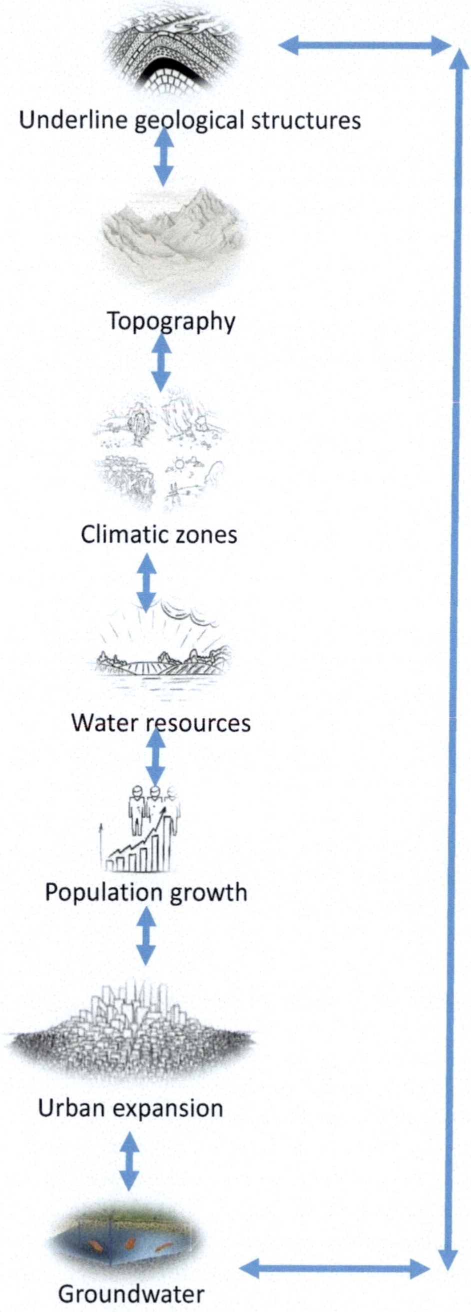

Fig. 5.10 A conceptual model of the interplay between geology, water resources, and urbanization

towns grow, the demand for water rises, further depleting already scarce resources. This urban growth often encroaches on agricultural lands, reducing the area available for food production and compounding the impact on water resources.

The intensive geological structures and high variations in lithological units create a high relief topography that generates zonations of water resources. This topography, while contributing to the complexity of water resource management, also offers potential for identifying and harnessing new water sources. The varied terrain and geological formations mean that water resources are not uniformly distributed, necessitating a tailored approach to water management that takes into account local conditions.

Acknowledgements The authors are grateful to The Hashemite University for supporting this research.

References

Al-Hasani, I., Al-Qinna, M., & Hammouri, N. A. (2023). Potential impacts of climate change on surface water resources in arid regions using downscaled regional circulation model and soil water assessment tool, a case study of Amman-Zerqa Basin Jordan. *Climate, 11*(3), 51.

Appleyard, S. (1995). The impact of urban development on recharge and groundwater quality in a coastal aquifer near Perth, Western Australia. *Hydrogeology Journal, 3*, 65–75.

Bender, F. (1974). Geology of Jordan. In *Contribution of the regional geology of the earth* (p. 196) Gebrüder Borntraeger.

Chou, Y. (1997). *Exploring spatial analysis in GIS* (p. 500). Onword Press.

Christoforidou, M., Borghuis, G., Seijger, C., van Halsema, G. E., & Hellegers, P. (2023). Food security under water scarcity: A comparative analysis of Egypt and Jordan. *Food Security, 15*(1), 171–185.

Jyrkama, M., Sykes, J., Normani, S. (2005). Recharge estimation for transient ground water modeling. *Ground Water 40*, 638–648.

Kazemi Garajeh, M., Haji, F., Tohidfar, M., Sadeqi, A., Ahmadi, R., & Kariminejad, N. (2024). Spatiotemporal monitoring of climate change impacts on water resources using an integrated approach of remote sensing and Google Earth Engine. *Scientific Reports, 14*(1), 5469.

Mohammady, M. (2021). Land use change optimization using a new ensemble model in Ramian County. *Iran Environment Earth Science, 80*, 780.

Odeh, T., Mohammad, A. H., Hussein, H., Ismail, M., & Almomani, T. (2019). Over-pumping of groundwater in Irbid governorate, northern Jordan: a conceptual model to analyze the effects of urbanization and agricultural activities on groundwater levels and salinity. *Environmental Earth Sciences, 78*, 1–12.

Odeh, T., Mohammad, A. H., Pradhanang, S. M., Ismail, M., & Rödiger, T. (2022). GIS-based analytical modeling on evaluating impacts of urbanization in Amman water resources Jordan. *Environmental Earth Sciences, 81*(5), 160.

Odeh, T., Boulad, N., Abed, O., Abu Yahya, A., Khries, N., Abu-Jaber, N. (2017). The influence of geology on landscape typology in Jordan: Theoretical understanding and planning implications. *Land, 6*(3), 51.

Qiu, J., Shen, Z., & Xie, H. (2023). Drought impacts on hydrology and water quality under climate change. *Science of the Total Environment, 858*, 159854.

Rödiger, T., Geyer, S., Odeh, T., & Siebert, C. (2023). Data scarce modelling the impact of present and future groundwater development on Jordan multiaquifer groundwater resources. *Science of the Total Environment, 870*, 161729.

Salameh, E., Bannayan, H. (1993). *Water resources of Jordan—Present status and future potentials* (p. 183). Friedrich Ebert Stiftung.

Stephenson, D. (2003). *Water resources management* (p. 32). Taylor and Francis.

Tsakiris, G. P., & Loucks, D. P. (2023). Adaptive water resources management under climate change: An introduction. *Water Resources Management, 37*(6), 2221–2233.

Zurqani, H. A. (2024). High-resolution forest canopy cover estimation in ecodiverse landscape using machine learning and Google Earth Engine: validity and reliability assessment. *Remote Sensing Applications: Society and Environment, 33*, 101095.

Chapter 6
Evapotranspiration Dynamics: A Comprehensive Review Amidst Water Challenges and Scope of Research in Egypt

Mohamed Elkollaly, Tamer A. Gado, Ahmed Sefelnasr, Faisal Baig, Abdel Azim Ebraheem, and Mohsen Sherif

Abstract Evapotranspiration (ET) is a key ingredient of the water cycle, which makes it an indispensable part of water resources management and irrigation scheduling, especially in arid and semiarid environments. Countries, such as Egypt, grappling with limited water resources, impose constraints on water utilization, often discerned through diverse modelling methodologies. Ensuring the precision of model outputs is contingent upon accurate Evapotranspiration (ET) data, as numerous models leverage ET for input or validation, and the imperative alignment with field-scale applications necessitates obtaining ET data at a corresponding scale. Several methods have been developed for estimating ET and they vary widely in procedures, devices, complexity, and purpose. Compared to Egypt, this study will be set up to review the quantified ET under local environmental conditions, evaluate weather-based reference Evapotranspiration (ET_0) estimation methods, and investigate the developed novel methods for estimating ET_0 using a gene expression programming (GEP) model. Recently, Remote Sensing data have been used to estimate daily evapotranspiration values, overcoming the limitations of climate parameters and field data scarcity. The objective of this study is to evaluate the various dynamics employed to define the spatial and temporal variation in evapotranspiration in response to climate change through the application of Artificial Intelligence and Remote Sensing techniques. Considering the intricate water challenges in Egypt, the main objective of this review is to pinpoint lacunae in the comprehension of evaporation in Egypt and

M. Elkollaly (✉) · F. Baig · M. Sherif
National Water and Energy Center, United Arab Emirates University, P.O. Box 15551, Al Ain, UAE
e-mail: mohamed_elkollaly@f-eng.tanta.edu.eg; 700043011@uaeu.ac.ae

M. Elkollaly · A. Sefelnasr · A. A. Ebraheem · M. Sherif
Civil and Environmental Engineering Department, College of Engineering, United Arab Emirates University, P.O. Box 15551, Al Ain, UAE

M. Elkollaly · T. A. Gado
Irrigation and Hydraulics Engineering Department, Faculty of Engineering, Tanta University, Tanta 31733, Egypt

A. Sefelnasr et al. (eds.), *Water Resources Management and Sustainability*, Water Science and Technology Library 114, https://doi.org/10.1007/978-3-031-80520-2_6

for estimating evapotranspiration, comparing them with studies conducted in Egypt, and conclude with necessary recommendations.

6.2 Historical Perspective and Definitions of Evapotranspiration

Precise evaluation of evapotranspiration (ET) is essential for calculating crop water requirements, managing water resources, and determining the water budget, particularly in arid conditions where water resources are scarce and fresh water is a limited resource (Shiri, 2017). Potential evapotranspiration (PET) is frequently used to calculate actual evapotranspiration, especially in cases where direct measurement through lysimeters or water balance approaches under field conditions is challenging (Lang et al., 2017). In contemporary terms, potential evapotranspiration (PET) aligns with reference crop evapotranspiration (ET) according to Allen's definition, representing hypothetical evapotranspiration based on assumed conditions, including a crop height of 0.12 m, a fixed surface resistance of 70 s/m, an albedo of 0.23, and a reference surface mimicking actively growing, well-watered, and uniformly tall green grass that completely shades the ground (Allen et al., 1998; Lang et al., 2017). Various empirical and theoretical equations are available to estimate potential evapotranspiration under specific weather conditions (El-Nashar & Hussien, 2013). Estimation of evapotranspiration (E) relies on fundamental principles such as the Bowen ratio method, the Monin–Obukhov similarity theory (MOST), and the Penman–Monteith equation. The latter two theoretical expressions merge surface energy balance with MOST. Discrepancies among these approaches in evapotranspiration estimates arise due to utilizing different input data. While surface and satellite-based measurement systems offer accurate estimates of diurnal, daily, and annual E variability, their ability to estimate longer-term variability remains largely unestablished. Although a global mean estimate of E can be reasonably derived using a surface water budget method, uncertainties persist in its regional distribution. Current land surface models exhibit significant variability in the ratio of vegetation transpiration to total E, limiting their ability to accurately assess the sensitivity of E to factors such as precipitation deficits and land cover changes (Wang & Dickinson, 2011).

The accurate estimation of evaporation in models necessitates the consideration of both soil evaporation and vegetation transpiration, accounting for various weather parameters such as air temperature, humidity, solar radiation, and wind speed. Models like the Penman–Monteith (PM) model, characterized by a single-layer resistance approach, are particularly valuable when vegetation extensively covers the soil (Abdrabbo & El Afandi, 2015; Allen et al., 2011; Lang et al., 2017). In cases where vegetation cover is less than 60–70%, the use of a multilayer or multi-source model becomes necessary. Such models incorporate separate surface and aerodynamic functions within the Penman–Monteith (PM) equation, allowing for the accurate representation of evaporation from both soil and the interaction between soil and canopy. Multi-layer models integrate aerodynamic resistances in series between layers, encompassing soil and one or more layers of vegetation canopy. On the other hand, multi-source models often involve coupling aerodynamic resistances from soil and canopy in parallel and may include multiple sources of soil, such as shaded and sunlit or wet and dry areas. (Allen et al., 2011; Lang et al., 2017; Shiri, 2017).

Terrestrial evapotranspiration (E) denotes the water transferred from the land surface to the atmosphere, involving a phase change from liquid to gas, absorbing energy and cooling the land. The latent heat associated with E is represented by λE, with λ as the latent heat of vaporization. These terms are crucial for short-term numerical weather prediction models, long-term climate simulations, and climate change assessments. In these models, E is typically parameterized as the sum of soil evaporation, and vegetation transpiration, the latter coupling with carbon uptake through photosynthesis. Assuming negligible energy storage by the canopy, λE can be calculated as a residual of surface net radiation (R_n), sensible heat flux (H), and ground heat flux (G), expressed as $\lambda E = R_n - H - G$. The determination of R_n involves the sum of incident downward and upward shortwave and longwave radiation,

$R_n = R_s - R_{su} + R_{ld} - R_{lu}$, where R_s is surface incident solar radiation, R_{su} is surface reflected solar radiation, and R_{ld} and R_{lu} are surface downward and upward longwave radiation (Wang & Dickinson, 2011).

Over the past 50 years, numerous scientists globally have developed various empirical methods to estimate evapotranspiration based on different climatic variables. These relationships often required meticulous local calibrations and demonstrated limited global applicability. Testing the accuracy of these methods under new conditions is challenging, time-consuming, and costly, while the demand for evapotranspiration data on short notice remains high for project planning or irrigation scheduling design (Allen et al., 1998; Ragheb et al., 2015). Temperature-based methods rely on the assumption that temperature is a reliable indicator of the evaporative capacity of the atmosphere. These methods are valuable due to the lack of meteorological parameters; however, the outcomes are generally less reliable compared to methods that consider additional climatic factors (McKenney & Rosenberg, 1993). In cases where meteorological data is limited, there are few models available to estimate reference evapotranspiration (ET_o). To address this constraint, data-driven machine learning techniques, specifically genetic approaches like genetic algorithm (GA) and gene expression programming (GEP), offer a solution. Notably, GEP provides an advantage over artificial neural networks (ANN) as it generates an explicit model structure that is easily understandable and adaptable (Mattar, 2018; Valipour et al., 2019).

Variations among methods can be ascribed to differences in their responsiveness to climate conditions, as well as disparities in the climatic factors they take into account (McKenney & Rosenberg, 1993). Recent advancements in research and more precise assessments of crop water utilization have exposed weaknesses in existing methodologies. Various researchers have scrutinized the performance of several methods across diverse locations, revealing variations in their behaviors (McMahon et al.,

2016; Ouda et al., 2016; Ragheb et al., 2015; Shiri, 2017). Although factors like site conditions and weather data collection bias could influence results in the mentioned studies, it has become apparent that these methods exhibit divergent performance in different global locations. Often, deviations from computed to observed values surpass the ranges suggested by the FAO. This analysis underscores the necessity for developing a standardized method for computing reference evapotranspiration (ET_0) (Allen et al., 1998). The FAO Penman–Monteith method is recommended as the exclusive standard approach (Lang et al., 2017). The FAO Penman–Monteith method is highly reliable for predicting reference evapotranspiration (ET_0) across various locations and climates. It is no longer recommended to use older FAO or alternative reference ET methods (Allen et al., 1998). The Food and Agriculture Organization of the United Nations (FAO) advises the adoption of the Penman–Monteith equation, which directly incorporates crucial meteorological variables governing evapotranspiration (ET). However, in numerous locations, limited data availability hinders the application of Penman–Monteith, leading to the use of simplified, empirical methods that demand fewer input data (Kingston et al., 2009; Lang et al., 2017). Consequently, it is essential to identify an alternative method with comparable efficiency to the Penman–Monteith (PM) approach (Lang et al., 2017). Other simple and effective methods were deployed for PET evaluation, such as Thornthwaite (1948), Abtew (1996) methods for humid climate conditions in east-central parts of America and south Florida, respectively. Hargreaves and Samani (1985), Blaney et al. (1950) techniques were established for arid and semi-arid climates; the former originated in the northwestern United States while the latter was famous in the western United States and has been widely used in other areas. Priestley and Taylor (1972) and Makkink (1957) proposed other formulas suitable for humid climatic conditions (Lang et al., 2017). Recently, two approaches have been employed in modelling evapotranspiration, one hinges on climatic factors and measurements, and the other is based on remote sensing theory.

6.3 Methods

6.3.1 Climatic Models

Climatic models can be classified into three primary clusters, i.e., temperature, radiation, and combination models. This classification is rooted in the predominant parameter influencing the climatic model, whether it pertains to temperature, radiation, or a combination of multiple factors.

6.3.1.1 Temperature-Based Models

The first effective temperature-based models appeared in 1948 when Thornthwaite (1948) developed a technique to estimate potential evapotranspiration using climatological data gathered from controlled experiments conducted in the humid Northeastern United States (Cruff & Thompson, 1967). The correlation between mean monthly temperature and potential evapotranspiration, standardized to a 30-day month with 12 h of potential sunshine, was examined in four specific regions (Kissimmee River Basin in Florida, Mesilla Valley in New Mexico, Skunk River Basin in Iowa, and Coshocton in Ohio). In each area, the relationship within the temperature range is effectively represented by an equation of the form $e = ct^a$, in which e is monthly evapotranspiration in °C. The coefficients c and a represent the variations across different locations (Thornthwaite, 1948). Three years later, Blaney et al. (1950) introduced a method for estimating water requirements for irrigated lands where limited or no data, except climatological information, is accessible. An empirical formula has been derived from experimental studies across the western United States. This formula delineates the relationship between temperature, length of growing season, monthly percent of annual daytime hours, and consumptive use of water (Doorenbos et al., 1977; Gotardo et al., 2016; Zhao et al., 2013). This method should be used with skepticism, particularly in equatorial regions with stable temperatures and variance weather parameters, small islands and coastal areas where sea temperature influences air temperature with little response to seasonal radiation changes, and high altitudes where low mean daily temperatures, especially during cold nights, contrast with high daytime radiation levels. In climates with significant sunshine variability during transition months, like monsoon and mid-latitude climates in spring and autumn, skepticism is advised. Under these conditions, prefer the Radiation Method, even if obtaining sunshine or radiation data from regional or global maps due to the absence of actual measured data (Doorenbos et al., 1977). In 1985, one of the simplest and most widely used equations that estimate Reference Crop Evapotranspiration from ambient air temperature was developed by Hargreaves and Samani (1985). This temperature-based formula was presented to estimate Reference Crop Evapotranspiration from values of maximum and minimum temperature.

$$ET_o = 0.0023 R_a \sqrt{T_{max} - T_{min}} \left(T_{mean}^{\circ C} + 17.8\right) \tag{6.1}$$

In this expression, R_a is the extraterrestrial radiation; Units for both ET_o and R_a in previous Equation are mm day^{-1}, T_{max} is the maximum air temperature (°C), T_{min} is the minimum air temperature (°C), T_{mean} is calculated as the average of T_{max} and T_{min} (°C).

6.3.1.2 Radiation-Based Models

In conjunction with temperature-based models, a set of equations emerged that relied mainly on radiation in their calculations. Makkink (1957) proposed a simple method to estimate ET in millimeters per day for grassed lands under the cool climatic conditions of the Netherlands, using only temperature and radiation parameters:

$$ET_o = 0.61\frac{\Delta}{\Delta + \gamma} * \frac{R_s}{\lambda} - 0.12 \tag{6.2}$$

where Δ is the slope of the saturation vapour pressure curve, λ is vaporization latent heat (MJ kg^{-1}) and γ is psychrometric constant (kPa $°C^{-1}$) (Ragheb et al., 2015). This approach is recommended for regions with available climatic data, including measured air temperature and sunshine, cloudiness, or radiation, but lacking measurements for wind and humidity. Outside equatorial zones, climatic conditions vary from year to year for each month or shorter period, leading to variations in Reference Crop Evapotranspiration (ET_o). To ensure water requirements are reliably met, calculations should be performed individually for each month or period, considering the specific climatic conditions of each year of record, rather than relying on mean radiation and mean temperature data over several years. (Doorenbos et al., 1977). Another radiation-based equation is Turc (1961) which was presented in Western Europe to estimate Evapotranspiration. It was used moderately in the United States (Mattar, 2018). This equation is written as:

$$ET_o = a_T 0.013\frac{T}{T + 15}\frac{23.8856R_s + 50}{\lambda} \tag{6.3}$$

The unit of ET_o is in mm d^{-1}, and λ is vaporization latent heat (MJ kg^{-1}). The value of a_T equals unity when the relative humidity is higher than 50%, otherwise, a_T is given by (Mattar, 2018):

$$a_T = 1 + \frac{50 + RH}{70} \tag{6.4}$$

About four decades later, Irmak et al. (2003) developed two formulas to detect the reference evapotranspiration ET_o in mm d^{-1} using the FAO56 Penman–Monteith method as an indicator. The first equation calculates the ET_o based on incoming solar radiation R_s along with maximum and minimum air temperature, while the second equation, net radiation (R_n) based, utilizes R_n. The equations were formulated based on 15 years (1980–1994) of ET_o values estimated using the FAO56-PM method, utilizing weather data near Gainesville, Florida. The performance of these equations was examined over six validation years, covering dry and wet periods, both at the original site and in other humid locations in the Southeast United States. The derived equations provide relatively accurate ET_o estimations compared with prevalent ET_o methods. The R_s-based equation is detailed below:

$$ET_o = -0.611 + 0.149R_s + 0.079T_m \tag{6.5}$$

where ET_o is the grass reference ET (mm day^{-1}); R_s is the incoming solar radiation (MJ m^{-2} day^{-1}); and T_m is the mean daily air temperature computed as the average of daily maximum and minimum air temperatures (°C). The second equation (R_n based, net radiation (MJ m^{-2} day^{-1}) is expressed as follows:

$$ET_o = 0.489 + 0.289R_n + 0.023T_m \tag{6.6}$$

6.3.1.3 Combination-Based Models

Combination-based models pioneered emerging in the early 1980s when Penman and Keen (1948) began formulating their first equations. Penman and Keen (1948) estimated Reference Evapotranspiration using the following equation:

$$ET_o = 0.35\left(1 + \frac{0.98}{100}U_2\right)(e_s - e_a) \tag{6.7}$$

where U_2 is the daily wind speed at 2 m (ms^{-1}), ($e_s - e_a$) is the vapor pressure deficit (kPa), e_s is the air saturation vapor pressure (kPa), and e_a is the actual vapor pressure (kPa) (Penman & Keen, 1948; Zhao et al., 2013). Many years later, Priestley and Taylor (1972) evaluated surface heat flux and evaporation using large-scale parameters. Targeting a simplification of Penman and Keen (1948) (Gotardo et al., 2016).

After five years, Doorenbos et al. (1977) suggested a proposed slight modification to the Penman equation to determine ET_o, involving a revised wind function form. The procedure utilizes mean daily climatic data considering the substantial influence of both daytime and nighttime weather conditions on evapotranspiration levels, an adjustment has been incorporated to account for this impact. This method appeared to be complex due to the components that the formula contains need to be derived from measured related climatic data when no direct measurements of required variables are available (Doorenbos et al., 1977). In May 1990, FAO, in association with the International Irrigation and Drainage Commission and the World Meteorological Organization, convened a gathering of specialists and researchers to review FAO methodologies regarding crop water requirements and provide guidance on the revision and update of procedures. The recommendation was to adopt the Penman–Monteith model as a new standard for reference evapotranspiration and produce procedures for calculating the various parameters. First, the reference crop was defined with specific attributes, and then the FAO Penman–Monteith method was formulated based on this hypothetical crop. This crop is assumed to have a height of 0.12 m, a surface resistance of 70 s m^{-1}, and an albedo of 0.23, closely resembling the evaporation of a vast surface of green grass with uniform height, active growth,

ET_o Reference evapotranspiration [mm day^{-1}],
R_n Net radiation at the crop surface [MJm^{-2}day^{-1}],
G Soil heat flux density [MJ m^{-2} day^{-1}],
T Mean daily air temperature at 2 m height [°C],
u_2 Wind speed at 2 m height [m s^{-1}],
e_s Saturation vapor pressure [kPa],
e_a Actual vapor pressure [kPa],
e_s-e_a Saturation vapour pressure deficit [kPa],
Δ Slope vapor pressure curve [kPa °C^{-1}],
γ Psychrometric constant [kPa °C^{-1}].

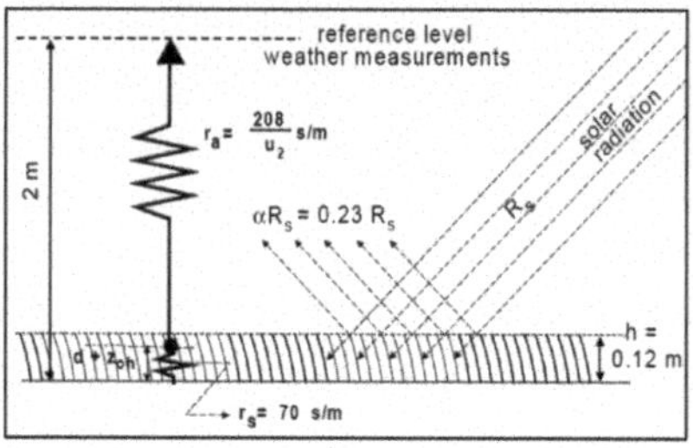

Fig. 6.1 Characteristics of the hypothetical reference crop (Allen et al., 1998)

and sufficient watering (Fig. 6.1). Unlike the previous FAO Penman method, this approach addresses shortcomings and produces values more stable with actual crop water use data worldwide (Allen et al., 1998; Lang et al., 2017; Shiri, 2017). The PMF-56 equation is defined by (Allen et al., 1998) as:

$$ET_o = \frac{0.408\Delta(R_n - G) + \gamma\frac{900}{T+273}\,u_2(e_s - e_a)}{\Delta + \gamma(1 + 0.34\,u_2)} \tag{6.8}$$

ET_o Reference evapotranspiration (mm day^{-1})
R_n Net radiation at the crop surface (MJm^{-2}day^{-1})
G Soil heat flux density (MJ m^{-2} day^{-1})
T Mean daily air temperature at 2 m height (°C)
u_2 Wind speed at 2 m height (m s^{-1})
e_s Saturation vapor pressure (kPa)
e_a Actual vapor pressure (kPa)
$e_s - e_a$ Saturation vapour pressure deficit (kPa)
Δ Slope vapor pressure curve (kPa °C^{-1})
γ Psychrometric constant (kPa °C^{-1})

The equation relies on standard climatological records of solar radiation (sunshine), air temperature, humidity, and wind speed. For accurate computations, it is essential that weather parameters are measured at 2 m above an extensive surface of green grass, ensuring shade over the ground and sufficient water availability (Allen et al., 1998). Recently, reference evapotranspiration has been widely used in the computation of actual evapotranspiration using different remote sensing-based models (Allen et al., 2007; Senay et al., 2007; Wagle et al., 2017).

6.3.2 *Remote Sensing-Based Models*

The Earth system operates in close proximity to an energy balance, where the amount of energy entering and leaving the system is approximately equal. Consequently, alterations in surface conditions, such as land or water coverage and the presence of snow or ice, influence the retention and distribution of energy within the Earth system. These variations in surface conditions contribute to changes in the surface energy balance. In 1993, the Surface Energy Balance Index (SEBI) technique was deployed by Menenti (1993), utilizing the evaporative fraction to estimate evapotranspiration, based on the Crop Water Stress Index (CWSI) (Jackson et al., 1981). In the twentieth century, Su (2002) introduced the Surface Energy Balance System (SEBS) for estimating atmospheric turbulent fluxes and evaporative fraction by integrating satellite earth observation data with meteorological information at appropriate scales. SEBS includes tools for determining land surface physical parameters, such as albedo, emissivity, temperature, and vegetation coverage, from spectral reflectance and radiance measurements. It also incorporates a model for determining the roughness length for heat transfer and a new formulation for calculating evaporative fraction based on energy balance at limiting cases. Evaluation with four experimental data sets demonstrates SEBS's capability to estimate turbulent heat fluxes and evaporative fractions at various scales with acceptable accuracy.

6.3.3 Machine Learning-Derived Models

In regions with limited data availability, machine learning methods become essential as the Penman–Monteith equation demands extensive input data, which is often lacking in many locations. Hence, achieving comparable outcomes with a reduced set of input data is preferable (Valipour et al., 2019). In recent years, heuristic data-driven models, such as gene expression programming (GEP), have been extensively used as an alternative to empirical and semi-empirical ET_0 models (Shiri, 2017). GEP offers several merits such as the straightforward entities of its chromosomes are linear, compact, relatively small, and easily subject to genetic manipulations like replication, mutation, recombination, and transposition (Ferreira, 2001). GEP, or genetic programming models, excels over other heuristic models by providing explicit formulations that illustrate relationships between input and target parameters. Unlike other methods, GEP doesn't rely on predefined functions; it randomly generates and selects functions that best fit experimental results. Additionally, GEP imposes no restrictions on the complexity and structure of the randomly formed functions (Guven et al., 2008).

Different applications of heuristic data-driven models in water resources engineering have been reported by researchers. In the ET_0 modeling context, Landeras et al. (2008) assessed the behavior of neural networks and different empirical ET_0 models for the humid region in Spain. Pour-Ali Baba et al. (2012) examined different models of neural networks and neuro-fuzzy for estimating daily ET_0 utilizing both gauged and estimated solar radiation data. Moreover, in Turkey, Kisi (2016) tested three heuristic regression methods. In 2016, Traore et al. (2016) deployed neural networks to predict near future ET_0 values using limited climate data. In the same year, Feng et al. (2016) employed different neural network algorithms for estimating

ET_0 in humid regions of Southwest China. Certain studies have indicated superior accuracy of GEP compared to ANN (Eslamian et al., 2012; Izadifar & Elshorbagy, 2010; Kim & Kim, 2008; Kisi & Guven, 2010) and Support Vector Regression (SVR) (Kisi & Guven, 2010).

Despite its development in water resources studies, the utilization of the GEP technique for ET_0 modeling remains constrained, with some notable successful applications. Irmak and Kamble (2009) explored evapotranspiration data assimilation using the GA and the SWAP model for on-demand irrigation. The assimilation methodology, derived from this research, serves as a practical tool at the field scale for scheduling irrigation based on remotely sensed evapotranspiration. Traore and Guven (2012) created regional-specific numerical models of ET_0 utilizing GEP. Statistically, GEP proved to be an effective modeling tool for accurate estimation of ET_0. Furthermore, in 2017, Shiri (2017) asserted that GEP surpasses empirical models in calculating ET_0 in hyper-arid regions in Iran, and Traore et al. (2017) discovered that GEP is an effective technique for modeling ET_0 in China.

6.4 Evapotranspiration Research and Bibliometric Analysis in Egypt

Egypt faces numerous challenges regarding water security, along with significant water losses due to the evapotranspiration process. Therefore, an accurate estimation of evapotranspiration rates would be a positive step toward an effective plan for sustainable water management.

This research conducted a bibliometric analysis to explore the trends, innovations, and applications of evapotranspiration research in Egypt. This study utilized the Scopus database to collect the publication information. Scopus is recognized for its extensive coverage of scientific literature. Keywords were precisely chosen to capture the core elements of evapotranspiration research in Egypt. A logical search strategy was implemented to combine these keywords effectively. These keywords include "evapotranspiration" and "Egypt" and "transpiration" or "Climatic models" or "Temperature" or "Radiation" or "Climate change" or "Heuristic Data-Driven Models" or "remote sensing" or "satellite data" or "Satellite" or "machine learning" or "deep learning" or "neural networks" or "AI". The R environment was chosen for its robust performance analysis, data visualization, and statistical modeling capabilities.

Data analysis commenced with keyword analysis to identify the frequency and distribution of the selected keywords within the retrieved articles. This provided insights into prevalent topics and technologies in the literature. Following keyword analysis, research constituents were analyzed to examine key entities contributing to the field. This included identifying influential Affiliations, and the number of publications. The dataset includes over 100 documents spanning from 1984 to 2024, sourced from various scholarly outlets including journals, books, and conference papers, with an annual growth rate of 3.53%. These publications cover a wide range of topics as reflected in the extensive list of Keywords. The total number of published papers ranged from 1 to 3 during the period from 1984 to 2008, then there was a significant positive trend until reached the peak at 14 and 18 publications in 2022 and 2023, respectively.

The analysis of the most relevant affiliations contributing to research in evapotranspiration in Egypt reveals a diverse array of academic institutions and research centres from Egypt. The National Research Centre tops the list with 19 articles, followed closely by Cairo University and the National Authority for Remote Sensing and Space Science (NARSS) with 14 and 11 articles, respectively. These institutions, alongside 6 others top like Ain Shams University, Mansoura University, and Water and Environment Research Institute are delineated in Fig. 6.2b. The frequency analysis of terms within the dataset reveals focal points in evapotranspiration research, with predominant terms such as evapotranspiration and remote sensing underscoring the advanced technologies used for monitoring and assessing evapotranspiration (Fig. 6.3). Specific parameters like climate change, irrigation, water supply, and crops highlight the importance of water sustainability in Egypt and its struggles with climate change, prompting researchers to estimate water consumption and evapotranspiration losses. Additionally, the absence of "machine learning" reflects a deficiency in the use of sophisticated computational methods for data analysis and modeling, indicating a reliance on traditional approaches in research methodologies.

Several studies have been conducted to estimate evapotranspiration in Egypt, employing both traditional methods and remote sensing techniques. One of the first studies in this regard was Faulkner (1988) that utilized the Energy Balance method to determine the actual evapotranspiration occurring over crops in Egypt. This study was carried out during the period from May to September 1987. Two sites were selected in Middle Egypt, at Giza and Bahtim, where meteorological data were available. The reference crop evapotranspiration (ET_0) was calculated using Blaney-Criddle, Radiation, original Penman, FAO modified Penman, and Evaporation Pan. Plus, the original Penman equation. It was found that the modified form of the Penman equation recommended by the FAO overestimated evapotranspiration by approximately

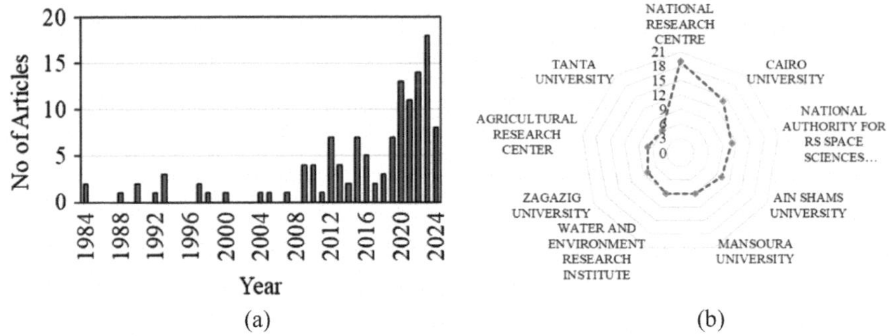

Fig. 6.2 **a** Annual scientific production, and **b** most relevant affiliations

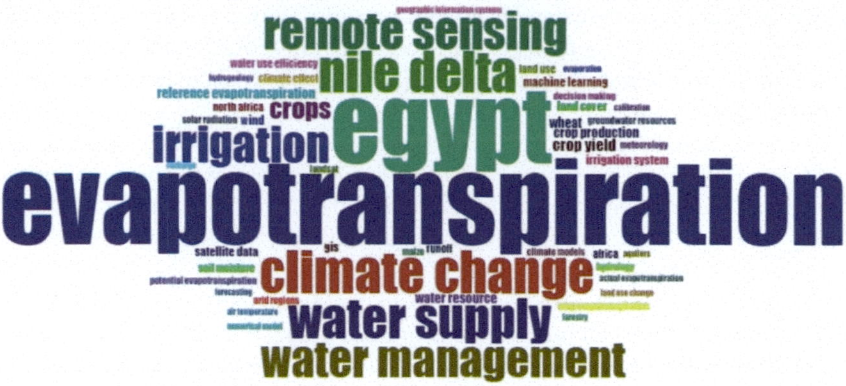

Fig. 6.3 Cloud map of keywords

30%. Alternatively, the use of an earlier form of the Penman equation yielded acceptable estimates. After twenty-four years, Abou El Hassan (2011) who identified the effect of climate change on reference crop evapotranspiration (ET_0), confirmed the previous outcomes that the FAO Penman Monteith method gives the highest values. This paper estimated the ET_0, using different models, recommended the Pan Evaporation Rate method as the most simple and precise method. Ragheb et al. (2015) calibrated and validated some ET_0 models using a lysimeter under the New Valley conditions. Nine drainage lysimeters were installed to compare with twelve ET_0 models from February 2013 to January 2015. The comparative study indicated that the locally calibrated FAO-24 pan model was the best model and gave excellent coinciding as compared with the lysimeter observations under the New Valley conditions. Another comparative study was executed by El Afandi and Abdrabbo (2015) to evaluate four reference evapotranspiration (ET_0) methods namely, Hargreaves, Thornthwaite, Blaney-Criddle, and FAO-56 Penman–Monteith (PM), based on Class-A pan or E-pan over different agro-climatic regions in Egypt. In this study, Egypt has been divided into several agro-climatic clusters according to the average temperature and the reference evapotranspiration. The results revealed that there were no significant differences between E-Pan and PM. The Hargreaves equation reported the highest ET_0 value at all regions while Thornthwaite was the lowest one. Proposing empirical equations for estimating the potential evapotranspiration was another approach that was investigated in the Middle Delta of Egypt by El-Nashar and Hussien (2013). The climatic data were obtained from the Tanta climatic station from 1990 to 2006. The result of the proposed equation is more accurate in calculating the potential evapotranspiration, relative to the measured values. Many other studies were implemented in Egypt to investigate the most accurate and suitable formula for evapotranspiration estimation, while others delineated the spatial distribution of evapotranspiration across Egypt using different models especially FAO Penman–Monteith (El-Shirbeny, 2017; Gad & El-Gayar, 2010; Kamel, 2015; Ouda & Norledin, 2017; Yassen et al., 2020). Overall, the FAO-56 Penman–Monteith method has proved its capability in the estimation of reference evapotranspiration over different agro-climatic regions in Egypt. Tables 6.1 and 6.2 encapsulates a summary of some studies, encompassing key information such as objectives, methodologies, utilized data, and significant findings derived from the research.

6.4.1 Evapotranspiration Under Climate Change Conditions

Temperature is considered one of the most significant factors influencing evaporation, and it is a key parameter in most equations. Therefore, there is a close relationship between climate change and evapotranspiration rates, which necessitated the presence of several research that investigated this trend. Based on the IPCC's SRES scenarios (A1, A2, B1, and B2), Attaher et al. (2006) aimed to investigate the effect of climate change on evapotranspiration ET_0 in Egypt. This study collected the data from 31 stations and applied the FAO-Peneman equation to calculate the current ET_0 trends. Future projections were estimated for the years 2025s, 2050s, and 2100s, using the HadCM3 climate model. Under current climate conditions, Results indicated that about 70% of the Egyptian planted area has an average ET_0 value of more than 4 mm per day. There was an increase in ET_0 values due to climate change, this increase will be uneven between regions and seasons. The future climatic changes will increase the potential irrigation demands by about 11% due to the upsurge in ET_0 only by 2100s. Khalil (2013) used the same Scenarios (IPCC SRES) projected by the GCM (HadCM3) and the same equation to discuss the spatial variation of reference evapotranspiration (ET_0) under climate change in Egypt. In this study, agrometeorological data were assembled from 20 stations in the Nile Delta and Nile Valley. Moreover, this study relied on the Penman–Monteith equation to determine reference evapotranspiration (ET_0) according to the agrometeorological data. The results show that under the current climate Aswan region suffers from a higher ET_0. Similarly, to the previous outcome, this research revealed a general increase in evapotranspiration under future climate for scenarios (A1, A2, and B1 and B2). A2 scenario in 2100 gave the highest ET_0 and B1 scenario in 2040 gave the lowest ET_0 in comparison with other scenarios. Likewise, Farag et al. (2014) utilized the same previous approach and scenarios to delineate the climate change implications on the ET_0 in Egypt.

In 2015, two studies were conducted regarding climate change utilizing the concerned RCPs scenarios: RCP2.6, RCP4.5, RCP6.0, and RCP8.5 (Abdrabbo & El Afandi, 2015; Abdrabbo et al., 2015). Abdrabbo and El Afandi (2015) compared the reference evapotranspiration (ET_0) estimations using FAO-56 Penman–Monteith and different equations namely, Blaney-Criddle, Hargreaves, Thornthwaite under future climatic conditions. In the other research, Abdrabbo et al. (2015) used the FAO 56 Penman–Monteith equation to estimate ET_0 based on the predicted maximum and minimum temperature under the RCPs scenarios. Both studies concluded that the expected climate changes in Egypt according to the RCP scenarios will cause an increase in the ET_0. In general, RCP 8.5 recorded the highest ET_0 values under climate

Table 6.1 Summary of state of art of evapotranspiration in Egypt

Study	Data		Models	Main objectives	Main findings
	No. of stations	Study period			
Mattar (2018)	32 (FAO) CLIMWAT	2013–2015	PMF-56 HS IR TR	Creating and assessing a gene expression programming (GEP) model for the estimation of mean monthly reference evapotranspiration (ET_o)	The GEP models exhibit superior performance in contrast to the empirical Hargreaves and Samani, Irmak, and Turc models. The accuracy of the GEP model experienced a substantial enhancement when incorporating either mean relative humidity (RH) or wind speed at 2-m height (u_2) as additional input variables
Ouda and Norledin (2017)	The Nile Delta and Valley	1985–2014	BISm	comparing the agro-climatic zones derived from weather data spanning a 10-year interval (2005–2014) with those generated over a 20-year period (1995–2014), both analyzed within the context of a 30-year interval	The agro-climatic zoning based on 20-year ET_o values closely resembled the zones created using 30-year ET_o values, albeit with distinct average ET_o values within each zone. The average ET_o value across the three classifications exhibited a close similarity
El-Shirbeny (2017	45 modeled data stations [NCEP, CFSR] 2 ground stations (calibration and validation)	1979–2014	PMF-56	Defining and delineating the boundaries of minimum and maximum reference evapotranspiration (ET_o) on both a spatial and temporal scale	Modeled data proves invaluable in situations where actual data is lacking, but it is crucial to undertake validation and calibration processes. According to the ET_o maps of Egypt, the northern region demonstrates higher water consumption efficiency compared to the southern counterpart

Table 6.2 Satellite based state of art of evapotranspiration in Egypt

Study	Remote Sensing			Models	Main objective	Main findings
	Satellite	Products	Spatial resolution			
Ayyad et al. (2019)	Landsat	EEFlux	30 m	RWS	Assessing the performance of three satellite-based Actual Evapotranspiration (ET_a) products, evaluating irrigation efficiency, and identifying agricultural areas in Egypt	The EEFlux Actual Evapotranspiration (ET_a) estimations tend to result in overestimations of the ET_a values. Conversely, both SSEBop and MOD16A2 products yielded closely aligned estimates for ET_a across the Nile Delta and Valley. Among the evaluated products, SSEBop demonstrated the highest performance. The irrigation efficiency indicator, represented by the Relative Water Stress (RWS), pointed to a high level of irrigation efficiency in Egypt
	MODIS	MOD16A2	1 km			
	MODIS	SSEBop	1 km			
	MODIS (Terra and Aqua)	Land Cover-MCD12Q1	500 m			
	MODIS (Terra)	NDVI-MOD13A3	1 km			
Wahab et al. (2018)	MERRA (Era)	Monthly evaporation "Lake Nasser"	2/3 long., 1/2 lat. degrees	SS Hs	Estimating the water loss of evapotranspiration and Lake Nasser evaporation in Egypt utilizing satellite and climate datasets Studying the effect of climate change on water loss	The total annual water loss is expected to increase in the future, with a noticeable rise. The highest quantity of total water loss during the study period was observed in the year 2021
	MODIS (Terra)	Land Surface Temperature LST	1 km			

(continued)

Table 6.2 (continued)

Study	Remote Sensing			Models	Main objective	Main findings
	Satellite	Products	Spatial resolution			
	MODIS (Terra)	NDVI	500 m			
Abdel Kader et al. (2015)	Landsat ETM + 7	SEBS	30 m Multispectral 15 m Panchromatic	SEBS MK PMF-56 PMF-24 HS	Estimating evapotranspiration utilizing remote sensing data and climatic models	Additional studies and verification efforts are essential to assess all factors that could impact the quality of data influencing the surface energy balance in arid land conditions. It is recommended to conduct further verification across multiple consecutive seasons and for various crops to estimate actual evapotranspiration at the field level. The values of Actual Evapotranspiration (ET_a) estimated by SEBS, P, R, and HS methods were found to be lower than those estimated by the PM (Penman–Monteith) method

change conditions. A year later, Morsy et al. (2016) used the Penman–Monteith equation for estimating the potential evapotranspiration (PET) for five governorates in Egypt (North Sinai, Marsa Matrouh, El-Gharbia, El-Sharkia and Sohag) from 1985 to 2014. RCPs scenarios developed from four global models were compared to detect the impact of climate change on PET. Results showed that the PET values projected by 2030 using this methodology were small compared to the values projected by Ouda et al. (2016) who used the A1B climate change scenario of the ECHAM5 model for El-Gharbia and Sohag. Furthermore, the values of the PET were slightly high in North Sinai and Marsa Matrouh.

6.4.2 Remote Sensing-Based Evapotranspiration

In recent years, remote sensing has played a pivotal role, particularly in hydrological research and environmental monitoring. Due to the lack of climatic and meteorological data in most parts of the world, most studies have turned to remote sensing as an attempt to monitor evapotranspiration and overcome these obstacles. These endeavors in Egypt date back to 2014 when El-Shirbeny et al. (2014) studied the potential crop Evapotranspiration in Egypt using satellite images of Landsat 8. Bands 4 and 5 were analyzed to calculate the Normalized Difference Vegetation Index (NDVI) and to classify the cultivated areas. The cultivated land area was 3,277,311 ha in August 2013. The ET_0 was then calculated from the meteorological data using the Penman-Montieth formula. The ET_0 increased towards the southern part of Egypt as a result of high radiation, air temperature, and low relative humidity. In 2015, Khalil et al. (2015) applied the remote sensing technique to estimate the potential evapotranspiration (ET_0) over Egypt and to estimate the water loss from agricultural lands. MODIS/TERRA—LST (Land Surface Temperature) and NDVI (Normalize Difference Vegetation Index) were used as index to calculate the ET_0 by Hargreaves potential evapotranspiration equation during the period from 2001 to 2013. The findings suggest that combining remotely sensed information with supplementary meteorological data has an efficient performance and enables the estimation of evapotranspiration at a pixel level, making it suitable for mapping reference crop evapotranspiration (ET_0) across the entirety of Egypt. The lowest and highest values were delineated in the results. The amount of water lost annually by evapotranspiration ranged from about 34.4–42.4 billion m^3 of water by the year 2012.

Another remote sensing approach called the surface energy balance system (SEBS model) was adopted in 2015 by Abdel Kader et al. (2015) to estimate values of evapotranspiration. This remote sensing-based model was used along with four climatic formulas including Hargreaves-Samani (HS), Penman–Monteith (FAO 56-PM), Penman (FAO 24-P), and radiation (R). Nine Landsat ETM + 7 satellite images were exploited to calculate daily evapotranspiration values for Sugar beet, wheat, and green onion crops for the season 2012–2013. Results detected variations in the estimated ET_a for each of the tested climatic and remote sensing models. The ET_a

values estimated by SEBS, P, R, and HS methods were lower than those calculated by the PM method. Comparison between climate models and those based on remote sensing continued until 2019 when Omar et al. (2019) decided to compare different models to find the superior method for estimating the evapotranspiration in the Nile Delta governorates. The methods were the FAO (33) method, the Irrigation Water Balance Calculation (IWBC) method, and the Remote Sensing method by Moderate Resolution Imaging Spectroradiometer (MODIS) ET_a-product (MOD16A2). These models were used in five governorates, where the data for IWBC were available. In the case of ET_a/ET_p estimation, there was a significant difference between FAO (33) and MOD16A2. However, the three methods gave near values for the ET_a showing a significant relationship. Generally, both MOD16A2 and IWBC can be utilized for ET_a estimation and FAO (33) cannot be utilized for estimating both ET_a and ET_a/ET_p ratio. Two other studies were implemented in 2019, and 2020 to identify superior ET_a products in agricultural areas and to map the spatial and temporal variation of ET_o for Egypt using remote sensing, respectively (Ayyad et al., 2019; Ezz & Abdelwares, 2020). In 2018, Wahab et al. (2018) combined remote sensing and climate change implications to estimate water loss based on Lake Nasser evaporation and agricultural evapotranspiration in Egypt using the ECHAM 6, and RCP4.5 scenario. Evapotranspiration has been calculated from climate data of ECMWF (ERA-Interim) and MODIS satellite data using the Hargreaves method. The results pointed out that the water loss over agricultural lands is about 38.4 billion m^3 and about 46.3 billion m^3 on average by utilizing data from Land Surface Temperature (MODIS) and ECMWF (ERA-Interim), respectively.

6.4.3 Applications of Heuristic Data-Driven Models

A few studies have employed artificial intelligence in simulating evapotranspiration in Egypt. One of the pioneers in this research is Mattar (2018), who employed and assessed a gene expression programming (GEP) model for estimating mean monthly evapotranspiration (ET_o) utilizing minimal climatic parameters from 32 weather stations in Egypt. The performance of the GEP was identified, relative to FAO-56 Penman–Monteith, revealing high accuracy detection when mean relative humidity (RH) or wind speed at 2-m height (u_2) were included as additional input variables. The GEP model, utilizing maximum and minimum air temperature, RH, and u_2 as inputs, exhibited the lowest root mean square error, highest coefficients of determination, and superior overall model performance and agreement indices for both training and testing sets. The results suggest that GEP is a more accurate and successful technique for modeling ET_o compared to other empirical models.

6.5 Conclusion

Potential evapotranspiration (PET) is commonly employed to calculate actual evapotranspiration, especially when direct measurements are challenging like lysimeter gauges and water balance quantities under field conditions. PET helps measure the atmospheric water demand of a region, making it valuable for applications like drought monitoring, irrigation scheduling, and assessing climate change impacts. Various empirical methods for estimating evapotranspiration from climatic variables have been developed over the past 50 years. Globally, the FAO56 Penman–Monteith method developed in 1998 is recommended as the standard PET method by the Food and Agriculture Organization (FAO) and World Meteorological Organization (WMO). It can be applied without extra parameter adjustments and is widely accepted. Recent studies have utilized Remote Sensing data to estimate daily evapotranspiration, examining spatial and temporal variations under the influence of climate change. Recently, Remote Sensing data and machine learning have been used to estimate evapotranspiration values and overcome the obstacle of data limitation.

In Egypt, the previous literature review detected several evapotranspiration studies. All the listed studies were through the last decade (2006–2020), except one study was in 1988. Only a few numbers of studies combined Remote sensing and climate change to estimate evapotranspiration in Egypt. This literature pointed out several studies that obtained evapotranspiration using a Remote sensing tool. other research papers focused on the investigation of the climate change impact on evapotranspiration. The remaining papers studied evapotranspiration using climatic-based models. Most of these studies used a pan instrument for evapotranspiration estimation and only one study used the lysimeter for this purpose.

The developed novel methods for representing the ET_0 using a gene expression programming (GEP) model were investigated. In Egypt, a limited number of studies employed gene expression programming (GEP) to create models for calculating the ET_0 (Mattar, 2018). It was concluded that the performance of the GEP models was more reliable and efficient than other climatic models, such as the empirical Hargreaves and Samani, Irmak, and Turc models. Therefore, the GEP model represents a promising and powerful approach.

This review provides a comprehensive overview of various methods and techniques employed for the precise estimation of evapotranspiration. Besides, this literature reviewed and analyzed the studies of evapotranspiration in Egypt to identify research gaps and formulate the necessary recommendations. From this perspective, the study emerges as an indispensable tool for decision-makers in the field of water resource management, particularly in arid regions like Egypt. It provides valuable guidance on the procedures necessary for accurately estimating evapotranspiration and playing a crucial role in reducing water losses, mitigating water deficits, and ensuring the sustainability of water resources.

6.6 Recommendations

In Egypt, the previous review showed that there are some research gaps relative to evapotranspiration. One of these gaps is the deficiency of machine learning studies in the context of evapotranspiration. Therefore, it is recommended to evaluate the different techniques of the heuristic data-driven models (ARIMA, SVR, ANN, WNN, and GEP) for evapotranspiration estimation. Besides, utilizing the gene expression programming (GEP) modelling under the impact of climate change should be conducted for further evapotranspiration evaluation more rigorously. In Egypt, the lack of field evapotranspiration observations represents a major flaw in verifying the accuracy of the different approaches, so several lysimeter instruments should be installed in different well-spatial distributed stations to detect the best-fit model regarding Evaluating evapotranspiration, whether climate or remote sensing models. Furthermore, the new generation of socioeconomic scenarios in climate change research based on the Shared Socio-Economic Pathways (SSPs) has not been studied yet in Egypt, so determining the influence of climate change on evapotranspiration under the new scenarios is an urgent issue that may change the current situation for evapotranspiration estimation.

References

Abdel Kader, M., Khalifa, H., Sheta, A., & Ibrahim, A. (2015). Evapotranspiration estimation using remote sensing data and some climatic models. *Journal of Soil Sciences and Agricultural Engineering, 6*(11), 1341–1354. https://doi.org/10.21608/jssae.2015.43926

Abdrabbo, M., & El Afandi, G. (2015). Comparison of reference evapotranspiration equations under climate change conditions. *Global Journal of Advanced Research, 2*, 556–568.

Abdrabbo, M., Farag, A., & El-Desokey, W. (2015). Implementing of RCPs scenarios for the prediction of evapotranspiration in Egypt. *International Journal of Plant Soil Science, 6*(1), 50–63.

Abou El Hassan, W. H. (2011). Impact assessment of long term climate change on reference evapotranspiration and water management in north delta. *Journal of Soil Sciences and Agricultural Engineering, 2*(6), 623–634. https://doi.org/10.21608/jssae.2011.55648

Abtew, W. (1996). Evapotranspiration measurements and modeling for three wetland systems in South Florida1. *JAWRA Journal of the American Water Resources Association, 32*(3), 465–473. https://doi.org/10.1111/j.1752-1688.1996.tb04044.x

Allen, R. G., Pereira, L. S., Howell, T. A., & Jensen, M. E. (2011). Evapotranspiration information reporting: II. Recommended documentation. *Agricultural Water Management, 98*(6), 921–929. https://doi.org/10.1016/j.agwat.2010.12.016

Allen, R. G., Pereira, L. S., Raes, D., & Smith, M. (1998). Crop evapotranspiration: Guidelines for computing crop water requirements. *FAO Irrigation and Drainage Paper, 56*. https://books.google.com.eg/books?id=42hRAAAAMAAJ

Allen, R. G., Tasumi, M., & Trezza, R. (2007). Satellite-based energy balance for mapping evapotranspiration with internalized calibration (METRIC)—model. *Journal of Irrigation and Drainage Engineering, 133*(4), 380–394. https://doi.org/10.1061/(ASCE)0733-9437(2007)133:4(380)

Attaher, S., Medany, M., Abdel Aziz, A., & El-Gindy, A. (2006). Irrigation-water demands under current and future climate conditions in Egypt. *Misr Journal of Agricultural Engineering, 23*(4), 1077–1089.

Ayyad, S., Al Zayed, I. S., Ha, V. T. T., & Ribbe, L. (2019). The performance of satellite-based actual evapotranspiration products and the assessment of irrigation efficiency in Egypt. *Water, 11*(9), 1913.

Blaney, H. F., Criddle, W. D. (1950). *Determining water requirements in irrigated areas from climatological and irrigation data.* U.S. Soil Conservation Service.

Cruff, R. W., & Thompson, T. H. (1967). *A comparison of methods of estimating potential evapotranspiration from climatological data in arid and subhumid environments* [Report](1839M). (Water Supply Paper, Issue). U.S. G.P.O. http://pubs.er.usgs.gov/publication/wsp1839M

Doorenbos, J., Pruitt, W. O., Aboukhaled, A., Damagnez, J., Dastane, N. G., Van den berg, C., Rijtema, P. E., Ashford, O. M. & Frere, M. (1977). Guidelines for predicting crop water requirements. *FAO Irrigation and Drainage Paper, 24*, 1–179. https://ci.nii.ac.jp/naid/10008232709/en/

El-Nashar, W. Y., & Hussien, E. A. (2013). Estimating the potential evapo-transpiration and crop coefficient from climatic data in Middle Delta of Egypt. *Alexandria Engineering Journal, 52*(1), 35–42. https://doi.org/10.1016/j.aej.2012.12.001

El-Shirbeny, M. A. (2017). Reference evapotranspiration borders maps of Egypt based on kriging spatial statistics method. *International Journal of Geomate, 13.*

El-Shirbeny, M. A., Ali, A., & Saleh, N. H. (2014). Crop water requirements in Egypt using remote sensing techniques. *Journal of Agricultural Chemistry and Environment, 3*(2), 57–65.

El Afandi, G., & Abdrabbo, M. (2015). Evaluation of reference evapotranspiration equations under current climate conditions of Egypt. *Turkish Journal of Agriculture-Food Science Technology, 3*(10), 819–825.

Eslamian, S. S., Gohari, S. A., Zareian, M. J., & Firoozfar, A. (2012). Estimating penman-monteith reference evapotranspiration using artificial neural networks and genetic algorithm: A case study. *Arabian Journal for Science and Engineering, 37*(4), 935–944. https://doi.org/10.1007/s13369-012-0214-5

Ezz, H., & Abdelwares, M. (2020). Spatial and temporal variation of eto for Egypt using remote sensing. *ARPN Journal of Engineering and Applied Sciences, 15.*

Farag, A., Abdrabbo, M., & Ahmed, M. (2014). GIS tool for distribution reference evapotranspiration under climate change in Egypt. *International Journal of Plant & Soil Science*, 2320–7035.

Faulkner, R. D. (1988). Estimates of crop evapotranspiration in Egypt. *Water and Environment Journal, 2*(6), 651–653. https://doi.org/10.1111/j.1747-6593.1988.tb01354.x

Feng, Y., Cui, N., Zhao, L., Hu, X., & Gong, D. (2016). Comparison of ELM, GANN, WNN and empirical models for estimating reference evapotranspiration in humid region of Southwest China. *Journal of Hydrology, 536*, 376–383. https://doi.org/10.1016/j.jhydrol.2016.02.053

Ferreira, C. (2001). *Gene expression programming: A new adaptive algorithm for solving problems.* arXiv preprint cs/0102027.

Gad, H., & El-Gayar, S. (2010). Climate parameters used to evaluate the evapotranspiration in delta central zone of Egypt. In *Fourteenth international water technology conference*, IWTC14.

Gotardo, J. T., Rodrigues, L. N., & Gomes, B. M. (2016). Comparison of methods for estimating reference evapotranspiration: An approach to the management of water resources within an experimental basin in the Brazilian cerrado. *Journal of the Brazilian Association of Agricultural Engineering, 36*(6), 1016–1026.

Guven, A., Aytek, A., Yuce, M. I., & Aksoy, H. (2008). Genetic programming-based empirical model for daily reference evapotranspiration estimation. *CLEAN – Soil, Air, Water, 36*(10–11), 905–912. https://doi.org/10.1002/clen.200800009

Hargreaves, G. H., & Samani, Z. A. (1985). Reference crop evapotranspiration from temperature. *Applied Engineering in Agriculture, 1*(2), 96–99. https://doi.org/10.13031/2013.26773

IPCC. (1990). *Climate change: The IPCC scientific assessment.* Mass, Cambridge.

Irmak, A., & Kamble, B. (2009). Evapotranspiration data assimilation with genetic algorithms and SWAP model for on-demand irrigation. *Irrigation Science, 28*(1), 101–112.

Irmak, S., Irmak, A., Allen, R. G., & Jones, J. W. (2003). Solar and net radiation-based equations to estimate reference evapotranspiration in humid climates. *Journal of Irrigation and Drainage Engineering, 129*(5), 336–347. https://doi.org/10.1061/(ASCE)0733-9437(2003)129:5(336)

Izadifar, Z., & Elshorbagy, A. (2010). Prediction of hourly actual evapotranspiration using neural networks, genetic programming, and statistical models. *Hydrological Processes, 24*(23), 3413–3425. https://doi.org/10.1002/hyp.7771

Jackson, R. D., Idso, S. B., Reginato, R. J., & Pinter, Jr., P. J. (1981). Canopy temperature as a crop water stress indicator. *Water Resources Research, 17*(4), 1133–1138. https://doi.org/10.1029/WR017i004p01133

Kamel, S. T. (2015). Variation of evapotranspiration during 2000 to 2010 in egypt due to consequences of climate change In *Eighteenth international water technology conference, IWTC18, Sharm ElSheikh.*

Khalil, A. (2013). Effect of climate change on evapotranspiration in Egypt. *Researcher, 5*(1), 7–12.

Khalil, A., Essa, Y., & Abdel-Wahab, M. (2015). Evapotranspiration mapping over Egypt using MODIS/Terra satellite data. *International Journal of Advanced Research, 15*(12), 512–522.

Kim, S., & Kim, H. S. (2008). Neural networks and genetic algorithm approach for nonlinear evaporation and evapotranspiration modeling. *Journal of Hydrology, 351*(3), 299–317. https://doi.org/10.1016/j.jhydrol.2007.12.014

Kingston, D. G., Todd, M. C., Taylor, R. G., Thompson, J. R., & Arnell, N. W. (2009). Uncertainty in the estimation of potential evapotranspiration under climate change. *Geophysical Research Letters, 36*(20).

Kisi, O. (2016). Modeling reference evapotranspiration using three different heuristic regression approaches. *Agricultural Water Management, 169*, 162–172. https://doi.org/10.1016/j.agwat.2016.02.026

Kisi, O., & Guven, A. (2010). Evapotranspiration modeling using linear genetic programming technique. *Journal of Irrigation and Drainage Engineering, 136*(10), 715–723. https://doi.org/10.1061/(ASCE)IR.1943-4774.0000244

Landeras, G., Ortiz-Barredo, A., & López, J. J. (2008). Comparison of artificial neural network models and empirical and semi-empirical equations for daily reference evapotranspiration estimation in the Basque Country (Northern Spain). *Agricultural Water Management, 95*(5), 553–565. https://doi.org/10.1016/j.agwat.2007.12.011

Lang, D., Zheng, J., Shi, J., Liao, F., Ma, X., Wang, W., Chen, X., & Zhang, M. (2017). A comparative study of potential evapotranspiration estimation by eight methods with FAO penman–monteith method in Southwestern China. *Water, 9*(10), 734. https://www.mdpi.com/2073-4441/9/10/734

Liu, Q., & Yang, Z. (2010). Quantitative estimation of the impact of climate change on actual evapotranspiration in the Yellow River Basin, China. *Journal of Hydrology, 395*(3), 226–234. https://doi.org/10.1016/j.jhydrol.2010.10.031

Makkink, G. (1957). Examination of Penman's revised formula [Esperanto]. *NJAS Wageningen Journal of Life Sciences, 5*(4), 290–305.

Mattar, M. A. (2018). Using gene expression programming in monthly reference evapotranspiration modeling: A case study in Egypt. *Agricultural Water Management, 198*, 28–38. https://doi.org/10.1016/j.agwat.2017.12.017

McKenney, M. S., & Rosenberg, N. J. (1993). Sensitivity of some potential evapotranspiration estimation methods to climate change. *Agricultural and Forest Meteorology, 64*(1), 81–110. https://doi.org/10.1016/0168-1923(93)90095-Y

McMahon, T. A., Finlayson, B. L., & Peel, M. C. (2016). Historical developments of models for estimating evaporation using standard meteorological data. *WIREs Water, 3*(6), 788–818. https://doi.org/10.1002/wat2.1172

Menenti, M. (1993). Parameterization of land surface evaporation by means of location dependent potential evaporation and surface temperature range. In *Exchange processes at the land surface for a range of space and time scales.* https://cir.nii.ac.jp/crid/1572543025224597760

Morsy, M., El-Sayed, T., & Ouda, S. A. H. (2016). Potential evapotranspiration under present and future climate. In *Management of climate induced drought and water scarcity in Egypt: Unconventional solutions* (pp. 5–25). Springer International Publishing. https://doi.org/10.1007/978-3-319-33660-2_2

Omar, M. E. M., Abd Elhamid, A. M. I., & Al Zayed, I. S. (2019). Estimating actual evapotranspiration over a large and complex irrigation system of the Nile Delta in Egypt. *International Journal of Recent Technology and Engineering (IJRTE), 8*(4).

Ouda, S., Noreldin, T., & Hosney, M. (2016). Evapotranspiration under changing climate. In *Major crops and water scarcity in Egypt: Irrigation water management under changing climate* (pp. 1–23). Springer International Publishing. https://doi.org/10.1007/978-3-319-21771-0_1

Ouda, S. A., & Norledin, T. A. (2017). Evapotranspiration data to determine agro-climatic zones in Egypt. *Water and Land Development, 32*(1), 79–86.

Penman, H. L., & Keen, B. A. (1948). Natural evaporation from open water, bare soil and grass. *193*(1032), 120–145. https://doi.org/10.1098/rspa.1948.0037

Pour-Ali Baba, A., Shiri, J., Kisi, O., Fard, A. F., Kim, S., & Amini, R. (2012). Estimating daily reference evapotranspiration using available and estimated climatic data by adaptive neuro-fuzzy inference system (ANFIS) and artificial neural network (ANN). *Hydrology Research, 44*(1), 131–146. https://doi.org/10.2166/nh.2012.074

Priestley, C. H. B., & Taylor, R. J. (1972). On the assessment of surface heat flux and evaporation using large-scale parameters. *Monthly Weather Review, 100*(2), 81–92. https://doi.org/10.1175/1520-0493(1972)100%3c0081:Otaosh%3e2.3.Co;2

Ragheb, H. M. A., Gomah, H. H., Hassanein, M. K., & Hassan, A. M. A. (2015). Validation of reference evapotranspiration models using lysimeters under arid climatic conditions. *Assiut Journal of Agricultural Sciences, 46*(2), 210–228. https://doi.org/10.21608/ajas.2015.569

Senay, G. B., Budde, M., Verdin, J. P., & Melesse, A. M. (2007). A coupled remote sensing and simplified surface energy balance approach to estimate actual evapotranspiration from irrigated fields. *Sensors, 7*(6), 979–1000. https://www.mdpi.com/1424-8220/7/6/979

Shiri, J. (2017). Evaluation of FAO56-PM, empirical, semi-empirical and gene expression programming approaches for estimating daily reference evapotranspiration in hyper-arid regions of Iran. *Agricultural Water Management, 188*, 101–114. https://doi.org/10.1016/j.agwat.2017.04.009

Su, Z. (2002). The surface energy balance system (SEBS) for estimation of turbulent heat fluxes. *Hydrology and Earth System Sciences, 6*(1), 85–99. https://edepot.wur.nl/33624

Thornthwaite, C. W. (1948). An approach toward a rational classification of climate. *Geographical Review, 38*(1), 55–94. https://doi.org/10.2307/210739

Traore, S., & Guven, A. (2012). Regional-specific numerical models of evapotranspiration using gene-expression programming interface in Sahel. *Water Resources Management, 26*(15), 4367–4380. https://doi.org/10.1007/s11269-012-0149-3

Traore, S., Luo, Y., & Fipps, G. (2016). Deployment of artificial neural network for short-term forecasting of evapotranspiration using public weather forecast restricted messages. *Agricultural Water Management, 163*, 363–379. https://doi.org/10.1016/j.agwat.2015.10.009

Traore, S., Luo, Y., & Fipps, G. (2017). Gene-expression programming for short-term forecasting of daily reference evapotranspiration using public weather forecast information. *Water Resources Management, 31*(15), 4891–4908. https://doi.org/10.1007/s11269-017-1784-5

Turc, L. (1961). Estimation of irrigation water requirements, potential evapotranspiration: A simple climatic formula evolved up to date. *Ann. Agron, 12*(1), 13–49.

Valipour, M., Gholami Sefidkouhi, M. A., Raeini-Sarjaz, M., & Guzman, S. M. (2019). A hybrid data-driven machine learning technique for evapotranspiration modeling in various climates. *Atmosphere, 10*(6), 311.

Wagle, P., Bhattarai, N., Gowda, P. H., & Kakani, V. G. (2017). Performance of five surface energy balance models for estimating daily evapotranspiration in high biomass sorghum. *ISPRS Journal of Photogrammetry and Remote Sensing, 128*, 192–203. https://doi.org/10.1016/j.isprsjprs.2017.03.022

Wahab, M. M. A., Essa, Y. H., Khalil, A. A., Elfadli, K., & Panegrossi, G. (2018). Water loss in Egypt based on the Lake Nasser evaporation and agricultural evapotranspiration. *Environment Asia, 11*(2), 192–204.

Wang, K., & Dickinson, R. E. (2011, December 01, 2011). *A review of global terrestrial evapotranspiration: Observation, modeling, climatology, and climatic variability*. https://ui.adsabs.harvard.edu/abs/2011AGUFM.B33F0532W. https://doi.org/10.1029/2011RG000373?download=true

Yassen, A. N., Nam, W.-H., & Hong, E.-M. (2020). Impact of climate change on reference evapotranspiration in Egypt. *CATENA, 194*, 104711. https://doi.org/10.1016/j.catena.2020.104711

Zhao, L., Xia, J., Xu, C.-Y., Wang, Z., Sobkowiak, L., & Long, C. (2013). Evapotranspiration estimation methods in hydrological models. *Journal of Geographical Sciences, 23*(2), 359–369. https://doi.org/10.1007/s11442-013-1015-9

Chapter 7
Predictive Modeling of Water Levels Using Univariate and Multivariate Time Series Approaches

Rehmana Younis, Aqib Rehman Pir Zada, Malik Muhammad Saad Missen, Muhammad Naeem Iqbal, Nadeem Iqbal Kajla, Mohamed Elkollaly, and Faisal Baig

Abstract This paper explores the dynamics of water levels in the Aquifer Petrignano dataset through comprehensive time series analysis. By examining temporal variations and seasonal patterns, we gain insights into the underlying trends affecting water levels. We address the impact of missing values and evaluate various methods for handling them, ensuring the robustness of our analysis. Data smoothing and resampling techniques are investigated to enhance predictive accuracy. Emphasis is placed on achieving stationarity using transformations and differencing methods. We analyze the influence of cyclical features, such as the time of year, on water level predictions. Time series decomposition is utilized to understand the fundamental components of the dataset. The inclusion of lagged variables is explored to improve forecasting models. Autocorrelation analysis guides the selection of appropriate models for water level prediction. We compare the performance of univariate models,

R. Younis
College of Letters and Sciences, Columbus State University, Columbus, GA, USA

A. R. P. Zada
COMSATS University Islamabad, Sahiwal, Pakistan

M. M. S. Missen
Department of Software Engineering, Islamia University of Bahawalpur, Bahawalpur, Pakistan

M. N. Iqbal
Department of Agricultural Engineering, MNS University of Agriculture Multan, Multan, Pakistan

N. I. Kajla
Institute of Computing, MNS University of Agriculture Multan, Multan, Pakistan

M. Elkollaly · F. Baig (✉)
National Water and Energy Center, UAE University, Al Ain, UAE
e-mail: faisalbaig@uaeu.ac.ae

M. Elkollaly
Irrigation and Hydraulics Engineering Department, Faculty of Engineering, Tanta University, Tanta 31733, Egypt

A. Sefelnasr et al. (eds.), *Water Resources Management and Sustainability*, Water Science and Technology Library 114, https://doi.org/10.1007/978-3-031-80520-2_7

such as Prophet, ARIMA, Auto-ARIMA, and LSTM, evaluating their strengths and limitations. The study extends to multivariate models, such as Multivariate Prophet, to assess improved prediction accuracy by incorporating multiple variables. Challenges associated with applying multivariate models are addressed, and solutions are proposed to overcome these obstacles. The findings provide valuable insights into water level dynamics and offer robust methodologies for accurate forecasting.

Keywords Predictive modeling · Time series decomposition · ARIMA · LSTM · Water level prediction

7.1 Introduction

Smart Water Analytics (Bejarano et al., 2018) represents a transformative approach in the field of water resource management, leveraging cutting-edge technologies such as advanced data analysis techniques, Internet of Things (IoT) (Madakam et al., 2015) devices, and artificial intelligence (AI) (Boden, 1996). The core concept revolves around the continuous collection and processing of real-time data from a variety of sources, including sensors, meters, and satellites. This data is then analyzed to generate actionable insights that can significantly enhance the efficiency and effectiveness of water distribution, consumption, and conservation efforts.

One of the primary advantages of smart water analytics is its ability to shift water management strategies from reactive to proactive (Rehman et al., 2023). Traditionally, water resource management has been characterized by a reactive approach, where issues are addressed only after they arise. However, with the integration of smart water analytics, it becomes possible to predict and prevent problems before they occur. For instance, predictive analytics can identify potential leaks (Farley & Trow, 2003) or inefficiencies in the water distribution network, allowing for timely maintenance and repair, thereby minimizing water loss and ensuring a more reliable supply.

As global pressures on water resources intensify due to factors such as population growth, industrial expansion, and climate change, the need for efficient and sustainable water management (Haie & Keller, 2008) becomes increasingly critical. Smart water analytics plays a pivotal role in addressing these challenges. By utilizing real-time data and sophisticated machine learning algorithms, these systems can enhance the accuracy of water level predictions, optimize water distribution schedules, and ensure that water usage is aligned with availability (Mustafa et al., 2024). This not only supports the sustainable use of water resources but also helps in mitigating the impacts of water scarcity and ensuring the resilience of water systems.

Moreover, the implementation of smart water analytics extends beyond mere operational improvements. It has the potential to facilitate better decision-making at both the policy and community levels. For example, detailed data on water consumption patterns can inform policymakers about the most effective water-saving measures, while also empowering communities with the knowledge they need to manage their water use more efficiently. This holistic approach ensures that the benefits of smart water analytics are felt across the entire spectrum of water management, from individual households to large-scale utilities (Iqbal et al., 2019).

In short, smart water analytics represents a significant advancement in the quest for a sustainable water future. By harnessing the power of IoT, AI, and advanced data analysis (Cramer, 2003), it offers a proactive, data-driven approach to water resource management (Ullah et al., 2023). This not only improves operational efficiency but also supports informed decision-making and promotes the sustainable use of water resources. As we continue to face growing pressures on our water systems, the adoption of smart water analytics will be crucial in ensuring the resilience and sustainability of our water infrastructure for the benefit of both communities and ecosystems alike (Bejarano et al., 2018). The followings are the contributions of this work summarized:

- The paper explores the dynamics of water levels in the Aquifer Petrignano dataset through time series analysis.
- Examines temporal variations and seasonal patterns in water levels.
- Addresses the impact of missing values and evaluates methods for handling them.
- Investigates data smoothing and resampling techniques to enhance predictive accuracy.
- Emphasizes the importance of achieving stationarity using transformations and differencing methods.
- Analyzes the influence of cyclical features, such as the time of year, on water level predictions.
- Utilizes time series decomposition to understand the underlying components of the dataset.
- Explores the inclusion of lagged variables to enhance forecasting models.
- Conducts autocorrelation analysis to guide the selection of appropriate models for water level prediction.
- Compares the performance of univariate models like Prophet, ARIMA, Auto-ARIMA, and LSTM, evaluating their strengths and limitations.
- Extends to multivariate models, such as Multivariate Prophet, to assess improved prediction accuracy by incorporating multiple variables.
- Addresses challenges associated with applying multivariate models and proposes solutions to overcome these obstacles.

Through this comprehensive approach, the paper aims to contribute to the body of knowledge on effective water level prediction and management in the Aquifer Petrignano dataset.

The paper is structured as follows: We begin by providing an overview of the data collection process, detailing the sources and types of data gathered. Next, we describe the methodology employed for data preprocessing, feature selection, and model development. We then present the results of our predictive models, highlighting their accuracy and potential applications. Finally, we discuss the implications of our findings for water resource management and outline future directions for research and development in this field.

7.2 Experiments

7.2.1 Data Collection

Water is an essential resource for human survival, ecosystem health, and economic activities (Bogardi et al., 2020). Effective management of water resources is crucial for ensuring their sustainable availability and quality. The Acea Group, a leading Italian multiutility operator, plays a pivotal role in managing water services for approximately 9 million inhabitants across Lazio, Tuscany, Umbria, Molise, and Campania. Predicting water levels in various waterbodies—such as springs, lakes, rivers, and aquifers—is essential for optimizing water distribution and ensuring its availability throughout the year.

Given the distinct characteristics of different waterbodies, influenced by geographical location, climatic conditions, and human activities, tailored approaches are necessary for accurate water availability forecasting. Among these waterbodies, aquifers, which are underground layers of water-bearing rock that supply a significant portion of drinking water, pose a unique challenge. Their hidden nature and the complexity of factors affecting their recharge and discharge rates make their management particularly difficult. The Data Collection is done by Acea Group which is one of the leading Italian multiutility operators. Listed on the Italian Stock Exchange since 1999, the company manages and develops water and electricity networks and environmental services. Acea is the foremost Italian operator in the water services sector supplying 9 million inhabitants in Lazio, Tuscany, Umbria, Molise, Campania. They uploaded this data on kaggle. We used dataset of Aquifer petrignano dataset from the Kaggle.

7.2.2 Data Visualisation

Data Visualization in Smart Water Analytics Using Time Series

In our research on Smart Water Analytics using Time Series, we examined several key features to illustrate significant trends and relationships within the Aquifer Petrignano dataset.

Rainfall, measured in millimeters, was depicted to visualize precipitation patterns over time, crucial for understanding its impact on water resources. Temperature, represented in degrees Celsius, provided insights into seasonal variations and potential correlations with water consumption and groundwater levels. Volume, measured in cubic meters, was graphed to show the quantity of water extracted from the treatment plant, essential for monitoring usage trends and ensuring sustainable resource management. Finally, hydrometry data, indicating groundwater levels in meters, was visualized to assess aquifer health and inform decisions regarding water extraction and conservation strategies. These visualizations not only enhance comprehension of the intricate dynamics within the water system but also support informed decision-making for efficient water resource management.

Variation of Water Levels Over Time

In analyzing the Aquifer Petrignano dataset, we sought to understand how water levels vary over time. By plotting water levels over different periods, such as monthly and yearly, we could observe long-term trends. These visualizations revealed consistent patterns, such as a gradual decline in water levels over several years, potentially due to increased water extraction or climatic changes. Anomalies, such as sudden spikes or drops in water levels, were identified, indicating unusual events like heavy rainfall or drought.

Seasonal Patterns in Water Levels

Seasonal patterns in the water levels of the Aquifer Petrignano dataset were also a focal point of our analysis. By applying Seasonal and Trend decomposition using Loess (STL), we broke down the water levels into trend, seasonal, and residual components. This decomposition helped isolate recurring seasonal patterns from overall trends and random noise. We identified that water levels were generally higher during the rainy season and lower during dry periods, highlighting the influence of seasonal precipitation on groundwater levels.

Furthermore, we examined correlations between water levels and other variables. Rainfall data was compared with water levels to understand how precipitation impacts groundwater. Increased rainfall typically correlated with rising water levels. Temperature data was also analyzed to see its effect on water levels; higher temperatures were associated with increased evaporation and reduced groundwater levels. Additionally, we analyzed the volume of water extracted from the treatment plant and its impact on groundwater levels. Increased extraction often led to lower water levels, emphasizing the importance of sustainable resource management.

Example Visualization and Analysis

In Fig. 7.1, plot of Depth and Temperature illustrates the relationship between the depth of groundwater levels and temperature. The top plot shows the variation in the depth to groundwater over time, highlighting long-term trends and fluctuations. The bottom plot depicts the temperature variation, displaying clear seasonal patterns with regular peaks and troughs.

In Fig. 7.2, plot of Drainage and River Hydrometry further supports our analysis: • The top graph in Fig. 7.2 shows the drainage volume over time. It illustrates the variability and trends in the volume of water being drained, with notable fluctuations and a slight increasing trend over the years. • The bottom graph in Fig. 7.2 presents river hydrometry data, which indicates the water levels in the river. This data highlights seasonal peaks and troughs, correlating with periods of higher and lower rainfall.

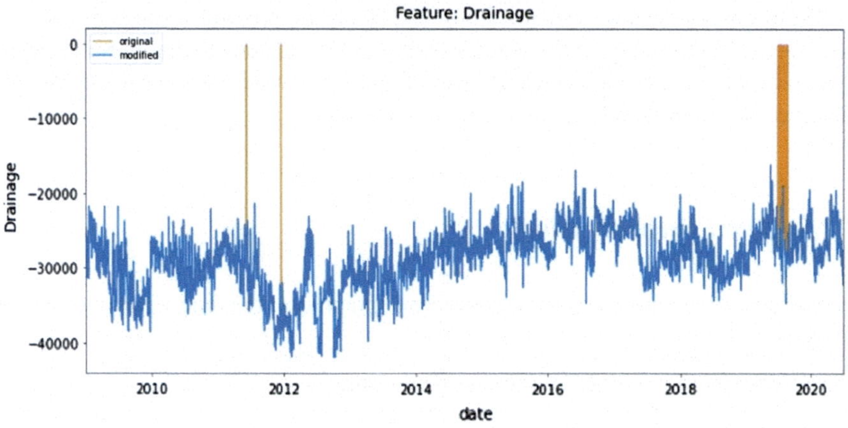

Fig. 7.4 Plot of drainage null values

Smoothing Data/Resampling

To adjust the frequency of samples, smoothing techniques and resampling were utilized (Fig. 7.5):

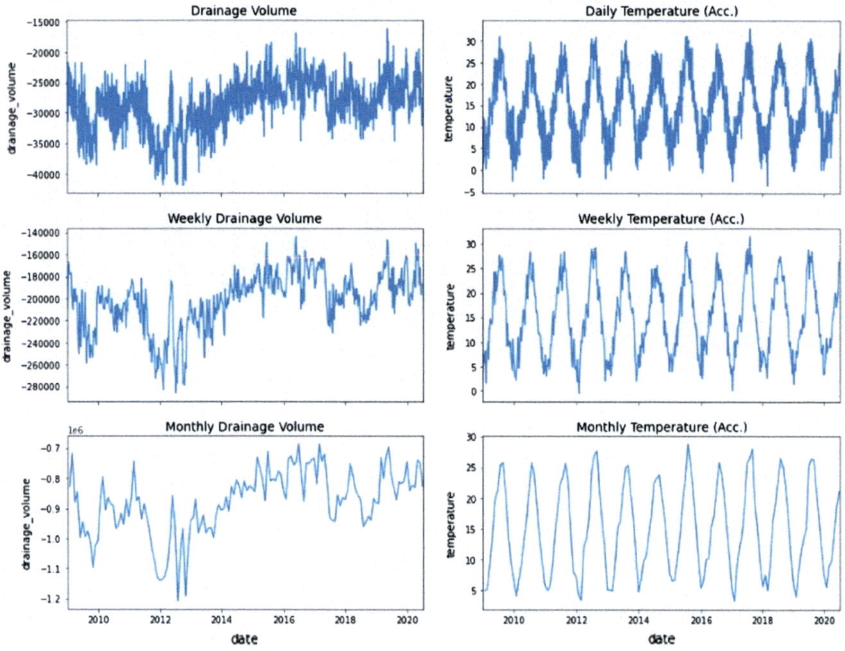

Fig. 7.5 Plot of smoothing feature

Upsampling: Increasing the frequency of samples (e.g., from days to hours).

Downsampling: Decreasing the frequency of samples (e.g., from days to weeks).

Stationarity

Certain time series models, like ARIMA, require the data to be stationary. Stationarity was assessed using multiple approaches:

- **Visual Inspection**: Time series plots were visually inspected for trends or seasonality (Figs. 7.6 and 7.7)
- **Basic Statistics**: Statistical metrics such as mean and variance were analyzed across partitions of the time series.
- **Statistical Tests**: The Augmented Dickey-Fuller (ADF) test, a unit root test, was conducted.

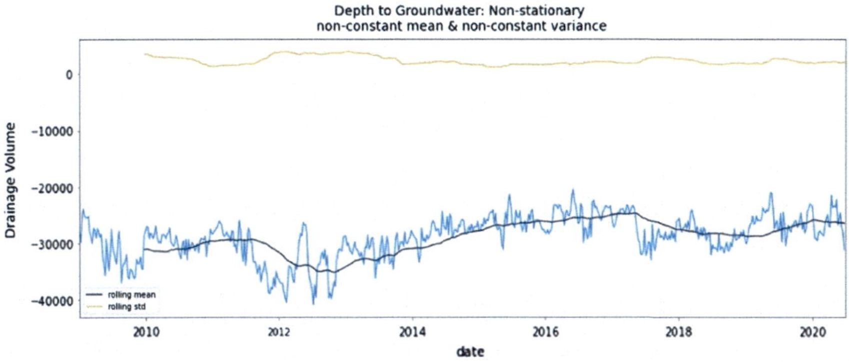

Fig. 7.6 Stationary 1

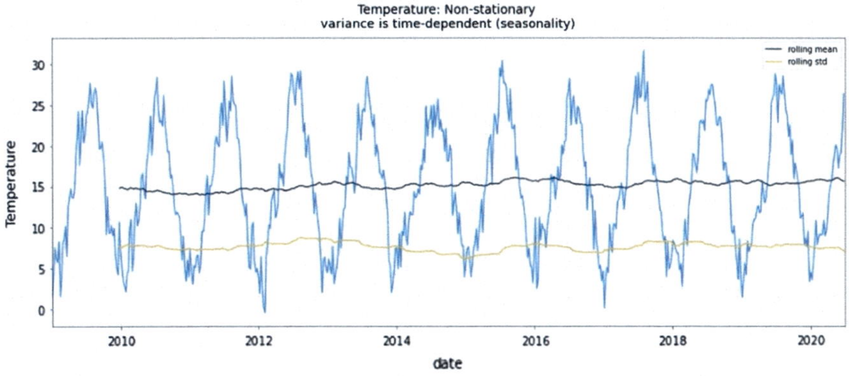

Fig. 7.7 Stationary 2

7.2.4 Feature Engineering

In this subsection, we detail the feature engineering techniques applied to enhance the time series data for analysis and modeling.

Encoding Cyclical Features

Time features such as months or days of the week are cyclical in nature. For instance, months cycle from 1 to 12 within a year, with December (12) transitioning to January (1) of the following year. Directly encoding these features using numerical values (e.g., 1 to 12 for months) can lead to misinterpretation by models due to the abrupt change from December to January. To address this, cyclical features are encoded using sine and cosine transformations, converting them into continuous variables that capture the periodic nature of time (Fig. 7.8).

Time Series Decomposition

Time series decomposition (Zarnowitz & Ozyildirim, 2006) involves breaking down a time series into its fundamental components:

- **Level**: The average value of the series.
- **Trend**: The increasing or decreasing pattern over time.
- **Seasonality**: The repeating cycle or pattern within a fixed time period.
- **Noise**: The random variation or irregular component.

Decomposition provides a structured framework to understand and model time series data, facilitating better analysis and forecasting. Depending on the nature of the data, not all components (trend and seasonality) may be present or significant (Fig. 7.9).

The decomposition model can be expressed additively or multiplicatively:

- **Additive Model:** $y(t) = \text{Level} + \text{Trend} + \text{Seasonality} + \text{Noise}$
- **Multiplicative Model:** $y(t) = \text{Level} \times \text{Trend} \times \text{Seasonality} \times \text{Noise}$

Each component contributes differently based on whether the additive or multiplicative model is chosen, impacting how the time series is interpreted and modeled.

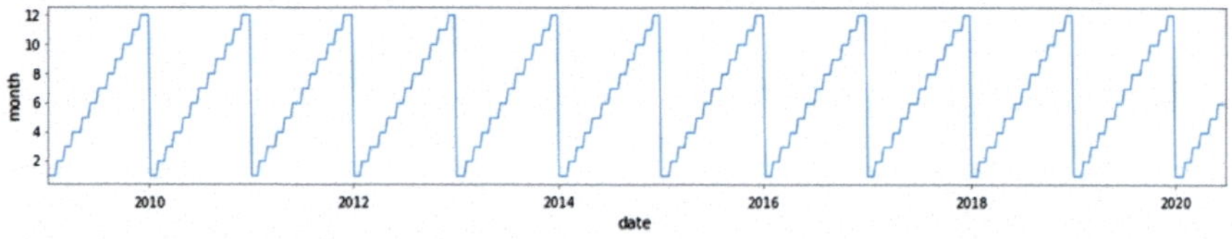

Fig. 7.8 Encoding cyclical features

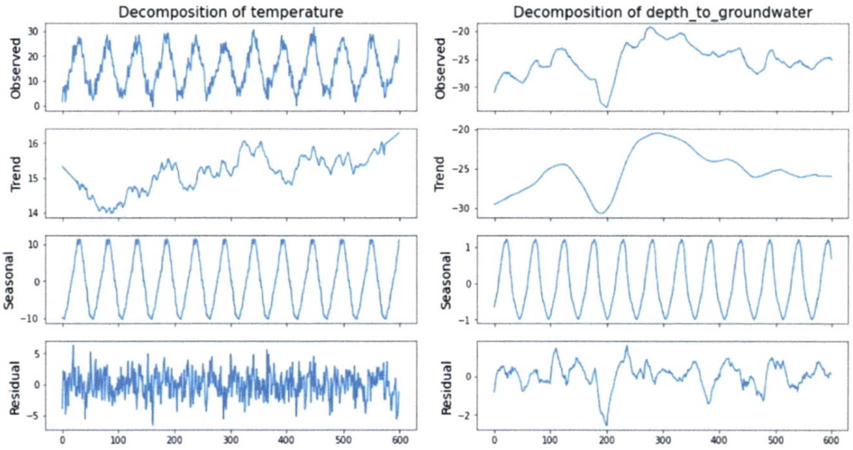

Fig. 7.9 Timeseries decomposition

Lag

To explore relationships and dependencies between variables, lagging (shifting) each variable by a certain time step is employed. This involves creating new variables that represent the values of existing variables at previous time points. Lagged variables allow for the analysis of how past values of one variable correlate with present or future values of another variable. Such analysis helps in uncovering patterns, trends, and potential predictors within the time series data.

These feature engineering techniques transform the raw time series data into structured features that improve the effectiveness of subsequent analysis and modeling tasks.

7.2.5 Exploratory Data Analysis (EDA)

In this subsection, we conduct exploratory data analysis (Lim & Zohren, 2021) to visualize the time series data and extract meaningful insights (Fig. 7.10).

Autocorrelation Analysis

Autocorrelation analysis (Anselin, 2003) involves examining the correlation of a variable with a lagged version of itself. It is crucial for time series modeling, especially for identifying patterns like seasonality and determining appropriate ARIMA model parameters.

After stationarizing the time series through differencing, the next step in ARIMA modeling is to assess whether Autoregressive (AR) or Moving Average (MA) terms are necessary. This is achieved through Autocorrelation Function (ACF) plots of the differenced series.

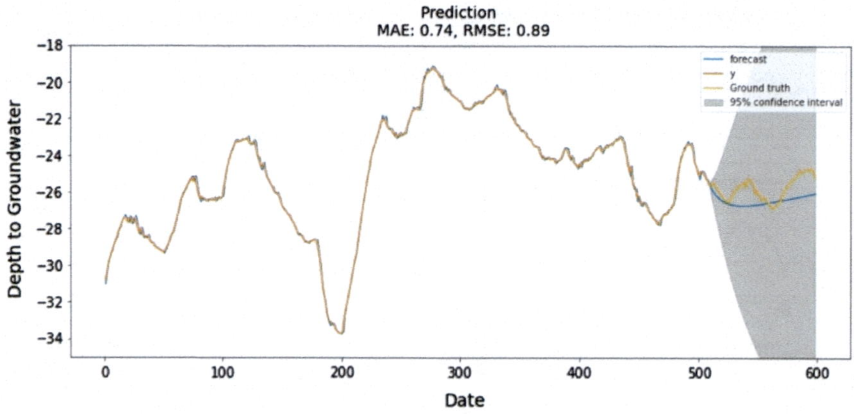

Fig. 7.14 Univariate ARIMA

- Auto-Regressive (AR) terms (p): Capture dependencies between an observation and a lagged value.
- Integrated (I) term (d): Represents the number of differencing required to make the time series stationary.
- Moving Average (MA) terms (q): Depict the dependency between an observation and a residual error from a moving average model.

Steps for ARIMA analysis include checking stationarity, differencing, model selection based on ACF and PACF plots, model building, and validation.

Auto-ARIMA Auto-ARIMA is an automated version of ARIMA that selects the optimal parameters (p, d, q) based on statistical tests and information criteria (Fig. 7.15).

LSTM Long Short-Term Memory (LSTM) (Graves & Graves, 2012) networks are a type of recurrent neural network (RNN) (Grossberg, 2013) particularly effective for sequence prediction tasks. For univariate time series forecasting, LSTM models can capture complex patterns and long-term dependencies in the data (Fig. 7.16).

Steps for LSTM modeling involve data preprocessing (normalization, reshaping), model training, and validation.

Model for Multivariate Time Series

- How does the inclusion of multiple variables in multivariate models (like Multivariate Prophet) improve the accuracy of water level predictions in the Aquifer Petrignano dataset?
- What challenges arise when applying multivariate models to the Aquifer Petrignano dataset, and how can these challenges be addressed?

Multivariate time series involve multiple variables that vary simultaneously over time, such as sensor data from multiple sources.

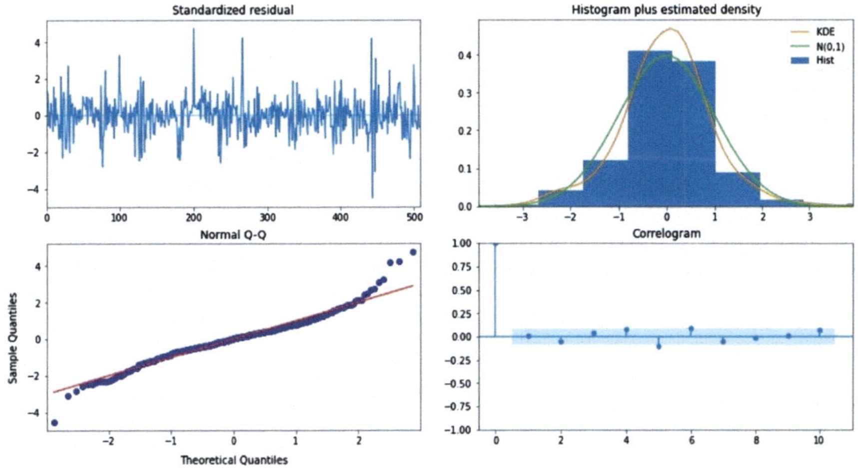

Fig. 7.15 Auto-ARIMA plot of results

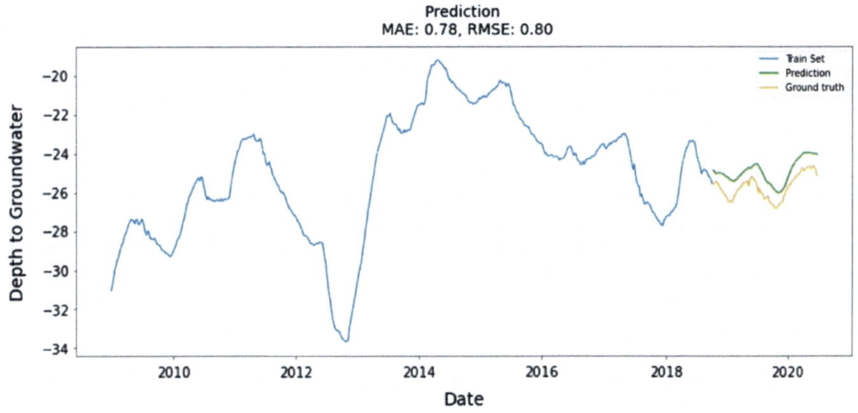

Fig. 7.16 LSTM model for univariate modelling

Multivariate Prophet

Prophet can also handle multivariate time series forecasting by incorporating multiple related variables into its forecasting framework. It extends the capabilities of Prophet to capture interactions and dependencies between variables. Further exploration into multivariate time series forecasting would involve techniques like Vector Autoregression (VAR), Dynamic Factor Models, and other approaches capable of modeling complex dependencies and interactions between variables (Fig. 7.17).

This subsection provides a foundation for applying various modeling techniques to different types of time series data, ensuring robust forecasting and analysis.

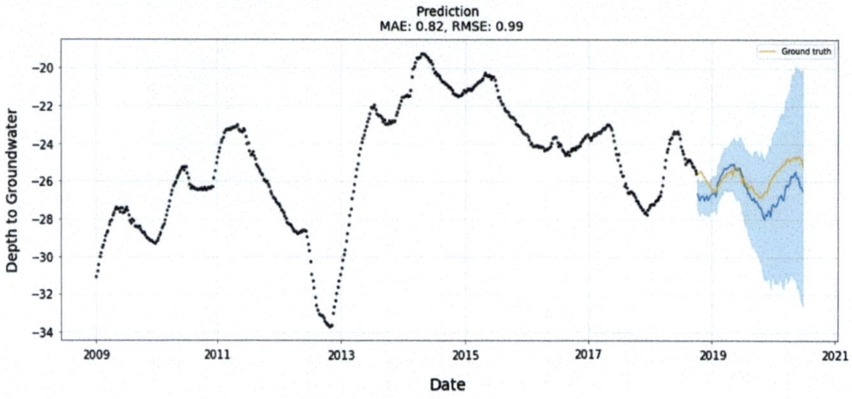

Fig. 7.17 Prophet model for multivariate modelling

7.2.7 Model Evaluation

In this subsection, we discuss the evaluation of time series forecasting models to assess their accuracy and effectiveness. When evaluating time series forecasting models, several key performance indicators (KPIs) are commonly used:

- **Mean Absolute Error (MAE)**: Measures the average magnitude of errors in a set of forecasts, without considering their direction. It is calculated as:

$$\text{MAE} = \frac{1}{n}\sum_{i=1}^{n}\left|Y_i - \widehat{Y}_i\right| \tag{7.1}$$

- **Mean Squared Error (MSE)**: Measures the average of the squares of the errors—that is, the average squared difference between the estimated values and the actual value. It is calculated as:

$$\text{MSE} = \frac{1}{n}\sum_{i=1}^{n}(Y_i - \widehat{Y}_i)^2 \tag{7.2}$$

- **Root Mean Squared Error (RMSE):** RMSE is the square root of the MSE, which provides a measure of the average magnitude of the errors in the same units as the target variable. It is calculated as:

$$\text{RMSE} = \sqrt{\frac{1}{n}\sum_{i=1}^{n}(Y_i - \widehat{Y}_i)^2} \tag{7.3}$$

- **Mean Absolute Percentage Error (MAPE):** MAPE expresses the forecast accuracy as a percentage of the absolute error relative to the actual values. It is calculated as:

$$\text{MAPE} = \frac{1}{n}\sum_{i=1}^{n}\left(\frac{\left|Y_i - \widehat{Y}_i\right|}{Y_i}\right) \times 100\% \tag{7.4}$$

- **Symmetric Mean Absolute Percentage Error (SMAPE):** SMAPE is an alternative percentage error measure that addresses the issue of scale dependence in MAPE. It is calculated as:

$$\text{SMAPE} = \frac{1}{n}\sum_{i=1}^{n}\left(\frac{2 \times \left|Y_i - \widehat{Y}_i\right|}{|Y_i| + \left|\widehat{Y}_i\right|}\right) \times 100\% \tag{7.5}$$

- **Forecast Bias**: Measures the average tendency of the model to overestimate or underestimate the actual values.
- **Forecast Accuracy**: Provides an overall measure of how well the model performs in terms of correctly predicting the direction and magnitude of future values.

Evaluation Metrics for Univariate Models

Among the univariate models, ARIMA demonstrates the best overall performance with the lowest Mean Absolute Error (MAE) of 0.861 and Mean Squared Error (MSE) of 1.130, indicating high accuracy. Deep Neural Prophet performs well in terms of Root Mean Square Error (RMSE) and Forecast Bias, but its high MAPE and SMAPE values suggest it struggles with proportional accuracy. Auto ARIMA has the highest Forecast Accuracy at 95.270%, but this comes with a higher Forecast Bias and other errors. LSTM shows the worst performance across most metrics, with particularly high MAE, MSE, and RMSE values, indicating poor fit for this specific univariate forecasting task. As shown in Table 7.1.

Evaluation Metrics for Multivariate Model

In the multivariate context, Deep Neural Prophet is the only model evaluated and demonstrates strong performance. It achieves a low Mean Absolute Error (MAE) of 0.824 and Mean Squared Error (MSE) of 0.988, with an exceptionally high Forecast

Table 7.1 Evaluation metrics for univariate time series models

Model	MAE	MSE	RMSE	MAPE	SMAPE	Forecast Bias	Forecast accuracy
Deep neural prophet	0.892	1.188	1.090	187.797	167.445	0.233	− 87.797
ARIMA	0.861	1.130	1.063	116.841	178.051	0.109	− 16.841
Auto ARIMA	1.200	1.963	1.402	4.730	4.580	− 1.200	95.270
LSTM	25.261	638.432	25.267	98.201	192.932	25.261	1.799

Table 7.2 Evaluation metrics for multivariate time series model

Model	MAE	MSE	RMSE	MAPE	SMAPE	Forecast Bias	Forecast accuracy
Deep neural Prophet	0.824	0.988	0.994	3.223	3.152	− 0.705	96.777

Accuracy of 96.777%. The low MAPE and SMAPE values further suggest that this model provides reliable and proportionally accurate forecasts, making it a robust choice for multivariate time series analysis. As shown in Table 7.2.

7.3 Conclusion

In this paper, we conducted an in-depth analysis of the Aquifer Petrignano dataset, revealing key insights into the temporal variations and seasonal patterns of water levels. Our investigation into handling missing values and employing data smoothing and resampling techniques has highlighted effective methods to enhance predictive accuracy. By achieving stationarity through transformations and differencing, we established a robust foundation for accurate modeling. The analysis of cyclical features and time series decomposition provided a comprehensive understanding of the dataset's underlying components.

The inclusion of lagged variables and autocorrelation analysis significantly improved our forecasting models, guiding the selection of the most appropriate techniques. Our comparative study of univariate models, including Prophet, ARIMA, Auto-ARIMA, and LSTM, demonstrated their respective strengths and limitations. Extending the analysis to multivariate models, such as Multivariate Prophet, underscored the benefits of incorporating multiple variables for improved prediction accuracy.

We addressed the challenges associated with applying multivariate models and proposed practical solutions, contributing to the broader field of time series analysis. Our findings offer valuable methodologies for accurate water level forecasting and can inform future research and practical applications in hydrology and water resource management.

References

Anselin, L. (2003). An introduction to spatial autocorrelation analysis with geoda. *Spatial Analysis Laboratory, University of Illinois, Champagne-Urbana, Illinois.*

Bejarano, G., Jain, M., Ramesh, A., Seetharam, A., & Mishra, A. (2018). Predictive analytics for smart water management in developing regions. In *2018 IEEE International Conference on Smart Computing (SMARTCOMP)* (pp. 464–469). IEEE.

Boden, M. A. (1996). *Artificial intelligence*. Elsevier.
Bogardi, J. J., Leentvaar, J., & Sebesvari, Z. (2020). Biologia futura: integrating freshwater ecosystem health in water resources management. *Biologia futura 71*(4), 337–358.
Chen, J. (2019). Autoregressive integrated moving average (arima). *Advanced Technical Analysis Concept*.
Cramer, D. (2003). *Advanced quantitative data analysis*. McGraw-Hill Education (UK).
Farley, M., & Trow, S. (2003). *Losses in water distribution networks*. IWA publishing.
Graves, A., & Graves, A. (2012). Long short-term memory. *Supervised Sequence Labelling with Recurrent Neural Networks*, 37–45.
Grossberg, S. (2013). Recurrent neural networks. *Scholarpedia, 8*(2), 1888.
Haie, N., & Keller, A. A. (2008). Effective efficiency as a tool for sustainable water resources management 1. *JAWRA Journal of the American Water Resources Association, 44*(4), 961–968.
Iqbal, N., Saad Missen, M. M., Salamat, N., & Prasath, V. S. (2019). On video based human abnormal activity detection with histogram of oriented gradients. *Handbook of Multimedia Information Security: Techniques and Applications*, 431–448.
Lim, B., & Zohren, S. (2021). Time-series forecasting with deep learning: A survey. *Philosophical Transactions of the Royal Society A, 379*(2194), 20200209.
Madakam, S., Ramaswamy, R., & Tripathi, S. (2015). Internet of things (IoT): A literature review. *Journal of Computer and Communications, 3*(5), 164–173.
Mustafa, M., Baig, M. F., Badar, H. M. S., Missen, M. M. S., & Kajla, N. I. (2024). An artificial intelligence and IoT based approach for the Pink bollworm detection in Cotton crop. *Journal of Computing & Biomedical Informatics*.
Rehman, D., Jamil, M. A., Badar, H. M. S., Awan, M. D. A., Chaudhry, M. U., & Kajla, N. I. (2023). Enhancing crop production and water conservation through IoT-based smart irrigation systems. *Journal of Computing & Biomedical Informatics, 5*(01), 96–104.
Toharudin, T., Pontoh, R. S., Caraka, R. E., Zahroh, S., Lee, Y., & Chen, R. C. (2023). Employing long short-term memory and Facebook prophet model in air temperature forecasting. *Communications in Statistics-Simulation and Computation, 52*(2), 279–290.
Ullah, S., Iqbal, N., Khan, A. H., Sajid, M., Ahmad, Z., Ahmad, H., & Hussain, M. (2023). Empowering agriculture: A green revolution with internet of energy-driven farm energy management for sustainable and eco-friendly practices. *Journal of Population Therapeutics and Clinical Pharmacology, 30*(19), 975–992.
Zarnowitz, V., & Ozyildirim, A. (2006). Time series decomposition and measurement of business cycles, trends and growth cycles. *Journal of Monetary Economics, 53*(7), 1717–1739.

Chapter 8
Flood Risk Assessment in Fujairah Emirate Under Climate Change

Ahmed Sefelnasr, Mohamed Abdel-Fattah, Mohamed Elkollaly, Mohsen Sherif, Dalal Alshamsi, Fouad Lamghari Ridouane, Shaher Bano, and Abdel Azim Ebraheem

Abstract A flash flood represents the most prevalent natural disaster that endangers the lives of people, as well as the stability of the economy and infrastructure. To effectively address this threat, watershed planning and management become crucial tasks, especially in identifying areas prone to flash floods and reducing their impact, particularly in regions like the Fujairah Emirate. These sudden floods can result in severe consequences for both individuals and infrastructure, emphasizing the importance of proactive planning. The utilization of geographic information systems (GIS) and a weighted overlay technique serves as a robust approach to predicting flash flood-prone zones (FFVZ). This approach integrates various factors, such as rainfall, elevation, slope, land use/land cover, drainage density, geology, geomorphology, and soil characteristics, derived from satellite data. This comprehensive framework helps in comprehending the complexities of flash flood susceptibility. Additionally, a classification system categorizes vulnerability levels, offering a clear understanding

A. Sefelnasr · M. Elkollaly · M. Sherif · D. Alshamsi · A. A. Ebraheem (✉)
National Water and Energy Center, United Arab Emirates University, Al Ain, United Arab Emirates
e-mail: abdelazim.aly@uaeu.ac.ae

M. Abdel-Fattah
Civil Engineering and Quantity Surveying Department, Military Technological College, P.O. Box 262, Muscat 111, Oman

M. Elkollaly · M. Sherif
Civil and Environmental Engineering Department, College of Engineering, United Arab Emirates University, P.O. Box 15551, Al Ain, United Arab Emirates

M. Elkollaly
Irrigation and Hydraulics Engineering Department, Faculty of Engineering, Tanta University, Tanta 31733, Egypt

D. Alshamsi
Geosciences Department, College of Science, United Arab Emirates University, P.O. Box 15551, Al Ain, United Arab Emirates

F. L. Ridouane · S. Bano
Fujairah Research Center, Fujairah, United Arab Emirates

A. Sefelnasr et al. (eds.), *Water Resources Management and Sustainability*, Water Science and Technology Library 114, https://doi.org/10.1007/978-3-031-80520-2_8

of the varying degrees of risk within the studied area. This classification can be a valuable tool for prioritizing mitigation efforts and allocating resources to the most vulnerable areas. Identifying critical sites with high vulnerability ratings is essential for effective emergency planning and response strategies. Furthermore, historical data from Landsat and Sentinel 2 imagery, which track changes in vegetation and urban development over time, shed light on the evolving environment and its potential link to flash flood susceptibility. The 2D Rainfall-Runoff-Inundation (RRI) model was utilized to simulate flash flood events under varying return periods. The simulation results were used to evaluate the effectiveness of the existing flood mitigation strategies under climate change and the potential for their augmentation.

Keywords Wadi flash floods · Flood risk management · Flood mitigation · Fujairah emirate

8.1 Introduction

Flood risk reduction is a prime concern for flood-vulnerable countries around the world and despite all international efforts and technological advancements, flood risk management remains a challenging task for many developed and developing nations (Shah et al., 2018). In the wadi system, flash floods are among the most devastating natural hazards in terms of human losses and economic damage. However, floods are an important source of water for such arid environments as wadi system. Recently in the arid region, Wadi Flash Floods (WFFs) are frequently occurring. Integrated management of wadi flash floods requires the development of strategic methodologies for evaluating the risk, mitigation, and water resource management (Sumi et al., 2013). The flood risk reduction can be achieved by decreasing the magnitude of the flood or the vulnerability of the flood-prone area using non-structural and structural measures (Hansson et al., 2008). Non-structural measures refer to non-engineering actions such as the use of insurance, increasing preparedness through early warnings, restricted development and land use planning and operation of flood control reservoirs, etc. (Hansson et al., 2008; Shah et al., 2018). UAE and Fujairah Emirate is one example where various challenges to manage this devastating natural hazard existed. The first step in any disaster risk management activities is to assess the potential hazard and risk resulting from this disaster as "What cannot be measured, cannot be managed".

Across the hyperacid Arabian Peninsula, preparing to face the evolving risks of flash-flood hazards in light of projected climate change scenarios requires both an understanding of the extreme rainfall events that cause them and the implementation of effective mitigation and adaptation measures. In July 2022, February 2024, and April 2024, torrential storms impacted northeastern UAE. The severe damage from these storms has sparked a broader consideration of flash-flood risks, particularly in rapidly urbanizing arid regions. The July 2022 and April 2024 storms, which occurred unexpectedly at the beginning and mid-summer respectively when rainfall is typically absent, were triggered by a combination of climatic and oceanic factors: the position of the Intertropical Convergence Zone (ITCZ), an unusual westward extension of a humid air mass from the Indian monsoonal trough, a cool sea-surface temperature anomaly that intensified onshore moisture advection, and orographic precipitation enhancement by the coastal mountain ranges.

The northern part of the UAE, including the study area (Fig. 8.1), has recently experienced several flash flooding incidents. Notably, Fujairah Port, which lies within the study area, recorded the highest rainfall at 234.9 mm, while Fujairah Airport saw 153 mm between July 25 and July 28, 2022. Additionally, Kalba received 85 mm of rainfall between February 10 and April 12, 2024, and 173 mm between April 14 and April 16, 2024, according to reports from the National Meteorology Center in Abu Dhabi and The Gulf Today National News.

Due to the unusually heavy rains on July 28, 2022, and the mountainous terrain of Fujairah Emirate, including the presence of seven wadis in Fujairah City—Wadi Hayl, Wadi Saham, Wadi Farfar, Wadi Ham, Wadi Yabsah, Wadi Madhab, and Wadi Safad (Fig. 8.2a)—a total of 4225 people were affected. Of these, 1638 personnel were involved in evacuating those impacted by flooding within these wadis and by water spilling over the southern side of the Wadi Ham Dam, located in Fujairah City (Fig. 8.2b). Approximately 870 individuals were rescued, and another 3897 were

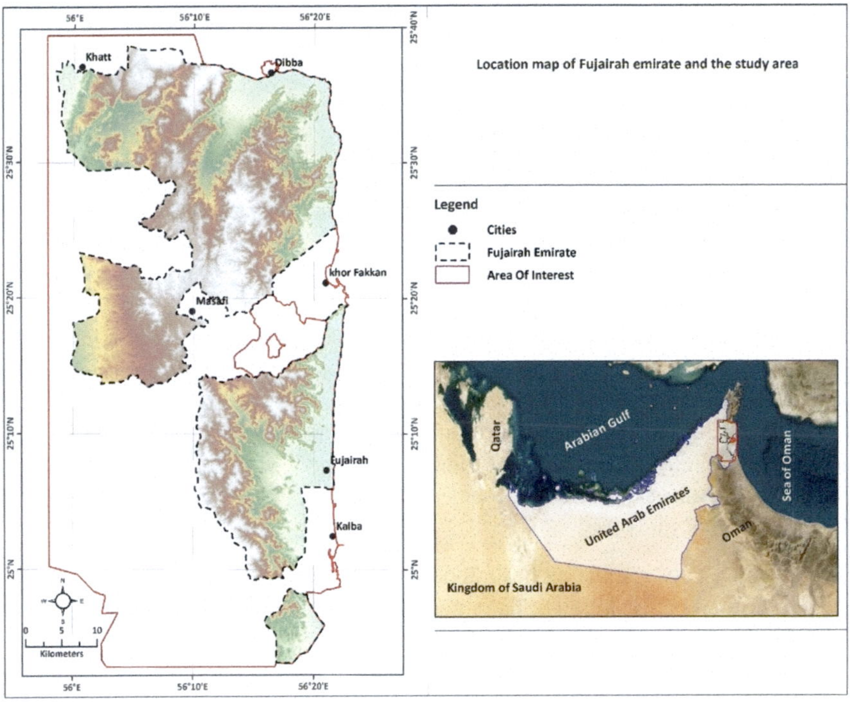

Fig. 8.1 Location map of the study area

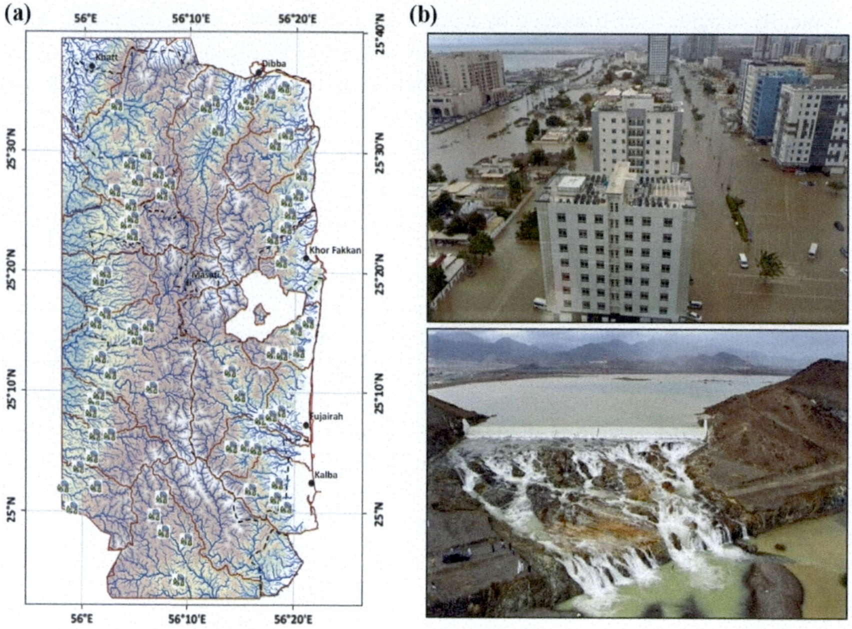

Fig. 8.2 **a** Location of dams, catchments and main streamlines in area of interest (modified from Sefelanasr et al., 2024), **b** Floods of July 2022 in Fujairah city when the water flowed over the Wadi Ham South Dam. *Source* Terry et al. (2023)

provided with shelter (Alhefeiti et al., 2022). Additionally, Fujairah Port, the world's third-largest fuel bunker center, experienced widespread flooding.

In addition to assessing the effectiveness of current flood emergency management strategies in the Fujairah emirate, UAE, the Government of Fujairah supported this research to enhance these strategies by focusing on studying technologies of greater capture of floodwaters for infiltration and groundwater replenishment. This effort aligns with national initiatives to develop sustainable solutions for addressing freshwater scarcity. The study underscores the need to increase monitoring of mountain wadi systems to complement meteorological forecasts with hydrological data and improve modeling to better predict runoff-generation processes in ungauged catchments. Additionally, it explores the significant impact of urbanization, landscape alterations, and infrastructure development on the emirate's hydrology. The specific objectives of this study are:

1. To identify gaps in existing flash flood mitigation measures and propose alternative strategies to address the ongoing risk, particularly in the context of climate change, which is altering precipitation intensity and patterns.
2. To estimate the flood hazard across the target area of interest under different rainfall events of 5, 10, 20-, 50-, 100-, and 200-years return periods, considering the applied mitigation measures (Dams).

Fig. 8.3 Examples of flash flood in the study area on 4th and 17th November 2016. *Source* https://twitter.com/Stormcenter; Gulf News 17 November 2023; https://gulfnews.com/uae/weather/rain-in-uae-fujairah-police-clarifies-on-vehicle-swept-away-in-the-floods-in-valley-1.99451683

8.2 Study Area

Fujairah emirate (Fig. 8.1), while renowned for its breathtaking beauty, confronts a vulnerability to flash floods during the rainy season. Intense rainfall has the potential to overwhelm the local drainage systems, leading to sudden flash floods in low-lying areas and wadis, which serve as natural water collectors and conduits directing water toward populated zones. Figure 8.3 shows the flooding of Fujairah City on the 4th, and 17th November, 2023. Starting April 14, 2024, severe weather hit the Arabian Gulf, most severely affecting Oman and the UAE. Extreme amounts of rain were brought by the storms. The scale of this event is exceptional, with the National Center of Meteorology announcing the UAE experienced the largest amount of rainfall since 1949, with more than 250 mm in a single day. The emirate's rugged terrain, characterized by wadis and valleys, amplifies the likelihood of flash floods (Subraelu et al., 2023). Consequently, even popular tourist destinations within Fujairah emirate, such as its picturesque beaches, wadis, and historical sites, are not exempt from the risks posed by flash floods. As an example, Al Aqah Beach, one of the emirate's most popular beach destinations, lies in a low-lying area that is susceptible to flash floods during the rainy season. Similarly, cherished wadis can become perilous for tourists when heavy rainfall occurs, as they are prone to flash floods. Importantly, the region has encountered significant storm events in recent years, including Tropical Cyclones Shaheen and Goni, which inflicted extensive damage and disruption.

8.3 Data and Methods

There are several hydrological models developed over the years to predict rainfall-runoff processes considering the interaction with the subsurface layers and simulate flood routing through the river networks. Rainfall-runoff modeling in wadi system is a complicated and challenging problem since we must consider various interconnected variables and processes. The unique hydrological characteristics of the arid wadis

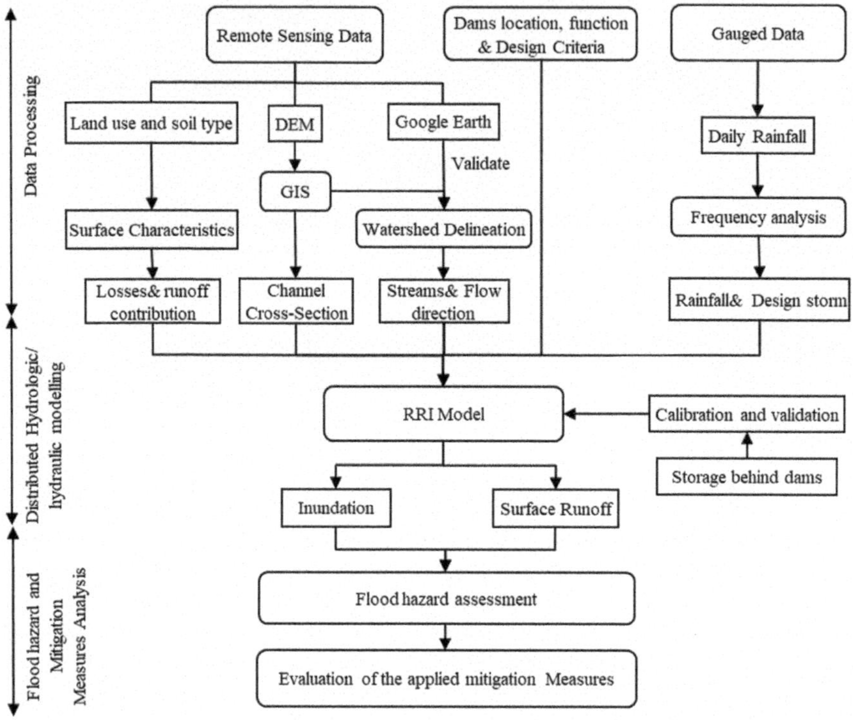

Fig. 8.4 Illustration of the developed approach for (1) input data processing, (2) hydrological modeling using Hydro-BEAM and (3) flood hazard and mitigation measures assessment

include high spatial and temporal storm variability, high evaporation rate, lack of vegetation cover, low infiltration capacity of the poorly vegetated soils and exposed rocks, absence of base flow, intermittent channel flow and significant transmission losses through the wadi channels. Therefore, the adopted methodology needs to be suitable for all these challenges (Fig. 8.4). Remote sensing data (e.g., Digital Elevation Model (DEM) data) as well as Geographic Information Systems (GIS) techniques (e.g., spatial analysis, interpolation, and watershed delineation tools) have been used for input data preprocessing and visualization of the results. RRI model has been utilized to simulate several flood events.

8.3.1 Rainfall Data

Fujairah Environment Authority has provided the available daily rainfall data between 1980 and 2023 that further have been used to assess the flood hazard using different scenarios. There are 9 rainfall gauges (as indicated in Fig. 8.5a) with maximum daily rainfall ranges from 86.2 mm at Al Tawiyen station to 220.9 mm at

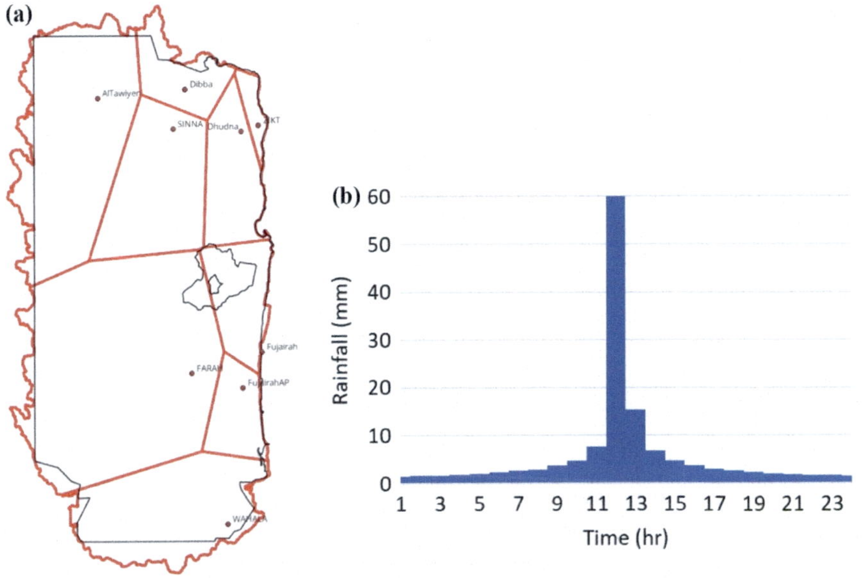

Fig. 8.5 **a** The available rainfall station and its areal coverage using Theissen polygon, **b** A 100 design storm for Dhudna station using SCS Type II method

Fujairah station. Where the storm events in the arid regions are typically short, lasting from less than 1 h to a few hours (Böer, 1997; Gheith & Sultan, 2002; Ghoneim & Foody, 2013), therefore the design storm duration was 24 h but its peak in the middle 3–4 h and distributed all over the basin based on each station areal coverage according to Theissen polygon method (as indicated in Fig. 8.5a). The rainfall data have been analyzed to the rainfall depths for a design storms of 5, 10, 20, 50, 100 and 200 years return period using Extreme Value Gumpel Type I distribution (as indicated in Table 8.1. For the design storms and rainfall return period analysis a method of SCS-Type II 24 h (see Fig. 8.5b) distribution has been used which fits the rainfall characteristics in the arid region (Chow, 1988).

8.3.2 Geomorphic and Topographic Setting

Topographic data for the study area derived from the Shuttle Radar Topography Mission (SRTM) 3arc-second (~90 m) resolution DEM data. The study area total area is 3138.3 km^2 and characterized by a high range of relief with elevation ranges from 1506 m above sea level (with slopes reached to 54.5% at the hill slope), to sea level at the downstream. DEM data (Fig. 8.6a) were utilized for watershed delineation, flow direction, flow accumulation and slope detection.

Table 8.1 The estimated design storms for the available rainfall station in the study area

Station/return period rainfall (mm) in years	5	10	20	50	100	200
Fujairah	87.4	116.4	144.2	180.1	207.1	233.9
Fujairah Airport	93.4	126.0	157.2	197.7	228.0	258.1
Dibba	53.0	67.7	81.8	100.0	113.7	127.3
Farah	72.9	98.9	123.8	156.1	180.2	204.3
Sinna	51.4	64.4	76.7	92.8	104.8	116.8
Al Tawiyen	48.5	62.0	75.1	91.9	104.6	117.1
Wahala	64.1	85.3	105.6	131.8	151.5	171.0
Dhudna	66.7	85.1	102.9	125.8	143.0	160.1
Zikt	67.9	89.8	110.8	137.9	158.3	178.6
Arithmetic average	67.3	88.4	108.7	134.9	154.6	174.2
Weighted average	60.3	79.8	98.6	122.8	141	159.1

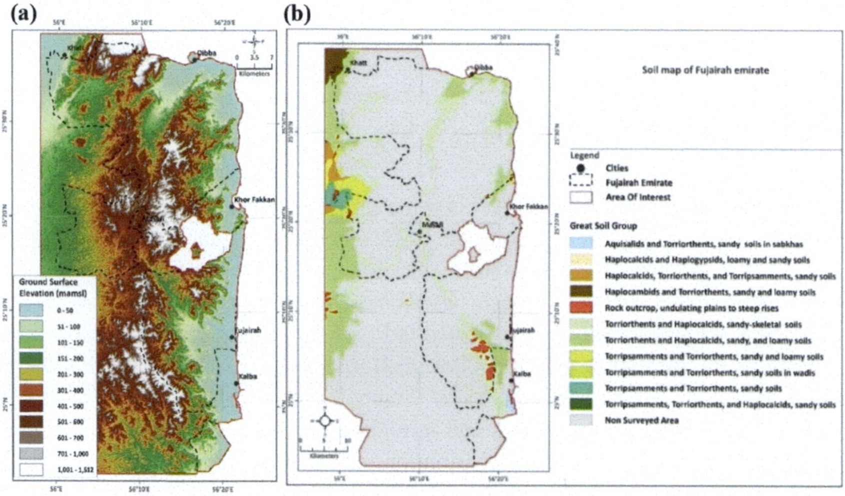

Fig. 8.6 **a** DEM model for the study area, **b** Geomorphology and topography of the study area

8.3.3 Soil/Landuse Data

Based on the available land use/cover and soil maps (Fig. 8.6b), where it is classified to 7 classes. Those classes include mountainous, urban and 5 sandy soil types. The mountainous land cover is the major class which covers 75% of the total area. The hydrological model parameters setting is linked to that soil classification.

8.3.4 Watershed Delineation

One of the preliminary steps is watershed delineation to detect the mainstream network, watershed, and sub-basins boundaries by using spatial analysis tools (i.e. GIS) and Digital Elevation Model (DEM) data. The generated streams networks were cross-correlated and checked with the available satellite images, Google Earth, and the local geological maps (Horton, 1932). Results of such comparisons confirmed that a reliable streams network for hydrologic modeling was obtained. The delineated basin is shown in Fig. 8.2a.

8.3.5 Dams Data

71 dams have been considered in the study area as indicated in Fig. 8.7a and the submitted geodatabase. Dam specifications, geometries, and capacities have been incorporated in the model. It is worth mentioning that there are other available dams, but their design capacity or function are not available. As will be indicated in the model section, the dam has been modeled to store all coming water until it reaches its maximum capacity then starts to release all extra water.

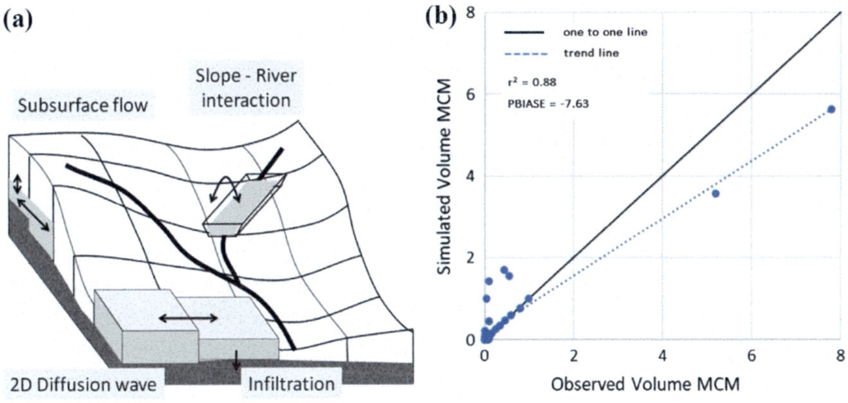

Fig. 8.7 **a** Schematic diagram of RRI model (Sayama et al., 2012), **b** The simulated volumes behind dams compared to the measures ones using the calibrated RRI model

8.4 Hydrological/Hydraulic Model

The used model is 2D Rainfall-Runoff-Inundation (RRI) model, which is capable of simulating rainfall–runoff and flood inundation simultaneously (Sayama et al., 2012). Figure 8.7a shows a schematic diagram of the RRI model. The model deals with slopes and river channels separately. A channel is discretized as a single vector along the center line of the overlying slope grid cell. The channel represents an extra flow path between grid cells lying over the actual river course (Sayama et al., 2012). RRI model, can be used for both humid areas and dry wadi systems (Abdel Fattah et al., 2018). The model deals with slopes and river channels separately. The flow on the slope grid cells is calculated with the 2D diffusive wave model, while the channel flow is calculated with the 1D diffusive wave model.

The inflow–outflow interaction between the slope and river is calculated based on different overflowing formulae depending on water-level and levee-height conditions (Sayama et al., 2012). Green–Ampt infiltration model (Sayama et al., 2012) is used by RRI model to consider vertical infiltration. As for the dam model, it has two parameters: outflow discharge and maximum storage volume where the storage continues to be updated based on simulated inflow and outflow. The outflow is maintained at a certain discharge rate until the storage volume reaches the dam's maximum storage level. RRI model has been applied for flash floods simulation in the area of interest after its calibration.

8.5 Model Calibration

Calibration is the process of adjusting parameter values of a model to obtain a better fit between observed and predicted variables. Volume accumulated behind the dams was used to calibrate the model to tackle the absence of runoff data. The initial parameter settings of RRI model were based on literature review and model manuals. Figure 8.7b compares the simulated total runoff volume with the available records of the accumulated storage volume at dam locations. To evaluate their effects on the predicted river discharges, we used the following five indices: Coefficient of determination or squared correlation coefficient (r2, describe the degree of collinearity between simulated and measured data), Percent bias (PBIAS, measures the average tendency of the simulated data to be larger or smaller than their observed counterparts) (Gupta et al., 2009). The simulated volums using the calibrated parameter setting (indicated in) show good agreements with the measured data (as shown in Fig. 8.7b) where PBIAS and r2 indices equal to -7.63 and 0.88, respectively. This model performance might be acceptable where any model due to variant simplifications of the real complex hydrological system in terms of scale and rainfall.

8.6 Simulation Results

RRI model has been applied for flash floods simulation in the study area after its calibration. Flood estimation was performed on an hourly basis to be able to reproduce the rapid WFFs events. In the following maps (Figs. 8.8, 8.9, and 8.10), and to avoid a lengthy report, only the simulated peak inundation (in m) and peak discharge (in m^3/s) throughout the target area of interest and some selected wadis (e.g., Wadi Ham). Wadi flash flood special characteristics can be figured out from the results where it takes a few hours to reach to the peak discharge and then gradually decreasing until the end of the event. Variation of hydrograph peak and spatial distribution can be noticed from one event to another. This is due to the different rainfall intensity and pattern for each event.

8.6.1 Inundation and Water Depth in the Area of Interest

The maximum predicted inundations for 5, 10, 20, 50, 100, and 200 years storms are 7.8 m, 9.4 m, 10.7 m, 12.1 m, 13.1 m, and 13.9 m, respectively. Figure 8.8 provides an overview of the inundation and water depth patterns in the Fujairah Emirate, illustrating how most streams flow from the middle mountainous region and towards the outskirts of the study area. The maps highlight that deep inundation

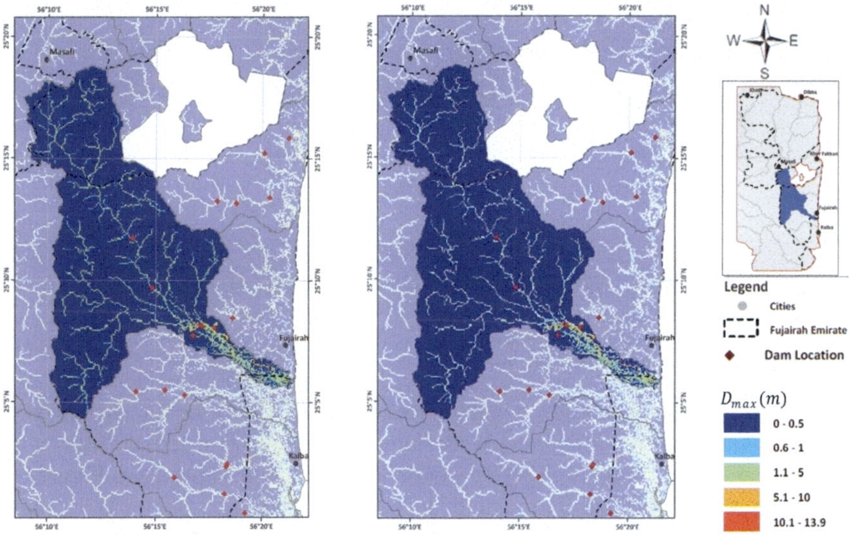

a)100 years return periods b) 200 years return periods

Fig. 8.8 Simulated floodwater depth in Wadi Ham for the different return periods (Maximum floodwater depth $D_{\max}(\text{m})$)

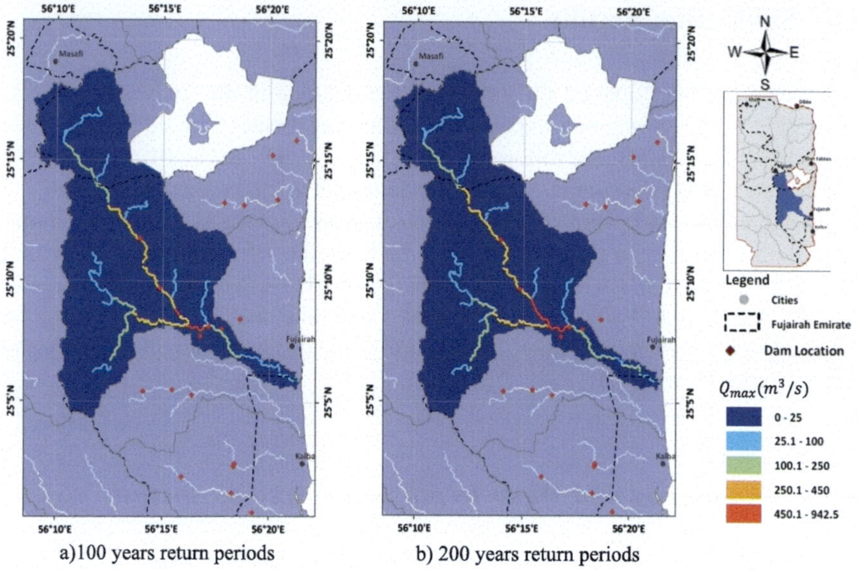

Fig. 8.9 Simulated floodwater discharge in Wadi Ham for the different return periods (Maximum floodwater discharge $Q_{max}(m^3/s)$)

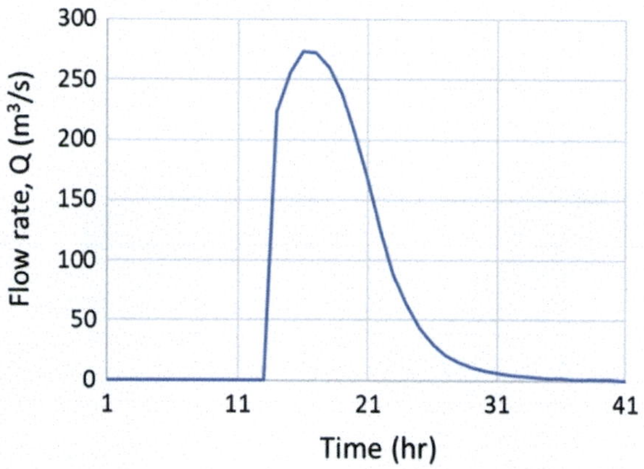

Fig. 8.10 The simulated outflow rate from Ham main dam for the 200 years return period storm

primarily occurs along the eastern and western boundaries, with significant depths in the southwestern regions, including Wadi Ham. Wadi Ham encompass the major city within the study area, underscoring its economic and strategic importance. Figure 8.8 depicts a more detailed delineation of the inundation and water depth for Wadi Ham.

For Wadi Ham, flash flood events result in substantial inundation, particularly in the southeastern plains. Inundation depths exceed 1.1 m for storms of 20 and 50 return periods in years. For extreme events with return periods of 100 and 200 years, significant inundation depths surpassing 10.1 m constitute, spreading over larger areas. This pattern is influenced by the topography and slight slopes of the valley, which facilitate water accumulation. The findings of Wadi Ham indicate the critical need to increase the storage capacity of the main dam to handle these extreme flood events. Additionally, implementing measures such as enhancing green infrastructure, improving storm drainage systems, and utilizing the excess stormwater to recharge groundwater would mitigate the impacts of flash floods.

Under the application of a 200 years return period storm, most of the dams (greater than 90%) of the investigated dams (71 dams) are expected to be totally filled that threatens the downstream areas of those dams and the safety of the dam structure itself. For example, Ham main dam is expected to have an outflow after exceeding the maximum storage volume of 273 m^3/s, under 200 years return period storm as indicated in Figs. 8.9 and 8.10.

The simulation results indicated the need for more gauging stations for wadi flow and rainfall measurements should be installed at the target area for accurate hydrological model calibration and precise prediction of the design storm. Table 8.2 shows the recommended minimum densities of hydrological stations according to WMO (2008) guidelines.

Alongside structural measures, non-structural measures provide flexible flood management options for adapting to the ever-changing river basins, socio-economic and climate scenarios, and are in line with the spirit of environment-friendly and sustainable development (Kundzewicz, 2002; Shah et al., 2018). Therefore, non-structural measures such as the use of insurance and restricted development planning, regulating human activities to mitigate flooding increasing preparedness through early warnings, and reducing the consequence of flooding by reducing both exposure and vulnerability (Hansson et al., 2008; Richards et al., 2008; Shah et al., 2018; Tapsell et al., 2002).

A flood disaster preparedness plan is urgently needed in the study area (if it does not already exist). The purpose of the preparedness stage is to provide the necessary decision support system for the case that the existing flood protection system has failed (Plate, 2002). It is evident that no technical solution to flooding is safe (Plate,

Table 8.2 Recommended minimum densities of stations (area in km^2 per station)

Physiographic unit	Precipitation		Evaporation	Streamflow	Sediments	Water quality
	Non-recording	Recording				
Coastal	900	9000	50,000	2750	18,300	55,000
Mountains	250	2500	50,000	1000	6700	20,000
Interior plains	575	5750	5000	1875	12,500	37,500
Hilly/ undulating	575	5750	50,000	1875	12,500	47,500
Small islands	25	250	50,000	300	2000	6000
Urban areas	–	10–20	–	–	–	–
Polar/arid	10,000	100,000	100,000	20,000	200,000	200,000

2002). Even if the system always does what it is supposed to do, it is hardly ever possible to offer protection against any conceivable flood (Plate, 2002). There is always a residual risk, due to failure of technical systems, or due to the rare flood which exceeds the design flood (Shah et al., 2018). Long-term prediction of climate change and the anthropogenic (as land use change) impact on wadi flash foods.

Sediment transport evaluation during wadi flash flood is also recommended, where sediment has severe impact on the mitigation structure, which increases the flash flood risk and decreases the efficiency of flood mitigation structures. Therefore, it is important to study this factor and to be linked with the wadi characteristics such as geomorphometry and soil type.

By sharing knowledge and experiences throughout the world, stakeholders have multiple choices of technologies to apply to flood-prone communities (Shah et al., 2015). Therefore, we highly recommended to have a database contains all the available data from all wadi systems and to be accessible for any researcher.

8.7 Conclusion

The absence of gauged surface runoff data (wadi/channel flow rate records) and inundation records (water depths) posed a significant limitation in the present study. These records are essential for further validating the model outputs and for incorporating other hydrological and groundwater processes in the study area, such as groundwater recharge, seepage to the channels, and wadi channel transmission losses. Given the anticipated impact of climate change on the magnitude and pattern of precipitation, there is also a need to increase the resolution of rainfall records.

With the recent dramatic changes in precipitation patterns and magnitude, it is crucial to reassess and update the geometry of channels and major wadis, including modeling and periodically updating sedimentation and sediment transport. During field visits, the research team observed that many dams have lost storage capacity due to sediment accumulation. Therefore, implementing a maintenance plan to restore the maximum storage volumes of all dams is recommended.

The existing flood mitigation measures, such as protection dams, are insufficient to handle the significant increase in runoff volume. As a result, it is recommended to evaluate flood mitigation strategies by considering additional measures, such as channel enhancement, groundwater recharge techniques, or non-structural measures like early warning systems. Groundwater recharge structures can serve as an effective approach for water resource management in this arid and dry region.

n the UAE, there are currently no established guidelines, codes, or standards related to hydrological or flood modeling, as well as the associated infrastructures such as drainage systems, dams, and highways. This lack of standardized practices makes it challenging to determine the approved methods for hydrological modeling, including the design of rainfall storms. However, distribution maps of inundation and discharge can play a crucial role in managing flash floods, water resources, urban planning, and land use. Identifying surface runoff zones can provide valuable insights into potential groundwater locations, as transmission losses and groundwater recharge are directly related to surface runoff.

Considering a 200-year return period design storm, it is anticipated that more than 90% of the investigated dams (71 dams) will be completely filled, posing a significant threat to downstream areas and the structural integrity of the dams themselves. As a result, it is crucial to conduct individual studies for each wadi as soon as possible to implement additional mitigation measures in the upstream areas, thereby enhancing the effectiveness of the existing dams in mitigating potential flood hazards. Additionally, downstream areas could experience flooding due to local rainfall without any contribution from upstream sources, emphasizing the need for efficient stormwater management and drainage systems in conjunction with the protection provided by upstream dam.

References

Abdel-Fattah, M., Kantoush, S. A., Saber, M., & Sumi, T. (2018). Rainfall-runoff modeling for extreme flash floods in wadi samail, Oman. *Journal of Japan Society of Civil Engineers, Ser. B1 (Hydraulic Engineering)*, *74*(5), I_691–I_696. https://doi.org/10.2208/jscejhe.74.5_I_691

Alhefeiti, F., Alblooshi, M., Abdalla, R., Kalathingal, M., Bano Mirza S., & Ridouane, F. (2022). Investigation of the effects of heavy rain and flood in the emirate of Fujairah and construction and compilation of flood map and elevation of flood map and elevation chart using state of the art GIS technologies. *International Journal of Development Research*. *12*(09), 58827–58831. https://doi.org/10.37118/ijdr.25356.09.2022

Böer, B. (1997). An introduction to the climate of the United Arab Emirates. *Journal of the Arid Environment, 35*, 3–16.

Chow, V. T. (1988). *Applied hydrology, book*. https://ponce.sdsu.edu/Applied_Hydrology_Chow_1988.pdf.

Gheith, H., & Sultan, M. (2002). Construction of a hydrologic model for estimating Wadi runoff and groundwater recharge in the Eastern Desert Egypt. *Journal of Hydrology, 263*(1–4), 36–55. https://doi.org/10.1016/S0022-1694(02)00027-6

Ghoneim, E., & Foody, G. M. (2013). Assessing flash flood hazard in an arid mountainous region. *Arabian Journal of Geosciences, 6*(4), 1191–1202. https://doi.org/10.1007/s12517-011-0411-7

Gupta, H. V., Kling, H., Yilmaz, K. K., & Martinez, G. F. (2009). Decomposition of the mean squared error and NSE performance criteria: Implications for improving hydrological modelling. *Journal of Hydrology, 377*(1–2), 80–91. https://doi.org/10.1016/j.jhydrol.2009.08.003

Hansson, K., Danielson, M., & Ekenberg, L. (2008). A framework for evaluation of flood management strategies. *Journal of Environmental Management, 86*(3), 465–480. https://doi.org/10.1016/j.jenvman.2006.12.037

Horton, R. E. (1932). Drainage basin characteristics. *Transactions, American Geophysical Union, 13*(1), 350–361

Kundzewicz, Z. W. (2002). Non-structural flood protection and sustainability. *Water International, 27*(1), 3–13. https://doi.org/10.1080/02508060208686972

Plate, E. J. (2002). Flood risk and flood management. *Journal of Hydrology, 267*(1–2), 2–11. https://doi.org/10.1016/S0022-1694(02)00135-X

Richards, J., White, I., & Carter, J. (2008). Local planning practice and flood risk management in England: Is there a collective implementation deficit? *Environnement Urbain, 2*, 11–20. https://doi.org/10.7202/019218ar

Sayama, T., Ozawa, G., Kawakami, T., Nabesaka, S., & Fukami, K. (2012). Rainfall–runoff–inundation analysis of the 2010 Pakistan flood in the Kabul River basin. *Hydrological Sciences Journal, 57*(2), 298–312. https://doi.org/10.1080/02626667.2011.644245

Shah, M. A. R., Rahman, A., & Chowdhury, S. H. (2018). Challenges for achieving sustainable flood risk management. *Journal of Flood Risk Management, 11*(S1). https://doi.org/10.1111/jfr3.12211

Subraelu, P., Ahmed, A., Ebraheem, A. A., Sherif, M., Mirza, S. B., Ridouane, F. L., & Sefelnasr, A. (2023). Risk assessment and mapping of flash flood vulnerable zones in Arid region, Fujairah city, UAE-using remote sensing and GIS-based analysis. *Water, 15*(15), 2802. https://doi.org/10.3390/w15152802

Sumi, T., Saber, M., & Kantoush, S. A. (2013). Japan-Egypt hydro network: Science and technology collaborative research for flash flood management. *Journal of Disaster Research, 8*(1), 28–36. https://doi.org/10.20965/jdr.2013.p0028

Tapsell, S. M., Penning-Rowsell, E. C., Tunstall, S. M., & Wilson, T. L. (2002). Vulnerability to flooding: Health and social dimensions. *Philosophical Transactions of the Royal Society of London. Series A: Mathematical, Physical and Engineering Sciences, 360*(1796), 1511–1525. https://doi.org/10.1098/rsta.2002.1013.

Terry, J. P., Al Ruheili, A., Almarzooqi, M. A., Almheiri, R. Y., & Alshehhi, A. K. (2023). The rain deluge and flash floods of summer 2022 in the United Arab Emirates: Causes, analysis and perspectives on flood-risk reduction. *Journal of Arid Environments, 215*, 105013. https://doi.org/10.1016/j.jaridenv.2023.105013

Part II
Droughts

Chapter 9
Impacts of Drought in India and Preparedness for Drought Under Global Warming

Jeongwoo Han and Vijay P. Singh

Abstract Global warming and climate change are detrimentally impacting the social environment and are increasing the death toll due to more severe natural disasters worldwide. For example, droughts in California, USA, caused economic losses amounting to US $2.2 billion and $3 billion in 2014 and 2015, respectively, and drought that occurred in India from 2000 to 2002 caused the deaths of 20 people and economic losses of US $1.5 billion. Besides, global arid regions are expected to increase ~ 10% by 2100 under a high-emission scenario. The more arid background witnesses the more frequent occurrence of severer, longer, and a really more extensive droughts that are likely to lead to water scarcity. Central northeast and west central India showed an increasing trend of the percentage (%) of the area under drought in the past (1950–2010), and % of the area under "above moderate drought" across India is expected to increase up to 70% by 2095 under the Representative Concentration Pathway (RCP) 8.5 scenario. Warming global temperature is also affecting the change in the location of deep convection over oceans and, in turn, the southwest monsoon rainfall, which contributes to 70–90% of the annual rainfall in India, tending to be dormant due to the amplified El Nino-Southern Oscillation (ENSO) and the stronger equatorial Indian Ocean positive anomalies. More frequent occurrence of Rossby waves, which stagnate atmospheric flow, and the intensified land-atmospheric feedback under warming climate make an already initiated drought more severe, longer, and spatially more extensive. Since drought occurs naturally in hydrologic processes, its occurrences are beyond human control. Thus, what humans can do is to be prepared for the anticipated droughts to mitigate drought impacts. To bolster relief efforts and

J. Han (✉)
Department of Biological and Agricultural Engineering, Texas A&M University, College Station, TX 77843-2117, USA
e-mail: jwhan82@hanyang.ac.kr

Research Institute of Engineering Technology, Hanyang University (ERICA), Ansan, South Korea

V. P. Singh
Department of Biological and Agricultural Engineering and Zachary Department of Civil and Environmental Engineering, Texas A&M University, College Station, TX 77843-2117, USA
e-mail: Vijay.Singh@ag.tamu.edu

National Water and Energy Center, UAE University, Al Ain, UAE

A. Sefelnasr et al. (eds.), *Water Resources Management and Sustainability*, Water Science and Technology Library 114, https://doi.org/10.1007/978-3-031-80520-2_9

mitigate drought impacts, a proactive drought management plan needs to be prepared, based on drought early warning, forecasting, and risk analysis. Application of deep learning, such as long short-term memory (LSTM) networks, and entropy spectral analysis have extended forecast lead time with higher accuracy than traditional statistical or machine learning models and have shown their applicability for proactive mitigation planning based on risk analysis. However, more improved monitoring systems for drought indicators, drought indices, which can consider implications of human activities and hydrologic processes explicitly, and more accurate drought forecasting is required to be prepared for severe droughts. Besides, historical drought impact data should be archived and organized for reliable assessment of vulnerability and risk to drought. Therefore, this study discusses the impacts of drought in the past and future and causal mechanisms based on drought events in India and suggests ways to be prepared for the expected more severe droughts from technical and governmental machinery perspectives.

Keywords Drought early warning · Drought forecasting · Drought impacts · Drought management · Global warming · Risk analysis

9.1 Introduction

Drought, as one of the natural disasters, has brought humongous impacts on socioeconomics, environment, and human mortality or well-being, once it occurred. Since drought is not an imminent natural phenomenon, unlike flooding and tsunami, timely drought management is hindered by the obscured detection of its tangible occurrence, rendering as the costliest natural disaster. In the U.S. alone, for example, annual economic losses of drought have been US$ 6–8 billion (Han & Singh, 2021). India also has a long history of social and economic disruption due to droughts. During British rule of India (1765–1947), 12 major famines occurred in India that led millions of people to death (Mishra et al., 2019). Besides the famines associated with the summer monsoon failure that brought crop failure during that era, large-scale droughts in the second half of the 20th and early twenty-first centuries exerted devastating impacts on India when they came with the rising population and lack of irrigation (Gadgil & Gadgil, 2006; Mishra et al., 2016).

As part of the hydrologic process, the occurrence of drought and evolution thereof are closely associated with precipitation, evapotranspiration, and antecedent terrestrial water storage (Han & Singh, 2020; Mishra & Singh, 2010). Under global warming, the rising temperature increases evapotranspiration and accelerates hydrological processes such that when a rainfall intermittent period is longer, arid regions are expected to expand, facilitating a higher frequency of and more severe droughts. Under a high-emission scenario (e.g., Shared Socio-economic Pathway scenario [SSP-8.5]), the global aridity areas are projected to expand up to ~ 10% by 2100 (Pascale et al., 2020; Sherwood & Fu, 2014). Depending on the projected evapotranspiration and precipitation, the trend ending up either drier or wetter by the end of the twenty-first century regionally varies under the SSP scenario. However, under SSP1-2.6, significant drier conditions are projected over North Africa, Saudi Arabia, Central America, and the southwest U.S., which showed significant negative SPEI (Balting et al., 2021). Furthermore, under SSP5-8.5, a more significant drying trend is expected and dry hotspots are expanding over the Mediterranean region, Central Asia, Central Europe, and the Middle East, including the drought hotspots mentioned above under SSP1-2.6 (Balting et al., 2021). Under the Representative Concentration Pathway (RCP) 8.5 scenario, for Central Northeast and West Central India, the percentage of the area under drought is expected to increase up to 70% by 2095 compared to the past, 1950–2010 (Bisht et al., 2019).

Since more severe and areally more extensive droughts are expected to occur more frequently in India under global warming, without systematic proactive drought mitigation measures and well-oiled government machinery, unprecedented drought impacts will occur. Thus, to be prepared for the upcoming droughts to mitigate their impacts, an understanding of causative mechanisms and developing techniques for drought forecasting and risk analysis are important (Han & Singh, 2020, 2023a, 2023b). The summer monsoon precipitation is a critical moisture source for India, that has a significant impact on developing drought (Kumar et al., 2013). El Niño-Southern Oscillation (ENSO) and Indian Ocean Dipole are important patterns of sea surface temperature (SST) anomalies affecting the Indian monsoon rainfall that significantly contributes to the annual rainfall in India (Kumar et al., 2013; Rajbanshi & Das, 2021). While SST anomalies govern ocean-atmospheric circulation at a large scale, atmospheric internal variability, such as Rossby waves, also significantly contributes to extreme weather at a synoptic scale (Han & Singh, 2020; Kumar et al., 2013; Mahto & Mishra, 2023; Rajbanshi & Das, 2021). Once drought is initiated by SST anomalies and/or atmospheric internal variability, land-atmospheric feedback under a warming climate can make drought more severe, longer, and spatially more extensive (Han & Singh, 2020, 2023a, 2023b; Mahto & Mishra, 2023).

The drought occurrence is beyond human control and human activities can make drought more severe and evolve to water shortage. Thus, to minimize drought impacts, humans should be prepared for the upcoming drought by harnessing accurate drought forecasting and proactive management plans (Han & Singh, 2023a). Drought forecasting has been done by stochastic, physical, and machine learning modeling approaches, and technological advancement for drought forecasting has been remarkable (Han & Singh, 2020; Mishra & Singh, 2010). Besides the forecasting methods, reliable drought mitigation planning should be achieved in advance of the drought occurrence. Thus, to contribute to understanding drought occurrences and developing a system for preparing drought management measures, this study discusses drought impacts in India, the projected droughts under global warming, drought causative mechanisms, and drought forecasting and risk analysis.

9.2 Impacts of Drought in India

Drought is a recurring natural disaster and India has witnessed several severe droughts in its history (Gadgil & Gadgil, 2006; Mishra et al., 2016). Since agriculture is a primary sector that supports the Indian economy, drought has often caused crop failures leading to economic and social devastation (Mishra et al., 2019). Although modern food distribution and irrigation systems have been developed, the Indian agricultural system heavily relies on monsoon rainfall as an essential water source for the agricultural sector (Mishra et al., 2019). Thus, the failure of monsoon rainfall and its consecutive failure affected human well-being and mortality due to famine, besides economic losses. The earliest famine, which occurred due to the consecutive failures of monsoon rainfall, is traceable back to the seventeenth century (Cook et al., 2010; Maharatna, 1996). The worst famine, referred to as the Great Famine that occurred due to drought during 1876–78, caused deaths of people between 5 and 10 million (Singh et al., 2018). Besides, during 1943–1944, severe droughts brought up the Bengal famine that affected 1.5 million people from starvation and led to ~ 2.5 million deaths (Padmanabhan, 1973). Between the years 1871 and 2002, India experienced 22 major/mega droughts (Samra, 2004). Some recent major drought impacts, that caused great economic losses and deprived human life and well-being, are listed in Table 9.1.

As drought impacts in India exemplified heretofore, the resulting brutal socio-economic fallouts highlight the importance of preparing drought mitigation plans and water resources management. Furthermore, to prepare long-term sustainable water supply and water-infrastructure, we may need to have a fore-view of future drought under global warming and climate change. Thus, before discussing drought causative mechanisms that can provide clues to improve drought forecasting, droughts under global warming are discussed in the following section.

9.3 Global Warming and Drought

9.3.1 Global Terrestrial Drought under Global Warming

Globally, drought characteristics, such as severity, duration, and area, are projected to increase in the future due to a warming climate (Han & Singh, 2020; Ji et al., 2023; Fig. 9.1). Ji et al. (2023) studied the change in drought characteristics, which were obtained using the Standardized Precipitation Evapotranspiration Index (SPEI) at a 3-month time scale (SPEI-3) under global warming as shown in Fig. 9.1. The figure shows the global median values (solid lines) of drought severity, duration, and area under different climate change scenarios. From the 2020s to the 2050s, drought characteristics under different scenarios consistently show an increasing trend. After the 2050s, while the lowest radiative forcing scenario showed a more or less plateaued or even slightly decreasing trend, other climate change scenarios

Table 9.1 Recent drought impacts in India

Years	Regions	Damages
2002	Uttar Pradesh, Madhya Pradesh, Rajasthan, Haryana, Delhi, Karnataka, Kerala, Nagaland, Orissa, Chhattisgarh, Himachal Pradesh, Gujarat, Maharashtra, Andhra Pradesh, and Tamil Nadu states	• 300,000,000 people affected from starvation and epidemic • Economic losses of US$ 910,721,000
2000–2001	New Delhi, Rajasthan, Gujarat, Orissa, Mahasamund, Raipur, Kawardha, Rajnandgaon, Maharashtra, Madhya Pradesh, and Andhra Pradesh	• 90,000,000 people affected and 20 deaths • Economic losses of US$ 588,000,000
1993	Bihar, Orissa, Andhra Pradesh, Madhya Pradesh, Uttar Pradesh, and Karnataka	• 1,175,000 people affected from starvation and epidemic
1987	Orissa, Gujarat, Madhya Pradesh, Andhra Pradesh, and Maharashtra	• 300,000,000 people affected and 410 deaths
1972–1973	Central India	• 100,000,000 people affected from starvation and epidemic • Economic losses of US$ 50,721,000
1942–1944	Kolkata and Bengal	• 1,500,000 deaths

induced a higher increasing rate as radiative forcing increased (Fig. 9.1). Drought durations are expected to be 0.34, 1.01, 1.25, and 1.44 months longer under SSP1-2.6, SSP2-4.5, SSP3-7.0, and SSP5-8.5, respectively, compared to the reference period. Vertical lines of Fig. 9.1 show central years among years with 20-year moving averaged global mean surface temperature exceeding the specific warming levels (i.e., 1.5 °C, 2 °C, 3 °C and 4 °C) relative to the mean temperature of the preindustrial levels of 1850–1900 (Ji et al., 2023). Compared to the baseline period (1995–2014) with an increasing rate of severity of 1.33×10^3 km^2month $decade^{-1}$, the period of warming level of 4 °C has a considerably steeper increasing rate of severity of $3.04 \times 10^3 km^2$month $decade^{-1}$. Thus, Fig. 9.1 shows that global terrestrial is expected to experience a significantly increasing trend of drought severity, duration, and areal extent during the twenty-first century. Besides, when the global mean temperature exceeds 3 °C and 4 °C compared to preindustrial temperature, drought characteristics will be getting severe rapidly. Thus, to slow down drought, which is getting worse, and to be capable of adapting to future droughts, global cooperation to achieve the well below 2 °C target of global warming was set by the Paris Agreement. In the following section, drought characteristics under global warming are discussed.

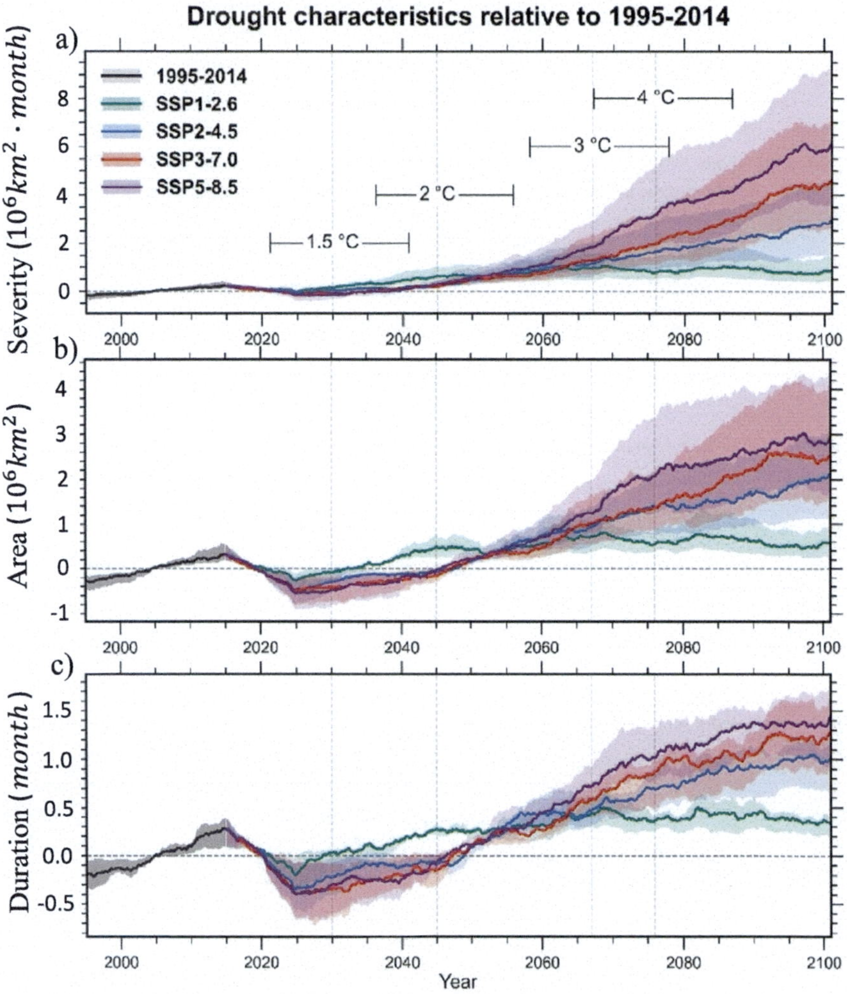

Fig. 9.1 Anomalies of drought characteristics based on the baseline period of 1995–2014. The solid line represents the median value of the multi-model outputs of the Coupled Model Intercomparison Project Phase 6 (CMIP6). The shading represents the interquartile range of the outputs of CMIP6. The projected warming was derived by the average 20-year period across simulations. The vertical dashed line indicates the corresponding central year among years exceeding the specified warming temperature (i.e., 1.5 °C, 2 °C, 3 °C, and 4 °C) (Ji et al., 2023)

9.3.2 Drought in India Under Global Warming

Although global terrestrial droughts show an increasing trend in their severity, area, and duration under global warming, drought characteristics may have regionally different trends. Thus, this section investigates drought characteristics in India under global warming. Based on the historical data from 1982 to 2013, meteorological drought, which is more severe than moderate drought, mostly occurred from December to April in most regions in India (Fig. 9.2). Figure 9.2 shows the occurrence of drought, which had a severity greater than or equal to that of moderate drought, was counted during 1982–2013 for each month at each grid of 0.5° resolution. While hydrological and soil moisture drought did not show significant monthly variation in their occurrences, vegetation drought appraised based on vegetation condition index (VCI) showed a higher occurrence from March to July (Poonia et al., 2021; Fig. 9.3). When considering meteorological drought-prone months (from December to April) and vegetation drought-prone months, it is confirmed that meteorological drought has a high likelihood to be propagating to crop failure.

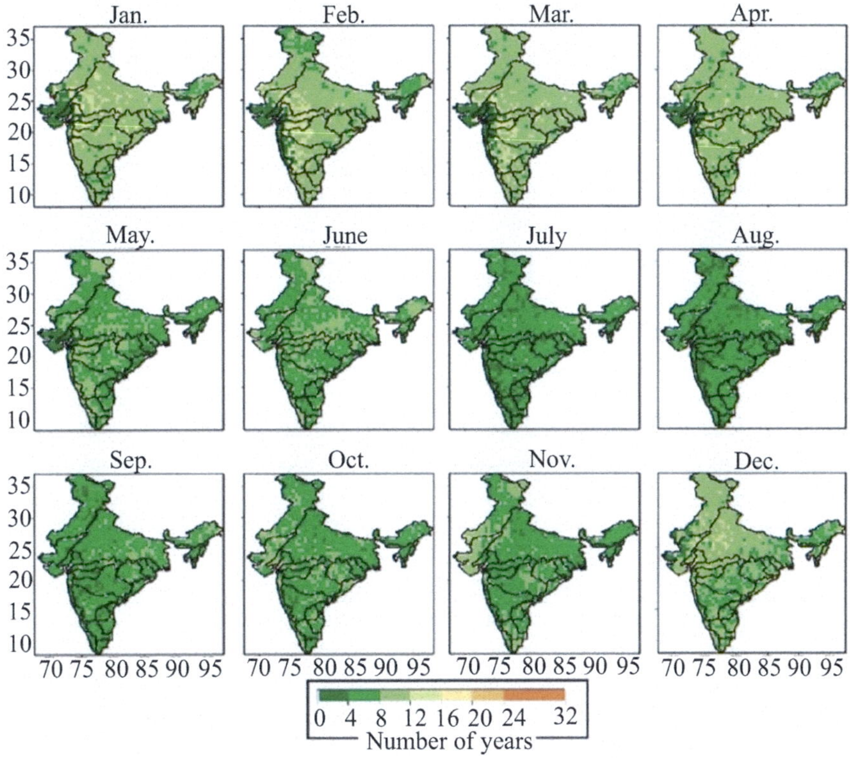

Fig. 9.2 Temporal variation of meteorological drought occurrence during 1982–2013 (Poonia et al., 2021)

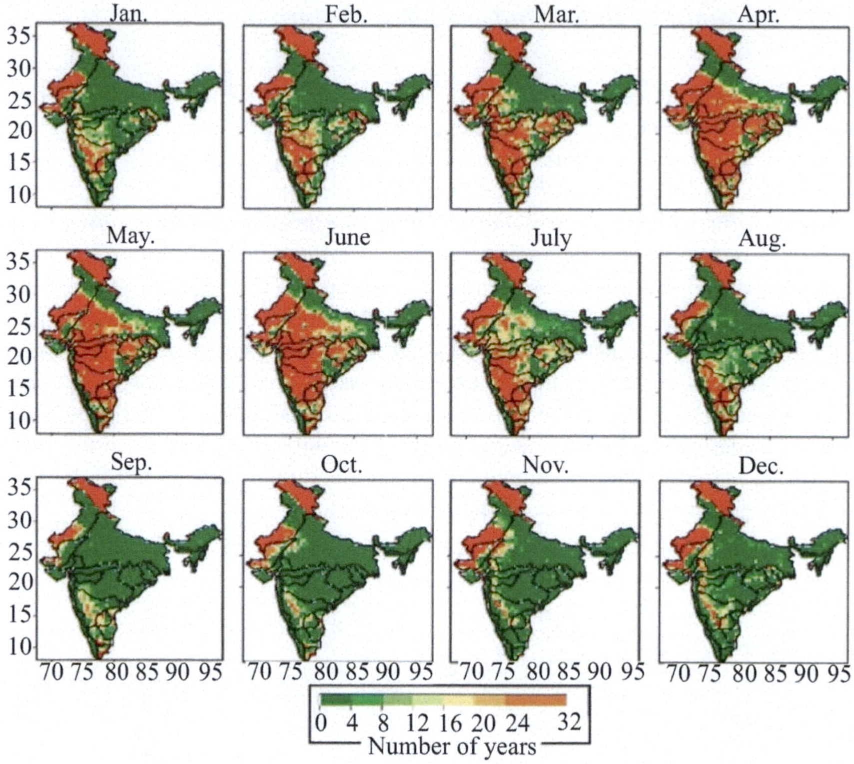

Fig. 9.3 Temporal variation of vegetation drought occurrence during 1982–2013 (Poonia et al., 2021)

To investigate the regional trend of the areal extent of meteorological drought in India, India was divided into 6 homogenous regions, including hilly regions of Northeast hills and Jammu and Kashmir Mountain regions, based on rainfall characteristics (Sharma & Mujumdar, 2017; Fig. 9.4). In each region, areal extents (%) under moderate droughts [Standardized Precipitation Index (SPI < -1.3] are shown in Fig. 9.5. During 1951–2010, after the Mann–Kendall trend test, monotonically increasing trends in the areal extent of droughts were observed in Central Northeast India and West Central India (Fig. 9.5). Besides, Mishra et al. (2016) and Kumar et al. (2013) showed a significant decrease in monsoon rainfall over Indo-Gangetic plains and an increase in the percentage area affected by moderate droughts of the plains of India.

Under the climate change scenarios (i.e., RCP 4.5 and 8.5), trends of drought severity and duration, which were calculated using SPEI-3, in near (2010–2039), mid (2040–2069), and far (2070–2099) futures are shown in Figs. 9.6 and 9.7, respectively. For each 30-year time frame, severity corresponding to 25-year return period was calculated and compared to that of the reference period (1976–2005). An increase

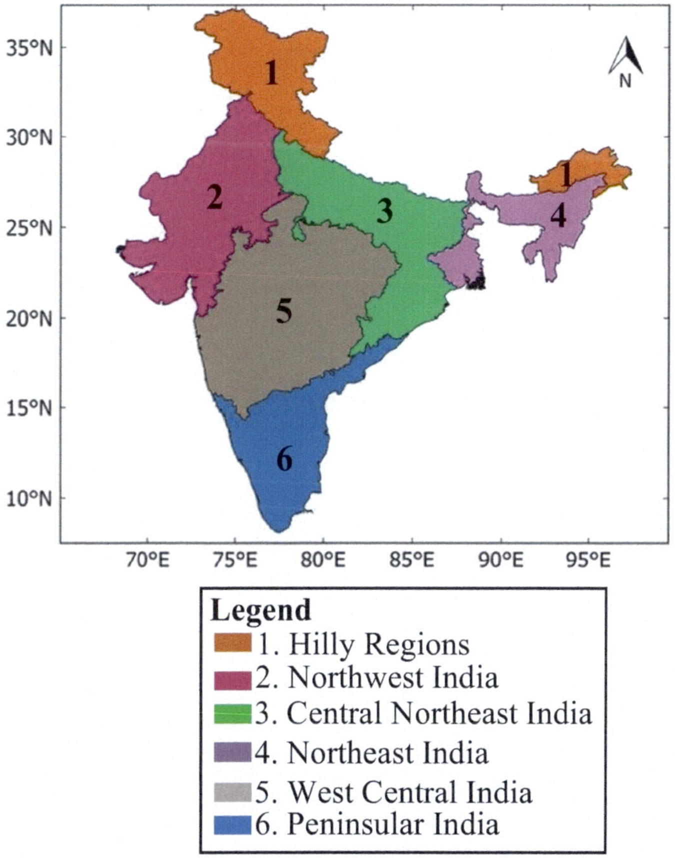

Fig. 9.4 Meteorologically homogeneous regions in India (Sharma & Mujumdar, 2017)

in magnitude of severity was found to be maximum for mid- and far-future in the northwest, west-central, and central northeast India (Fig. 9.6). Figure 9.7 compares the average duration of droughts in each time frame to that in the reference period. The average duration is expected to be longer in most of India and their maximum was found in the mid- and far-future period for both climate change scenarios.

Besides, the area subjected to moderate drought was found to be decreasing for both scenarios, except for the far future under RCP 4.5 which showed an insignificant increasing trend (Fig. 9.8). In the long term (2010–2099), the areal extent under moderate drought was found to be decreasing significantly at a 5% significance level under RCP 4.5 and 8.5 scenarios (Fig. 9.8). On the other hand, an increasing trend of areal extent affected by the above moderate drought (i.e., more severe than moderate drought) was found under both scenarios in near-, mid-, and far-futures and long term (Fig. 9.8). As more severe and longer droughts are expected and there are chances that moderate droughts are replaced by the droughts more severe than moderate,

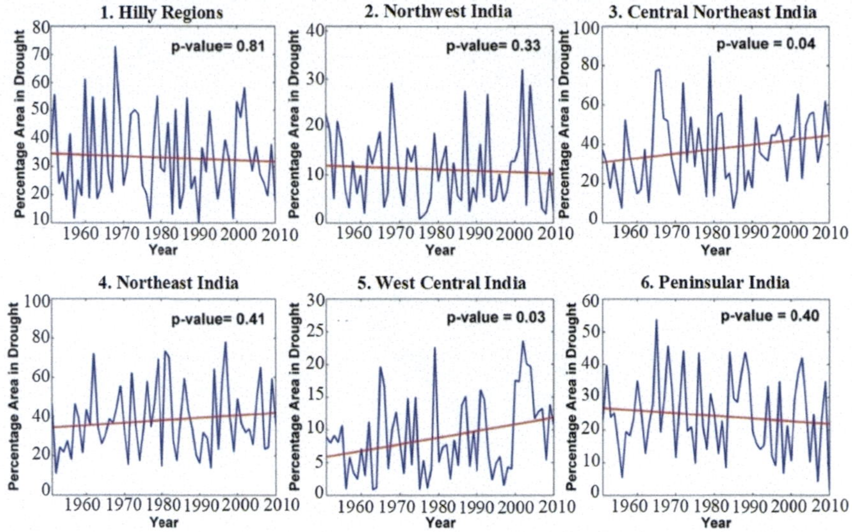

Fig. 9.5 Percentage area under meteorological drought in each meteorologically homogenous regions defined in Fig. 9.4 (Sharma & Mujumdar, 2017)

larger areas can suffer from droughts in the future. Thus, to be well-prepared for the upcoming severe droughts, Indian agriculture and water resources systems should be equipped with improved drought early warning and forecasting techniques and well-organized government machinery.

9.4 Mechanisms Causing Drought in India

As more severe, longer, and areally extensive droughts are expected to occur in India, an understanding of drought causative mechanisms may be required to enhance drought early warning, forecasting, and long-term water management policy (Han & Singh, 2020). Drought causative mechanisms can be discussed in three fields: sea surface temperature (SST) teleconnections, atmospheric internal variability, and land–atmosphere interaction (Han & Singh, 2020, 2021). The majority of atmospheric moisture is supplied from the ocean by the teleconnection of global SST anomalies and/or atmospheric internal variability (Han & Singh, 2020; Hoerling et al., 2009; Mo & Schemm, 2008; Newman et al., 2003; Sheffield & Wood, 2012; Wang et al., 2014). Once the meteorological drought is initiated, drought can become more severe, longer, and spatially extensive by land-atmospheric feedback (Cook et al., 2009; Hoerling et al., 2014; Sheffield & Wood, 2012; Williams et al., 2015). Since drought causative mechanisms are intertwined via ocean–atmosphere-land circulation, dichotomization of the causal mechanisms is not possible in the real

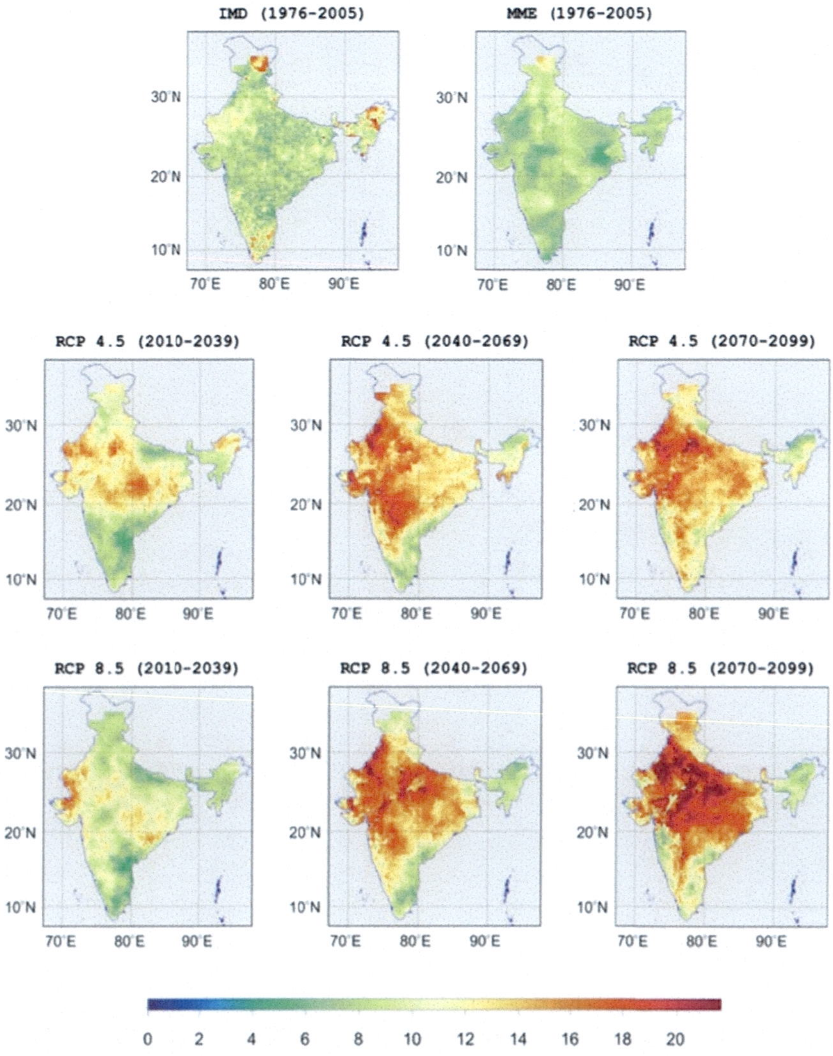

Fig. 9.6 Comparison of the severity of SPEI-3 drought corresponding to the 25-year return period to that in the reference period for different future time frames under different climate change scenarios. India Meteorological Department (IMD) and Multi-Model Ensemble (MME) represent data sources used for the reference period (Bisht et al., 2019)

world. However, to understand the process that brings up and develops droughts, this study discusses the three aforementioned processes separately.

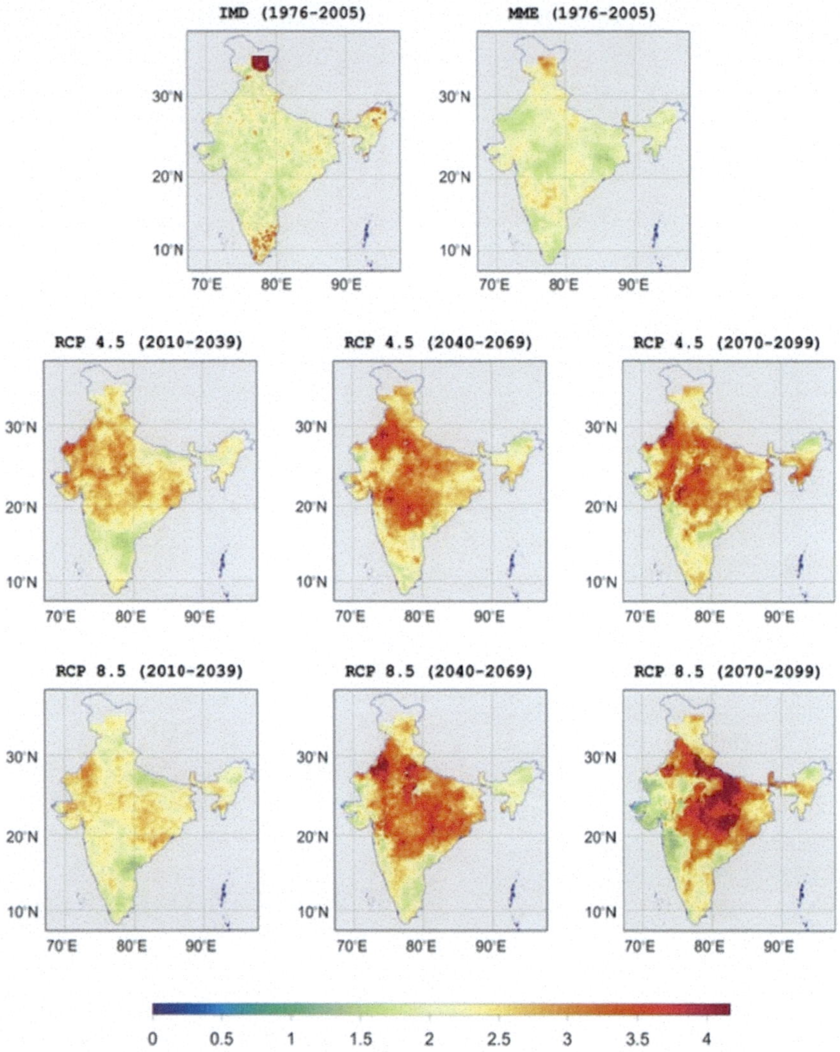

Fig. 9.7 Comparison of the average duration (months) of SPEI-3 drought to the reference period for different future time frames under different climate change scenarios. India Meteorological Department (IMD) and Multi-Model Ensemble (MME) represent data sources used for the reference period (Bisht et al., 2019)

9.4.1 Sea Surface Temperature Teleconnections

Precipitation in India is concentrated between June and September and takes up 70–90% of the annual mean rainfall (Jain et al., 2019). The interannual variable monsoon rainfall is linked to the El Niño-Southern Oscillation (ENSO), the equatorial Indian

9.4.2 Atmospheric Internal Variability and Land-Atmospheric Feedback

Besides the implications of SST teleconnections on droughts, drought can be more severe and longer due to atmospheric internal variability, which can shift the location of the jet stream and mid-latitude storm track and affect zonal atmospheric anomalies (Han & Singh, 2020; Liu & Alexander, 2007; Wang et al., 2014). A Rossby wave, known as an atmospheric bridge, is initially formed by SST anomalies and propagates energy waves from west to east inside wave guards/boundaries by meandering between ridge and trough of atmospheric pressure (Hoskins & Ambrizzi, 1993; Wang et al., 2014). When the amplitude of Rossby waves is amplified and propagation speed slows down due to atmospheric blocking, extreme weather can stagnate at the synoptic scale (i.e., a horizontal scale of the order of 1000 km or more). Thus, when a deep trough locates over India, dry and cold southerly air comes down such that dry weather occurs. The dry weather can dwell longer when atmospheric blocking stagnates Rossby waves and/or the East Asian Jet stream. Joseph and Srinivasan (1999) analyzed the relationship between the Indian Summer Monsoon Rainfall (ISMR) and Rossby waves and found a phase difference in Rossby waves between the weak and strong monsoon years. For the weak monsoon years, anomalous southerly winds were dominant, potentially leading to droughts when they lasted longer.

Large-scale atmospheric circulation modulates upper-level atmospheric internal variability and monsoons that contribute to the land-atmospheric interactions deriving local and regional extreme events (Mahto & Mishra, 2023). Land-atmospheric interactions/feedback can be referred to as a process modulating the cycles of water and energy fluxes over land areas through antecedent terrestrial water storages, meteorological conditions, and land use conditions. The mechanism of land-atmospheric feedback that can affect drought is summarized as follows: an increase in sensible heat, when precipitation is deficient, raises the atmospheric water demand (Han & Singh, 2023a, 2023b, 2023c). Thus, when the soil or terrestrial water does not meet the atmospheric water demand, drought becomes more severe and flash drought can occur.

As an example, the simultaneous occurrence of extremely high temperature and precipitation deficit may facilitate soil moisture depletion, potentially propagating meteorological drought to agricultural, hydrological, and/or flash drought. When drought is elongated and temperature increases, vapor pressure deficit (VPD) occurs due to not meeting AWD. Mahto and Mishra (2023) showed that soil moisture (SM) and VPD coupling, which represents the case of SM decrease and VPD increase, corresponds to the area under flash drought in India. More intense flash droughts contributed by the SM-VPD coupling are projected in the future due to an increase in temperature and VPD along with a more frequent monsoon rainfall failure (Mahto & Mishra, 2023). Besides the case of flash drought, the land-atmospheric feedback can contribute to the development of megadroughts, such as the Dust Bowl that occurred in the U.S. during the 1930s.

9.5 Drought Forecasting and Management

In order to mitigate the impact of drought, timely management, including crisis management, risk management, vulnerability assessment, and proactive mitigation plan, is needed. To that end, better monitoring and more accurate and reliable forecasting are essential, which will facilitate integrated drought management. Thus, this section discusses widely used drought forecasting methods and drought management to reduce the impacts of drought.

9.5.1 Drought Forecasting Methods

Drought forecasting methods are broadly categorized into statistical forecasting methods and dynamical forecasting methods. Advances in understanding drought causative mechanisms and computational efficiency have led to an increase in the accuracy at a longer lead time for drought forecasting. Although dynamical drought forecasting methods, such as a coupled hydrologic model with a general circulation model, can handle the nonlinear nature of hydrometeorological phenomena, their lack of stochastic nature hinders long-term forecasting (Han & Singh, 2020). On the other hand, statistical methods, such as time series and regression models, use temporal dependence, periodicities, and/or a statistical relationship between predictors and predictand for forecasting (Han & Singh, 2020; Mishra & Singh, 2010). However, statistical methods sometimes have limitations as not efficient in their applications for the nonstationary system (Han & Singh, 2020). To compensate for downsides of dynamic and statistical methods, hybrid dynamic-statistical methods have been developed (Han & Singh, 2020; Hao et al., 2018). However, hybrid dynamic-statistical methods do not have an objective merging approach to combine the two forecasting methods and have not had much attention compared to individual methods (Han & Singh, 2020). Rather than hybrid dynamic-statistical methods, along with the advances in computational efficiency, machine learning and/or deep learning models have achieved great improvement. In this section, entropy spectral analysis, which is a novel approach belonging to statistical methods, and deep learning methods are further discussed.

Entropy spectral analysis for forecasting time series can be expressed with linear prediction models (i.e., autoregressive models) (Han & Singh, 2023c). However, its distinction from the original linear prediction models comes from how entropy spectral analysis estimates prediction coefficients for the forecasting process. Entropy spectral analysis defines entropy based on the spectral density and maximizes or minimizes entropy constrained on the autocorrelation to estimate prediction coefficients (Singh, 2013, 2014). Thus, entropy spectral analysis uses spectral density as a key component that links entropy theory to time series analysis for forecasting. Entropy spectral analysis has two pillars (Singh, 2013, 2014; Singh & Cui, 2015): the principle of maximum entropy (POME) and the principle of minimum cross entropy

(POMCE). While POME finds spectral density with the highest entropy among spectral densities that satisfy the posed constraints, POMCE finds spectral density as does POME but as close to the given prior spectral density as possible. Han and Singh (2023a, 2023b, 2023c) applied Burg entropy spectral analysis (BESA) for forecasting meteorological drought across the continental U.S. and showed the applicability of BESA for long-term forecasting (longer than 12 months) and drought management with high accuracy. However, application of entropy spectral analysis for drought forecasting has been rarely found. Besides, the entropy spectral analysis can be extended to multichannel entropy spectral analysis to incorporate multiple predictors. The entropy spectral analysis has been mainly developed under the assumption of stationarity. If it can consider second-order nonstationarity, more reliable and accurate long-term forecasts are expected to be achieved under climate change.

During the last decades, machine learning has had a prominent advance and deep learning has been invented to be applied for various sectors in sciences and engineering. Since deep learning has non-linearity and non-regularization in constructing input–output relations underlying larger feature layers, it showed an advantage in finding patterns and forecasting time series by overcoming algorithmic limitations that shallow machine learning had (Nearing et al., 2021). For forecasting time series, long short-term memory (LSTM) neural networks have been widely used, since they improved recurrent neural networks (RNNs) to use longer memory dependencies in time series (Dikshit et al., 2021; Xiang et al., 2020). Furthermore, traditional deep learning, which relies solely on extensive data, has improved to incorporate process-based models such that the running process is understood and more accurate forecasting can be achieved with less data usage. Thus, explanatory artificial intelligence or scientific machine learning has emerged in the hydrologic research community. Although scientific machine learning has not been widely applied for drought forecasting, its potential to improve forecasting accuracy and lead time is promising.

9.5.2 Drought Management

Most of the countries' initial drought management plans were crisis management, which means post-disaster actions aimed at relief. However, more frequent and severe droughts at larger spatiotemporal scales under global warming require proactive drought management measures (Ding et al., 2011; Wilhite et al., 2000; Zhang & Delworth, 2016). Whereas crisis drought management is a myopic strategy to maintain the function of the impacted regions by emergency supplies from the government, a risk management or proactive management plan is to improve resilience to drought by integrating governance, scientific analysis and measures, and public participation (Ding et al., 2011; Wilhite et al., 2000; Zhang & Delworth, 2016). Proactive management starts with the early warning or forecasting of drought occurrence and characteristics thereof (Zhang & Delworth, 2016). Since drought monitoring and forecasting can foresee the trend of evolution and transition of drought, they are essential components for risk management. For example, long-lasting meteorological drought and co-occurrence of high temperatures increase the probability of the transition of meteorological drought to agricultural drought and other drought types. However, early warnings based on only a single drought indicator (e.g., precipitation, snowpack, soil moisture, and streamflow) cannot provide a realistic risk assessment of a region, since a drought's transition and its impacts on social and environmental sectors are affected by biophysics, population, the susceptibility of values at risk, and the capacity of a drought mitigation system (Wilhite et al., 2000; Zhang & Delworth, 2016). For comprehensive drought risk assessment, developing a risk index, which accounts for hydroclimatic monitoring, the susceptibility of industries and environment, and political support, is necessary from basin to national scales (Zhang & Delworth, 2016).

To minimize the susceptibility of industries and the environment, watersheds or water-supplying systems should be resilient to drought. Water conservation/management, soil management, diversification of crops and industries, risk-sharing systems, and public education can enhance resilience to drought.

9.6 Conclusions

Drought is one of the hydrological extreme events caused by persistent precipitation deficit over a region such that abnormally lower precipitation, streamflow, groundwater, and snowpack, resulting in a loss in industrial and agricultural productivity and depriving human well-being and life. Thus, long-lasting severe drought not only causes the costliest economic losses but also leads to social fabric disruption due to the non-productive land attributed to soil degradation and water scarcity. So, drought impacts are detrimental. For instance, California droughts in 2014 and 2015, respectively, caused economic losses of US $2.2 billion and $3 billion. Besides, droughts that occurred in India from 2000 to 2002 caused the deaths of 20 people and economic losses of US $1.5 billion. However, under global warming, arid regions are expected to increase and more severe, longer, and a really more extensive droughts are expected to occur more frequently across the globe. Thus, to minimize the drought impacts, well-prepared drought mitigation plans are required. India also has witnessed a number of severe droughts that caused huge economic losses and human deaths associated with famines and epidemics, such as the Bengal famine during 1943–1944 that affected 1.5 million people from starvation and led to ~ 2.5 million people deaths.

In most regions in India, meteorological droughts mostly occurred from December to April and vegetation droughts mostly occurred from March to July. Besides, during 1951–2010, areas under drought showed an increasing trend in Central Northeast India and West Central India. Indo-Gangetic plains had a significant decrease in monsoon rainfall and an increase in drought-impacted areas over the past period. In the future, under climate change scenarios (e.g., RCP 4.5 and 8.5), an increase in magnitude of severity was found to be maximum for 2040–2099 in Northwest, West

Central, and Central Northeast India. Besides, under global warming, an increasing trend of areal extent affected by the above moderate drought (i.e., more severe than moderate drought) is projected across India.

Since the major moisture source for India is the Indian Summer Monsoon Rainfall (ISMR), ISMR failure is a primary trigger for the occurrence of drought. The development of ISMR can be attributed to ENSO, the equatorial Indian Ocean anomalies, and the Atlantic Ocean climate anomalies. Among the SST teleconnections, ENSO explained the most variability (~ 26.2%) of drought, and the equatorial Indian Ocean and South Atlantic Ocean anomalies explained the second most variability (~ 8.6%) of drought in India. The strength of SST teleconnection to precipitation variability may change due to warming SST under global warming. Under a higher carbon emission scenario, droughts are expected to be more severe, since stronger El Niño events are expected to be associated. Once drought initiates with precipitation shortfall, a drought can be more severe when the trough of Rossby waves is located toward India and dwells for a long period due to the stagnation of atmospheric flow. While SST teleconnections and atmospheric internal variability, such as Rossby waves, provide the background that drought evolves, land-atmospheric feedback can make drought more severe and longer when temperature increases and atmospheric water demand is not satisfied. Under global warming, an increase in temperature and a longer rainfall intermittent period can accelerate drought evolution.

Amid droughts expected to become more extreme, mitigation of drought impacts should be achieved by systematically applying crisis management, risk management, vulnerability assessment, and proactive mitigation plan. To establish timely management, better monitoring and forecasting techniques are essential. Among the advanced forecasting methods, entropy and deep learning models showed unsurmountable forecasting accuracy at a longer lead time. To prepare a reliable proactive drought management plan based on improved forecasting and early warning systems, it is vital that more observations, location-specific forecasting technology, records of drought impacts, and well-oiled governmental machinery are developed, and a proactive drought management plan needs to be prepared.

References

Balting, D. F., AghaKouchak, A., Lohmann, G., & Ionita, M. (2021). Northern Hemisphere drought risk in a warming climate. *NPJ Climate and Atmospheric Science, 4*(1), 61. https://doi.org/10.1038/s41612-021-00218-2

Bisht, D. S., Sridhar, V., Mishra, A., Chatterjee, C., & Raghuwanshi, N. S. (2019). Drought characterization over India under projected climate scenario. *International Journal of Climatology, 39*(4), 1889–1911. https://doi.org/10.1002/joc.5922

Cook, E. R., Anchukaitis, K. J., Buckley, B. M., D'Arrigo, R. D., Jacoby, G. C., & Wright, W. E. (2010). Asian monsoon failure and megadrought during the last millennium. *Science, 328*(5977), 486–489. https://doi.org/10.1126/science.1185188

Cook, B. I., Miller, R. L., & Seager, R. (2009). Amplification of the North American "dust bowl" drought through human-induced land degradation. *Proceedings of the National Academy of Sciences, 106*(13), 4997–5001. https://doi.org/10.1073/pnas.0810200106

Dikshit, A., Pradhan, B., & Alamri, A. M. (2021). Long lead time drought forecasting using lagged climate variables and a stacked long short-term memory model. *Science of the Total Environment, 755*, 142638. https://doi.org/10.1016/j.scitotenv.2020.142638

Ding, Y., Hayes, M. J., & Widhalm, M. (2011). Measuring economic impacts of drought: A review and discussion. *Disaster Prevention and Management: An International Journal, 20*(4), 434–446. https://doi.org/10.1108/09653561111161752

Gadgil, S., & Gadgil, S. (2006). The Indian monsoon, GDP and agriculture. *Economic and Political Weekly*, 4887–4895. https://www.jstor.org/stable/4418949

Han, J., & Singh, V. P. (2020). Forecasting of droughts and tree mortality under global warming: A review of causative mechanisms and modeling methods. *Journal of Water and Climate Change, 11*(3), 600–632. https://doi.org/10.2166/wcc.2020.239

Han, J., & Singh, V. P. (2021). Impacts of Rossby wave packets and atmospheric rivers on meteorological drought in the continental United States. *Water Resources Research, 57*(12), e2021WR029966. https://doi.org/10.1029/2021WR029966

Han, J., & Singh, V. P. (2023a). Understanding drought: Definitions, causes, assessments, forecasts, and management. *Integrated Drought Management, 1*, 1–42.

Han, J., & Singh, V. P. (2023b). A review of widely used drought indices and the challenges of drought assessment under climate change. *Environmental Monitoring and Assessment, 195*(12), 1438. https://doi.org/10.1007/s10661-023-12062-3

Han, J., & Singh, V. P. (2023c). Long-lead drought forecasting across the continental United States using burg entropy spectral analysis with a multiresolution approach. *Water Resources Research, 59*(4), e2022WR034188. https://doi.org/10.1029/2022WR034188

Hao, Z., Singh, V. P., & Xia, Y. (2018). Seasonal drought prediction: Advances, challenges, and future prospects. *Reviews of Geophysics, 56*(1), 108–141. https://doi.org/10.1002/2016RG000549

Hoerling, M., Quan, X. W., & Eischeid, J. (2009). Distinct causes for two principal US droughts of the 20th century. *Geophysical Research Letters, 36*(19). https://doi.org/10.1029/2009GL039860

Hoerling, M., Eischeid, J., Kumar, A., Leung, R., Mariotti, A., Mo, K., Schubert, S., & Seager, R. (2014). Causes and predictability of the 2012 great plains drought. *Bulletin of the American Meteorological Society, 95*(2), 269–282. https://doi.org/10.1175/BAMS-D-13-00055.1

Hoskins, B. J., & Ambrizzi, T. (1993). Rossby wave propagation on a realistic longitudinally varying flow. *Journal of Atmospheric Sciences, 50*(12), 1661–1671. https://doi.org/10.1175/1520-0469(1993)050%3c1661:RWPOAR%3e2.0.CO;2

Huang, J., Zhang, G., Zhang, Y., Guan, X., Wei, Y., & Guo, R. (2020). Global desertification vulnerability to climate change and human activities. *Land Degradation & Development, 31*(11), 1380–1391. https://doi.org/10.1002/ldr.3556

Jain, S., Salunke, P., Mishra, S. K., Sahany, S., & Choudhary, N. (2019). Advantage of NEX-GDDP over CMIP5 and CORDEX data: Indian summer monsoon. *Atmospheric Research, 228*, 152–160.

Ji, Y., Fu, J., Lu, Y., & Liu, B. (2023). Three-dimensional-based global drought projection under global warming tendency. *Atmospheric Research, 291*, 106812. https://doi.org/10.1016/j.atmosres.2023.106812

Joseph, P. V., & Srinivasan, J. (1999). Rossby waves in May and the Indian summer monsoon rainfall. *Tellus A, 51*(5), 854–864. https://doi.org/10.3402/tellusa.v51i5.14497

Kumar, K. N., Rajeevan, M., Pai, D. S., Srivastava, A. K., & Preethi, B. (2013). On the observed variability of monsoon droughts over India. *Weather and Climate Extremes, 1*, 42–50. https://doi.org/10.1016/j.wace.2013.07.006

Liu, Z., & Alexander, M. (2007). Atmospheric bridge, oceanic tunnel, and global climatic teleconnections. *Reviews of Geophysics, 45*(2). https://doi.org/10.1029/2005RG000172

Maharatna, A. (1996). *The Demography of famines: An Indian historical perspective*. Oxford University Press.

Mahto, S. S., & Mishra, V. (2023). Flash drought intensification due to enhanced land-atmospheric coupling in India. *Journal of Climate*, 1–31. https://doi.org/10.1175/JCLI-D-22-0477.1

Mishra, A. K., & Singh, V. P. (2010). A review of drought concepts. *Journal of Hydrology, 391*(1–2), 202–216. https://doi.org/10.1016/j.jhydrol.2010.07.012

Mishra, V., Aadhar, S., Asoka, A., Pai, S., & Kumar, R. (2016). On the frequency of the 2015 monsoon season drought in the Indo-Gangetic Plain. *Geophysical Research Letters, 43*(23), 12–102. https://doi.org/10.1002/2016GL071407

Mishra, V., Tiwari, A. D., Aadhar, S., Shah, R., Xiao, M., Pai, D. S., & Lettenmaier, D. (2019). Drought and famine in India, 1870–2016. *Geophysical Research Letters, 46*(4), 2075–2083. https://doi.org/10.1029/2018GL081477

Mishra, V., Thirumalai, K., Singh, D., & Aadhar, S. (2020). Future exacerbation of hot and dry summer monsoon extremes in India. *Npj Climate and Atmospheric Science, 3*(1), 10. https://doi.org/10.1038/s41612-020-0113-5

Mo, K. C., & Schemm, J. E. (2008). Relationships between ENSO and drought over the southeastern United States. *Geophysical Research Letters, 35*(15). https://doi.org/10.1029/2008GL034656

Nearing, G. S., Kratzert, F., Sampson, A. K., Pelissier, C. S., Klotz, D., Frame, J. M., Prieto, C., & Gupta, H. V. (2021). What role does hydrological science play in the age of machine learning? *Water Resources Research, 57*(3), e2020WR028091. https://doi.org/10.1029/2020WR028091

Newman, M., Compo, G. P., & Alexander, M. A. (2003). ENSO-forced variability of the Pacific decadal oscillation. *Journal of Climate, 16*(23), 3853–3857. https://doi.org/10.1175/1520-0442(2003)016%3c3853:EVOTPD%3e2.0.CO;2

Padmanabhan, S. Y. (1973). The great Bengal famine. *Annual Review of Phytopathology, 11*(1), 11–24. https://doi.org/10.1146/annurev.py.11.090173.000303

Pascale, S., Kapnick, S. B., Delworth, T. L., & Cooke, W. F. (2020). Increasing risk of another Cape Town "day zero" drought in the 21st century. *Proceedings of the National Academy of Sciences, 117*(47), 29495–29503. https://doi.org/10.1073/pnas.2009144117

Poonia, V., Goyal, M. K., Gupta, B. B., Gupta, A. K., Jha, S., & Das, J. (2021). Drought occurrence in different river basins of India and blockchain technology based framework for disaster management. *Journal of Cleaner Production, 312*, 127737. https://doi.org/10.1016/j.jclepro.2021.127737

Rajbanshi, J., & Das, S. (2021). The variability and teleconnections of meteorological drought in the Indian summer monsoon season: Implications for staple crop production. *Journal of Hydrology, 603*, 126845. https://doi.org/10.1016/j.jhydrol.2021.126845

Samra, J. S. (2004). Review and analysis of drought monitoring, declaration and management in India.

Sharma, S., & Mujumdar, P. (2017). Increasing frequency and spatial extent of concurrent meteorological droughts and heatwaves in India. *Scientific Reports, 7*(1), 15582. https://doi.org/10.1038/s41598-017-15896-3

Sheffield, J., & Wood, E. F. (2012). *Drought: Past problems and future scenarios*. Routledge.

Sherwood, S., & Fu, Q. (2014). A drier future? *Science, 343*(6172), 737–739. https://doi.org/10.1126/science.124762

Singh, V. P. (2014). *Entropy theory in hydraulic engineering: An introduction*. American Society of Civil Engineers. https://doi.org/10.1061/9780784412725

Singh, D., Seager, R., Cook, B. I., Cane, M., Ting, M., Cook, E., & Davis, M. (2018). Climate and the Global Famine of 1876–78. *Journal of Climate, 31*(23), 9445–9467. https://doi.org/10.1175/JCLI-D-18-0159.1

Singh, V. P. (2013). *Entropy theory and its application in environmental and water engineering*. Wiley.

Singh, V. P., & Cui, H. (2015). Entropy theory for streamflow forecasting. *Environmental Processes, 2*(3), 449–460. https://doi.org/10.1007/s40710-015-0080-8

Wang, S. Y., Hipps, L., Gillies, R. R., & Yoon, J. H. (2014). Probable causes of the abnormal ridge accompanying the 2013–2014 California drought: ENSO precursor and anthropogenic warming footprint. *Geophysical Research Letters, 41*(9), 3220–3226. https://doi.org/10.1002/2014GL059748

Wilhite, D. A., Hayes, M. J., Knutson, C., & Smith, K. H. (2000). Planning for drought: Moving from crisis to risk management 1. *JAWRA Journal of the American Water Resources Association, 36*(4), 697–710. https://doi.org/10.1111/j.1752-1688.2000.tb04299.x

Williams, A. P., Seager, R., Abatzoglou, J. T., Cook, B. I., Smerdon, J. E., & Cook, E. R. (2015). Contribution of anthropogenic warming to California drought during 2012–2014. *Geophysical Research Letters, 42*(16), 6819–6828. https://doi.org/10.1002/2015GL064924

Xiang, Z., Yan, J., & Demir, I. (2020). A rainfall-runoff model with LSTM-based sequence-to-sequence learning. *Water Resources Research, 56*(1), e2019WR025326. https://doi.org/10.1029/2019WR025326

Zhang, L., & Delworth, T. L. (2016). Simulated response of the Pacific decadal oscillation to climate change. *Journal of Climate, 29*(16), 5999–6018. https://doi.org/10.1175/JCLI-D-15-0690.1

Chapter 10
Advancing Hydrological Drought Prediction in the Topl'a River, Slovakia: A Deep Learning Approach with SMOTE Enhancement

Wael Almikaeel, Lea Čubanová, and Alexandra Vidová

Abstract Water resource management faces significant challenges due to hydrological droughts, emphasizing the critical need for accurate forecasting. This study meticulously analyzes hydrological droughts in the Topl'a River, Slovakia, employing an innovative deep-learning model based on the Streamflow Drought Index (SDI) spanning from 1989 to 2020. The assessment of hydrological droughts involves a comprehensive data preparation phase, transforming and aligning daily discharge, water level, and temperature data with SDI requirements. This process categorizes hydrological years into dry or non-dry, laying the foundation for subsequent modeling. At the core of this research is the development of a sophisticated deep-learning model that harnesses the predictive power of SDI for precise forecasting. To overcome the challenge of limited training data, the study employs the Synthetic Minority Over-sampling Technique (SMOTE) for data augmentation, ensuring a balanced representation of both dry and normal hydrological years. The model architecture is carefully designed to predict SDI for an entire hydrological year, drawing on input features from the initial six months, which include water level, discharge, and temperature data. Implemented through TensorFlow and Keras libraries, the model incorporates strategic measures to prevent overfitting, enhancing its robustness. Through extensive training and evaluation, the model exhibits exceptional performance, achieving a remarkable 100% accuracy on both training and validation datasets. Impressively, this high level of accuracy extends to the testing dataset spanning 2010–2020, underscoring the model's ability to generalize effectively to unseen data. This study significantly contributes to the advancement of hydrological drought

W. Almikaeel (✉) · L. Čubanová · A. Vidová
Department of Hydraulic Engineering, Faculty of Civil Engineering, Slovak University of Technology in Bratislava, Radlinského 11, 810 05 Bratislava, Slovak Republic
e-mail: wael.almikaeel@stuba.sk

L. Čubanová
e-mail: lea.cubanova@stuba.sk

A. Vidová
e-mail: alexandra.vidova@stuba.sk

A. Sefelnasr et al. (eds.), *Water Resources Management and Sustainability*, Water Science and Technology Library 114, https://doi.org/10.1007/978-3-031-80520-2_10

prediction through the integration of deep learning and innovative data augmentation techniques. The model's exceptional accuracy and generalizability underscores its potential as a valuable tool for drought assessment and water resource management. The research highlights the importance of an accurate data transformation process, ensuring compatibility between diverse data types and the chosen model architecture. The study's results not only validate the model's robust performance but also signal a promising avenue for further exploration in the realm of hydrological variability. The potential applications across diverse hydrological contexts and geographical regions emphasize the far-reaching implications of this research in contributing to our understanding of and ability to mitigate the impacts of hydrological extremes on society and ecosystems.

Keywords Climate change · Deep learning · Drought · Neural network · SDI · SMOTE

10.1 Introduction

Drought is a complex meteorological and hydrological problem that affects many regions worldwide, posing major challenges to ecosystems, agriculture, water resources, and human societies. Drought manifests in several distinct types, each with its own set of consequences, including meteorological drought, agricultural drought, hydrological drought, and socioeconomic drought. One of the most significant types of droughts is hydrological drought, which directly affects water resource availability. This can have a major impact on critical areas such as drinking water, irrigation, and industrial processes. (Almikaeel et al., 2022a; European Environment Agency, 2007).

Hydrological drought is characterized by prolonged periods of prolonged low streamflow, groundwater levels (Koczka Bara et al., 2014), or reservoir storage, leading to deficits in water availability and quality. To assess and comprehend the complexities of hydrological drought, numerous drought indices have been developed (Almikaeel et al., 2022a). Two of the most important indices are the Streamflow Drought Index (SDI) and the Water-Bearing Coefficient method. SDI is particularly valuable as it leverages historical streamflow data to gauge drought severity and duration, thereby facilitating the formulation of early warning systems and strategies for drought mitigation. The Water-Bearing Coefficient method compares the average water flow each year to the long-term average (Almikaeel & Čubanová, 2023; Nalbantis, 2008).

One of the most notable features of drought is its strong regional variability. Different regions around the world have unique climates and vulnerabilities, which affect the characteristics and impacts of droughts (Touma et al., 2015). For instance, Europe and the Arabian Gulf represent two distinct regions with unique climate and water resource dynamics. Europe, with its diverse climate zones, experiences a range of drought types, from mild to severe, impacting agriculture, ecosystems, and water supply. In contrast, the Arabian Gulf, characterized by arid and hyper-arid conditions, faces extreme water scarcity challenges, with droughts severely affecting water security, agriculture, and infrastructure (Alsumaiei & Alrashidi, 2020).

Climate change has significantly affected both Europe and arid regions like the Arabian Gulf, but its impact on drought severity varies. In Europe, climate change has resulted in more frequent and longer-lasting droughts, alongside increased occurrences of extreme weather events like heavy rainfall and floods. These changes can be attributed to shifts in atmospheric circulation patterns, which have reduced the amount of rainfall reaching the region. Furthermore, rising temperatures have increased evaporation rates, exacerbating drought conditions (Dankers & Feyen, 2008; Spinoni et al., 2017; Sýs et al., 2021).

In contrast, the Arabian Gulf has witnessed a notable rise in drought severity attributed to climate change. Higher temperatures have driven up evaporation rates, reducing available water resources for human use, agriculture, and ecosystems. Additionally, shifts in precipitation patterns have brought about reduced rainfall, exacerbating the severity of droughts. The combination of decreased rainfall and increased evaporation rates has depleted the region's water reserves, posing significant risks to agriculture, the economy, and public health. (Alsumaiei & Alrashidi, 2020; Shahid & Behnassi, 2014).

Recent advances in artificial intelligence (AI) and deep learning have opened new horizons in the realm of drought prediction and management. These cutting-edge technologies offer the potential to model and forecast hydrological drought events more accurately than traditional methods. AI and deep learning models, including neural networks, recurrent neural networks (RNNs), and convolutional neural networks (CNNs), have demonstrated promising capabilities in capturing complex relationships within hydrological systems. Such models can leverage large datasets to forecast drought events with greater precision, enhancing our ability to mitigate the consequences of these natural disasters (Almikaeel et al., 2022a; Sit et al., 2020).

This study aims to introduce and demonstrate the use of neural networks to model hydrological catchments with limited data, focusing on flow and temperature observations. The study shows that deep learning models can effectively predict droughts in various geographic regions, with the potential to achieve high predictive accuracy.

10.2 Methods

The hydrological drought evaluation within the selected geographic region of Slovakia was assessed using the Streamflow Drought Index (SDI). The analysis utilized data from a single hydrological gauging station, containing essential parameters such as water levels, hourly discharge measurements, and temperature data, over the period of 31 hydrological years.

10.2.1 Study Area

The study area focuses on the Topľa River (a part of the Bodrog River basin), located in eastern Slovakia (Fig. 10.1). This river, with a catchment area of 1544 km^2 and a length of 129.8 km, experiences a temperate climate, with varying rainfall patterns across different elevations. The region can be classified into three climatic regions, ranging from warm to cool, based on temperature and summer days. Precipitation data analysis highlights July as the month with the highest rainfall in the eastern region of Slovakia during the period from 1997 to 2017 (Almikaeel & Čubanová, 2023; Almikaeel et al., 2022b).

The Topľa River's hydrological regime exhibits seasonality, with peak discharge during spring snowmelt and summer rainfall events. Conversely, low flows and occasional ice cover characterize the winter months. This study utilizes the Bardejov gauging station, situated at 103.5 rkm with a catchment area of 325.8 km^2, reporting a mean annual discharge of 2.978 $m^3\ s^{-1}$ for the years 1961–2000, according to the Slovak Hydrometeorological Institute (Almikaeel & Čubanová, 2023). Understanding the hydrological characteristics of the Topľa River is critical for effective water management and mitigating extreme events like floods and droughts.

Over the last six decades, human activities and regulatory measures have notably narrowed the river's channel width. In Bardejov, channelization efforts have straightened the meandering channel and fostered the growth of riparian vegetation. Recent flood events occurred in previous years, including 2008, 2010, and 2018, and have also been recorded in 2023. These events were interspersed with drought incidents noted in 2003, 2011, 2015, and 2022.

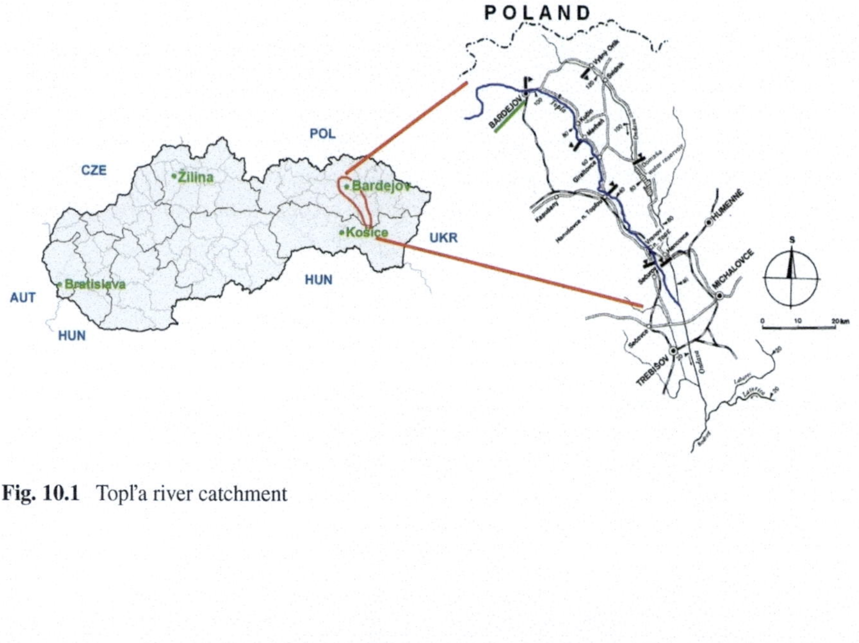

Fig. 10.1 Topľa river catchment

10.2.2 The Streamflow Drought Index Method

The Streamflow Drought Index (SDI), introduced by Nalbantis in 2008 (Nalbantis, 2008), represents a significant advancement in the field of hydrological monitoring and drought assessment. It is a powerful tool for evaluating drought conditions based on streamflow discharge data. Its simplicity, adaptability to varying data conditions, and alignment with standardized intervals make it a valuable asset in the realm of hydrological monitoring and drought assessment. The incorporation of SDI into drought studies contributes to more effective decision-making processes and a deeper comprehension of hydrological drought dynamics, thereby aiding in the mitigation of water-related risks and the sustainable management of water resources (Almikaeel & Čubanová, 2023).

The Streamflow Drought Index (SDI), introduced by Nalbantis in 2008 (Nalbantis, 2008), is a powerful tool for evaluating drought conditions using streamflow discharge data. SDI is based on the Standardized Precipitation Index (SPI), but it uses monthly runoff values instead of rainfall data. This makes SDI more sensitive to changes in streamflow, which can be a more accurate indicator of drought severity than rainfall alone. SDI can be used to assess both wet and dry periods, and it provides a quantitative measure of drought severity in relation to streamflow conditions. SDI relies on two main input parameters: monthly discharge values and a historical time series of streamflow data from a specific streamflow gauge. (Nalbantis, 2008; WMO & GWP, 2016).

SDI is considered a valuable index designed for the monitoring and assessment of drought events at specific streamflow gauges, regardless of basin size. Its user-friendly nature and adaptability distinguish it from more complex hydrological drought indices. SDI can handle missing data, making it robust even when there are gaps in the data. Additionally, its accuracy improves with the length of the streamflow record, making it particularly valuable for long-term hydrological assessments. SDI's flexibility is evident in its ability to analyze drought occurrences across various timescales, offering a comprehensive understanding of hydrological drought dynamics (Akbari et al., 2015).

Similar to SPI, SDI utilizes standardized intervals to evaluate the magnitude of drought events. This standardized categorization ensures consistent and easily understandable assessments of drought conditions. SDI values effectively quantify drought severity, as detailed in Table 10.1 (Akbari et al., 2015; Nalbantis, 2008). This standardized classification simplifies the comparison of drought severity, facilitating a comprehensive understanding of hydrological droughts across diverse regions and time periods.

To compute the Streamflow Drought Index, it is essential to have a time series of monthly streamflow volumes denoted as $Q_{i,j}$, where 'i' represents the hydrological year ($i = 1, 2, \ldots$), and 'j' corresponds to the month within that hydrological year ($j = 1, 2, \ldots, 12$, e.g., $j = 1$ for October, and $j = 12$ for September). The cumulative streamflow volume denoted as $V_{i,k}$, is calculated as follows (Akbari et al., 2015):

Table 10.1 Drought severity measured using SDI values according to the standard intervals (Akbari et al., 2015; Nalbantis, 2008)

State	Drought level	Values
0	Non-drought	SDI ≥ 0.0
1	Mild drought	− 1.0 ≥ SDI < 0.0
2	Moderate drought	− 1.5 ≥ SDI < − 1.0
3	Severe drought	− 2.0 ≥ SDI < − 1.5
4	Extreme drought	SDI < −2.0

$$V_{i,k} = \sum (Q_{i,j}) \tag{10.1}$$

where $V_{i,k}$ represents the cumulative streamflow volume for the *i*-th hydrological year and the *k*-th reference period. The reference periods are defined as follows: *k* = 1 for October until December, *k* = 2 for October until March, *k* = 3 for October until June, and *k* = 4 for October until September (*k* = 1, 2, 3, 4).

SDI is subsequently derived from cumulative streamflow volumes $V_{i,k}$ for each reference period *k* (*k* = 1, 2, 3, 4) of the i-th hydrological year (*i* = 1, 2, ...) and is calculated as follows (Akbari et al., 2015):

$$SDI_{i,k} = \frac{V_{i,k} - v}{S_k} \tag{10.2}$$

In the equation, $SDI_{i,k}$ represents the Streamflow Drought Index calculated for a specific time period (*i*) and location (k). $V_{i,k}$ corresponds to the actual observed streamflow value for the same time period (*i*) and location (*k*) for which the SDI is being computed. Additionally, v and S_k stand for the mean and standard deviation of cumulative streamflow volumes for the reference period denoted as k. These statistical parameters are calculated over an extended time frame.

It is essential to note that v typically represents the long-term mean or average streamflow for the same location (*k*) over the reference period, serving as a baseline or reference point for comparison. Essentially, it indicates what streamflow is ordinarily expected for that specific location. On the other hand, S_k signifies the standard deviation of streamflow values during the same reference period for the same location (*k*). This metric quantifies the variability or spread of streamflow values in relation to the mean (v). A higher S_k value implies more substantial variability in streamflow, as observed from the mean (v) (Akbari et al., 2015; Tareke & Awoke, 2022). These statistical components collectively contribute to the computation of the Streamflow Drought Index, enabling a comprehensive assessment of hydrological drought conditions.

Equation (10.2) essentially calculates the difference between the observed streamflow ($V_{i,k}$) for a specific time and place and the long-term mean streamflow (v) for that location, standardized by the standard deviation (S_k) of streamflow values over the reference period. This standardization allows to assess how unusual or extreme the observed streamflow is compared to historical norms (Akbari et al., 2015; Tareke & Awoke, 2022).

Positive SDI values indicate wet conditions, while negative values signify hydrological drought. The SDI is used to categorize hydrological drought into five states, ranging from 0 (non-drought) to 4 (extreme drought), based on predefined criteria as outlined in Table 10.1 (Akbari et al., 2015; Nalbantis, 2008).

10.2.3 Neural Networks

Neural networks have gained prominence in the field of hydrology and drought prediction due to their capacity to capture complex patterns and dependencies in hydrological data, which is often characterized by sophisticated, non-linear relationships. These networks are part of the broader field of machine and deep learning, where they stand out as versatile tools for modeling and forecasting hydrological variables (Almikaeel et al., 2022a).

Neural networks are a subset of machine learning algorithms inspired by the structure and function of the human brain and consist of layers of artificial neurons, inspired by the neurons in the human brain. These layers include an input layer, hidden layers, and an output layer. Each neuron receives inputs from neurons in the previous layer, computes a weighted sum of these inputs, and then applies an activation function to introduce non-linearity into the model. The network's parameters, including the weights and biases associated with each neuron, are iteratively adjusted during a training process using historical data to minimize prediction errors. This adaptability enables neural networks to detect complex relationships and nuances in hydrological data (Almikaeel et al., 2022a; Alsumaiei & Alrashidi, 2020; Rozos et al., 2021; Sit et al., 2020).

The advantages of neural networks in hydrology and drought prediction include their capability to model highly complex and non-linear interactions between variables. Hydrological systems are influenced by numerous factors, including rainfall, temperature, topography, and land use, which often exhibit complicated interdependencies. They can also learn from historical data patterns and make predictions, making them valuable tools for time-series forecasting in the hydrological domain. Additionally, they can process large and multi-dimensional datasets, which is common in hydrology due to the high number of parameters involved (Almikaeel et al., 2022a; Alsumaiei & Alrashidi, 2020; Rozos et al., 2021; Sit et al., 2020).

However, neural networks are not without limitations. They often require a substantial amount of data for training, which may be a constraint in some hydrological applications. Additionally, they are considered "black-box" models, meaning that understanding the inner workings of the model can be challenging, potentially limiting their utility in scenarios where interpretability is crucial (Almikaeel et al., 2022a; Alsumaiei & Alrashidi, 2020; Rozos et al., 2021; Sit et al., 2020).

An example of neural network application in hydrology and drought prediction involves the use of historical weather data, river discharge information, and remote sensing data as inputs to predict future drought conditions. By training a neural network on past drought events and their associated meteorological and hydrological parameters, researchers can develop predictive models to forecast potential drought occurrences, aiding in early warning systems and effective water resource management strategies, particularly in regions vulnerable to drought events (Almikaeel et al., 2022a; Alsumaiei & Alrashidi, 2020).

The neural network method employed in this study represents a powerful computational framework used for predictive modeling in a variety of fields, including hydrology and drought prediction. In the context of this study, the neural network architecture is tasked with predicting SDI based on key hydrological variables such as water level, discharge, and temperature for the initial six months of each year.

10.3 Results and Discussion

10.3.1 Drought Classification Using SDI

In the analysis part, discharge data spanning from 1989 to 2020 is employed as the primary dataset. To evaluate hydrological drought conditions, the average monthly discharge values for each month within the hydrological year, which begins in October and ends in September, are analyzed. This analysis is conducted at two distinct levels: firstly, from October to March, covering the initial half of the hydrological year, and secondly, encompassing the whole hydrological year.

Figure 10.2 illustrates the fluctuating Streamflow Drought Index values when evaluating the situation up to the midpoint of the year compared to the assessment of the entire hydrological year. The figure demonstrates the variability in the severity of drought conditions, which ranges from slight to significant, or, interestingly, transitional shifts from drought to normal situations. These fluctuations in SDI values mark the potential significance of the hydrological situation during the mid-year as a valuable indicator for forecasting the annual hydrological condition.

However, it is essential to acknowledge that the severity of drought conditions is vulnerable to transformations over the course of the entire year. This vulnerability is highlighted by instances where the hydrological situation transitions from a dry state to a normal state, reflecting the complex interaction of various hydrological factors.

Given these findings, the primary aim of this study is to develop a predictive model. This model utilizes data on daily average discharge, temperature, and water levels for the first six months of the hydrological year to predict whether the overall hydrological condition for the year will be classified as normal or dry, based on the SDI index. Such a predictive model has the potential to increase our ability to effectively address hydrological drought events, thereby enhancing water resource management and mitigation strategies.

For a better understanding of the discharge patterns, the hydrological years are classified into dry and non-dry. Figure 10.3 illustrates the contrasting mean monthly discharges observed in both dry-assessed and non-dry-assessed years. The distinction is clear: the red bars representing dry years prominently display mean monthly

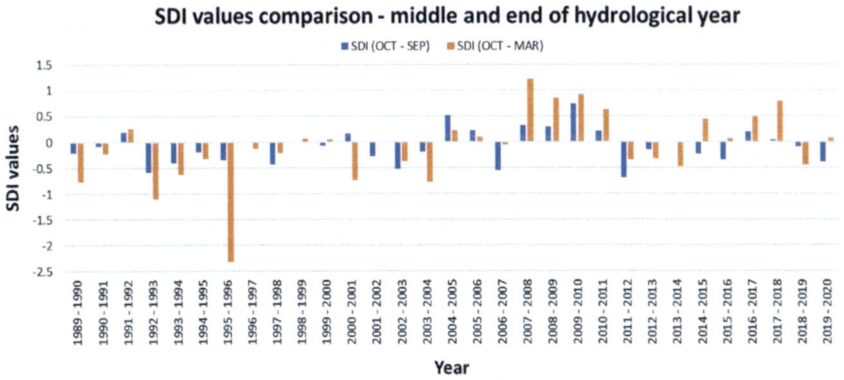

Fig. 10.2 SDI values comparison—evaluation until the mid-year point versus the evaluation until the year's end

discharges that are significantly lower than those observed in non-dry years, except for February and April, where the difference in discharges is relatively smaller.

Additionally, Fig. 10.3 includes a green line, which represents the long-term average discharge of the Topľa River spanning from 1961 to 2000. Notably, in non-dry years, a distinct pattern emerges as the mean discharges for the months of March, April, May, June, and July consistently exceed the long-term average. In contrast, during the dry-classified years, mean discharges surpass the long-term average primarily in February, March, April, and May. These variations in discharge magnitude across different hydrological classifications suggest the presence of discernible patterns. Examining these patterns holds significance for classification and forecasting dry hydrological conditions. This understanding, rooted in the complex relationships between mean discharges in individual months and their

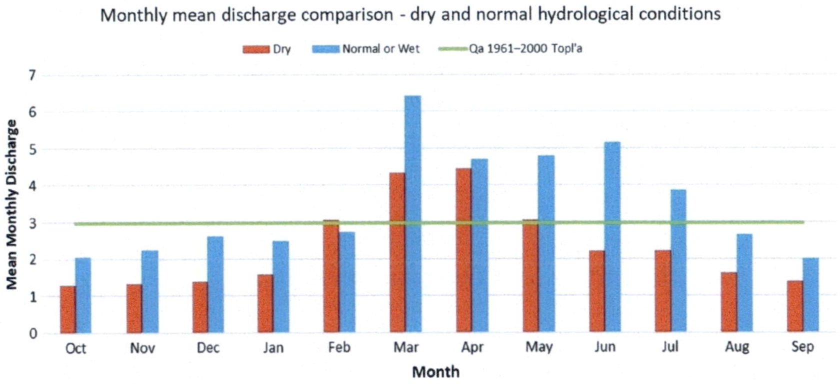

Fig. 10.3 Monthly mean discharge comparison—assessment for dry and normal hydrological conditions

cumulative impact on the river's overall hydrological state, improves the foundation for predictive modeling, promising high accuracy in dry condition forecasts.

10.3.2 Data Preparation

Data preparation for modeling involves a series of essential steps aimed at refining raw data to make it compatible with the models utilized in the study. In this study, the dataset comprises daily average discharge, water level, and temperature data over 1989–2020. The initial step of data processing involves the derivation of monthly average discharge values, which is crucial for assessing drought conditions using the Streamflow Drought Index (SDI). Consequently, each hydrological year in the dataset is evaluated and categorized as either "1" for normal or "0" for dry, representing the target for the modeling process. The primary objective of this study is to predict the hydrological status of the Topl'a River, primarily relying on daily average discharge data from the initial six months. As such, the inputs for the modeling consist of daily average discharge values for the first 183 days, while the desired outputs correspond to the accurate hydrological status denoted as "1" for normal and "0" for dry.

The modeling process within machine learning typically contains two key phases: training and testing. Consequently, the dataset is split into two distinct groups. The first group, held for training, comprises 60–80% of the data, while the second group, sorted for testing, represents the remaining 20–40%. Since the general dataset exhibits a relatively balanced distribution of dry and normal years, with dry years accounting for approximately 60% of the 31-year period and normal years constituting 40%, this balance is preserved within the training and testing sets. Specifically, data from the period 1989–2010 is dedicated to training, while data from 2010 to 2020 is set aside for testing.

In summary, the training set includes daily average discharge, water level, and temperature data for the first 183 days (representing the initial 6 months) spanning from 1989 to 2010, along with the corresponding hydrological statuses as target values. However, the testing set consists of daily average discharge, water level, and temperature data for the first 183 days from 2010 to 2020, accompanied by the associated hydrological statuses as target values. This essential data preparation process forms the foundation for subsequent modeling and predictive analysis.

10.3.3 Data Balancing and Data Augmentation

Given that the training dataset encompasses records from only 21 years, comprising three distinct data types (discharge, water level, temperature), which is conventionally considered a relatively limited dataset for training deep learning models, there arises a compelling need to augment this dataset. This augmentation must be done to expand the training dataset's size without compromising its quality or introducing any form of bias that may unintentionally influence the model's predictive accuracy. To address this challenge, the Synthetic Minority Over-sampling Technique (SMOTE) is employed, an approach widely recognized in the field of machine learning for its effectiveness in data augmentation and balancing imbalanced datasets (Chawla et al., 2002).

The SMOTE technique operates by generating synthetic data points that exhibit statistical characteristics similar to the original data. In essence, it augments the dataset by interpolating between existing data points, creating synthetic instances of the minority class to address class imbalance while preserving the underlying data distribution. This augmentation is achieved through an advanced algorithm that carefully balances the representation of different classes within the dataset (Chawla et al., 2002; Lemaître et al., 2017).

The utilization of SMOTE in this study notably expands the training dataset, which increases the number of training samples, elevating the count from its initial 21 instances to a more substantial 28 instances. This augmentation serves a twofold purpose: firstly, it balances the dataset, ensuring an equal representation of dry and normal years, and secondly, it increases the training dataset's size, providing the model with more data to learn from. This augmentation is a pivotal step in enhancing the model's capacity to capture the non-linearities of the hydrological system and improve the accuracy of its drought forecasting predictions.

10.3.4 Data Transformation

The preparatory phase of data transformation and preprocessing holds profound significance as a critical precursor to the training of the neural network in this study. The data contains three distinct variables: discharge, water level, and temperature, each characterized by unique scales, units of measurement, and, notably, diverse statistical parameters and distributions. It's important to highlight that while discharge and water level share some similarities in their distributions, the differences across these variables require a careful and detailed approach to data transformation and normalization.

The main goal of this data preprocessing is to normalize the input data with the specific requirements of the forecasting model, as illustrated in Fig. 10.4. Essentially, this involves transforming the data to align with the structure and functioning of the neural network. By addressing the differences in scale, units, and statistical characteristics, this transformation process serves multiple purposes. It enhances the model's capacity to interpret and extract significant patterns from the data while also reducing the likelihood of complications arising from variations in the input data.

In practical terms, this means applying techniques and methods systematically to standardize the data, making it compatible with the neural network's computation. Furthermore, it's crucial to perform these transformations carefully to avoid any misrepresentation or loss of vital information within the data, ensuring that the intrinsic hydrological insights are retained throughout the preprocessing phase. The

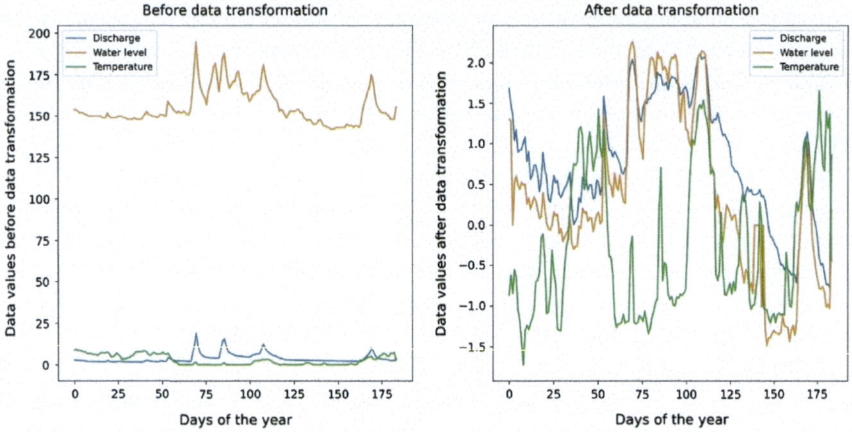

Fig. 10.4 Data processing (discharge, water level and temperature)—before (left one) and after (right one) transformation

effectiveness of this preprocessing phase significantly affects the following training process, eventually leading to the model's capability to offer precise and reliable drought predictions.

10.3.5 Model

The model developed in this study is designed for the prediction of the Streamflow Drought Index (SDI) for an entire hydrological year based on input features consisting of water level, discharge, and temperature data for the first 6 months of each year. This model leverages deep learning techniques, specifically utilizing the TensorFlow and Keras libraries (Abadi et al., 2016; Chollet et al., 2015). To explain this architecture to a wide range of audience, each layer's role and purpose will be simplified and explained.

At its core, the model features three distinct input features, each dedicated to a specific type of data: water level, discharge, and temperature information, covering the initial six months of the hydrological year. These input layers are corresponding to the channels through which the model receives essential information about the hydrological system. The utilization of separate input layers ensures that each type of data is considered independently before subsequent processing.

The model architecture is composed of the following layers:

- Input Layers—three input layers are defined, corresponding to the three types of data: water level, discharge, and temperature. Each layer has a shape of (183, 1), indicating that it accepts sequences of data with 183-time steps and a single feature at each time step. These inputs represent the historical data for the first half of the year.

- Flatten Layers—following the input layers, three flatten layers are employed to transform the sequence data into a one-dimensional format that can be processed by subsequent layers. This step simplifies the data for further analysis.
- Dense Layers (Fully Connected Layers)—the flattened data from each input (water level, discharge, and temperature) is separately passed through three dense layers with 1500 neurons each. These layers use the hyperbolic tangent (tanh) activation function (Eq. (10.3)), which introduces non-linearity to the data and enables the network to capture complex patterns in the input data.

$$\tan h(x) = (e^x - e^{(-x)})/(e^x + e^{(-x)}) \tag{10.3}$$

- Dropout Layers—to prevent overfitting and improve the model's generalization capability, dropout layers with a dropout rate of 0.05 are applied after the first set of dense layers. Dropout randomly deactivates a fraction of neurons during training, encouraging the network to learn more robust features.
- Additional Dense Layers with Regularization—three more dense layers with 500 neurons each are introduced, following the dropout layers. These layers utilize the rectified linear unit (ReLU) activation function (Eq. (10.4)) and incorporate L2 regularization with a coefficient of 0.001. L2 regularization helps prevent overfitting by adding a penalty term to the loss function, which discourages overly complex models.

$$\mathrm{ReLU(x) = x, \quad if\ x > 0, 0, \quad if\ x \le 0} \tag{10.4}$$

- Concatenate Layer—the outputs from the three sets of dense layers are concatenated together to combine the information extracted from the water level, discharge, and temperature data. This merged data is then used for further analysis.
- Additional Dense Layers—following the concatenation, three more dense layers with 250, 100, and 50 neurons are added. These layers continue to process the combined data, capturing higher-level abstractions in the input information.
- Output Layer—the final layer consists of a single neuron with a sigmoid activation function. This output layer is responsible for predicting the SDI for the entire hydrological year. The sigmoid activation function maps the model's output to a value between 0 and 1, which can be interpreted as the likelihood or severity of a drought event.

In summary, this model architecture is designed to take historical data on water level, discharge, and temperature for the first 6 months of a year and use deep learning techniques to predict the Streamflow Drought Index (SDI) for the entire year. The model's layers are carefully structured to capture relevant patterns in the input data while preventing overfitting through regularization and dropout layers. This approach allows for the automated assessment of drought conditions based on hydrological data, which can be valuable for water resource management and drought prediction.

10.3.6 Obtained Results

Upon conducting 50 iterations of model training, it is notable that the accuracy achieved on the training dataset reached an exemplary 100%. Such a result, though typically indicative of overfitting (a phenomenon where the model demonstrates high performance on training data but fails to generalize to unseen data) demands further investigation. In this context, it's essential to note that the accuracy of the model on the validation dataset, which is a subset of the training data used to evaluate the model's performance during training, also reaches an impressive 100%. This alignment of perfect accuracy on both the training and validation datasets prompts important considerations about the model's generalization capabilities.

Subsequently, the model's accuracy extends to the classification of unseen data, which is a crucial validation step. The test dataset covers daily discharge, water level, and temperature data for the first six months of the hydrological year, spanning from 2010 to 2020. Remarkably, the model was able to accurately classify the hydrological situation for entire test data set and performed very well identifying the dry hydrological years, achieving an outstanding 100% accuracy rate. Figure 10.5 represents the confusion matrix of the model performance on the testing dataset.

The specific outcomes of this model evaluation are significant. The model correctly identifies six years as drought-affected and four as non-drought, aligning seamlessly with the comprehensive analytical assessment previously presented. This concordance between model predictions and the analysis underscores the model's ability to classify hydrological conditions, thereby confirming its applicability as a valuable tool for drought assessment.

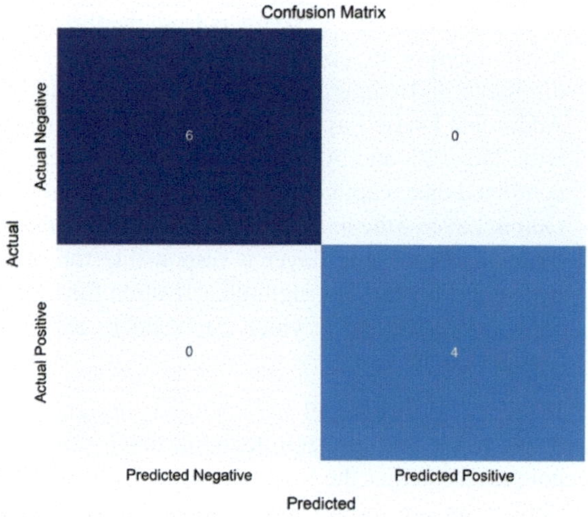

Fig. 10.5 Confusion matrix with achieved accuracy rate

10.3.7 Interpretation and Discussion

This study investigated the complexities of hydrological drought prediction, emphasizing the significant role of deep learning in this field. The design of the classification model, particularly its architecture, proved to be a critical step in achieving accurate and reliable results. Since the model needed to process daily average discharge, water level, and temperature data from the initial six months of hydrological cycles, choosing an appropriate architecture was essential for recognizing complex patterns within these distinct data types. Different hydrological contexts and datasets may require different model architectures. Therefore, it is important to carefully select and customize the model architecture to the specific dataset. This adaptation ensures that the model can effectively identify relevant features and relationships in the data, leading to better predictive accuracy.

This study emphasizes the importance of data transformation and preprocessing. Discharge, water level, and temperature data have different scales, units, and statistical distributions. Therefore, it is essential to ensure compatibility between these diverse data types and the chosen model architecture. Data transformation is crucial to this process, allowing for the conversion of disparate data sources into a unified scale that aligns seamlessly with the model's requirements. Choosing the right transformation methods is crucial for ensuring that the model can glean meaningful insights from the data. The transformative processes employed must preserve the inherent information contained within the data while enhancing its compatibility with the chosen model. By selecting the appropriate data transformation techniques, the model can achieve higher predictive accuracy by effectively accounting for the variations and nuances present within the hydrological datasets.

10.4 Conclusions

The Topľa River basin, exhibits a unique runoff regime, with the highest average monthly runoff occurring in March and April, while the lowest is observed in September and October (Fendeková et al., 2018). This runoff pattern is influenced by the below-average period (August till January), while monthly mean discharges of July and February reached nearly long-term average values for the period 1961–2000 ($Q_{a\ 1961\text{–}2000}$). Additionally, the Topľa River's flow dynamics are impacted by delayed spring runoff due to the slower snowmelt process at higher elevations. These characteristics contribute to a comprehensive understanding of the river basin's runoff patterns.

This study makes a significant contribution to hydrological drought prediction by leveraging deep learning techniques to improve the accuracy and reliability of forecasts. The model, which uses daily average discharge, water level, and temperature data, shows great promise in discerning hydrological situations. The incorporation of the Synthetic Minority Over-sampling Technique (SMOTE) addresses the challenge of limited training data, expanding the dataset without compromising quality or introducing bias. The model's exceptional performance, with a 100% accuracy rate on unseen data, highlights its potential as a valuable tool for drought assessment. Additionally, the analysis of discharge patterns during dry and non-dry years reveals a potential method for pattern-based classification. However, it is important to rigorously test the model's efficiency across diverse hydrological contexts and geographical regions to ensure its robustness and generalizability.

The study demonstrates the potential of deep learning models in hydrological research, providing a promising approach to address the challenges of hydrological variability. Including additional environmental variables and exploring refined model architectures offers opportunities to improve predictive accuracy and comprehensiveness. As hydrological variability continues to increase, studies like this one will play a crucial role in advancing our understanding and prediction of hydrological phenomena, ultimately helping to mitigate the impacts of hydrological extremes on society and ecosystems.

Acknowledgements This contribution was developed within the framework and based on the financial support of the APVV-19-0383 project, together with APVV-20-0023 project.

References

Abadi, M., Agarwal, A., Barham, P., Brevdo, E., Chen, Z., Citro, C., ... & Zheng, X. (2016). Tensorflow: Large-scale machine learning on heterogeneous distributed systems. https://www.tensorflow.org/

Akbari, H., Rakhshandehroo, G. R., Sharifloo, A. H., & Ostadzadeh, E. (2015). Drought analysis based on standardized precipitation index (SPI) and streamflow drought index (SDI) in Chenar Rahdar River basin, southern Iran. *Watershed Management.* https://doi.org/10.1061/9780784479322.002

Almikaeel, W., Čubanová, L., & Šoltész, A. (2022b). Comparison of mean daily discharge data for under-mountain and highland-lowland types of Rivers. *Acta Hydrologica Slovaca, 23*(1), 73–81. https://doi.org/10.31577/ahs-2022-0023.01.0008

Almikaeel, W., & Čubanová, L. (2023). Drought analysis using water-bearing coefficient and streamflow drought index—Topľa River case study. In *14th Conference of Civil and Environmental Engineering for PhD Students and Young Scientists*. Young scientists 2022 (YS 2022). https://doi.org/10.1063/5.0158671

Almikaeel, W., Čubanová, L., & Šoltész, A. (2022a). Hydrological drought forecasting using machine learning—Gidra river case study. *Water, 14*(3), 387. https://doi.org/10.3390/w14030387

Alsumaiei, A. A., & Alrashidi, M. S. (2020). Hydrometeorological drought forecasting in hyper-arid climates using nonlinear autoregressive neural networks. *Water, 12*(9), 2611. https://doi.org/10.3390/w12092611

Chawla, N. V., Bowyer, K. W., Hall, L. O., & Kegelmeyer, W. P. (2002). Smote: Synthetic minority over-sampling technique. *Journal of Artificial Intelligence Research, 16*, 321–357. https://doi.org/10.1613/jair.953

Chollet, F., et al. (2015). *Keras*. GitHub. https://github.com/fchollet/keras. Accessed September 25, 2023.

Dankers, R., & Feyen, L. (2008). Climate change impact on flood hazard in Europe: An assessment based on high-resolution climate simulations. *Journal of Geophysical Research, 113*(D19). https://doi.org/10.1029/2007jd009719

European Environment Agency. (2007). *Climate change and water adaptation issues; European environment agency: Copenhagen, Denmark* (p. 112).

Fendeková, M., Horvát, O., Blaškovičová, L., Danáčová, Z., Fendek, M., & Bochníček, O. (2018). Prognosis of climate change driven drought in the Poprad, Torysa, and Topľa River Basins. *Acta Hydrologica Slovaca, 19*(2), 234–243. ISSN 2644–6291.

Koczka Bara, M., Velísková, Y., Dulovičová, R., & Schügerl, R. (2014). Influence of surface water level fluctuation and riverbed sediment deposits on groundwater regime. *Journal of Hydrology and Hydromechanics, 62*(3), 177–185. https://doi.org/10.2478/johh-2014-0030

Lemaître, G., Nogueira, F., & Aridas, C. K. (2017). Imbalanced-learn: A python toolbox to tackle the curse of imbalanced datasets in machine learning. *Journal of Machine Learning Research, 18*(17), 1–5. http://jmlr.org/papers/v18/16-365.html

Nalbantis, I. (2008). Evaluation of a hydrological drought index. *European Water*, 67–77.

Rozos, E., Dimitriadis, P., Mazi, K., & Koussis, A. D. (2021). A multilayer perceptron model for stochastic synthesis. *Hydrology, 8*(2), 67. https://doi.org/10.3390/hydrology8020067

Shahid, S. A., & Behnassi, M. (2014). Climate change impacts in the Arab region: Review of adaptation and mitigation potential and practices. In *Vulnerability of agriculture, water and fisheries to climate change* (pp. 15–38). https://doi.org/10.1007/978-94-017-8962-2_2

Sit, M., Demiray, B., Xiang, Z., Ewing, G., Sermet, Y., & Demir, I. (2020). *A comprehensive review of deep learning applications in hydrology and water resources*. https://doi.org/10.31223/osf.io/xs36g

Spinoni, J., Vogt, J. V., Naumann, G., Barbosa, P., & Dosio, A. (2017). Will drought events become more frequent and severe in Europe? *International Journal of Climatology, 38*(4), 1718–1736. https://doi.org/10.1002/joc.5291

Sýs, V., Fošumpaur, P., & Kašpar, T. (2021). The impact of climate change on the reliability of water resources. *Climate, 9*(11), 153. https://doi.org/10.3390/cli9110153

Tareke, K. A., & Awoke, A. G. (2022). Hydrological drought analysis using streamflow drought index (SDI) in Ethiopia. *Advances in Meteorology, 2022*, 1–19. https://doi.org/10.1155/2022/7067951

Touma, D., Ashfaq, M., Nayak, M. A., Kao, S.-C., & Diffenbaugh, N. S. (2015). A multi-model and multi-index evaluation of drought characteristics in the 21st century. *Journal of Hydrology, 526*, 196–207. https://doi.org/10.1016/j.jhydrol.2014.12.011

World Meteorological Organization (WMO) & Global Water Partnership (GWP). (2016). In M. Svoboda & B. A. Fuchs (Eds.), *Handbook of drought indicators and indices* (p. 52). Integrated Drought Management Programme (IDMP), Geneva.

Chapter 11
Assessing the Impact of Long-Term Drought on Water Resources in Arid Regions Under Climate Change Conditions

Hany F. Abd-Elhamid, Mohsen Sherif, Martina Zeleňáková, Ahmed Sefelnasr, Tatiana Soľáková, and Mohamed Mahdy

Abstract Climate change is accelerating due to increased human activities and the associated increase in the greenhouse gas emissions. Climate change has impacted the parameters of the hydrologic cycle, including temperature, wind and precipitation patterns. Climate change might affect hydrological regimes and increase the likelihood of extreme events such as extended drought and flash floods. The Middle East is among the most vulnerable areas to climate change. The UAE, like other countries in the Middle East, is vulnerable to extreme weather events including drought and flood. Climate change represents a main challenge for water resources management in many countries. The drought is a natural hazard that might become more frequent under changing weather conditions. Drought analysis is usually conducted using different drought indices, but the Standardized Precipitation Index (SPI) is the most used method and is recommended by the World Meteorological Organization for monitoring drought conditions in various areas across the world. This study aims to assess the long-term drought in the northern area of UAE using SPI based on

H. F. Abd-Elhamid (✉)
Department of Water and Water Structures Engineering, Faculty of Engineering, Zagazig University, Zagazig 44519, Egypt
e-mail: hany.abdelhamid@tuke.sk

H. F. Abd-Elhamid · M. Zeleňáková · T. Soľáková
Department of Environmental Engineering, Faculty of Civil Engineering, Technical University of Kosice, 040 01 Košice, Slovak Republic
e-mail: martina.zelenakova@tuke.sk

M. Sherif · A. Sefelnasr
National Water and Energy Center, UAE University, Al Ain, UAE
e-mail: msherif@uaeu.ac.ae

A. Sefelnasr
e-mail: ahmed.sefelnasr@uaeu.ac.ae

M. Mahdy
Department of Geography, Faculty of Art, Zagazig University, Zagazig 44519, Egypt
e-mail: mmmahdy@arts.zu.edu.eg

A. Sefelnasr et al. (eds.), *Water Resources Management and Sustainability*, Water Science and Technology Library 114, https://doi.org/10.1007/978-3-031-80520-2_11

recorded monthly rainfall data from land stations for a period of 40 years from 1980 to 2020. SPI-12 months were calculated using DrinC to examine meteorological drought that may affect water resources. Four stations have been selected for rainfall trend and SPI analysis. Rainfall trends analysis showed that Masafi station had the highest annual rainfall (134.7 mm) and Um Ghafa station had the lowest (43.0 mm). Rainfall trend analysis showed a significant decline at most of the stations, possibly due to climate change in the last two decades. According to SPI-12 results, moderate droughts occurred in 2001 and 2005, severe droughts occurred in 2010 and 2011, while extreme droughts occurred in 2012 and 2016. The driest years were 2005, 2010, 2012, and 2016 and the estimated average return period is 3–4 years. All the studied stations had a decreasing trend in rainfall, indicating that climate change would have a significant impact on the UAE's water resources and more drought periods are expected. The findings of the study will allow for proper management of the UAE's water resources and adaptation to climate change.

Keywords Arid regions · Climate change · Meteorological drought · SPI · UAE

11.1 Introduction

Global warming is mainly attributed to the rapid increase of the emission of greenhouse gases (GHG) in the atmosphere as a result of increased human activities and high consumption of fossil fuels. Climate change has altered the hydrologic cycle parameters including temperature, wind, and precipitation. Climate change may alter hydrological regimes and increase the likelihood of extreme events like flood and drought. The risk of prolonged drought increases as temperatures rise and the precipitation quantities and patterns shift. Therefore, it is important to project future rainfall trends and droughts to develop effective management strategies.

Assessment of rainfall trends using various techniques has been the subject of numerous investigations in various areas. Haylock et al. (2006) used the daily rainfall in Latin America to assess changes in the total rainfall over the years (1960–2000). The findings revealed a significant decrease in rainfall in the studied area. Endo et al. (2009) examined precipitation changes in Southeast Asian countries using daily data from 1950 to 2000. The findings showed that stations with an upward tendency are more numerous than stations with a downward tendency. Sen's slope and Mann-Kendall were used by Alahacoon and Edirisinghe (2021) to analyze the seasonal rainfall changes in Sri Lanka from 1989 to 2019 in response to climate change. The findings revealed that the annual rainfall dramatically increased from 1989 to 2019. Moreover, Alahacoon et al. (2022) examined rainfall patterns in the African continent between 1983 and 2020.

Drought is a natural phenomenon that results from prolonged low rainfall patterns. Climate change has an aggravating effect on the frequency and severity of drought episodes. Negative effects of drought can be seen in different fields including agriculture, environment, ecosystems, economy, and society. Drought can be classified to four different types: agricultural, hydrological, meteorological, and socioeconomic. Different indicators, including SPI, RDI, and SPEI, have been utilized for drought analysis. McKee et al. (1993) developed SPI for drought analysis based on rainfall data. SPI is calculated monthly for n months, e.g. 1, 3, 6, 9, 12, or 24 months depending on the time of interest. SPI is recommended by the World Meteorological Organization (WMO) for monitoring drought conditions. SPI has been used for drought analysis in a variety of places, including Syria (Zeleňáková et al., 2022a), Greece (Karavitis et al., 2011), Iran (Mansouri et al., 2012), Serbia (Gocic & Trajkovic, 2014), Botswana (Batisani, 2011), and Slovakia (Zeleňáková et al., 2022b, c).

The Middle East is among the most vulnerable areas to climate change. UAE is one of the Middle Eastern countries that could be severely impacted by climate change. Recently, limited studies have been done to evaluate rainfall trends and drought in UAE. In three wadis in the northern region of UAE, Sherif et al. (2009) examined the rainfall distribution in space and time, its consistency, likelihood of occurring, and drought conditions. As a consequence of the drought characterization, the study region was found to be classed as arid. In four climate zones in UAE, Sherif et al. (2013) investigated the spatial and temporal characteristics of rainfall. They studied the patterns of precipitation, its likelihood to occur, its intensity-duration-frequency (IDF) relationship, its likely maximum precipitation (PMP), and possible drought scenarios. The East Coast region had the highest estimated PMP values, while the Gravel Plains region had the lowest. The estimated drought severity index revealed that the regions have experienced comparable patterns of drought over the years.

Drought conditions in Sharjah, UAE was assessed by Yilmaz et al. (2020) using the Rainfall Anomaly Index (RAI) and Aridity Index (AI) based on observed rainfall data. Decreasing trends in RAIs (drier future) were detected for February and March which may cause significant problems, particularly for agricultural activities. Oyounalsoud et al. (2022) used monthly precipitation and average temperature data collected over 35 years (1981–2015) to assess the drought in Sharjah, UAE. Future droughts in the area were predicted using the Standardized Precipitation Index (SPI) and Reconnaissance Drought Index (RDI). The statistical distribution functions (gamma and lognormal) for SPI and RDI are fitted across an annual time period, and the Mann-Kendal test is then used to analyze the trend of the index values. Both indices indicated significant drought frequency and a tendency to get drier over time that requires a proper drought management and monitoring.

Due to its dry climate and decreased annual rainfall, UAE faces extreme droughts, like many other Middle Eastern nations. The UAE has been subjected to scarce conventional water resources for prolonged periods of time as a result of these droughts. Over the past 30 years, a 20% decrease in rainfall has been recorded. Nevertheless, the UAE has achieved water security through the desalination of seawater and reuse of treated water for development of development of forests and green areas. This study aims to investigate changes in rainfall trends in UAE and analysis of drought due to climate change. SPI is used in this study to evaluate the drought in UAE using DrinC software.

11.2 Materials and Methods

In this section the study area, data collection and the methodology used in assessing the rainfall trend and drought in the northern area of the UAE will be described.

11.2.1 Study Area and Data Collection

UAE is located in the eastern of the Arabian Peninsula. Figure 11.1 shows the boundaries of the UAE, which are Oman to the east, Saudi Arabia to the south and west, and the Arabian Gulf to the north. The UAE latitude and longitude is 23.4241° N and 53.8478° E. The total area of UAE is 83 600 km^2 and the total population is about 9.5 million. The climate is arid and more than 75% of the area is covered by desert. The average minimum daily temperature is 14 °C in the winter and 45 °C in the summer (Aldababseh & Temimi, 2017). The average annual rainfall in the UAE is in the range of 80 and 140 mm.

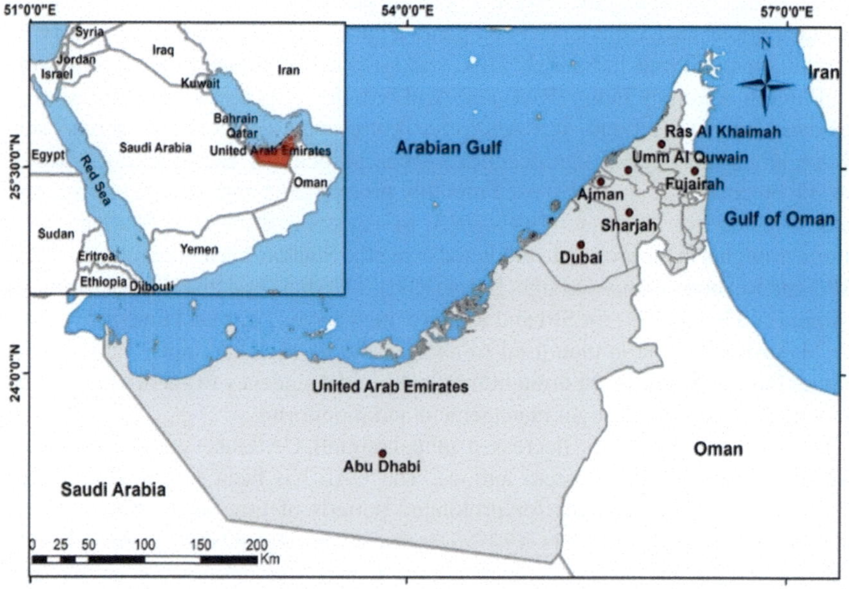

Fig. 11.1 Location map of the study area

The rainfall data from 20 stations were gathered from the National Center of Meteorology, Abu Dhabi, for the period from 1980 to 2020 (40 years). The stations are distributed across the northern area as shown in Fig. 11.2. The stations are concentrated in the north where the majority of the rainfall events are encountered. Four significant stations of highest rainfall have been selected for the analysis of SPI and rainfall trends (see Fig. 11.2).

The annual rainfall for the 20 stations for the period (1980–2020) is presented in Fig. 11.3. The maximum annual rainfall is about 134.7 mm at Masafi station and the minimum is about 43.0 mm at Um Ghafa station. The stations of high rainfall records are located on the Arabian Gulf and Gulf of Oman coasts. The stations in the south and western parts receive less annual rainfall.

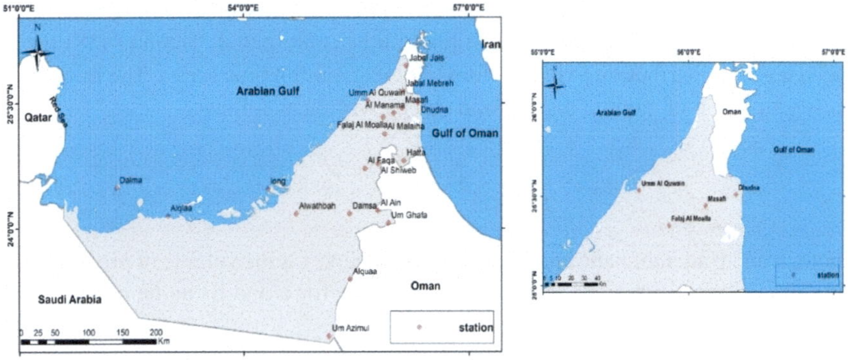

Fig. 11.2 Locations of the selected rain gauges stations

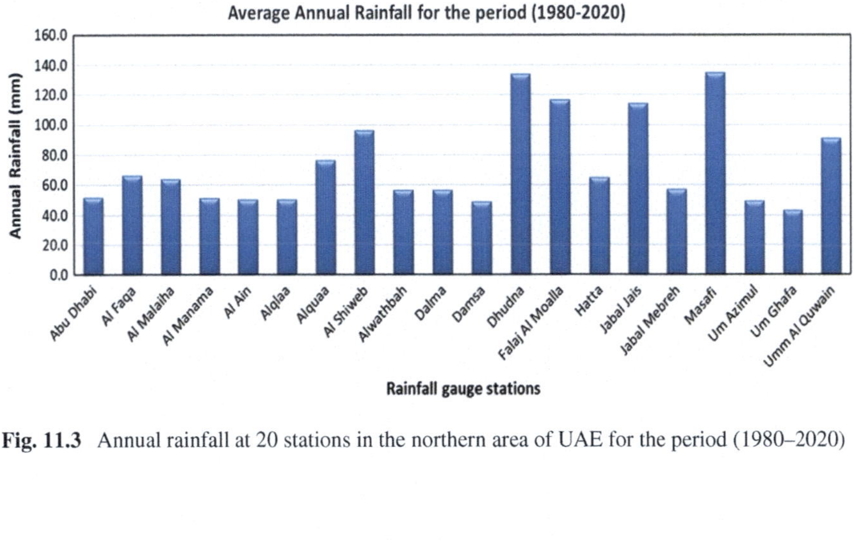

Fig. 11.3 Annual rainfall at 20 stations in the northern area of UAE for the period (1980–2020)

11.2.2 Methodology

Drought Indices Calculator (DrinC) program is used to assess the historical drought in UAE in the period (1980–2021). DrinC can be used to calculate the Reconnaissance Drought Index (RDI), the Stream flow Drought Index (SDI), the Standardized Precipitation Index (SPI), and the Precipitation Deciles. It can also be used to monitor, analyze, and evaluate droughts. It has many applications and hence is becoming more popular for drought analysis, especially in arid and semi-arid countries. It has a straightforward, approachable, and comprehensive graphical user interface (GUI) (Tigkas et al., 2015). The long-term rainfall records are fitted to a probability distribution for the selected time period to determine SPI. This probability distribution is then converted to normal distribution, with the result that the mean SPI for the location and desired period is zero (Edwards & McKee, 1997; McKee et al., 1993). When SPI is normalized, wetter and drier conditions can be represented. Thom (1958) introduced gamma distribution to fit the climatological rainfall time series. The frequency density function of gamma distribution can be written as follows:

$$g_x = \frac{1}{\beta^\alpha \Gamma(\alpha)} x^{\alpha-1} e^{-x/\beta} \quad \text{for x} > 0 \tag{11.1}$$

where, α and β are the shape and the scale parameters, x is the volume of precipitation and (Γ) is gamma function. Maximum probability for α and β can be estimated as following:

$$\alpha = \frac{1}{4A}\left(1 + \sqrt{1 + \frac{4A}{3}}\right), \beta = \frac{\overline{x}}{\alpha}, \quad \text{where } A = \ln(\overline{x}) - \frac{\sum \ln(x)}{n} \tag{11.2}$$

where, n is the number of observations.

The cumulative probability of a recorded rainfall event is then determined using the obtained parameters. Hence, the gamma function is unknown and precipitation distribution could include zeros, the cumulative probability is determined as following:

$$H(x) = q + (1 - q)G(x) \tag{11.3}$$

where, q is the likelihood of no precipitation, and G(x) is the cumulative probability of the incomplete gamma function. The formula (m/n) can be used to get q if m is the total number of zeros in a time series of precipitation. The cumulative probability H(x) is then converted to the standard normal random variable z, with mean zero and variance of one (Abramovitz & Stegun, 1965), to determine the SPI value. A drought event, as defined by the SPI value, takes place when the index consistently reaches an intensity of – 1.0 or less. Therefore, the start and end of each drought event define how long it lasts. The drought value of SPI for each month of the drought episode is

Table 11.1 Drought categories according to SPI

SPI value	Drought classification
≥ 2	Extremely wet
1.5–1.99	Very wet
1.0–1.49	Moderately wet
0.99 to − 0.99	Near normal
− 1.0 to − 1.49	Moderate drought
− 1.5 to − 1.99	Severe drought
≤ -2	Extreme drought

the positive sum (Hayes et al., 2007). As given in Table 11.1, there are seven different categories of drought (Blain, 2012; McKee et al., 1995; Sienz et al., 2012).

11.3 Results and Discussion

This study investigated the rainfall patterns in the northern region of UAE based on rainfall data from 20 stations for the period (1980–2020). SPI is used to monitor historical drought in the UAE considering climate change.

11.3.1 Rainfall Trends Analysis in the Period (1980–2020)

Rainfall is a significant meteorological parameter that must be carefully monitored and evaluated. Variations in rainfall values, including mean, maximum, and minimum seasonal values, must be taken into account when evaluating climate change. Every region requires mean rainfall analysis, but understanding the maximum and minimum values is crucial for a variety of applications such as agriculture and water resources management. The amount of water that will be available to meet diverse demands for agriculture, industry, and other human activities must be estimated based on real rainfall data. In the current study area, most of the rainfall occurs in the winter but the summer receives a small quantity of rainfall. The trends of rainfall at four significant stations in the northern area of UAE for the period (1980–2020) have been selected and presented in Fig. 11.4. The rainfall trends for the period (1980–2020) reveal that the utmost rainfall was recorded in 2003 (626.5 mm) at Masafi station and the smallest rate was recorded in 2012 (0.1 mm) at Um Ghafa station. Rainfall trend analysis showed a significant decline in rainfall at most of the stations possibly due to climate change in the last two decades (2000–2020) as shown in Fig. 11.4.

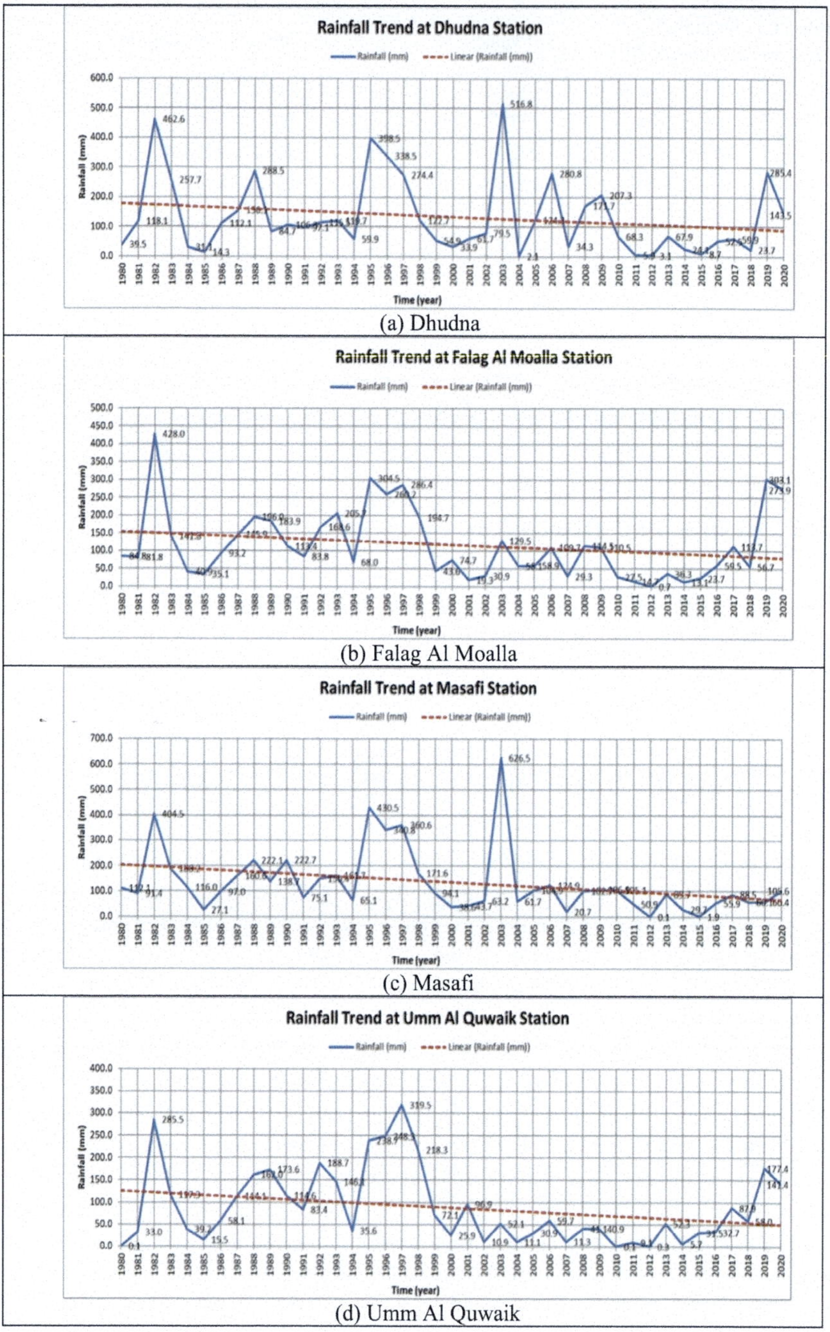

Fig. 11.4 Rainfall trend in the UAE for the period (1991–2009)

11.3.2 Drought Analysis Using SPI in the Period (1980–2020)

SPI is a useful index for analyzing historical drought. In the current study, SPI is calculated for a time scale of 12 months to assess the meteorological drought. Four significant stations have been chosen to present SPI results (see Fig. 11.5). SPI-12 results at Dhudna station show that severe drought occurred in 2005, 2012, 2013 and 2016. Also, severe drought was recorded in 2012 and 2013 at Falag Al Moalla station. Furthermore, Extreme drought occurred in 2012, 2013 and 2017 at Masafi, but at Umm Al Quwaik station extreme drought observed from 2010 to 2013. Long-term drought index revealed that the extreme drought occurred in 2012–2013 and 2016–2017.

According to SPI results, Umm Al Quwaik station has the lowest propensity of drought occurrence while Masafi station has the highest tendency, but the other stations gave different values between the two stations. Moreover, SPI-12 results showed that moderate droughts (1.0–1.5, SPI) occurred in the years 2001 and 2005, severe droughts (1.5–2.0, SPI) occurred in the year 2010 and 2011, while extreme droughts (SPI > 2.0) occurred in 2012 and 2016. The extreme drought occurred in the years 2012 and 2016 at most of the stations. The analysis of SPI-12 results from 1980 to 2020 showed that the driest years were 2012–2013 and very dry events were repeated in the years: 1998, 1999, 2007 and 2008. The average return period is estimated to be 3–4 years. Also, rainfall trend analysis showed decreasing trend at all the investigated stations which reveal that UAE is impacted by the climate change that occurred in the Middle East.

11.4 Conclusion

The parameters of hydrologic cycle, specifically precipitation pattern could be affected by climate change, which could increase the possibility of floods and droughts. Arid and semi-arid areas, like the Middle East is vulnerable to climate change. Recently, changes in temperature, wind, and precipitation patterns have become noticeable in many areas which could affect water resources. UAE is one of the Middle East countries that could be affected by climate change. This study assesses the impact of climate change on rainfall trends in UAE. The variation in rainfall caused by climate change between 1980 and 2020 was assessed based on data from 20 land stations located in the northern region of the UAE. Also, SPI is used to assess the historical drought during the same period. SPI is computed using DrinC software at 12 months' time scale. Four significant stations with the highest rainfall rates have been selected for rainfall trend and SPI analysis. From the SPI results, it can be concluded that severe drought and extreme drought have been repeated in different years and the driest periods were recorded in 2005, 2010, 2012, and 2016.

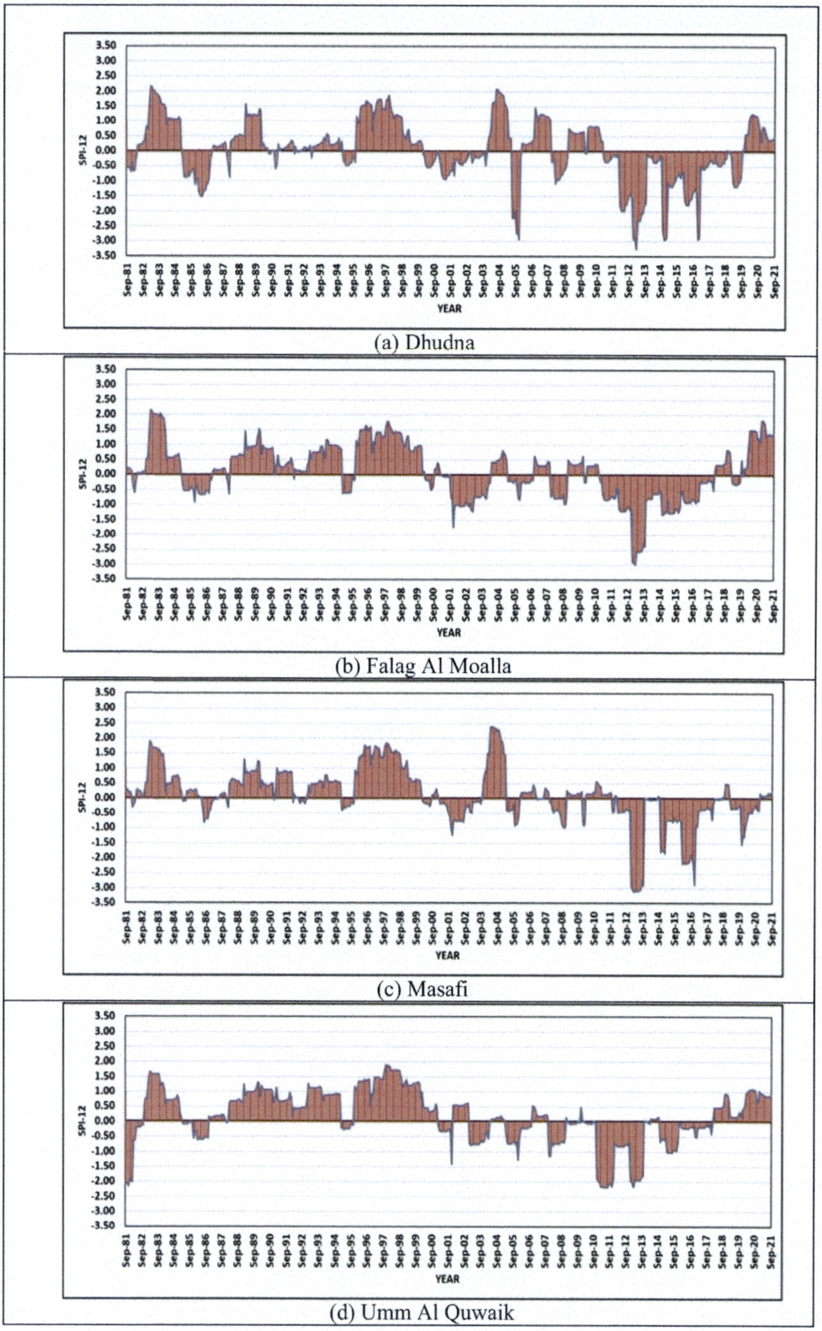

Fig. 11.5 SPI-12 at four significant stations for the period (1980–2020)

The results revealed the occurrence of severe drought and extreme drought conditions at most of the stations at different intervals. The extreme drought occurred in the years 2012 and 2013 at most of the stations and the average return period was estimated to be 3–4 years. All the studied stations showed a decrease in the rainfall trends which indicates that UAE would be extremely affected by the climate change. The results are consistent with the IPCC's 5th assessment report. Future shortage of the conventional water resources is expected in UAE due to climate change. Similar to other countries in the region, the water security might be achieved through desalination of seawater for potable and reuse of treated wastewater for developing of green lands. The findings of this study could aid policymakers in creating water resource management strategies that can adapt with the likely climate change conditions.

Acknowledgements This work was supported by the National Water and Energy Center, UAE University, and the Slovak Research and Development Agency under the Contract no. APVV-20-0281. The rainfall data was provided by the National Center of Meteorology, Abu Dhabi, UAE.

References

Abramovitz, M., & Stegun, I. (1965). Handbook of mathematical functions, National Bureau of Standards. In *Applied Mathematics Series* (Vol. 55). Washington, D.C.

Aldababseh, A., & Temimi, M. (2017). Analysis of the long-term variability of poor visibility events in the UAE and the link with climate dynamics. *Atmosphere, 8*, 242.

Alahacoon, N., & Edirisinghe, M. (2021). Spatial variability of rainfall trends in Sri Lanka from 1989 to 2019 as an indication of climate change. *ISPRS International Journal of Geo-Information, 10*, 84.

Alahacoon, N., Edirisinghe, M., Simwanda, M., Perera, E., Nyirenda, V. R., & Ranagalage, M. (2022). Rainfall variability and trends over the african continent using TAMSAT data (1983–2020): Towards climate change resilience and adaptation. *Remote Sensing, 14*, 96.

Batisani, N. (2011). The Spatio-Temporal-Severity dynamics of drought in Botswana. *Journal of Environmental Protection, 2*, 803–816.

Blain, G. C. (2012). Revisting the probabilistic definition of drought: Strengths, limitations and an agrometeorological adaptation. *Bragantia, 71*(1), 132–141.

Edwards, D.C., McKee T.B. (1997). Characteristics of 20th century drought in the United States at multiple time scales. Climatology Report Number 97–2, Colorado State University, Fort Collins, Colorado.

Endo, N., Matsumoto, J., & Lwin, T. (2009). Trends in precipitation extremes over Southeast Asia. *SOLA, 5*, 168–171.

Gocic, M., & Trajkovic, S. (2014). Spatiotemporal characteristics of drought in Serbia. *Journal of Hydrology, 510*, 110–123.

Hayes, M. J., Alvord, C., & Lowrey, J. (2007). Drought indices. *Feature Article, Intermountain West Climate Summary., 3*(6), 2–6.

Haylock, M. R., Peterson, T. C., Alves, L. M., Ambrizzi, T., Anunciação, Y. M. T., Baez, J., Barros, V. R., Berlato, M. A., Bidegain, M., Coronel, G., Corradi, V., Garcia, V. J., Grimm, A. M., et al. (2006). Trends in total and extreme South American rainfall in 1960–2000 and links with sea surface temperature. *Journal of Climate, 19*, 1490–1512.

Karavitis, C. A., Alexandris, S., Tsesmelis, D. E., & Athanasopoulos, G. (2011). Application of the standardized precipitation index (SPI) in Greece. *Water, 3*, 787–805.

Mansouri, D. R. M., Bagherzadeh, A., & Khosravi, M. (2012). Assessment of drought hazard impact on wheat cultivation using standardized precipitation index in Iran. *Arabian Journal of Geosciences, 6*, 4463–4473.

McKee, T. B., Doesken, N. J., & Kleist, J. (1993). The relationship of drought frequency and duration of time scales. In *Proceedings of the 8th Conference on Applied Climatology* (pp. 179–186). American Meteorological Society, Anaheim, CA, USA, 17–23 January.

McKee, T.B., Doesken, N.J., & Kleist, J. (1995) Drought monitoring with multiple time scales. In *9th Conference on Applied Climatology. American Meteorological Society, Boston* (pp. 233–236).

Oyounalsoud, M. S., Najah, A., Yilmaz, A. G., Abdallah, M., & Siddique, M. (2022). Meteorological drought assessment in Sharjah, UAE using drought indices. *International Journal of Environmental Science and Development, 13*(1), 16–20.

Sherif, M., Almulla, M., Shettya, A., & Chowdhurya, R. K. (2013). Analysis of rainfall, PMP and drought in the UAE. *International Journal of Climatology, 34*(4), 1318–1328.

Sherif, M., Akram, S., & Shetty, A. (2009). Rainfall Analysis for the Northern Wadis of United Arab Emirates: A case study. *Journal of Hydrologic Engineering, 14*(6), 535–544.

Sienz, F., Bothe, O., & Fraedrich, K. (2012). Monitoring and quantifying future climate projections of dryness and wetness extremes: SPI bias. *Hydrology and Earth System Sciences*, 2143–2157.

Thom, H. C. S. (1958). A note on gamma distribution. *Monthly Weather Review, 86*, 117–122.

Tigkas, D., Vangelis, H., & Tsakiris, G. (2015). DrinC: A software for drought analysis based on drought indices. *Earth Science Informatics, 8*(3), 697–709.

Yilmaz, A. G., Najah, A., Hussein, A., Khamis, A., Kayemah, N., & Atabay, S. (2020). Climate change effects on drought in Sharjah, UAE. *International Journal of Environmental Science and Development, 11*(3), 116–122.

Zeleňáková, M., Abd-Elhamid, H. F., Krajníková, K., Smetanková, J., & Alkhalaf, I. (2022a). Spatial and temporal variability of rainfall trends in response to climate change, a case study: Syria. *Water, 14*(10), 1670.

Zeleňáková, M., Soľáková, T., Purcz, P., & Abd-Elhamid, H. F. (2022b). Analysis of drought in the eastern part of Slovakia using standardized precipitation index. In *International Conference on Water Resources Management and Sustainability: Solutions for Arid Regions*, 22–24 March, 2022, Dubai, UAE.

Zeleňáková, M., Soľáková, T., Purcz, P., & Abd-Elhamid F. H. (2022c). Analysis of hydrological drought using Standardized Stream flow Index. In *Public recreation and landscape protection—with the environment hand in hand*, 9–11 May, 2022, Brno, Czech Republic.

Chapter 12
Assessing the Suitability of SPI and SPEI in Steppe Hot and Arid Climatic Zones in India

Siddhant Panigrahi and Vikas Kumar Vidyarthi

Abstract Steppe Hot and desert Climatic Zones present distinct problems and opportunities for managing water resources in the region. This study aims to compare and evaluate the suitability of the Standardized Precipitation Index (SPI) and Standardized Precipitation Evapotranspiration Index (SPEI) for some of the Steppe Hot and Arid Climatic Subdivisions (SHACS) in the Indian Meteorological Subdivisions. The $1^\circ \times 1^\circ$ resolution meteorological data for SHACS obtained from the Indian Meteorological Department (IMD), Pune was employed in this study. The results demonstrate that SPI and SPEI classify the majority of drought events in the moderate category across all evaluation periods. Both of these indices recognized Tamilnadu with maximum extreme drought events. The correlation values are observed to be maximum for Delhi & Haryana subdivision (0.86) and minimum for Gujarat subdivision (0.64) across 3-month timescales. It is also observed that the correlation values increase with increasing assessment periods for Delhi & Haryana and Gujarat subdivisions. However, for Tamilnadu subdivision correlation values continuously decline in the increasing assessment periods with the values of the correlation from 0.79 to 0.66. The results obtained from the Modified Mann–Kendall Test (MMKT) indicate that a significant decreasing trend exists for SPEI across all the assessment periods for all the subdivisions considered in this study as their p-values are less than 0.05. Further, considering SPEI, it is observed that the Tamilnadu Subdivision has high Z-Score values (in the range of − 17.25 to − 11.52) as compared to the Z-score values for other subdivisions (ranging from − 5.16 to − 3.05) which show that out of the three SHACS, Tamilnadu is emerging to be the hotspot region for meteorological drought over the years. The results obtained from the analysis based on identifying actual drought years show that the SPEI identified 13 drought years as compared to SPI which identified 11 out of the 16 actual drought years in the study region.

S. Panigrahi · V. K. Vidyarthi (✉)
Department of Civil Engineering, NIT Raipur, Raipur, Chhattisgarh, India
e-mail: vkvidyarthi.ce@nitrr.ac.in

S. Panigrahi
e-mail: spanigrahi.phd2022.ce@nitrr.ac.in

A. Sefelnasr et al. (eds.), *Water Resources Management and Sustainability*, Water Science and Technology Library 114, https://doi.org/10.1007/978-3-031-80520-2_12

Keywords Meteorological drought · Modified Mann–Kendall Test (MMKT) · Standardized Precipitation Evapotranspiration Index (SPEI) · Standardized Precipitation Index (SPI) · Steppe hot and arid regions

12.1 Introduction

Drought is a complex and pervasive natural phenomenon that poses substantial difficulties to ecosystems, cultures, and economies around the world. The hallmark of this phenomenon is an extended lack of precipitation in comparison to the statistical multi-year averages for a specific area, resulting in a scarcity of water supplies. Drought has far-reaching consequences for agriculture, water supplies, ecosystems, and human health (Panigrahi & Vidyarthi, 2024). Long-lasting droughts are starting to affect many countries, and the issue of water scarcity is growing more widespread (Sherif et al., 2023). For efficient resource management, risk reduction, and sustainable development, it is essential to comprehend and monitor drought. There are many ways that drought can appear, and it is frequently classified according to the particular element of the ecosystem that it impacts. Drought is classified into four categories: meteorological (MD), agricultural (AD), hydrological (HD), and socioeconomic (SD). MD is the most fundamental form, defined by a lengthy period of below-average precipitation (Singh et al., 2023). It provides a fundamental understanding of drought onset and severity by explicitly examining a region's precipitation deficit. MD provides a platform for evaluating the potential effects on water availability, soil moisture, and other hydrological variables. Meteorological drought is extremely important in arid and semi-arid regions because it has a direct impact on water supply, agriculture, ecosystems, and human activities (Devdatta & Pandit, 2018). Thus, analyzing MD in India's arid and semi-arid regions is crucial for solving the multifaceted difficulties posed by water scarcity, ensuring livelihoods, and conserving ecosystems. Vidyarthi and Jain (2020) assessed the meteorological drought in Gangetic West Bangal in India using SPI. Parthsarthi et al. (2023) assessed the proneness of drought in semi-arid regions of Gujarat in India using SPEI and developed severity-duration-frequency curves. Raja and Gopikrishnan (2022) studied dryness using the SPEI at several timescales in India's arid Barmer region, revealing growing patterns at 6, 9, and 12 months. Moghimi and Zarei (2019) evaluated the performance of precipitation-based indices such as Decile Index, SPI, China Z-Index (CZI), Modified (CZI), Z-Score, and Percent of Normal (PN) for modelling Reconnaissance Drought Index (RDI) in arid regions and determined that SPI was the best index. Kantamaneni et al. (2022) developed a framework to assess drought in Tamilnadu in India using SPI, Scaled Drought Condition Index (SDCI) and Standardized Water Level Index (SWI). Surendran et al. (2019) analysed drought in a few locations of India using PN, Rainfall Anomaly Index (RAI), SPI, and RDI, and discovered that RDI is a good indicator of drought severity and prediction. Both SPEI and RDI take precipitation and evapotranspiration into consideration while evaluating drought, although with different approaches. However, SPEI provides a more thorough and standardised approach to evaluate drought conditions, particularly for long-term analysis and climate change research. Pandey (2018) analysed drought characteristics in different climatic regions of India and concluded that arid regions in India are more prone to drought. However, it is unknown yet, as which drought index is most appropriate for India's dry and semi-arid regions based on climatic variables and their trend.

As SHACS are known to be more vulnerable to drought the analysis of such regions is of vital importance. SPI has grown in popularity and use as a result of its simplicity and versatility, whereas SPEI is becoming more recognised and employed for its capacity to provide a more thorough knowledge of droughts, particularly in light of evolving climate variables and water availability dynamics. Thus, in this work, SPI and SPEI were employed to measure drought circumstances across SHACS subdivisions. This study aims to present an accuracy evaluation between SPI and SPEI drought analysis over SHACS.

This study's precise aims include: (a) to find the drought severity in the SHACS using SPI, (b) to find the drought severity in the SHACS using SPEI, (c) to compare the SPI and SPEI based on drought severity in the SHACS. These drought indices were evaluated across three, six, nine, twelve, and twenty-four-month periods to establish the distinctive features of the MD in SHACS.

After a brief introduction, the study begins with an overview of the tools and techniques used in this investigation. Thereafter, the details with regard to study region and the data used are provided. Following that, a brief summary of the findings has been provided before giving concluding remarks.

12.2 Methodology

The drought indices have been selected such that different aspects of MD are taken into account. SPI measures precipitation (PRCN) deficits throughout several assessment periods, whereas SPEI considers temperature (TEMPR), integrates the impacts of surface evaporation variations and is mostly based on excess rainfall. SPI is popular because of its ease of calculation, and ability to offer a standardised measure of precipitation anomalies over many time scales (e.g., monthly, seasonal, annual). In recent years, SPEI has grown in prominence, particularly among researchers and academics interested in climate change implications, water resource management, and ecological studies. The advantage of SPEI over SPI is that it takes into account the dynamics of moisture supply and demand, allowing for a more thorough assessment of drought conditions. This is achieved by accounting for both precipitation and potential evapotranspiration. These metrics provide a quantitative computation of the severity of the drought. The interested readers can refer to "Mckee et al. (1993)" for SPI and "Vicente-Serrano et al. (2010)" for SPEI for a concise overview along with their mathematical formulations. In the present study the SPI has been evaluated across different timescales using the SPI Generator software developed by National Drought Mitigation Center (2018), University of Nebraska-Lincoln. The

Table 12.1 Drought severity classification based on SPI (Mckee et al., 1993)

SPI value	Description
> 2	Extremely wet (EW)
1.5–1.99	Very wet (VW)
1–1.49	Moderately wet (MW)
0.99 to – 0.99	Near normal
– 1 to – 1.49	Moderately dry (MD)
– 1.5 to – 1.99	Severely dry (SD)
< – 2	Extremely dry (ED)

SPEI has been calculated using the Python package made available by Vonk (2024). The drought severity classification based on SPI values given by "Mckee et al. (1993)" is presented in Table 12.1.

12.2.1 SPI

SPI is calculated by defining the time scale for estimation, such as monthly, seasonal, or annual SPI. The probability density function is then computed for the historical PRCN data using a gamma distribution. The recorded PRCN data are then transformed into a conventional normal distribution using the gamma distribution's inverse cumulative distribution function (CDF). Determine SPI values for each period using the converted PRCN data.

12.2.2 SPEI

SPEI is calculated by gathering historical data on monthly or seasonal PRCN and evapotranspiration in the region of interest. Next, calculate the difference between precipitation and evapotranspiration for each time period to get the climate moisture balance. The climate moisture balance data is then fitted with a probability distribution (gamma distribution), and the values are transformed into a conventional normal distribution using the inverse of the cumulative distribution function (CDF). Compute the SPEI values for each time scale using the converted climate moisture balance data.

12.2.3 Modified Mann–Kendall Test (MMKT)

The trend analysis has been done by the popular Modified Mann–Kendall Test (MMKT). The modified Mann–Kendall test is a statistical approach for identifying trends or patterns in time series data. It is an extension of the conventional Mann–Kendall test and is commonly employed when dealing with data having serial correlation or autocorrelation.

Typically, the Mann–Kendall test's variance and covariance estimators are adjusted to account for data autocorrelation. This improves trend detection accuracy by taking into consideration the reliance between subsequent data in a time series. To carry out MMKT "modifiedmk" package in R-Studio has been used. The MMKT statistics (S) is estimated as: -

$$sign(x_j - x_i) = \begin{cases} 1; & if \quad x_j > x_i \\ 0; & if \quad x_j = x_i \\ -1; & if \quad x_j < x_i \end{cases} \tag{12.1}$$

$$S = \sum_{i=1}^{N-1} \sum_{j=i+1}^{N} sign(x_j - x_i) \tag{12.2}$$

Interested readers can refer to Hamed and Rao (1998) and Dubey et al. (2023) for detailed descriptions along with their mathematical formulation.

12.3 Study Area and Data Preparation

India is a big and diversified country in South Asia, recognised for its rich cultural heritage, diverse landscapes, and climate (Fig. 12.1). Factors influencing India's climate include latitude, altitude, topography, and monsoon winds. The Indian Meteorological Subdivisions are the regions of India that have been divided based on meteorological and climatic factors. The India Meteorological Department (IMD) uses these subdivisions to monitor, analyse, and report the weather. The country is separated into meteorological subdivisions, each reflecting a distinct geographic area. Based on their climatic conditions these subdivisions are categorized into different climatic zones such as Montane Climate; Hot Desert & Arid; Steppe Hot & Arid; Humid Subtropical; Tropical Savannah, Wet & Dry, and Tropical Monsoon Area. The Steppe Hot and Arid Zone in India is predominantly connected with locations with steppe climate features, such as minimal rainfall and high temperatures. Steppe Hot and Arid Climate Subdivisions (SHACS) in India are hard but diverse regions with distinct biological and socioeconomic characteristics. Effective management techniques, sustainable practices, and adaption measures are critical for dealing with the issues posed by arid and semi-arid climates in these areas. Water scarcity is a major issue in this region and poses a barrier to agriculture in this zone. Farming operations are frequently tailored to dry environments, and crops that can withstand drought are grown. SHACS are subject to climate change consequences such as rising temperatures, shifting precipitation patterns, and water scarcity. These changes could have a profound impact on agriculture, ecology, and human livelihoods in these regions. Drought-resistant species such as prickly shrubs, cacti, and arid-adapted grasses make up the native vegetation in SHACS. This climatic zone is most prevalent in north-western India, encompassing Gujarat, Punjab, Delhi, and Haryana, as well as Tamil Nadu and Rayalseema. High temperatures are common in the SHACS, notably in the summer. Temperatures during the day may reach above 40 °C. The lack of cloud cover and dry air contribute to intense heat. Encouraging conservation of water, management of the soil, and cultivation of drought-tolerant crops are crucial in tackling the distinct circumstances of this region of India.

The PRCN and TEMPR data have been obtained from IMD at a resolution of 1° latitude × 1° longitude. The historic PRCN data has been obtained for the period 1901–2017 and the maximum and minimum TEMPR data has been obtained for the period 1951–2017. Out of all SHACS regions in India, three namely, Delhi & Haryana, Gujarat, and Tamilnadu subdivisions are taken into consideration in this study. The precipitation and temperature variations across these subdivisions are shown in Figs. 12.2 and 12.3, respectively.

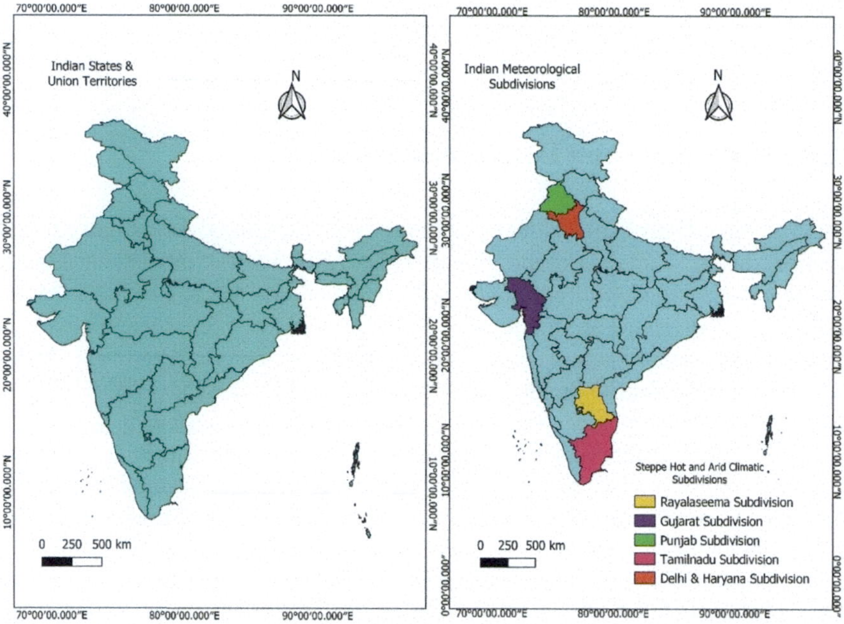

Fig. 12.1 Study area including Steppe Hot and arid climatic subdivisions

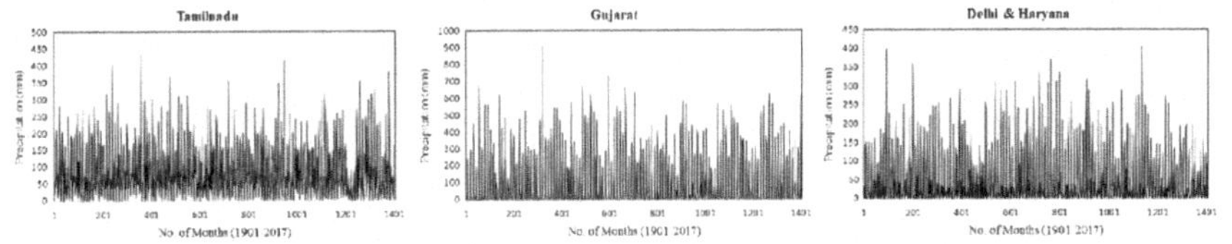

Fig. 12.2 Time series plot of PRCN over study regions

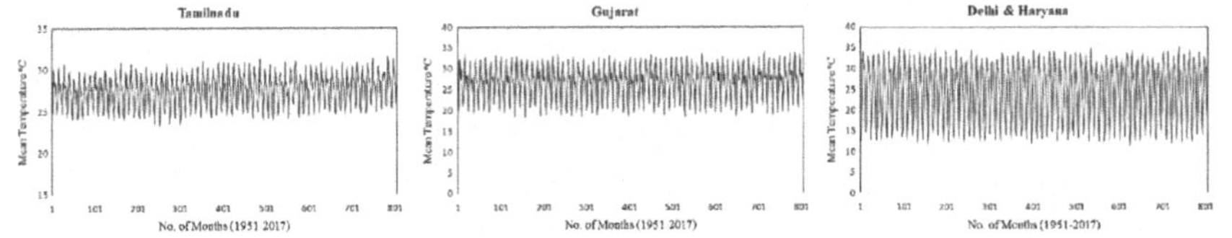

Fig. 12.3 Time series plot of TEMPR over study regions

12.4 Result and Discussions

In this study, for drought analysis of SHACS, mainly Delhi and Haryana subdivisions from Northern India, Gujarat subdivision from Western India, and Tamilnadu subdivision from Southern India have been considered. The drought events seized by SPI and SPEI across varying assessment periods (3, 6, 9, 12, and 24 months) with different severity for these subdivisions have been depicted in Table 12.2.

The Delhi and Haryana subdivisions demonstrate a consistent drop in ED occurrences across all evaluation years, with the highest number of drought events falling within the MD severity class. Except SPI-24, the Gujarat subdivision has identified a consistent rise in ED events across the assessment period. While the SD event increases from SPI-3 to SPI-6 and then gradually decreases over the following assessment periods until SPI-24. The Tamil Nadu subdivision reports that ED and MD incidents have been steadily falling during all assessment periods. The SD event increases from SPI-3 to SPI-6, then decreases for the following evaluation periods. Overall, SPI determines the most extreme drought occurrences for the Tamil Nadu subdivision.

The Delhi and Haryana subdivisions identify a similar number of ED occurrences across SPEI-6 and SPEI-12, with the highest number observed in SPEI-3. Maximum MD events are examined throughout SPEI-6, and SD events across SPEI-24. The number of SD occurrences in the Gujarat subdivision decreases from SPEI-3 to SPEI-9, then increases to SPEI-24. However, no clear pattern of rise and reduction in MD and ED occurrences is identified across different evaluation periods. Tamil Nadu subdivision examines the drop in ED and MD incidents from SPEI-6 to SPEI-24. SD occurrences are found to be highest across SPEI-3. Overall, SPEI identifies the most extreme drought occurrences for the Tamil Nadu subdivision.

Table 12.2 Total number of moderate, severe and extreme drought events seized by SPI and SPEI in study regions on the basis of Mckee et al. (1993) classification table

Metrological subdivision	Drought index	MD	SD	ED	Near normal	Drought index	MD	SD	ED	Near normal
Delhi and Haryana	SPI-3	136	55	41	942	SPEI-3	91	29	16	524
	SPI-6	115	69	38	937	SPEI-6	92	35	12	512
	SPI-9	131	64	33	932	SPEI-9	84	34	15	505
	SPI-12	133	73	28	908	SPEI-12	58	45	12	501
	SPI-24	137	91	22	914	SPEI-24	73	53	5	505
Gujarat	SPI-3	162	42	23	962	SPEI-3	92	43	9	514
	SPI-6	125	86	29	935	SPEI-6	86	28	15	515
	SPI-9	118	80	38	937	SPEI-9	95	27	14	512
	SPI-12	123	65	45	958	SPEI-12	93	30	13	518
	SPI-24	131	57	31	943	SPEI-24	74	50	14	500
Tamil Nadu	SPI-3	111	38	49	988	SPEI-3	72	40	19	544
	SPI-6	92	41	47	1030	SPEI-6	75	29	26	541
	SPI-9	91	32	47	1032	SPEI-9	70	30	25	541
	SPI-12	76	27	41	1087	SPEI-12	59	37	25	537
	SPI-24	76	20	30	1114	SPEI-24	48	36	23	526

The Correlation values among SPI and SPEI is observed to be increasing across the assessment periods Delhi and Haryana and Gujarat subdivision. The values are observed to be maximum for Delhi & Haryana subdivision (0.86) and minimum for Gujarat subdivision (0.64) across 3-month timescales. Across (6, 9, 12 and 24-month) timescales the correlation however increases for both these subdivisions. However, for Tamil Nadu subdivision the correlation values between SPI and SPEI is declining across the assessment periods. This may be due to variations in moisture supply and demand dynamics, especially in regions with complex climatic patterns. The correlation among the drought indices over different assessment periods have been shown in Fig. 12.4.

Further, the MMKT results for SPI and SPEI across all the assessment periods have been depicted in Table 12.3.

Statistical measurements such as Z-score, Sen's Slope, Kendall's Tau, and p-values are available for each time scale and index (SPI & SPEI). These measurements quantify the trend path and statistical relevance of the drought indices. It can be observed that a trend exists for SPEI across all the assessment periods for Delhi & Haryana, Gujarat, and Tamilnadu subdivisions as their p-value is less than 0.05. The null hypothesis was rejected, and the tau value was found to be negative, implying a negative trend for SPEI. However, for SPI negative trend is observed, across SPI-3 and SPI-6 for Gujarat subdivision and across SPI-3 for the Tamilnadu subdivision. Z-scores describe the magnitude of the trend, with larger absolute Z-scores indicating stronger trends. The magnitude of the trend seems to be significant for SPEI as

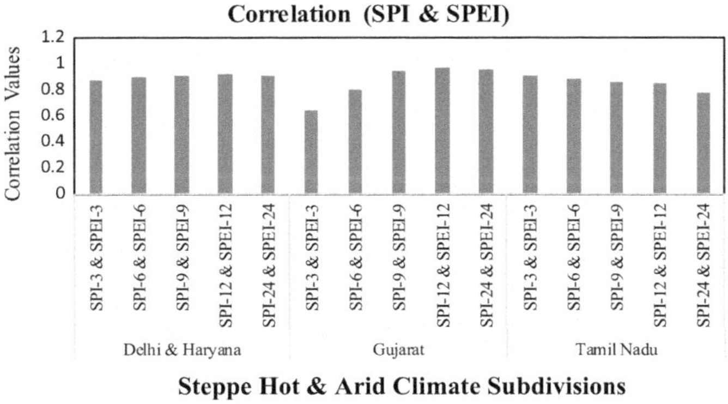

Fig. 12.4 Correlation plot of SPI and SPEI across all the assessment periods in study regions

compared to SPI for the study area indicating stronger trends. Further, considering SPEI, it is observed that the Tamilnadu Subdivision has high Z-Score values (in the range of − 17.25 to − 11.52) as compared to the Z-score values for other subdivisions (ranging from − 5.16 to − 3.05) which show that out of the three SHACS, Tamilnadu is emerging to be the hotspot region for meteorological drought over the years. Also, the Sen's slope measures how much the MMKT test slope is increasing or declining. For SPEI, it is observed to be a negative value (declining). The time series plots of SPI and SPEI across all the assessment periods have been depicted in Fig. 12.5.

It is observed that SPI and SPEI perform particularly well in the study regions in terms of seizing historical drought events. The actual drought years' data is obtained from "IMD (2015) Report" and "UNICEF India Report on Drought (2000)", and from the newspaper reports. Based on the analysis of these Drought indices and Actual Drought years post 1980's SPEI outperforms SPI. The Drought years seized by these indices and actual drought years in the study region have been depicted in Table 12.4.

Based on Table 12.4, the SPEI comes out to be the best drought index for the taken regions within SHACS in India as SPI correctly identifies 11 actual drought years whereas SPEI identifies 13 actual drought years in the study region.

Table 12.3 MMKT analysis for SPI and SPEI across all the assessment periods for study regions

Metrological subdivision	Drought index	p-value	Z	Sen's slope	Tau	Drought Index	p-value	Z	Sen's slope	Tau
Delhi and Haryana	SPI-3	0.55	− 0.6s0	-2.69×10^{-5}	− 0.007	SPEI-3	6.48×10^{-5}	− 3.99	-7.88×10^{-4}	− 0.115
	SPI-6	0.35	− 0.93	-6.35×10^{-4}	− 0.017	SPEI-6	9.7×10^{-5}	− 3.90	-9.75×10^{-4}	− 0.141
	SPI-9	0.61	− 0.52	-5.13×10^{-5}	− 0.013	SPEI-9	2.29×10^{-5}	− 4.23	-1.19×10^{-3}	− 0.176
	SPI-12	0.46	− 0.74	-9.5×10^{-5}	− 0.024	SPEI-12	5.46×10^{-6}	− 4.55	-1.41×10^{-3}	− 0.211
	SPI-24	0.41	− 0.83	-1.43×10^{-4}	− 0.037	SPEI-24	2.46×10^{-7}	− 5.16	-2.08×10^{-3}	− 0.310
Gujarat	SPI-3	1.05×10^{-11}	− 6.80	-2.49×10^{-4}	− 0.067	SPEI-3	0.002	− 3.08	-5.06×10^{-4}	− 0.073
	SPI-6	0.03	− 2.22	-1.07×10^{-4}	− 0.028	SPEI-6	0.002	− 3.05	-4.57×10^{-4}	− 0.067
	SPI-9	0.73	0.34	1.59×10^{-5}	0.005	SPEI-9	0.002	− 3.15	-4.6×10^{-4}	− 0.065
	SPI-12	0.80	0.25	9.62×10^{-6}	0.004	SPEI-12	6.5×10^{-4}	− 3.41	-5.61×10^{-4}	− 0.078
	SPI-24	0.63	0.49	3.33×10^{-5}	0.010	SPEI-24	8.22×10^{-5}	− 3.94	-8.27×10^{-4}	− 0.118
Tamil Nadu	SPI-3	0.03	− 2.22	-1.63×10^{-4}	− 0.045	SPEI-3	1.11×10^{-66}	− 17.25	-1.33×10^{-3}	− 0.201
	SPI-6	0.17	− 1.39	-1.32×10^{-4}	− 0.038	SPEI-6	4.50×10^{-45}	− 14.09	-1.5×10^{-3}	− 0.226

(continued)

Table 12.3 (continued)

Metrological subdivision	Drought index	p-value	Z	Sen's slope	Tau	Drought Index	p-value	Z	Sen's slope	Tau
	SPI-9	0.44	− 0.77	-8.46×10^{-5}	− 0.026	SPEI-9	4.62×10^{-35}	− 12.35	-1.58×10^{-3}	− 0.238
	SPI-12	0.61	− 0.51	-6.29×10^{-5}	− 0.020	SPEI-12	6.88×10^{-31}	− 11.56	-1.71×10^{-3}	− 0.253
	SPI-24	0.74	− 0.33	-4.56×10^{-5}	− 0.016	SPEI-24	1.09×10^{-30}	− 11.52	-2.26×10^{-3}	− 0.331

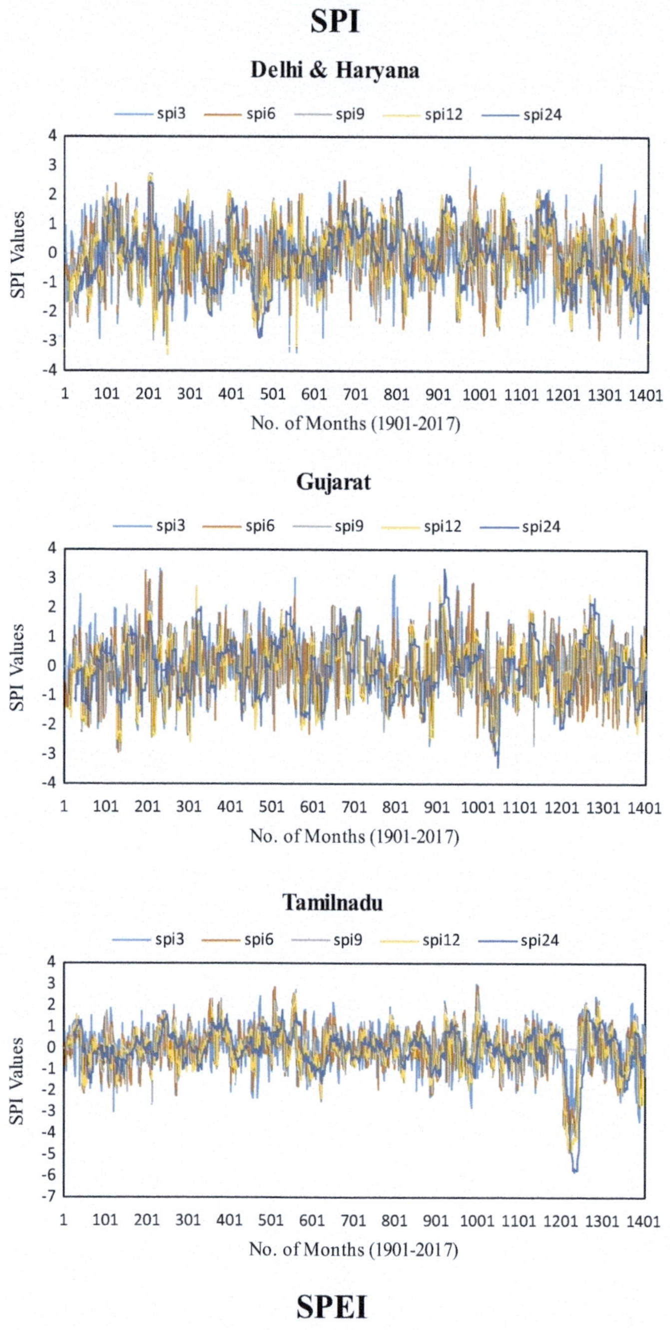

Fig. 12.5 SPI and SPEI analysis across various assessment periods in study regions

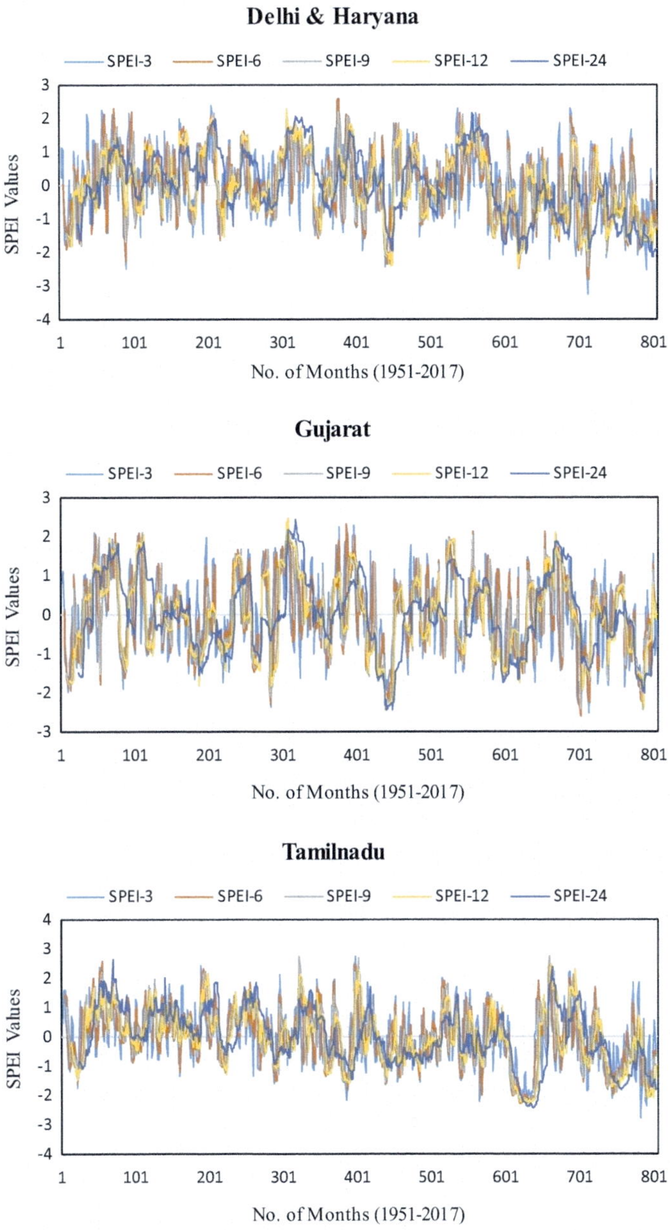

Fig. 12.5 (continued)

Table 12.4 Comparison of drought years seized by SPI and SPEI with actual drought years in the study region

Drought severity	SPI	SPEI	Actual drought years
SD and ED years	1918, 1936, 1941, 1965, 1972, 1979, 1981, 1985, 1987, 1990, 1993, 1995, 1999, 2002, 2010, 2012, 2016, 2018	1965, 1972, 1982, 1986, 1989, 1990, 1993, 1999, 2000, 2002, 2004, 2010, 2012, 2014, 2016, 2019	1918, 1936, 1941, 1965, 1972, 1982, 1986, 1990, 1999, 2000, 2002, 2004, 2010, 2012, 2016, 2019

12.5 Conclusion

The primary goal of this study is to ascertain the severity of historic drought events in few parts of India's Steppe Hot and Arid Climatic Subdivisions (SHACS) using SPI and SPEI drought indices, and to present an accuracy evaluation between SPI and SPEI drought analysis over SHACS for varying assessment periods of 3, 6, 9, 12, and 24 months. The results of these Drought Indices were validated with actual drought years' data taken from the "IMD (2015) Report" and the "UNICEF India Report on Drought (2000)" as well as newspaper sources. The analysis has led to the following observations:

1. SPI and SPEI indices identify most of the drought events under moderate class.
2. The Correlation values among SPI and SPEI are observed to be greater than 0.6 for the study regions that lie in SHACS indicating strong positive association between their values.
3. MMKT indicates a significant negative trend for SPEI across all the assessment periods for SHACS regions taken in the study.
4. The SPEI is capturing more number of actual drought events as compared to SPI in the SHACS regions taken in the study.

The originality of this work is that it compares the accuracy of SPI and SPEI drought evaluation over SHACS. Based on the validation of the results of SPI and SPEI with the actual drought years' data, it is concluded that SPEI is an apt index in comparison to SPI for analyzing meteorological drought (MD) conditions in the SHACS regions in India. MD analysis delivers scientific data and insights into water policy, drought management programs, and long-term water resource planning in SHACS regions.

In order to perform MD analysis over SHACS combining various meteorological factors, the current work uses, SPI and SPEI. The effectiveness of Multivariate Standardized Drought Index incorporating the joint dependence between SPI and SPEI for the study region can be explored. Similar studies can be conducted for various climatic zones to assess different types of droughts using appropriate drought indicators.

Acknowledgements The data utilised in this study were collected from IMD (https://www.imdpune.gov.in/) and are fully acknowledged. The SPI was calculated using SPI Generator [software] University of Nebraska-Lincoln (https://drought.unl.edu/Monitoring/SPI/SPIProgram.aspx), and the SPEI was estimated using a simple Python programme supplied by Vonk (2024) https://doi.org/https://doi.org/10.5281/zenodo.10816741, which is appropriately acknowledged. The "IMD (2015) Report" and the "UNICEF India Report on Drought (2000)" downloaded freely online for validation purposes are also acknowledged.

References

Dubey, V., Panigrahi, S., & Vidyarthi, V. K. (2023). Statistical trend analysis of major climatic factors over Chhattisgarh State India. *Earth Systems and Environment, 7*(3), 629–648. https://doi.org/10.1007/s41748-023-00345-1

Hamed, K. H., & Rao, A. R. (1998). A modified Mann-Kendall trend test for autocorrelated data. *Journal of Hydrology, 204*(1–4), 182–196. https://doi.org/10.1016/S0022-1694(97)00125-X

Indian Meteorological Department, Annual Report. (2015). https://metnet.imd.gov.in/docs/imdnews/ar2015.pdf.

Kantamaneni, K., Periasamy, T., & Roy, P., et al. (2022). Monitoring of multi-aspect drought severity and socio-economic status in the semi-arid regions of eastern Tamil Nadu, India. Water.https://doi.org/10.3390/w14132049

McKee, T. B., Doesken, N. J., & Kleist, J. (1993, January). The relationship of drought frequency and duration to time scales. In *Proceedings of the 8th Conference on Applied Climatology* (Vol. 17, no. 22, pp. 179–183).

Moghimi, M. M., & Zarei, A. R. (2019). Evaluating performance and applicability of several drought indices in arid regions. *Asia-Pacific J Atmospheric Sciences, 57*, 645–661.

National Drought Mitigation Center. (2018). *SPI Generator [software]*. University of Nebraska-Lincoln. https://drought.unl.edu/Monitoring/SPI/SPIProgram.aspx.

Pandit, D. V. (2018). Assessment of meteorological drought for semi-arid region of Maharashtra. *International Journal of Agriculture, Environment and Biotechnology*. https://doi.org/10.30954/0974-1712.08.2018.11

Pandey, R. P. (2018). Drought management in different climatic region in India. *International Journal of Engineering Research and Technology, 4*. https://api.semanticscholar.org/CorpusID:189418135.

Panigrahi, S., & Vidyarthi, V. K. (2024). Identification of the most suitable meteorological drought index for a region: A case study of Raigarh district in Chhattisgarh. *Natural Hazards*, 1–21.

Parthsarthi, A., Pandya, P., & Gontia, N. K., (2023). Development of drought severity–duration–frequency curves for identifying drought proneness in semi-arid regions. *Journal of Water and Climate Change*. https://doi.org/10.2166/wcc.2023.438

Raja, A., & Gopikrishnan, T. (2022). Drought analysis using the standardized precipitation evapotranspiration index (SPEI) at different time scales in an arid region. *Engineering, Technology & Applied Science Research, 12*(4), 9034–9037.

Sherif, M., Singh, V. P., Sefelnasr, A., & Abrar, M. (Eds.). (2023). *Water resources management and sustainability: Solutions for arid regions* (Vol. 121). Springer Nature.

Singh, V. P., Jhajharia, D., Mirabbasi, R., & Kumar, R. (Eds.). (2023). *Integrated drought management, volume 1: Assessment and spatial analyses in changing climate*. CRC Press.

Surendran, U., Anagha, B., Raja, P., Kumar, V., Rajan, K., & Jayakumar, M. (2019). Analysis of drought from humid, semi-arid and arid regions of India using DrinC model with different drought indices. *Water Resources Management*. https://doi.org/10.1007/S11269-019-2188-5

UNICEF India Report on Drought and Floods. (2000, September).

Vicente-Serrano, S. M., Beguería, S., et al. (2010). A multiscalar drought index sensitive to global warming: The standardized precipitation evapotranspiration index. *Journal of Climate, 23*(7), 1696–1718.

Vidyarthi, V. K., & Jain, A. (2020). Knowledge extraction from trained ANN drought classification model, *Journal of Hydrology, 585*, HYDROL_124804. https://doi.org/10.1016/j.jhydrol.2020.124804.

Vonk, M. A. (2024). *SPEI: A simple Python package to calculate and visualize drought indices (vX.X.X)*. Zenodo. https://doi.org/10.5281/zenodo.10816741.

Part III
Groundwater

Chapter 13
Managed Aquifer Recharge (MAR) on Rye Island, Slovakia

Michaela Červeňanská, Jakub Mydla, Andrej Šoltész, and Andrea Vranovská

Abstract Fast water transfer to rivers during floods causes a water deficit during dry periods. For a long-term retention of water in aquifers and its subsequent use in drier or heavier demand periods, the Managed Aquifer Recharge (MAR) techniques were studied and implemented in four pilot areas of the DEEPWATER-CE project funded by the Interreg CENTRAL EUROPE programme. In Slovakia, the Recharge Dam MAR type was investigated using the groundwater numerical modelling technique in order to retain the surface water in an aquifer and to recharge the groundwater for its subsequent use for irrigation during dry periods. The calibrated and verified groundwater flow model was developed using MODFLOW and was used for the prediction of two scenarios (prognosis) improving the water level regime in the channel system of Rye Island resulting in the improvement of the groundwater regime as well. By appropriate manipulation of the gate slides during flood situation, the amount of water recharged to the groundwater can be increased by 11,000 $m^3.d^{-1}$ without further technical measures. After the construction of three additional weirs, the amount of water recharged to the groundwater can be even higher.

Keywords DEEPWATER · Gabčíkovo-Topoľníky channel · Groundwater · MAR · Numerical modelling · Rye Island

M. Červeňanská (✉) · J. Mydla · A. Šoltész
Department of Hydraulic Engineering, Faculty of Civil Engineering, Slovak University of Technology in Bratislava, Radlinského 10, 810 05 Bratislava, Slovak Republic
e-mail: michaela.cervenanska@stuba.sk

J. Mydla
e-mail: jakub.mydla@stuba.sk

A. Šoltész
e-mail: andrej.soltesz@stuba.sk

A. Vranovská
Water Research Institute, Nábr. arm. gen. L. Svobodu 5, 812 49 Bratislava, Slovak Republic
e-mail: andrea.vranovska@vuvh.sk

A. Sefelnasr et al. (eds.), *Water Resources Management and Sustainability*, Water Science and Technology Library 114, https://doi.org/10.1007/978-3-031-80520-2_13

13.1 Introduction

The natural frequent phenomenon of interlacement of dry periods and heavy rainfalls causes droughts and floods, both of which harm the economy and have a general adverse effect on the quality of living. The current management of floods, which concentrates on the fast water transfer to rivers and, finally, to the sea, causes a water deficit during dry periods (Stachniak, 2020). Therefore, the project focused on the potential storage of surplus water from heavy rainfalls and from the snow melt and for retention of river water to store it for dry periods, was funded by the Interreg CENTRAL EUROPE program. Its name was DEEPWATER-CE: Development of an integrated implementation framework for managed aquifer recharge solutions to facilitate the protection of Central European water resources endangered by climate change and user conflict (Interreg Central Europe, 2021). The idea of the funded project consists of balancing water resources and flows using man-made measures, which replenish natural groundwater resources from short-lasting surface water surpluses, serve for long-term retention and use them in drier or heavier demand periods (Stachniak, 2020). The project aims to build a joint resource management strategy for Managed Aquifer Recharge (MAR) solutions (Fig. 13.1) to facilitate the protection of Central European water resources endangered by the climate change and user conflict. The focus was on the retention of excess water from rainfall and surface water that can be used to recharge the groundwater. The recharged water can then be harvested at times of need and used for drinking or irrigation purposes.

Fig. 13.1 MAR types verified in the framework of the ongoing research of the DEEPWATER-CE project

MAR is defined as the intentional infiltration of water into aquifers with the purpose of either later recovering that water for different purposes like agricultural, industrial, urban, or obtaining an environmental benefit (Dillon et al., 2009). It has replaced the term "artificial recharge" which maybe evoked something what seemed to be unnatural in the environment (Dillon, 2005).

The DEEPWTAER project included four pilot sites—Island of Vis, Croatia; Maros alluvial fan, Hungary; Podunajska lowland, Slovakia and Tarnów Waterworks, Poland—to investigate the potential of MAR in Central Europe.

13.2 Methods

Rye Island (Slovakian pilot area) is located in the Podunajska lowland, in the south-western part of the country (Fig. 13.2). Rye Island is an important agricultural area and an important source of quality drinking water in Slovakia. It consists of a vast network of drainage and irrigation channels. The pilot area is bounded by the left-hand side seepage channel of the supply channel of Gabčíkovo water structure (Fig. 13.2). There are four main drainage channels located in the investigated area—Gabčíkovo-Topoľníky channel (also known as S VII), Vojka-Kračany (A VII), Šuľany-Jurová (B VII) and Baka-Gabčíkovo (C VII).

The hydrological conditions of the area of interest are strongly affected by the construction of the Gabčíkovo water structure. After construction of the Gabčíkovo water structure, the decrease of the groundwater level in the adjacent area of the River Danube was observed. The groundwater regime, as well as the water level regime in drainage channels, changed. To improve and to control the groundwater regime

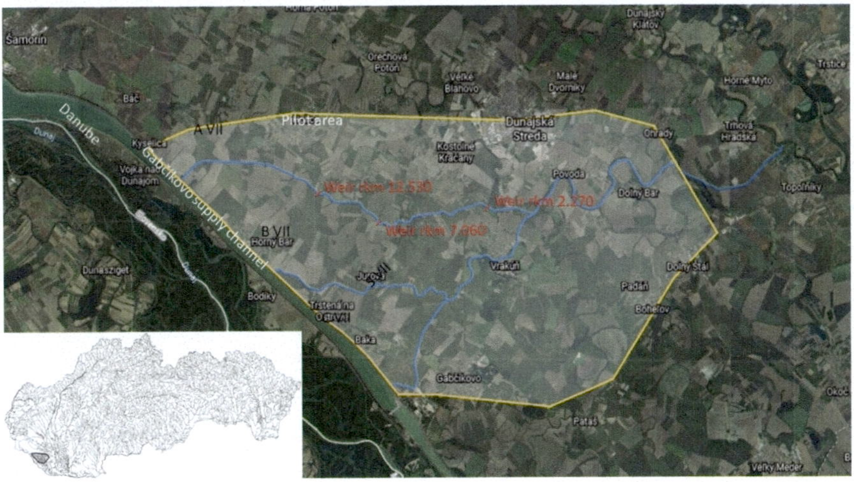

Fig. 13.2 The pilot area, Slovakia

in the floodplain area, a water supply structure situated on the left-hand side of the seepage channel was constructed. This structure allows an artificial water supply into the branch system and consequently into the groundwater (Červeňanská et al., 2016; Sikora & Slota, 1992; Šoltész, 2002).

The local geology determines the hydrogeological conditions. The entire Rye Island, and thus also the investigated part, lies in the Gabčíkovo depression. Fluvial Quaternary sediments form three complexes—lower, middle and upper. The lower complex is characterized by the cyclic alternation of layers of sandy-gravel-like sediments of smaller thickness with clay and loam layers of greater thickness. The thickness of the lower complex ranges from 10 to 350 m. The middle complex is formed by Danube gravels. It is separated from the lower complex by loam-clay layers. The thickness of this formation spatially varies, in the vicinity of Komárno it is about 8–12 m, in the middle of the depression about 160 m. The upper complex is formed by loam and sandy-loam flood sediments with a thickness from 0.5 to 3.0 m (Benková et al., 2013). The variation of the thickness of the Quaternary sediments led us to studying the final hydrogeological survey reports and reviews of the Geofond digital archive of the State Geological Institute of Dionýz Štúr (SGIDŠ, 2021). The bottom elevation was adopted from these reports and reviews.

Irrigation and drainage channels provide a background for technical possibilities for the water flow control, i.e. creating the Recharge Dam MAR type (Fig. 13.1) (Stachniak, 2020). For the retention of the surface water in an aquifer and for the recharge of the groundwater and its subsequent use for irrigation during dry periods, the Recharge Dam MAR type was investigated using the groundwater numerical modelling technique implemented by the Department of Hydraulic Engineering of the Slovak University of Technology in Bratislava in cooperation with the Water Research Institute, the main DEEPWATER-CE project partner from Slovakia.

Table 13.1 describes data needed for the numerical modelling of the groundwater flow.

Three hydrological years have been selected for the data analysis:

- 2008 as the year characterized as the precipitation normal year for the regions of western Slovakia (100–109% of the long-term normal);
- 2010 as the year extremely above normal in terms of precipitation (159% of the long-term normal);
- and 2018 with extremely low precipitation totals recorded in April, May, July and October.

A brief overview of minimum, maximum and median values of measured water levels on the S VII channel is presented in Table 13.2. For the purpose of a model design, the hydrological years 2010 and 2018 were selected.

Table 13.1 Data availability and sources of data

Data type	Description	Form	Source
Geomorphological	Digital relief model (DRM) = top elevation of the aquifer	Digital (raster)	Geodesy, Cartography and Cadaster Authority of the Slovak Republic (GCCA SR)
Geological and hydrogeological	Bottom elevation of the aquifer; hydraulic conductivity	Paper (final hydrogeological survey reports)	State Geological Institute of Dionýz Štúr (SGIDŠ)
Hydrological	Water levels in channels; daily precipitation; groundwater tables in boreholes	Digital	Slovak Hydrometeorological Institute (SHMI)
	Longitudinal profiles and cross sections of drainage channels	Paper	Slovak Water Management Enterprise (SWME)

Table 13.2 Extreme and median values of water levels in two stage discharges on the S VII drainage channel [m a.s.l.]

Stage discharge	2008	2010	2018
9914_Gabčíkovo	Min: 112.35 Max: 113.96 Median: 112.95	Min: 112.41 Max: 113.85 Median: 112.90	Min: 112.66 Max: 113.49 Median: 113.11
9924_Topoľníky	Min: 109.24 Max: 110.09 Median: 109.40	Min: 109.38 Max: 112.59 Median: 109.66	Min: 109.16 Max: 109.65 Median: 109.40

13.2.1 Data Sources

The data sources for this study include aeronautical laser scanning (ALS) data from the GCCA SR and DRM 5.0, which are available online through the Geoportal of Slovakia (accessed on February 14, 2021). Additional data were provided in cooperation with the Slovak Hydrometeorological Institute (SHMI) 2021 and the Slovak Water Management Enterprise (SWME) in 2020. Furthermore, hydrogeological survey reports and reviews were obtained from the Geofond digital archive, available through the Slovak Geological Survey's online services (accessed on January 12, 2021).

13.2.2 Numerical Modelling of the Groundwater Flow

A numerical model was developed using the MODFLOW program in the Groundwater Modeling System (GMS) environment. MODFLOW is a computer program that numerically solves the three-dimensional groundwater flow equation for a porous medium by using a finite-difference method (Harbaugh et al., 2000). The three-dimensional movement of groundwater of constant density through porous earth material may be described by the partial-differential Eq. (13.1) (McDonald & Harbaugh, 1988):

$$\frac{\partial}{\partial x}\left(K_{xx}\frac{\partial h}{\partial x}\right)+\frac{\partial}{\partial y}\left(K_{yy}\frac{\partial h}{\partial y}\right)+\frac{\partial}{\partial z}\left(K_{zz}\frac{\partial h}{\partial z}\right)-W=S_S\frac{\partial h}{\partial t} \tag{13.1}$$

where K_{xx}, K_{yy} and K_{zz} are values of hydraulic conductivity along the x, y and z coordinate axes, which are assumed to be parallel to the major axes of hydraulic conductivity (Lt^{-1}); h is the potentiometric head (L); W is a volumetric flux per unit volume and represents sources and/or sinks of water (t^{-1}); S_S is the specific storage of the porous material (L^{-1}); and t is time (t).

A modelling protocol, which was used for the creation of the numerical model, is described in Fig. 13.3. During the conceptual model stage, the system boundaries were identified, field data describing geomorphological (ALS: GCCA SR, 2021), geological (SGIDŠ, 2021), hydrogeological (SGIDŠ, 2021), hydrological (SHMI, 2021) and meteorological (SHMI, 2021) conditions were assembled, and field measurements were conducted. After that, the conceptual model was mapped to the designed grid, the boundary conditions were set and preliminary selection of values for aquifer parameters and hydrologic stresses was performed (the model design).

For defining the boundary condition as well as for the calibration of the model, data from 32 groundwater level monitoring objects (Fig. 13.4) of the State Hydrological Network of groundwater quantity monitoring of the SHMI (or observation wells) were evaluated. Data from 19 of these objects were used as the observation points, where the elevation of the groundwater level (the head) was set using the calibration targets. The Dirichlet (or the first-type) boundary condition was specified on the entire edge of the model boundary. The value of the boundary condition was computed as the mean value of groundwater levels measured in every individual observation well during the modelled period (SHMI, 2021). The drainage channels represent the Cauchy boundary condition, which was specified using the elevation of the bottom of the river (channel) bed (entered using the longitudinal profiles of drainage channels and the field-measured values) and the head in the river (the measured values in two stage discharges on the S VII drainage channels (SHMI, 2021)). Also, the value of the so-called conductance had to be specified. This value was one of the calibration parameters.

The calibration of the model was performed on the steady-state model using the period from May 26, 2010 to September 9, 2010 (Fig. 13.5) during the 2010 flood

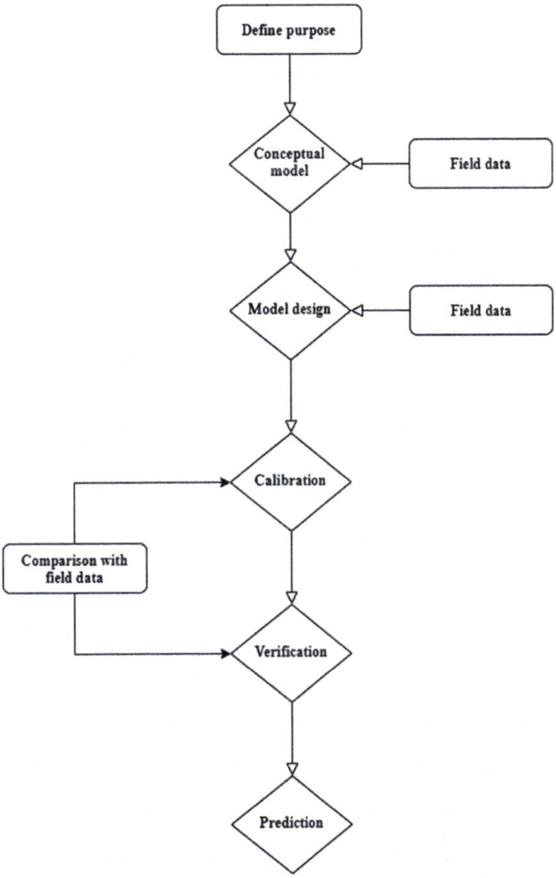

Fig. 13.3 Modeling protocol. Adapted from Anderson and Woessner (1992)

situation on the rivers Danube, Váh and the Little Danube. Calibration of the model was done using:

- The automated parameter estimation code PEST (Doherty, 2004) for the calibration of the hydraulic conductivity. The estimated range (from minimum to maximum value) was set from 90.72 to 683.64 m.d^{-1}. The first estimated value was 253.58 m.d^{-1}, which corresponds to the average value of the hydraulic conductance determined by the pumping tests located in the area of interest. The final value determined by PEST was 228.36 m.d^{-1}.;
- And by the trial-and-error adjustment of the conductance of the riverbed. A conductance is defined as the hydraulic conductivity of the riverbed materials divided by the vertical thickness (length of travel based on the vertical flow) of the riverbed materials, multiplied by the area (width times the length) of the river in the cell. The last term, area, is the hardest parameter to determine by hand since it varies from cell to cell. GMS can automatically calculate the lengths. Therefore,

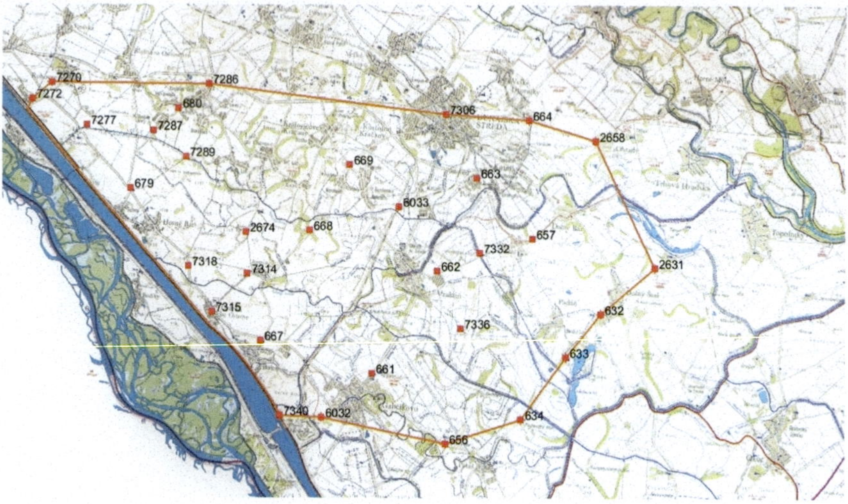

Fig. 13.4 Groundwater level monitoring objects of the State Hydrological Network of the groundwater quantity monitoring of the SHMI located in the modelled area

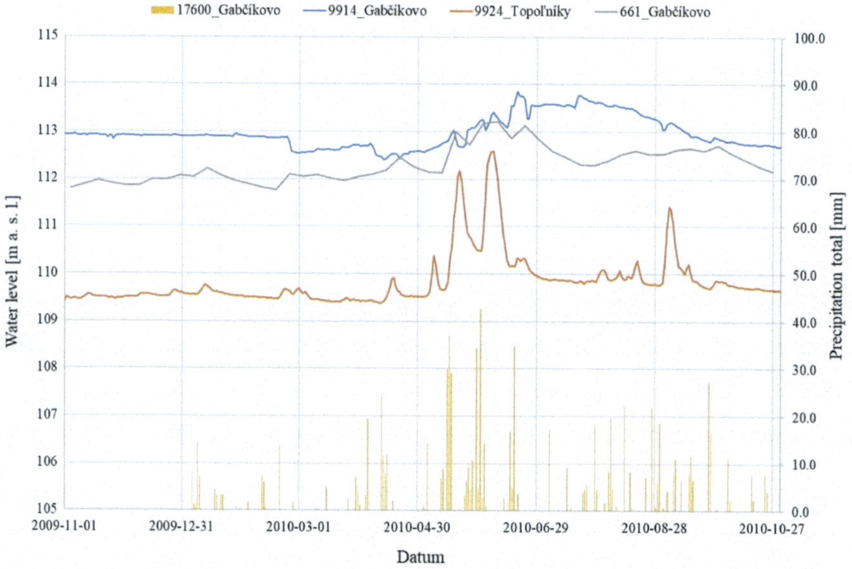

Fig. 13.5 The course of the water level at stage discharges on the S VII channel (9914 Gabčíkovo and 9924 Topoľníky) and the course of the groundwater level in well no. 661 in Gabčíkovo; daily total precipitation in the rain gauging station 17,600 Gabčíkovo—hydrological year 2010 (SHMI, 2021)

when a conductance is entered, it should be in GMS entered in terms of conductance per unit length. When GMS applies the boundary condition from the arc to the grid cell, it automatically multiplies the entered value of conductance by the length of the arc that intersects the cell to create an accurate conductance value for the cell.

The values of the conductance were calibrated as follows: for the S VII drainage channel $C = 2.0\ m^2.d^{-1}/m$; for the AVII and B VII drainage channels $C = 3.0\ m^2.d^{-1}/m$.

13.3 Results and Discussion

Figure 13.6 captures the contours of the calculated groundwater head for the steady-state model of the period from May 26, 2010 to September 9, 2010. Calibration targets display a calibration error for each object. The center of the target represents the observed value, the top of the target is equal to the observed value + the interval (0.10 m is our case) and the bottom of the target is the difference of the observed value and the interval. If the color of the calibration target is green, the bar lies entirely within the target. A yellow color means the bar is outside the target, but the error is less than 200%. A red color of the calibration target indicates that the error is greater than 200%. As can be seen from the Fig. 13.5, all targets (except from two targets) have the green color, which means that the difference between the observed and computed head is less than 0.10 m. The difference between the observed and calculated head in the two yellow calibration targets is 0.17 (12.04% of the difference between the maximum and minimum value in the observation well, or a variance) and – 0.11 m (12.94% of the variance). The model can be assumed calibrated in the accepted level of accuracy.

Figure 13.7 shows the depth to groundwater table. The average groundwater level is about 2.5 m bellows the terrain. The values of 6.0–8.0 m are errors caused by the defects on the digital relief model.

The verification of the model, which uses the set of calibrated parameter values and stresses to reproduce a second set of field data, was focused on the period from the March 13, 2018 to May 2, 2018 (Fig. 13.8). Figure 13.9 captures the contours of the calculated groundwater head for the steady-state model of the verification period.

As can be seen from the Fig. 13.9, several targets have the green color. The rest of the calibration targets have the yellow and red color, which means, that the difference between the observed and computed head is more than 0.10 m. The reason is, on the one hand, that the influence of the left-hand side seepage channel was not taken into account because of the missing data—the water level on the seepage channel is not measured or observed. On the other hand, it could be caused by the estimation of the water level courses in channels A VII and B VII. There are no records about the water levels on these channels, nor the records about the manipulation with gate slides of

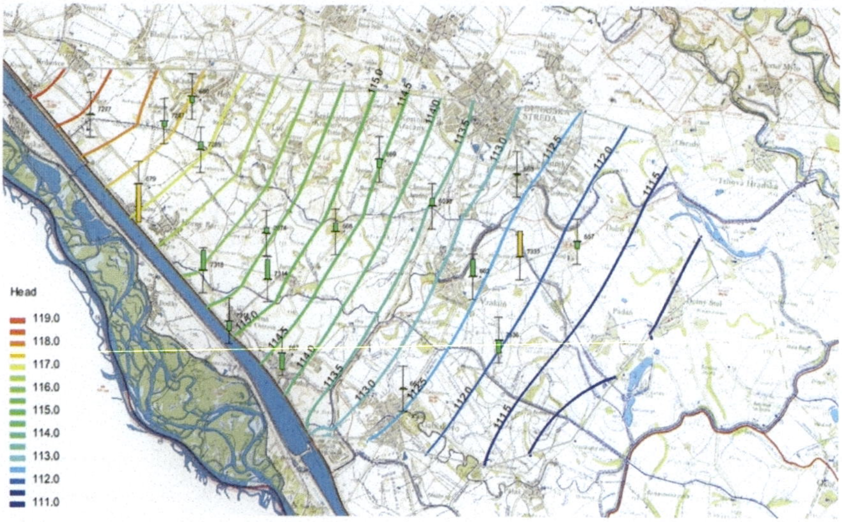

Fig. 13.6 The calculated groundwater head for the period from May 26, 2010 to September 9, 2010 [m a.s.l.]

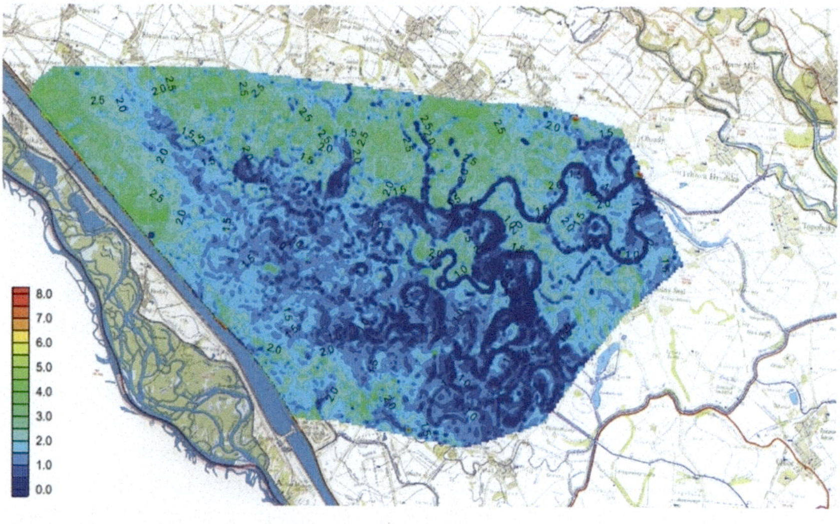

Fig. 13.7 Depth to groundwater table [m]

the weirs on the channels. The course of the water levels is therefore estimated based on the field measurements.

Figure 13.10 shows the depth to groundwater table. The average groundwater level is about 2.9 m bellows the terrain. The values of 6.0–8.0 m are errors caused by the defects on the digital relief model (Fig. 13.10).

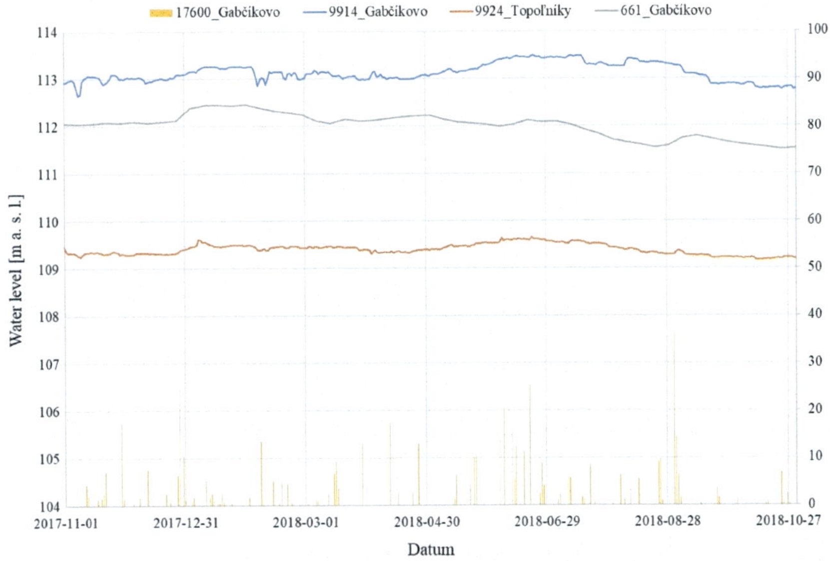

Fig. 13.8 The course of the water level at stage discharges on the S VII channel (9914 Gabčíkovo and 9924 Topoľníky) and the course of the groundwater level in well no. 661 in Gabčíkovo; daily total precipitation in the rain gauging station 17,600 Gabčíkovo—hydrological year 2018 (SHMI, 2021)

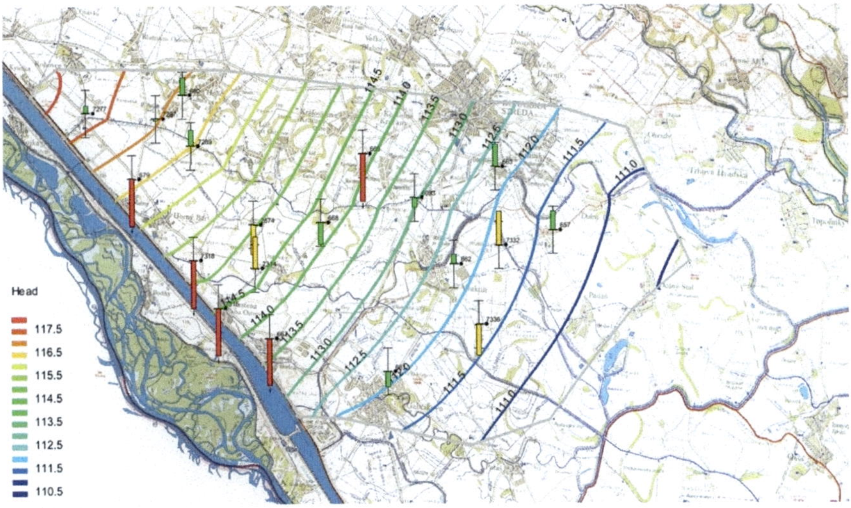

Fig. 13.9 The calculated groundwater head for the period from March 13, 2018 to May 2, 2018 [m a.s.l.]

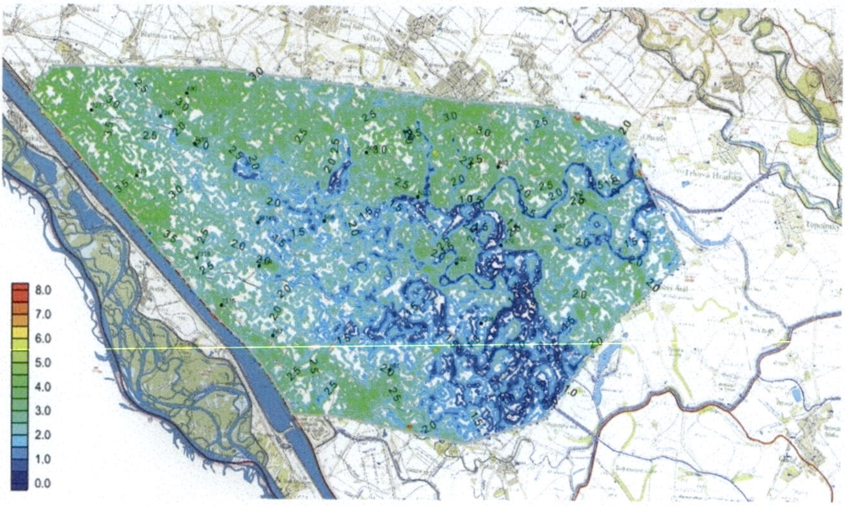

Fig. 13.10 Depth to groundwater table [m]

Calibrated and verified model was then used for the prediction for two scenarios (prognosis) examined on the A VII channel for both modelled periods:

- Prognosis 1—water level corresponds to the maximum levels at each weir (for the period from 2018, Fig. 13.11c) or to the maximum level during the flood situation (in 2010, Fig. 13.11a) + gate slides in the rkm 0.000 and 17.171 are closed;

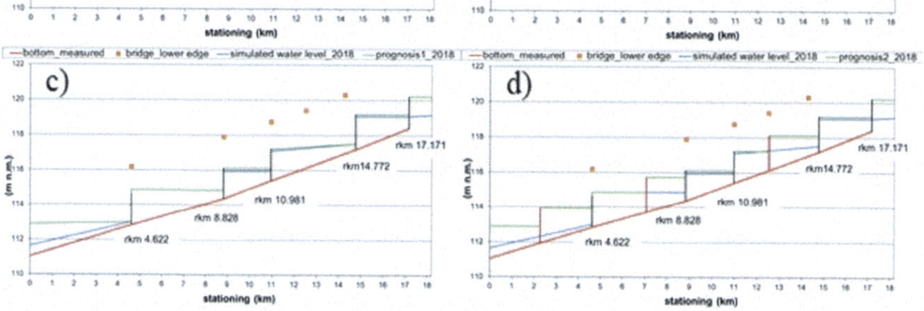

Fig. 13.11 Longitudinal profile of the A VII drainage channel—simulated water level and: **a** prognosis 1 (2010), **b** prognosis 2 (2010), **c** prognosis 1 (2018), **d** prognosis 2 (2018)

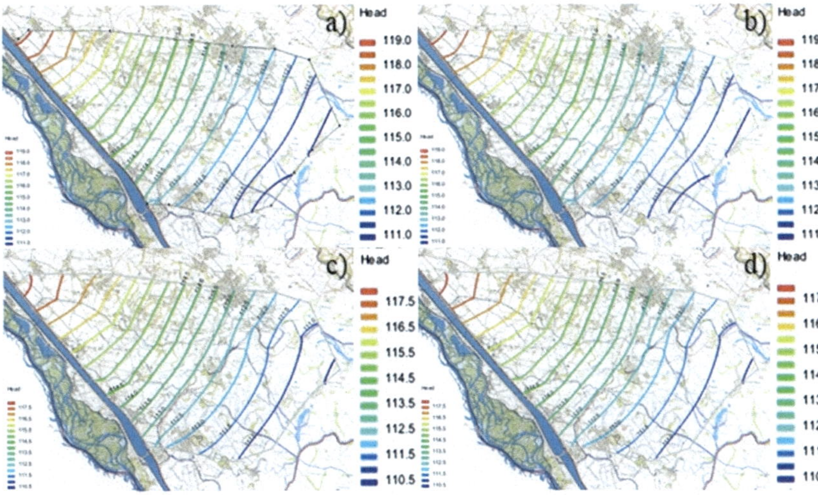

Fig. 13.12 The calculated groundwater head: **a** prognosis 1 (2010), **b** prognosis 2 (2010), **c** prognosis 1 (2018), prognosis 2 (2018)

- Prognosis 2—water level corresponds to the maximum levels at each weir (for the period from 2018, Fig. 13.11d) or to the maximum level during the flood situation (in 2010, Fig. 13.11b), the gate slides on the rkm 0.0 and 17.171 are closed + proposed weirs in the rkm 2.270, rkm 7.060 and rkm 12.530 are in operation (the height of each gate slide is 2.0 m). The location of proposed weirs can be seen in Fig. 13.2.

The calculated groundwater heads for the steady-state models capturing the prognosis 1 for the period of 2010 is in Fig. 13.12a and for the period from 2018 in Fig. 13.12b. The prognosis 2 of the period from 2010 is in Fig. 13.12c and for the period of 2018 in Fig. 13.12d.

Figure 13.13 shows the differences between prognosis and the so called 'zero variant'.

As can be seen from Fig. 13.13, the differences in groundwater levels vary from 0.0 to 0.25 m in the modelled period from 2010 and from 0.0 to 0.35 m in the modelled period from 2018. Increased groundwater level affects the volume of water infiltrated to the aquifer (Table 13.3) but does not cause the flooding of the adjacent area. Due to the slight increase in water level, the groundwater gets closer to the roots of cultivated crops as well.

The Table 13.3 shows that during the floods (in this case during the same hydrological conditions as were in 2010), closing the gate slides creates suitable conditions for water recharge into aquifer with more than 37,000 $m^3.d^{-1}$ (instead of previous more than 26,000 $m^3.d^{-1}$). With construction of the weirs in the rkm 2.270, rkm 7.060 and rkm 12.530, this number can be even higher (more than 41,000 $m^3.d^{-1}$).

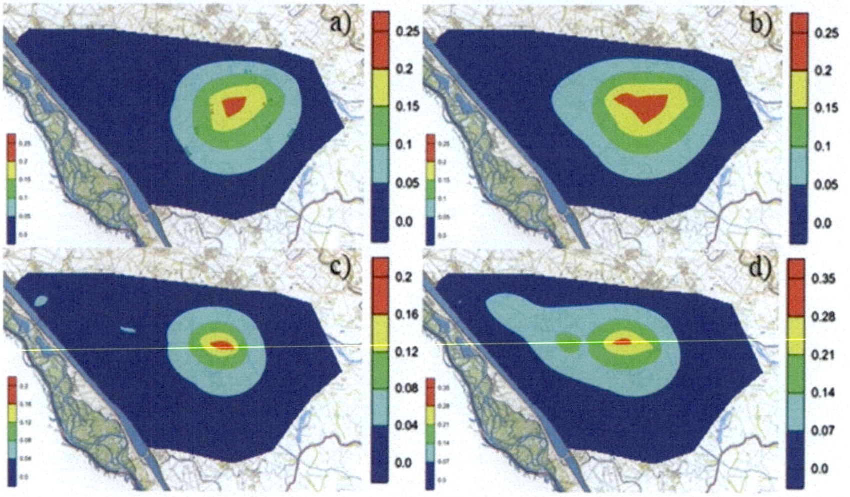

Fig. 13.13 Difference between the prognosis and the 'zero variant': **a** prognosis 1 (2010), **b** prognosis 2 (2010), **c** prognosis 1 (2018), **d** prognosis 2 (2018)

Table 13.3 River leakage [$m^3.d^{-1}$]

	2010			2018		
	In	Out	Total	In	Out	Total
'Zero variant'	46,648	– 19,767	26,881	45,318	– 30,959	14,359
Prognosis 1	48,053	– 10,143	37,910	50,784	– 27,186	23,598
Prognosis 2	49,946	– 8,733	41,213	58,029	– 25,240	32,789

13.4 Conclusions

The current management of floods, which concentrates on the fast water transfer to rivers and to the sea causes a water deficit during dry periods (Stachniak, 2020). For a long-term retention of water in aquifers and its subsequent use in drier or heavier demand periods, the Managed Aquifer Recharge (MAR) techniques were studied and implemented in four pilot areas of the DEEPWATER-CE project funded by the Interreg CENTRAL EUROPE programme. Irrigation and drainage channels of Rye Island, Slovakia, provides a background for technical possibilities for water flow control, i.e. creating the Recharge Dam MAR type. Rye Island is an important agricultural area and an important source of quality drinking water in Slovakia. The Recharge Dam MAR type on Rye Island channel system was investigated using an application of a modern mathematical methods of numerical modelling of the groundwater flow. A numerical model was developed using the MODFLOW program in the Groundwater Modeling System (GMS) environment. Multiple scenarios were explored relatively fast using the numerical modelling approach.

The calculated groundwater heads for the steady-state models capturing the prognosis 1 – water level corresponds to the maximum level during the flood situation in 2010 with closed gate slides in the rkm 0.000 and 17.171 are in Fig. 13.12a and for the period from 2018 with water level corresponding to the maximum levels at each weir with closed gate slides in the rkm 0.000 and 17.171 in Fig. 13.12b. The calculated groundwater heads for the steady-state models capturing the prognosis 2 (conditions from the prognosis 1 + proposed weirs in the rkm 2.270, rkm 7.060 and rkm 12.530) of the period from 2010 is in Fig. 13.12c and for the period of 2018 in Fig. 13.12d.

Figure 13.13. shows the differences between the prognosis and the so called 'zero variant'. As can be seen this figure, the differences in groundwater levels vary from 0.0 to 0.35 m. Increased groundwater level affects the volume of water infiltrated to the aquifer (Table 13.3) but does not cause the flooding of the adjacent area. By appropriate manipulation of the gate slides during flood situation, we can increase the amount of water recharged to the groundwater by 11,000 $m^3.d^{-1}$ without further technical measures. After construction of the weirs in the rkm 2.270, rkm 7.060 and rkm 12.530, the amount of water recharged to the groundwater can be even higher.

Acknowledgements The contribution was developed within framework and based on the financial support of the APVV-19-0383 project as well as VEGA project 1/0728/21.

References

Anderson, M. P., & Woessner, W. (1992). *Applied groundwater modeling. Simulation of flow and advective transport*. Academic press.

Benková, K., Bodiš, D., Nagy, A., Maglay, J., Švasta, J., Černák, R., & Marcin, D. (2013). *Explanations to the basic hydrogeological and hydrogeochemical map of the Danube Plain—Rye Island and the right bank of the Danube at a scale of 1: 50,000* (184 p) Bratislava: Štátny geologický ústav Dionýza Štúra. ISBN 978-80-89343-82-9.

Červeňanská, M., Šoltész, A., Baroková, D., & Janík, A. (2016). Evaluation of water management in regions affected by water structure construction and operation. In "*Ovidius*" *University Annals—Constantza, Year XVIII, Series: Civil Engineering*, Constantza, Romania (no. 18, pp. 17–24). ISSN 1584-5990.

Dillon, P. (2005). Future management of aquifer recharge. *Hydrogeolog Journal, 13*, 313–316. https://doi.org/10.1007/s10040-004-0413-6

Dillon, P., et al. (2009). *Managed aquifer recharge: An introduction* (Tech. Report, 13). Canberra: National Water Commission.

Doherty, J. (2004) PEST. Model-independent parameter estimation. (User Manual 5th ed., 336 p) Watermark Numerical Computing.

Harbaugh, A. W., Banta, E. R., Hill, M. C., & McDonald M. G. (2000). *MODFLOW-2000, the U.S. geological survey modular ground-water model-user guide to modularization concepts and the ground-water flow process* (p. 130) Open-File Report 00-92. Reston, Virginia: U.S. Geological Survey.

Interreg Central Europe. (2021). DEEPWATER-CE. Available on the internet: https://www.interreg-central.eu/Content.Node/DEEPWATER-CE.html

McDonald, M. G. & Harbaugh, A. W. (1988). A modular three-dimensional finite-difference ground-water flow model. In *Book 6. Modeling techniques* (Chap. A1, 588 p). Washington: U.S. Geological Survey.

Sikora, A. & Slota, R. (1992). *Research of discharge and water level regime in left-hand side branch system of the Danube by method of physical modelling* (42 p). Water Research Institute.

Šoltész, A. (2002). Water management in regions affected by water structure construction and operation. In *Advances in Hydro-Science and –Engineering. Abstracts. Proceedings of the 5th International Conference on Hydro-Science and –Engineering* (pp. 127) Warsaw: University of Technology.

Stachniak, K. (2020). *DEEPWATER-CE. Workpackage WPC, Activity D.C.2.1.* Press release. Final (summary from all PPS). PP4, (8 p) University of Selesia in Katowice.

Chapter 14
Garonne River Behavior Under Climate Change and the Potential Adaptation Strategies

Nazeer Asmael, Alain Dupuy, Sophie Leleu, Paul McLachlan, Francois Larroque, Michel Franceschi, Jean-François Bethoumieu, and Nicolas Cardot

Abstract According to the IPCC reports, the effects of climate change are well present from several decades ago, and it is expected to become even more pronounced during the remainder of the twenty-first century. The river basin system is defined as a complex of interaction between aquatic and terrestrial ecosystems. The interface between these ecosystems results in different responses of the river regime

N. Asmael (✉) · A. Dupuy · S. Leleu · F. Larroque · M. Franceschi
University of Bordeaux, EPOC UMR 5805, CNRS, Bordeaux INP-ENSEGID, 1 Allée Fernand Daguin, 33600 Pessac, France
e-mail: nazeer.asmael@ensegid.fr

A. Dupuy
e-mail: alain.dupuy@ensegid.fr; a.dupuy@brgm.fr

S. Leleu
e-mail: Sophie.Leleu@ensegid.fr

F. Larroque
e-mail: larroque@bordeaux-inp.fr

M. Franceschi
e-mail: michel.franceschi@ensegid.fr

A. Dupuy
French Geological Survey (BRGM), 3 Avenue Claude Guillemin, 45100 Orléans, France

P. McLachlan
Department of Geoscience, Aarhus University, Nordre Ringgade 1, 8000 Aarhus C, Danemark
e-mail: pm@geo.au.dk

J.-F. Bethoumieu
ACMG Association Climatologique de la Moyenne Garonne, 846 All. de la Seyncs, 47310 Sainte-Colombe-en-Bruilhois, France
e-mail: acmg@acmg.asso.fr

N. Cardot
SMEAG Syndicat Mixte d'Études et d'Aménagement de la Garonne, 61 rue Pierre Cazeneuve, 31200 Toulouse, France
e-mail: nicolas.cardot@smeag.fr

A. Sefelnasr et al. (eds.), *Water Resources Management and Sustainability*, Water Science and Technology Library 114, https://doi.org/10.1007/978-3-031-80520-2_14

ranging from floods to low flow. The instability of alluvial valley due to geomorphological and fluvial dynamics, represents an additional challenge for the riparian societies face to climate changes. The Garonne River, the third largest French Atlantic River in terms of discharge, flows down in the Garonne Valley, the largest occupied and cultivated alluvial plain in France. This area is influenced by a wide range of Oceanic, Mediterranean, and mountainous climates. Hence, the complex Garonne River regime is influenced by the precipitation generated over the Massif Central and the French and Spanish Pyrenees Mountains. The surrounding alluvial aquifer, which forms an important part of this system, maintains a strong hydraulic connection with the river and sustains its low flow. Consequently, this aquifer is extremely sensitive to meteorological changes and hence to recharge system variation. The inhabitants in this area are strongly depending to the river and its alluvial aquifer. However, the impact of climate changes on socio-economic activities could be significant in this area. The adaptation to these changes creates additional challenges for decision-makers and becomes a major objective of the watershed policy. The trends assessment of the Garonne River's discharge over the last fifty years has shown a decrease in average annual discharges and substantial decrease in low flows. The dry periods are starting earlier and becoming more severe and longer. Upstream, the snow cover shows a decrease in thickness and duration, which has a negative impact on the river's discharge and consequently on the groundwater recharge ratio. Groundwater measurements show a general decrease in water level. The observations show that most of the wells located close to the river have dried up during the dry periods since the alluvial aquifer drains into the river during these periods. The groundwater level reduction could have a negative impact on agriculture, wetlands, and river ecosystems. Thus, the implementation of adaptation measures has been initiated. The artificial recharge of the alluvial aquifer is considered as an adaptation strategy to address the effect of climate change and sustains the low flow of the river. In addition, this strategy will help to buffer the river temperature during the low flow periods. Several pilot sites in the Garonne Valley are under investigation for testing the artificial recharges. The first site, where artificial recharge has been tested since 2019, is located near the city of Agen, where the runoff is collected in a retention basin and is used to recharge the alluvial aquifer. Groundwater level in the wells near the retention basin have increased by about 1 m following rainfall events. The results of the groundwater modeling show (1) a similar magnitude aquifer response to the induced infiltration and (2) that the infiltrated water would take about 4 months to reach the Garonne River, which is an appropriate time to maintain the river's low flow, since recharge takes place in spring. Another site that has been tested is located near the city of Marmande, where the surface water from the Lateral Canal of the Garonne River has been used as a source to recharge the alluvial aquifer in May 2023. The infiltrated water has created a piezometric dome and the groundwater level has increased by 15 and 33 cm into two boreholes located about 40 m upstream and downstream of the infiltration site respectively. This experiment will be carried out on a larger scale, and further groundwater measurements and modeling will be carried out for this and other selected sites in the near future.

Keywords Climate change · France · Garonne River · Low-flow management · MAR—managed aquifer recharge

14.1 Introduction

In south-west France, the two main rivers Adour and the Garonne, drain one-fifth of the country's surface area and about 60% of this land is used for the agricultural purposes. These two river basins are managed together from an administrative point of view. In this area, about 15% of the agricultural area is irrigated, which requires about 264 M m^3 $year^{-1}$. The Garonne River is the main river with the larger catchment area, and it supplies about 2 million inhabitants with drinking water. In the Garonne Valley, three factors give it a particular geographical outline: a history marked by navigation along the river, river behavior which alters between extreme low flows and catastrophic flood, and almost the same landscape structure (Valette, 2002, 2011). The alternation of the river's flows between low flow and catastrophic floods is associated with sediment deposition, erosion, river backwaters, and wetlands. Consequently, the alluvial flood plain of the river extends throughout history and results in the formation of different terraces. As a result of climate change, urbanization, agricultural practices and the artificialization of watercourses, the river's discharge has decreased, particularly during low-flow periods.

According to recent studies (Lhuissier et al., 2016; Sauquet & Evin, 2022, etc.), the major effects of climate changes in the south-west of France during the coming 30 years will be: increase in average annual air temperature of about 2 °C, increase in annual evapotranspiration and extreme events such droughts and floods, decrease in watercourses average flows of about 20–40% and about 50% during the low-flow periods which will start earlier and will be more severe and longer, reduction in the duration and quantity of snow cover in the mountain ranges, and decrease in groundwater recharge ratio. This will result in water resource depletion, soil drying, forest fires, deterioration in water quality, and loss of biodiversity in both aquatic ecosystems and wetlands. Consequently, agriculture production, tourism, nuclear power stations, economic activities, will be impacted in this area. This impact and the demographic growth become a serious challenge for the decision makers in this region. Hence, they are starting to look to water resources as an indicator of climate change but also as a nature-based solution for future planning to fight these changes. This solution can be considered as an environmentally, economically, and socially beneficial and acceptable approach in the Garonne Valley which aims to improve the river's quality and quantity condition.

Historical meteorological records (https://meteofrance.com) show that the average temperatures in France have risen by 1.7 °C since 1900, with a considerable increase in warming over the last 3 decades. The records show also that the rainfall trends vary according to regions and seasons, the frequency of heat waves have increased, the extreme rainfall become more intense and more frequent especially in the southern region, the duration of snow cover in mid-mountain areas has decreased (Asmael et al., 2023a, 2023b; Lambert et al., 1990; Steiger et al., 1998). The soil becomes drier, and the droughts intensity has increased, consequently, it is estimated that 70% of wetlands have been lost over the last 50 years. The temperature contrasts were clearly visible at the end of August in summer 2023 in the Alps Mountain: the hottest day of this summer was recorded on Thursday 24 August, and 5 days later, on Tuesday 29 August, the temperature dropped by 20 degrees. After the peak temperatures, the snow returns as early at 2000 m.a.s.l in this area. In the same year, in the city of Bordeaux close to the Atlantic Ocean, the monthly average temperature recorded in September 2023 was 21.8 °C, the highest since records began, and the monthly averages temperature for August and October 2023 was also among the highest values since data has been recorded in Bordeaux.

According to the Adour and Garonne Water Agency (Richard et al., 2022), 1/3 of the watercourses were observed to be dry in August 2022, and around 60,000 inhabitants had been cut off from their drinking water supply in this area. During this year, the low-flow support program, initiated in 1993, continued for the first time until the end of November. This has resulted particularly from the changes in rainfall and snow patterns and water supply in the upstream mountain areas. Consequently, the maximum level of restriction on water use was applied, with significant consequences on the economic sectors. According to the same report, the main hydrological changes expected to occur in this area will be a decrease in the average annual flow of the Garonne River by 20 to 40% per year, and by about 50% during low-flow periods. Considering the groundwater, the recharge will also be reduced by between 30 to 50%. This will result in quantitative and qualitative changes in water resources and hydrological functioning of Rivers and related eco-systems.

Since 1993, the policymakers decided to mobilize surface water stored in hydropower reservoirs in mountainous areas to sustain the river's low flow. The amount of this water could represent about 50% of the river's flow during dry periods. The mobilized volume has more than doubled since the beginning of this program and increasing significantly from 2021 onwards. Water resources are a crucial issue in the context of climate change and an adaptive measure that can be implemented. Groundwater artificial recharge is one of the most economically attractive strategies to improve the quantitative and qualitative status of shallow alluvial aquifer and can also respond to local problems. In the study area, the molasse layer forms an impermeable barrier underlying the shallow alluvial aquifer and limiting downward groundwater flow. The natural flow direction of this aquifer is normally oriented towards the Garonne River, which acts as a natural drain. These two conditions make the artificial recharging of the alluvial aquifer a very interesting strategy to sustain River's low flow. In addition, this strategy, which is relatively low-cost, will participate in water resources development for collective interest and contribute to the biodiversity preservation in the study area. This study aims at: 1-study the evolution of the Garonne River discharge especially during the dry periods to show the effects of climate change on its regime, so that adaptation measures can be taken to limit these impacts., 2-Demonstrate the potential of artificial recharge of the surrounding alluvial aquifer as a nature-based solution to mitigate the effect of climate change and sustain the low flow of the River.

14.2 Physiography of the Garonne River System

The Garonne River in the South-West of France rises in Spain's central Pyrenees and reaches the French border after 50 km in the Haute-Garonne Department. It crosses the hills of the Pyrenean foreland and then the plain area where our study takes place. The river is joined by numerous tributaries along its course. It finally reaches the Atlantic Ocean after 529 km via the Gironde estuary, where it merges with the Dordogne River. The lower reach of the Garonne River is under tidal influence in the last 120 km before the onset of the Gironde estuary. The Garonne River is the main river of a tributary fluvial system that drains both the Pyrenees and the Massif Central highlands. The study area is located between the cities of Toulouse and Langon where the Garonne alluvial plain reaches its maximum (Fig. 14.1).

Fig. 14.1 Map of France, prepared by QGIS, free and open-source software, showing the location of the Garonne River between the cities of Toulouse and Langon

This alluvial plain area is considered one of the largest alluvial plains in France (Asmael et al., 2023a, 2023b; Danneville, 1995), which extends between 5 and 8 km and is asymmetrical as the Garonne River has sunk successively northwards. It is bordered by asymmetrical hillsides that play an important role in its hydro-geological system (Fig. 14.2). The northern hillsides are steep and wooded, while the southern hillsides are softer, and alternate between crops and forests. The altitude of the Garonne riverbed in the Garonne Valley steps between 140 and 1 m.a.s.l between Toulouse and Langon respectively.

In this tributary system, the hydrographic network in the Garonne Valley (Fig. 14.3), is organized around major watercourses. This network converges towards the Garonne River, the largest watercourse in the Valley and generally collects the surface water and groundwater that drains from the hillsides and alluvial plains.

Through time the Garonne Valley was formed in successive depositional and erosional phases that created 6 alluvial terraces upstream south of Toulouse named from T1 the youngest to T6 the oldest (Christophoul et al., 2014; Hubschman, 1975), and 4 terraces downstream close to Langon ranged from Fx the youngest to Ft the

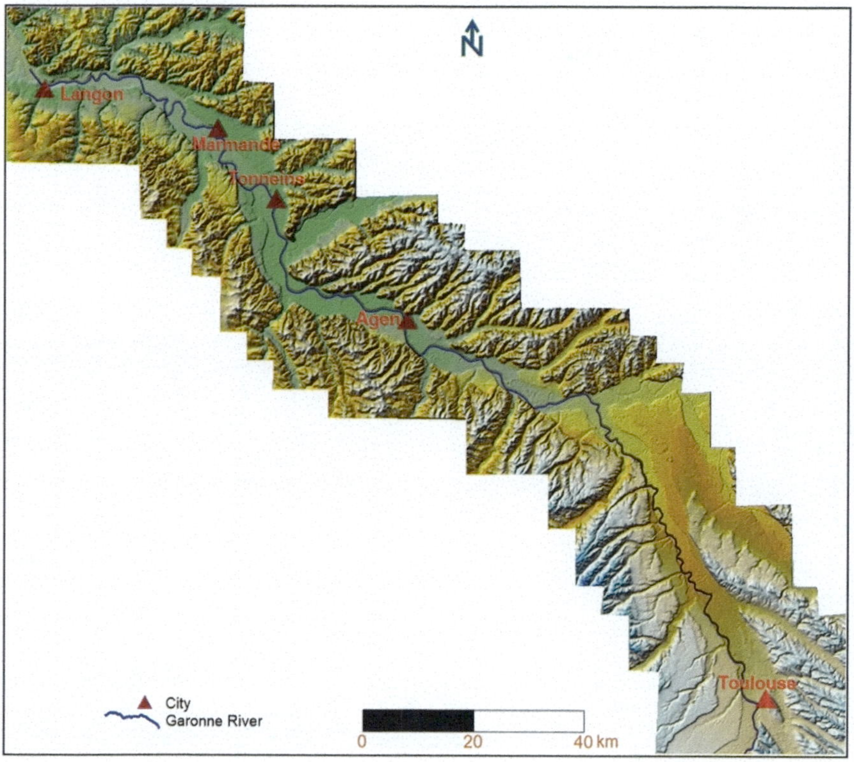

Fig. 14.2 Garonne Valley showing the Garonne River, main cities, and surrounding hills (based on IGN, French national geographic institute, https://geoservices.ign.fr/ spatial satellite data with a 5 m resolution)

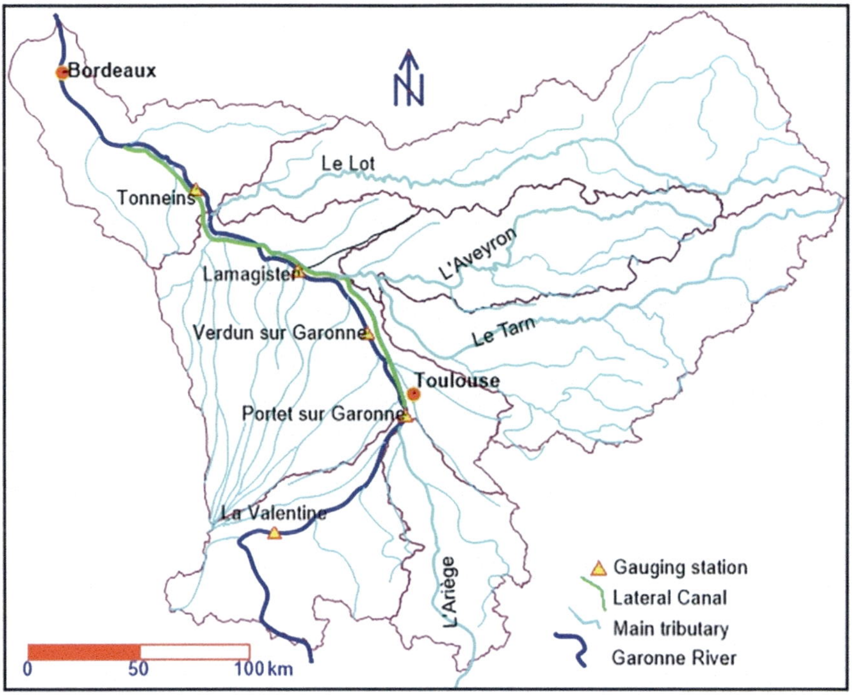

Fig. 14.3 Hydrographic network showing the Garonne River, the main tributaries, the lateral canal and the sub-basins

oldest (Dubreuilh & Alvinerie, 1978). The most recent terrace forms the alluvial plain deposited during the last glacial phase. Agriculture predominates in the Garonne Valley (33%) and forms, with the forests (32%) and semi-natural landscapes (25%), 90% of the land cover in this area (Fig. 14.4).

14.2.1 Geological Setting

The Garonne River is located in the Aquitaine Basin that forms the northern foreland basin of the Pyrenean orogen. The Aquitaine Basin is limited to the north by the Massif Central that corresponds to a large crustal terrane overlain by Mesozoic successions that recorded large scale uplift from the Late Cretaceous times with an increase of topography since about 40–30 My (Olivetti et al., 2020). From latest Cretaceous to early Miocene, the Pyrenean relief built up during orogenic phases while subsidence, folding and thrusting occurred in the Aquitaine foreland basin (Ford et al., 2016). The post-orogenic phase started during Miocene and triggered the major uplift phase (Bernard et al., 2020) during which the Pyrenean altitude increased creating a major erosional cycle that produced important volume of sediments trapped

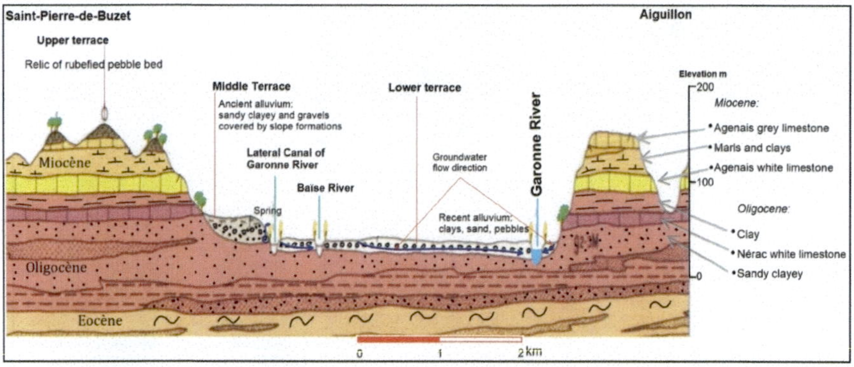

Fig. 14.6 Geological cross-section in the Garonne valley, after (BRGM and Asmael et al., 2023a, 2023b)

river and this aquifer can have significant quantitative and qualitative consequences on surface water and groundwater (Asmael et al., 2023a, 2023b). The processes that control these exchanges depend on the differences between the water levels in the river and the aquifer (Fig. 14.7), and the permeability of the alluvial deposits. The alluvial aquifer is recharged by the river during the flood periods, when the exchange surface increases, and it drains by the river during the dry periods. The recent alluvial deposits and the Garonne lower terrace consist of gravels or pebbles intercalated with sandy clay constitute an abundant aquifer of 4 to 10 m thick with transmissivity values change between 5.10^{-3} and 5.10^{-2} m^2s^{-1} and permeability values of between 10^{-1} and 6.10^{-3} ms^{-1}. The lateral and vertical heterogeneity of these formations can result in significant changes in hydrodynamic parameters of the aquifer. The alluvial deposits are often covered by silts and sand formations which become richer in clay resulted from the slops as far as from the river. The existence of meander bars consists of gravel sandstones induces preferential flows and makes the aquifer very productive locally. However, the presence of lenses filled up dominantly with finer grained deposits (silt, clays) makes the aquifer very impermeable in places. The Garonne River discharge changes in a wide range. According to the data measured in Tonneins station between 1989 and 2022, the mean annual discharge is 517 m^3s^{-1}, the lowest value measured in this period was 34.1 m^3s^{-1} in August 1989, and the highest was 6190 m^3s^{-1} during the floods of February 2021(https://www.hydro.eau france.fr). The river is usually flooding in winter or spring after the snow melts. This flood is sometimes followed by a significant low flow in summer.

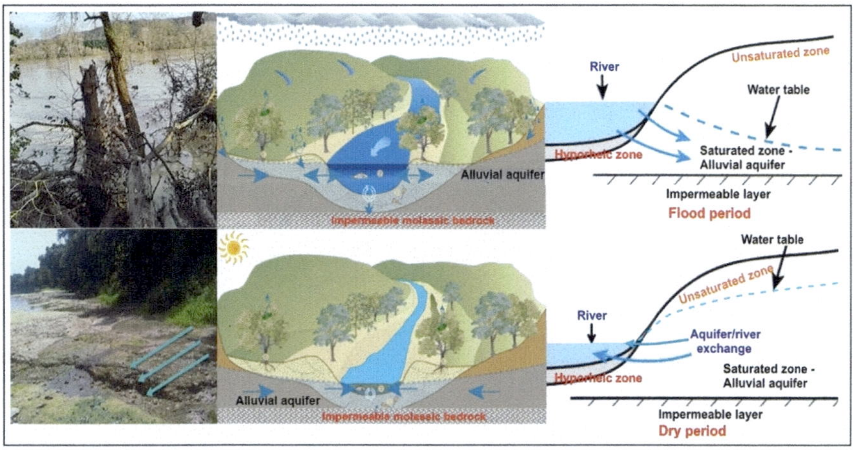

Fig. 14.7 Alluvial aquifer/Garonne River interface and exchange patterns during flood and dry periods, after (Asmael et al., 2023a, 2023b; Hendrickx & Sauquet, 2013; Queensland Government, 2017)

14.2.3 Hydropower Reservoirs to Sustain the Low-Flow of the Garonne River

Being among the first to be impacted by the climate change, water resources can be used as nature-based solutions to fight against these changes. Over the last 30 years, because of the effect of climate change and human activity on the river's low flow, a surface water supply programme has been initiated by SMEAG (Syndicat Mixte d'Etudes et d'Aménagement de la Garonne). This program aims at sustaining the river low flows and to satisfy the users' needs and the aquatic environment. Around 70 Mm^3year^{-1}, from 14 hydroelectric reservoirs in the Pyrenees Mountains, can be mobilized for this purpose. This supply can represent up to 50% of the river's discharge from mid-August to mid-September. This water is discharging directly by gravity into the river or through its tributaries. During the last year (2023) in the upstream part close to the city of Toulouse, 61.1 M m^3 were released to sustain the low flow of the river, which is the largest released amount since this program has been started in 1993. In mid-August 2023, the released amount was 21 m^3s^{-1} which represents more than 50% of the river discharge and allows to secure the agricultural productions. Without the low-flow support program, restrictions would have been imposed on the river, its alluvial aquifer, and its tributaries. Figure 14.8 shows the river discharge in four gauging stations from the upstream to the downstream between October and December 2023. From these four graphs, we can recognize that the river low flow is almost exceeded the objective low flow even with the support program. The effect of the precipitation on the river discharge is clearly observed at the beginning of the hydrological year.

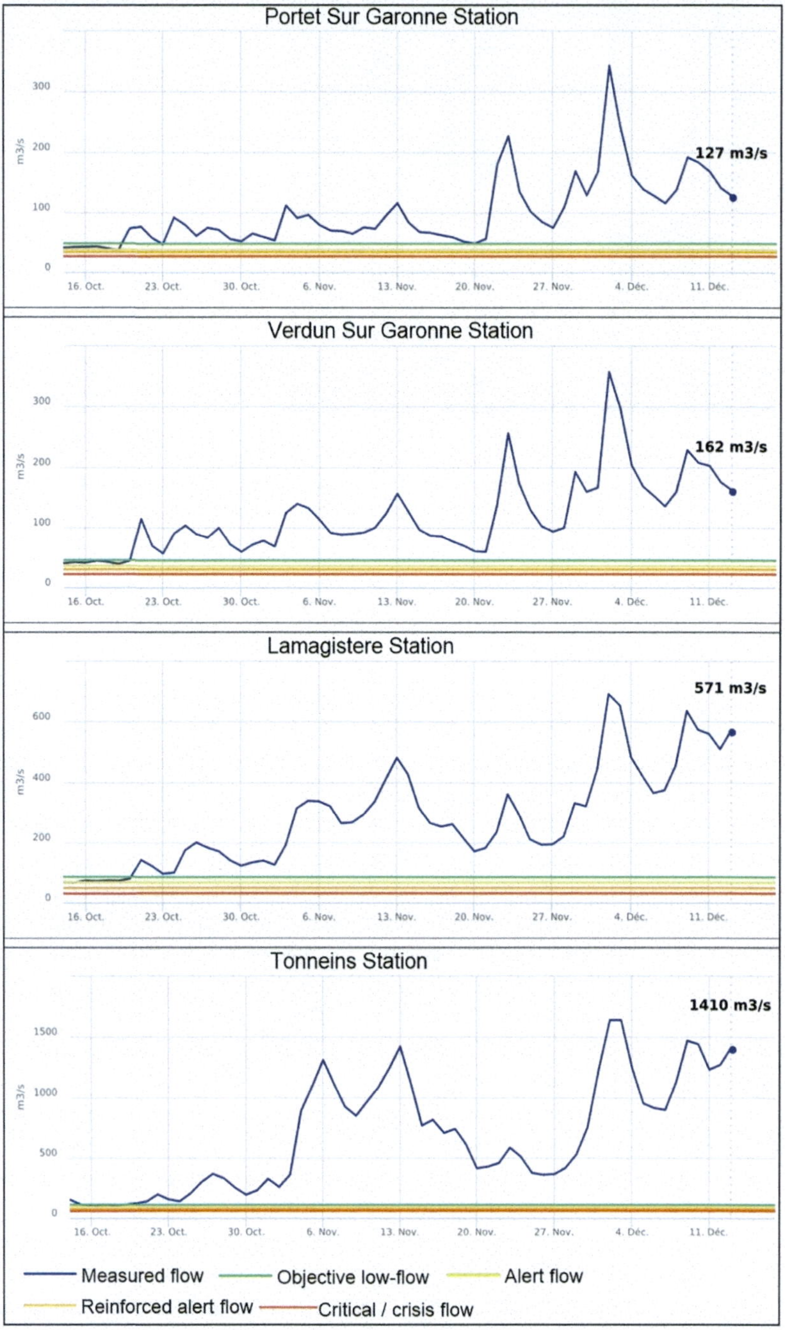

Fig. 14.8 Garonne River discharge measurement in four gauging stations from upstream toward the downstream between October and December 2023. After SMEAG (https://www.smeag.fr)

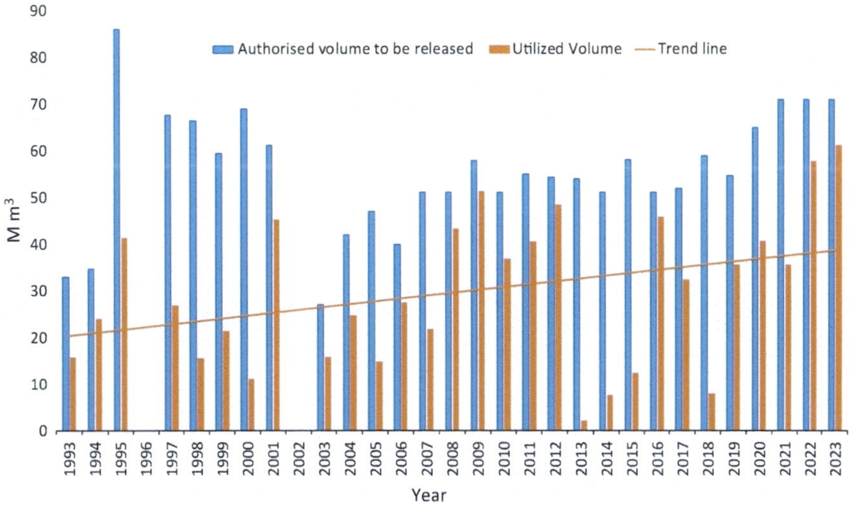

Fig. 14.9 Volume of surface water used to sustain the low flow of the Garonne River between 1993 and 2023, After SMEAG (https://www.smeag.fr)

The amount of released water since the beginning of the program is shown in Fig. 14.9 which shows an increasing trend with time. This operation has costed 4.33 million Euros in 2023.

14.3 Methodology

To evaluate the climate trend in the study area, the temperature data has been collected from Agen station (ACMG - Association Climatologique de la Moyenne-Garonne et du Sud-Ouest) for the last fifty years. The data show a warming trend with time resulted from the effect of the climate changes during this period. A decreasing precipitation trends with high confidence is also recognized in this area according to Meteofrance projected data (https://meteofrance.com) between 1961 and 2012. The river discharges show a clear response to these two parameters over time. Hence, a decrease in annual average discharge and a significant decrease in low flows have been projected (Sauquet & Evin, 2022).

To tackle the effects of climate change in the Garonne valley, new adaptation strategies has been implemented based on the exchange between the Garonne River and its surrounding alluvial aquifer. The artificial recharge of this aquifer is one of the management tools that being considered by water resource decision makers. Serval sites located between the cities of Agen and Marmande (Fig. 14.10) have been chosen for this purpose. For each site, the existing wells have been used to measure the groundwater level, at least two time by year, for two years. The groundwater level

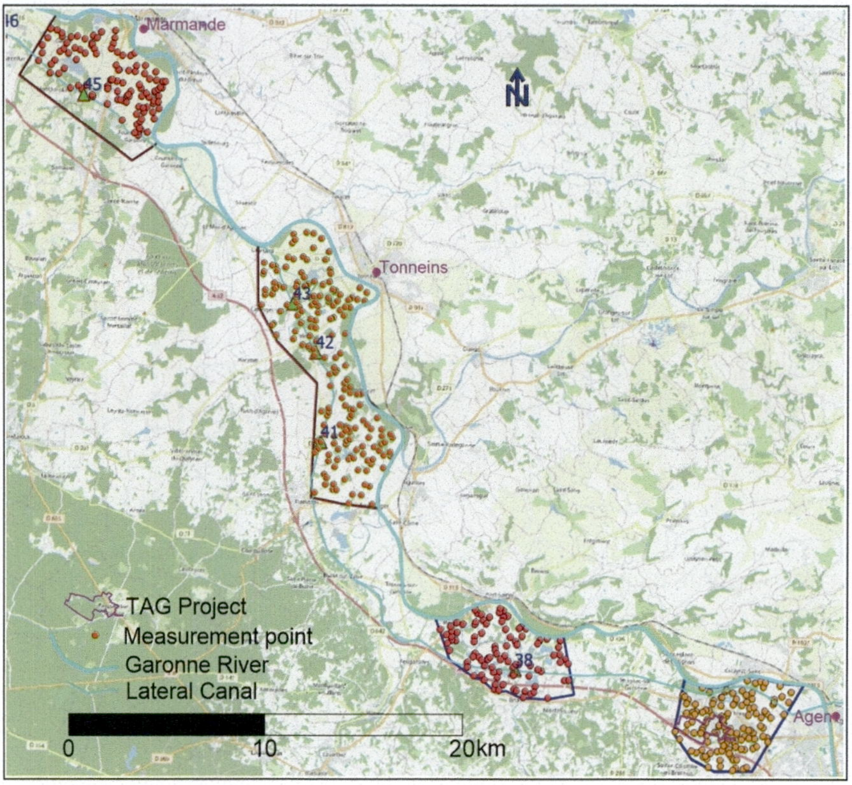

Fig. 14.10 Pilot sites selected in the Garonne valley for groundwater artificial recharge

changes, the relationship between the river and the aquifer based on groundwater levels fluctuation, the piezometric maps, and the lithological stratification at several points have been determined for each site.

In the first site close to the city of Agen, a large socio-economic project called TAG (Technopole Agen Garonne) has been developed in 2019 to increase the region's economic wealth while respecting the environmental conditions. Runoff water from this project area is collected into retention basins and is used as source to recharge the alluvial aquifer (Fig. 14.11).

Groundwater flow model has been constructed based on finite element method by using FEFLOW 7.1 by DHI WASY (Diersch, 2018) to simulate groundwater flow. The developed model covers an area of approximately 20 km^2 around the TAG project. Based on the measured and available data (groundwater levels, hydraulic conductivity, etc.…), the model has been calibrated in steady state and a good agreement is achieved between the measured and calculated groundwater level contours which has allowed us to turn the model in transient status. For the other three sites, the surface water from the artificial structure close to the Garonne River, known as "Canal Latéral à la Garonne" will be used to recharge the alluvial aquifer. After the

Fig. 14.11 Collection runoff ditch and retention basin after a storm event

field investigation, the first experiment of groundwater artificial recharge from this Canal was carried out on 17 May 2023 by using a ditch located on the left bank of the Canal close to the city of Marmande. The surface water was drained by gravity from the Canal to fill this ditch (Fig. 14.12). This experiment has allowed the determination of the infiltration capacity of the ditch and to observe the possible effect on the groundwater level.

The test was carried out intermittently over 4 weeks and the water level has measured continuously using piezometric sensors installed in the ditch and into two boreholes located about 40 m from the infiltration site. Borehole 1 is located in the upstream part of the infiltration site at the left bank of the Canal while borehole 2 is located in the downstream at the right bank of the Canal, in the groundwater flow direction.

Fig. 14.12 Geological investigation, cleaning up and filling in a ditch from the Lateral Canal close to the city of Marmande

14.3.1 Climate Change Effect

To identify the hydrological consequences of climate change in the Garonne Valley, three main parameters have been considered, the temperature, precipitation, and river discharge. According to the available data from the Agen meteorological station, the maximum temperatures measured between 1973 and 2023 clearly show a rising trend (Fig. 14.13).

The evolution of the number of days that exceeding the temperature thresholds between April and October from 1951 to 2023 in the same station is shown in

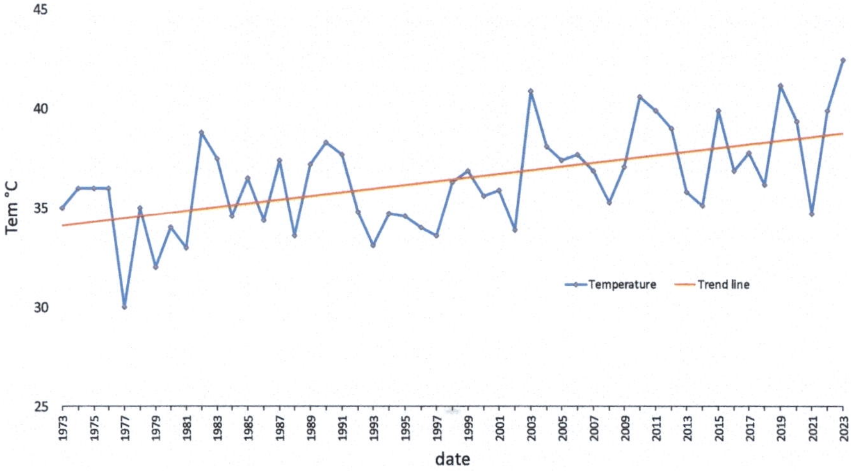

Fig. 14.13 Maximum temperature measured at the Agen station between 1973 and 2023

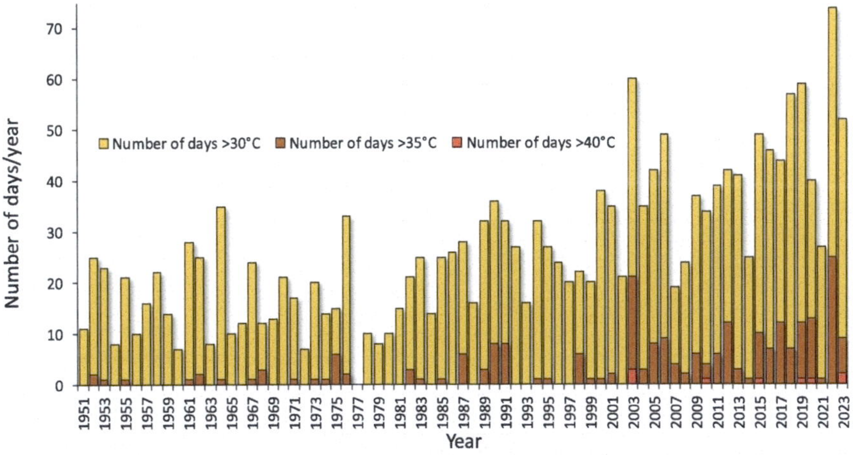

Fig. 14.14 The number of days above the threshold temperature measured at the Agen station

Fig. 14.14. This figure shows that the extreme high temperature is increasing with time accompanying with the increasing in the evapotranspiration and water demand in this agricultural area.

Considering the precipitation, the measured data from 1961 to 2012 over the whole country following by modelling projection (Fig. 14.15), highlights a difference impact of climate change between the north and south parts of France. A significant reduction in the precipitation amount has estimated in the southern half, while there was no change or even increase in the northern half where the snow cover plays an important role for the river's discharge. Consequently, a significant reduction in the watercourses discharge in the south and south-western parts of the country has been recognized.

In terms of rivers discharge, the model projection shows that the north part of France is likely to have higher river flows during the winter resulting from the increasing of precipitation amount in this area. However, in the southern parts, the Garonne River low-flows will significantly reduce by about 50% based on QMNA5 (minimum annual monthly flow rate with a 5-year return period). This effect will be less significant considering the average annual flow of the river (QA) (Fig. 14.16). (Sauquet & Evin, 2022).

14.3.2 Garonne River Regime and Climate Change Effect

In general, natural river regime reflects the effects of meteorological situation, different landscapes, associated altitudes and slopes and human impacts (Asmael et al., 2023a, 2023b; Grusson et al., 2018; Probst, 1983). Hence, the Garonne River discharge is always influenced by the melting of the Aneto glacier in the Spanish

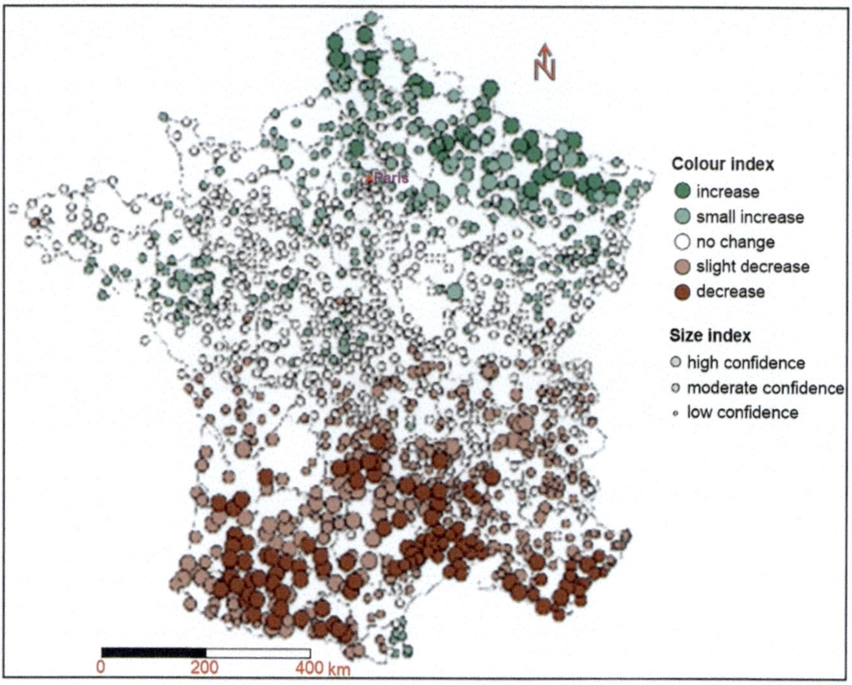

Fig. 14.15 Rainfall evolution patterns in France over the period from 1961 to 2012, data source: Meteofrance (https://meteofrance.com, accessed on 30 August 2023)

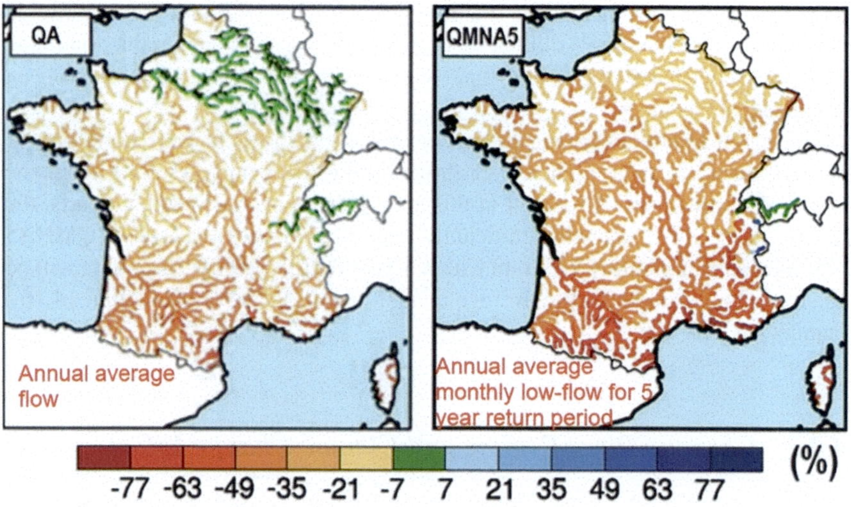

Fig. 14.16 Average river flows changes according to QA and QMNA5 modules for the period 2070–2100 with respect to the measured data between 1960 and 1990

Garonne side, snowmelt from the Pyrenees Mountains, rainfall patterns, and the condition of drainage of its accompanying alluvial aquifer. The historical measurements of the Garonne River discharge show that the so-called Oceanic Pyrenean flood, has generated a value up to 8500 $m^3\ s^{-1}$ in 1875, whereas The Oceanic Massif Central flood, has reached a discharge of 6700 $m^3\ s^{-1}$ in 1927 in Tonneins gauging station. The most significant flood that generated from the south of the Massif Central, from the Tarn River Tributary, was 8300 $m^3\ s^{-1}$ in 1930 and had a Mediterranean origin (Carozza et al., 2014). The historical river flow measurements also show a gradual decline trend in annual river discharge. This trend is consistent with the increase in temperature and evapotranspiration observed over the last 60 years and, a priori, is resulting from the effect of climate change.

To better identify the effects of climate change on the Garonne River, it is necessary to determine the river's natural flow by excluding the human influence such as low-flow support program or the extraction from the river for the different purposes. The Tonneins station is about 8 km after the junction point between the Garonne River and its tributary, the Lot River. It is located in the downstream area, represents the overall hydrological conditions in the Garonne Basin, and is the lowest station that does not influenced by the tidal action of the ocean. Figure 14.17 shows the difference between mean annual flow and reference natural flow calculated between 1970 and 2000 in this station. When the deviation is less than 0, this means that the annual flow for the corresponding year was less than the "normal" flow, and vice versa when the deviation is greater than 0. The signal of deviations shows a great decrease in River flow over the years resulting from the effect of climate change, which has induced a reduction in the natural resource.

The objective low-flow rate, below which neither the proper functioning of the environment nor the satisfaction of users is guaranteed, depends on station location. In Tonneins station, this flow is considered to be 110 m^3s^{-1}. From the Fig. 14.18 and according to the data measured in this station between 1960 and 2020, each point (green) represents a given year (x-axis), and the date on which the objective low-flow

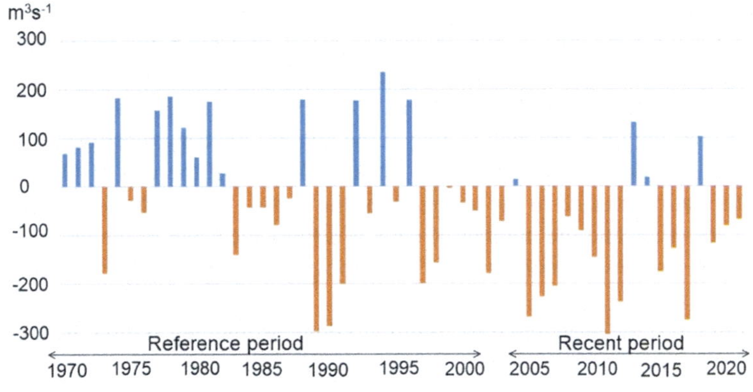

Fig. 14.17 Difference between mean annual flow and reference natural flow measured between 1970 and 2000 in the Tonneins station

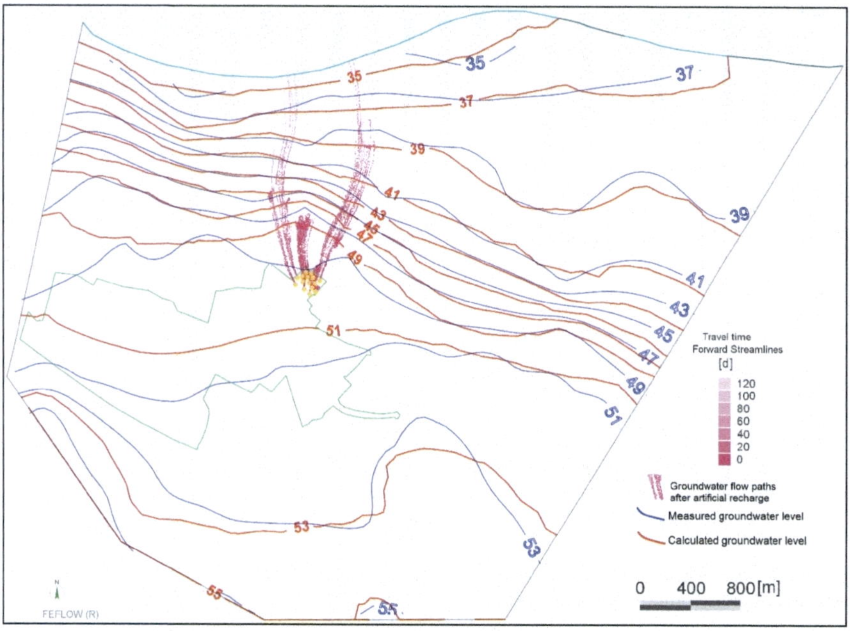

Fig. 14.22 The results of the groundwater modelling show the flow patterns, from the yellow nodes of infiltration basin selected by Feflow, towards the Garonne River after artificial groundwater recharge, after (Asmael et al., 2023a, 2023b)

This infiltrated water could be used for a wide range of agricultural, industrial, and drinking water purposes.

The initial result shows a clear positive correlation between the infiltrated water and the groundwater level measured into the two boreholes (Fig. 14.23). The groundwater level measured in the borehole1 shows that the infiltrated water takes about one day to reach this upstream borehole and the groundwater level increases according to dich filling. The groundwater level rose by about 0.15 m in this borehole when the ditch was filled up furthest on 6 June. However, the infiltrated water has reached the downstream borehole2, which was drilled later, after 14 h, and its level rose by about 0.33 m after the ditch was filled up furthest on 6 June.

A groundwater models are under developing to simulate the groundwater flow and estimate the potential of this method as an adaptation strategy against the effect of climate change on the Garonne River discharge.

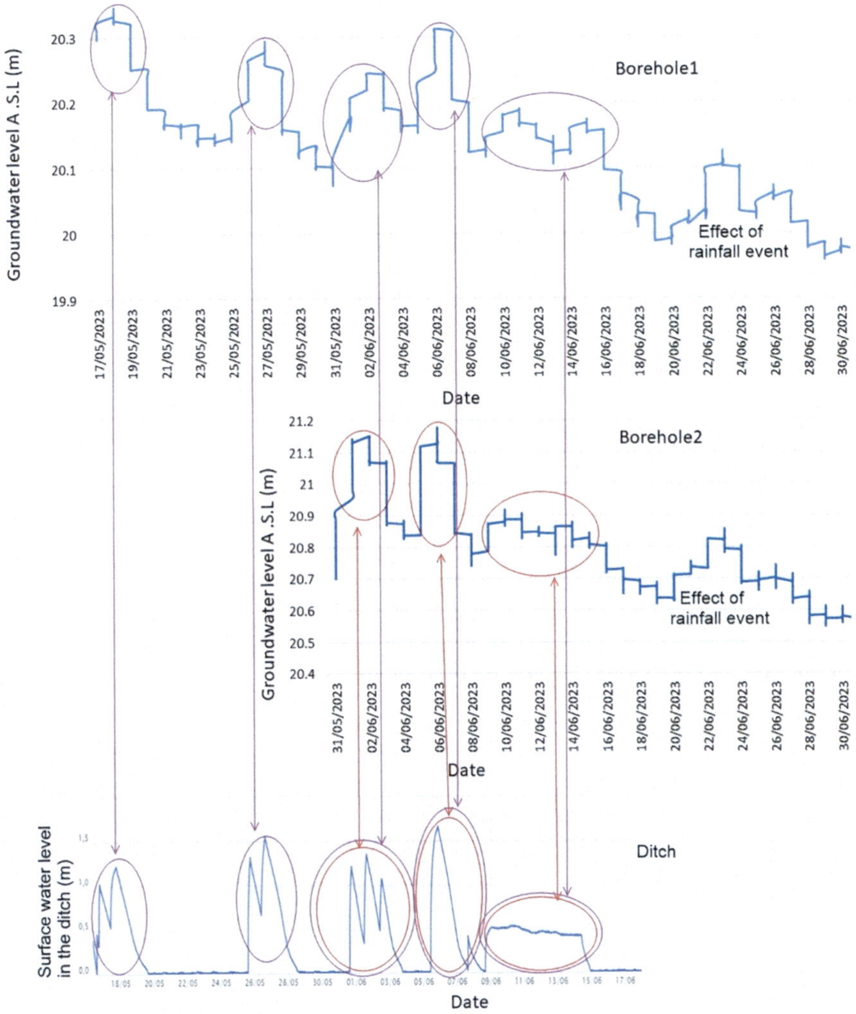

Fig. 14.23 Correlation between piezometric fluctuations measured in the boreholes and surface water level into the ditch during the infiltration experiment

14.5 Conclusion

At national scale, the existing modelling results show an increase in precipitation in the northern part of France and a significant decrease in the southern part. However, these results show an increase in the temperature across the whole country. Hence, the possible effects of climate change in the Garonne Valley will be the increase in the temperature, decrease of the rivers discharge, increase the vulnerability of the water resources availability, especially for the agriculture and drinking water in the downstream parts. The reduction in the snow cover thickness and the earlier melt resulting from temperature increase, will modify the regime of the Garonne River and its tributaries in the mountainous area. The result shows a decrease in the river's average seasonal flows and a significant decrease in its low-flows. Hence, the vulnerability of the aquatic and wetland environments will also increase.

On-site runoff water infiltration and then groundwater modelling were released to evaluate the effect of infiltrated runoff water on the groundwater level and to estimate its residence time before reaching the Garonne River. The obtained results show that the artificial recharge of the alluvial aquifer, which is well exploited and is maintained a strong hydraulic relationship with the river, can be potentially interesting as an adaptation strategy to fight against of climate changes in the study area. The preliminary result of the first experiment of using the surface water of the Lateral Canal as a source to recharge the alluvial aquifer allows us to determine the infiltration capacity of ditch located front of this Canal and to monitor the effect on groundwater level. Further measurements and groundwater modelling for each selected site will be carried out for a better understanding and assessment of this strategy to maintain the river's low flow. The unavoidable changes in the hydrological regime in the Garonne basin will increase the pressure on water resources. The adaptation to climate changes does not consider only the water management authorities but also private users such as industry, farmers and the individual users. Consequently, integrating strategies for adapting to climate change into integrated water management plans must be considering in this very important part of France in terms of economy, especially agricultural production. Hence, the nature-based solution can be considering for better and sustainable water resource management in maintaining the biodiversity of the corresponding ecosystems. It will help to strengthen the resilience of the stakeholders and the relevant ecosystems to climate change. Artificial recharge of the alluvial aquifer can therefore be considered in this alluvial plain where other methods such as reservoir creation are not feasible.

Acknowledgements The authors are grateful to SMEAG (Syndicat Mixte d'Etudes et d'Aménagement de la Garonne) and Water Agency of Adour-Garonne Basin (L'agence de l'eau Adour-Garonne) for their financial support and assistance in providing the data.

References

Asmael, N., Dupuy, A., & Leroy, B. (2023a). Artificial recharge of the shallow alluvial aquifer as an adaptation strategy in the Garonne valley, France. *IAHS, 100*, 1–6. Accepted 22 Jan 2023

Asmael, N., Dupuy, A., McLachlan, P., & Franceschi, M. (2023b). Hydro-Geochemical characteristics of the shallow alluvial aquifer and its potential artificial recharge to sustain the low flow of the Garonne River. *Water, 15*(16), 2972. https://doi.org/10.3390/w15162972

Bernard, T., Sinclair, H., Naylor, M., Weir, E., Christophoul, F., & Ford, M. (2020). Post-orogenic sediment drape and flux of mountain range-foreland basin systems: An example from the Northern Pyrenees (No. EGU2020-9079). Copernicus Meetings.

BRGM Infoterre. Available online: https://infoterre.brgm.fr Accessed on 22 Feb 2022.

Carozza, J. M., Carozza, L., Valette, P., Llubes, M., Py, V., Galop, D., et al. (2014). The subfossil tree deposits from the Garonne Valley and their implications on Holocene alluvial plain dynamics. *Comptes Rendus Geoscience, 346*, 20–27.

Christophoul, F., Regard, V., Martinod, J., & Darrozes, J. (2014). *Morphodynamics of the Upper Pleistocene Garonne River (SW France): Conditions of braiding/meandering transition* (p. 21).

Grusson, Y., Anctil, F., Sauvage, S., & Sánchez Pérez, J. M. (2018). Coevolution of hydrological cycle components under climate change: The case of the Garonne River in France. *Water, 10*, 1870.

Danneville. L. (1995). Eau-Aménagement-Environnement CONTRIBUTION DES EAUX SOUTERRAINES AUX DÉBITS ET À LA QUALITÉ DES EAUX DE SURFACE: exemple de la Garonne, de ses sources à la confluence du Tarn. Documents du BRGM 262 Editions BRGM. Ingénierie de l'environnement. Université Paris XI—Orsay, France.

Diersch, H. J. G. (2018). *FEFLOW. Finite element simulation system for subsurface flow and transport processes, Berlin, Germany.*

Dubreuilh, J., & Alvinerie, J. (1978). Notice explicative, Carte géol. France (1/50 000), feuille Langon (852). Rapport technique, BRGM.

Dubreuilh, J., Capdeville, J. P., Farjanel, G., Karnay, G., Platel, J. P., & Simon-Coinçon, R. (1995). Dynamique d'un comblement continental Néogène et Quaternaire: l'exemple du bassin d'Aquitaine. *Géologie de la France, 4*, 3–26.

Ford, M., Hemmer, L., Vacherat, A., Gallagher, K., & Christophoul, F. (2016). Retro-wedge foreland basin evolution along the ECORS line, eastern Pyrenees, France. *Journal of the Geological Society, 173*(3), 419–437.

Hendrickx, F., & Sauquet, E. (2013). Impact of warming climate on water management for the Ariège River basin (France). *Hydrological Sciences Journal, 58*, 976–993.

Hubschman, J. (1975). Les terrasses récentes de la Garonne et leur évolution. *Bulletin de l'association française pour l'étude du quaternaire, 12*(3–4), 137–147.

Lambert, R., Lami, J.M., & Senges, F. (1990). Etude Hydraulique de la Garonne. La Sécheresse de 1989 dans le Bassin de la Garonne (à L'amont du Mas d'A gênais); Université de Toulouse-le-Mirail, Institute de Géographie, et du Ministère de l'Environnement, Délégation de bassin Adour-Garonne: Toulouse, France (Vol. 2, 75p).

Lhuissier, L., Lamblin, V., Sauquet, E., Arama, Y., Goulard, F., & Strosser, P. (2016). Garonne 2050: A prospective study. *La Houille Blanche, 102*(6), 30–35. https://doi.org/10.1051/lhb/2016057

Olivetti, V., Balestrieri, M. L., Godard, V., Bellier, O., Gautheron, C., Valla, P. G., Zattin, M., Faccenna, C., Pinna-Jamme, R., & Manchuel, K. (2020). Cretaceous and late Cenozoic uplift of a Variscan Massif: The case of the French Massif Central studied through low-temperature thermochronometry. *Lithosphere, 12*(1), 133–149.

Ortiz, A., Guillocheau, F., Robin, C., Lasseur, E., Briais, J., & Fillon, C. (2022). Siliciclastic sediment volumes and rates of the North Pyrenean retro-foreland basin. *Basin Research, 34*(4), 1421–1439.

Probst, J. L. (1983). *Hydrologie du Bassin de la Garonne: Modèles de Mélange, Bilan de l'Erosion, Exportation des Nitrates et des Phosphates* (Ph.D. Thesis). University Toulouse, Toulouse, France.

Queensland Government. (2017). Groundwater dependent ecosystem pictorial conceptual model 'alluvia': version 1.5, Queensland Government, Brisbane. Available online: https://wetlandinfo.des.qld.gov.au/resources/static/pdf/ecology/gde/170303-gde-pcm-alluvia.pdf. Accessed on 22 Feb 2022.

Richard, B., Bayle, A., Legube, B., Enjalbal, D., Leurent, T., Sauquet, E., & Flipo, S. (2022). Hydrologie et changements climatiques : quelles tendances observées et à venir sur le bassin Adour-Garonne ? Report Agence de l'eau Adour-Garonne

Sauquet, E. & Evin, G. (2022). Rapport Explore2 synthese Connaissance Metropole—VF.pdf, Explore2. https://doi.org/10.57745/V4G4Y9.

Steiger, J., James, M., & Gazelle, F. (1998). Channelization and consequences on floodplain system functioning on the Garonne River, UK. *Regulated Rivers Research & Management: An International Journal Devoted to River Research and Management, 14*, 13–23.
Valette, P. (2002). *Les paysages de la Garonne: Les métamorphoses d'un fleuve (entre Toulouse et Castets-en-Dorthe)* (p. 554). Université de Toulouse Le Mirail.
Valette, P. (2011). Changements environnementaux et adaptation des sociétés dans la moyenne vallée de la Garonne (XVIIIe siècle à aujourd'hui), Sud-Ouest européen.

Chapter 15
Enhanced Understanding of Recharge Mechanisms in Brackish/Freshwater Sedimentary and Carbonate Aquifers

Amjad Aliewi, Muhammad Al-Rashed, Khalid Hadi, Ahmed Abdulhadi, and Fawzia Al-Ruwaih

Abstract Estimating recharge holds significant importance in water resources planning and management, particularly in arid and semi-arid regions, influencing both sedimentary and carbonate aquifers. Recharge plays a crucial role in determining the sustainable yield of aquifers, particularly within evidence of climate change marked by more frequent flash floods and fluctuations in annual rainfall quantities and patterns. This study examines the concepts surrounding two types of recharge for sedimentary and carbonate aquifers utilized in arid and semi-arid environments. Recharge to aquifers typically occurs through various mechanisms, including vertical recharge from rainfall and seepage from sewage and water supply networks. Additionally, lateral recharge or discharge, whether it's transboundary (regional recharge) or through submarine groundwater discharge (SGD), contributes significantly to the recharge dynamics. The mechanisms governing vertical and lateral recharge to brackish aquifers in arid/semi-arid environments differ from those of freshwater aquifers. This disparity stems from variations in the salt mass balance control when recharging water reaches the aquifer water levels, depending on whether the native aquifer water is fresh or brackish. In addition, the process of water recharge from the unsaturated zone to groundwater levels may exhibit variations between arid and semi-arid environments. This paper shows how the governing equations for recharge to brackish aquifers differ from those of freshwater aquifers in both arid and semi-arid environments. The mechanisms of transboundary recharge and submarine groundwater discharge are based on the development of contour maps depicting groundwater levels of the stressed aquifer under investigation. This study applies these mechanisms to Kuwait's sedimentary and carbonate aquifers, namely the Kuwait Group and Dammam aquifers. By accurately estimating recharge and SGD to/from the utilized aquifers, it is better Equipped to ensure the sustainable management and development of their groundwater resources. The results of applying the above mechanisms

A. Aliewi (✉) · M. Al-Rashed · K. Hadi · A. Abdulhadi
Water Research Centre, Kuwait Institute for Scientific Research, Kuwait City, Kuwait
e-mail: aaliewi@kisr.edu.kw

F. Al-Ruwaih
Department of Earth and Environmental Sciences, Kuwait University, Kuwait City, Kuwait

A. Sefelnasr et al. (eds.), *Water Resources Management and Sustainability*, Water Science and Technology Library 114, https://doi.org/10.1007/978-3-031-80520-2_15

show that recharge to brackish aquifers is in the range from 5 to 20% of rainfall and it is up to 25% of rainfall for freshwater aquifers.

Keywords Aquifer recharge · Arid and semi-arid environments · Salt mass balance · Sustainable yield of aquifers · Fresh/brackish aquifers · Sedimentary/carbonate aquifers

15.1 Introduction

Estimating recharge presents a challenge due to the inherent uncertainty in the process, as highlighted by Walker et al. (2019), particularly across diverse spatial and temporal scales, as noted by Li et al. (2023). Recharge to aquifers originates from various sources and occurs through different mechanisms, including: (1) vertical infiltration from rainfall, irrigation runoff, or leaks from water and sewage networks; (2) Lateral influxes from regional or transboundary sources. Recharge occurs in both the primary porosity of rocks and the secondary porosity of karstified rocks. In karst rock systems, the travel time of recharging water is shorter (Blake & Goldschmidt, 1947). Water moves from the unsaturated zone of lower permeability to the saturated zone of higher permeability (Hogan et al., 2004). Recharge rates are higher in the unsaturated zone when soil moisture is high, as less moisture allows water to accumulate in the unsaturated zone, increasing the travel time to recharge the aquifer's water table (Huet et al., 2016). In deep unsaturated zones, recharge may take years to occur and can be erratic (Izbicki et al., 2002). Temporal variations in recharge depend on the nature and frequency of rainfall storms and floods. Aquifer recharge is vital for water resource sustainability (Al-Rashed & Aliewi, 2018; Al-Rashed et al., 2023). Permeable soil and gentle ground slopes promote greater infiltration and thus greater recharge (Delin et al., 2000). In desert environments, depressions with higher soil moisture contribute significantly to aquifer recharge. Vegetation affects recharge processes by extracting infiltrated water for evapotranspiration, leaving only a fraction for recharge. Un-vegetated areas decrease total evapotranspiration, thereby increasing recharge (Gburek & Folmar, 1999). The presence of fractures and karst features near the ground surface enhances aquifer recharge spatially. Accurate recharge estimation is crucial for establishing a reasonable water budget and assessing sustainable yield, especially in areas where groundwater extraction is high. Different recharge assessment methods for the same aquifer can yield diverse results due to inherent limitations. Precise estimation requires accurate runoff, soil moisture, and evapotranspiration data, particularly under changing climate conditions, urbanization, and increasing water demand. Studies on various aquifers have shown different recharge percentages relative to total rainfall, emphasizing the importance of local conditions. The aim of this article is to enhance understanding of recharge mechanisms in brackish and fresh groundwater aquifers in arid and semi-arid environments.

15.2 Methods

The following methods are the most important methods that are applicable to estimate aquifer recharge in arid and semi-arid environments:

15.2.1 Water Budget Method

This method includes establishing a water budget for the aquifer area under consideration (Bredehoeft et al., 1982; Maurer & Thodal, 2000; Messerschmid & Aliewi, 2022). This water budget (Eq. 15.1) includes precipitation, irrigation returns, leakages from water and wastewater networks, evapotranspiration, runoff etc., plus change in aquifer storage.

$$R_P = P + L + TF - ET - R_{off} - Q \pm \Delta S \tag{15.1}$$

where

R_P recharge (m^3/day).
P Precipitation (m^3/day).
L Leakages from water supply and sewer systems (m^3/day).
TF Transboundary influx (m^3/day).
ET Evapotranspiration (m^3/day).
R_{off} Surface runoff (m^3/day).
Q Total abstraction from wells (m^3/day)
ΔS Change in storage (m^3/day).

However, this method suffers from several shortcomings (Levin et al., 2023) such as:

(a) The term "change in storage" requires accurate information about water levels time series for the utilized aquifers under investigation. Having a short length of time series of water levels will make it difficult to assess changes in the storage of aquifer to conclude an accurate water balance of the catchment.
(b) Accurate estimation of spatial and temporal values of precipitation, leakages, transboundary fluxes, evapotranspiration, surface runoff and abstraction from wells is always a major challenge to yield accurate recharge distribution in space and time.
(c) Equation 15.1 does not include the use of soil water budgets which are important to assess potential recharge to aquifers through the unsaturated zone.
(d) Although Eq. 15.1, accounts for evapotranspiration, but it does not connect it to changes in water storage. Clearly, water storage in the unsaturated zone is related to both components of evapotranspiration and aquifer recharge.

15.2.2 Water Levels Fluctuations Method

Fluctuations in groundwater levels are normally affected by aquifer recharge, therefore their variation in time over specific locations can be used to estimate aquifer recharge (Healy & Cook, 2002). This method (Eq. 15.2; Fig. 15.1) is normally used for unconfined aquifers.

$$R = S_y \frac{\Delta h}{\Delta t} \tag{15.2}$$

where

R	Recharh rate (m/day)
S_y	specific yield [dimensionless]
H	water table height (m)
t	time (day)
$\Delta h/\Delta t$	is the difference in water table (Δh) over time between measurements (Δt).

Equation 15.2 is applied only over periods of water-level rise (i.e. Δh is positive as seen in Fig. 15.1, Δh_1, Δh_2 and Δh_3) which show how water level was risen as a result of rainfall events that took place a few months prior to the maximum rise in the water levels. The method works best over short periods for shallow water tables that display sharp water table level rises and declines. The method inherently assumes that recharge occurs only as a result of brief events. No estimation of recharge under steady state conditions is possible.

The application of Eq. 15.2 suffers from the following shortcomings:

(a) Fluctuations in groundwater levels can be caused not only by recharge but by also other events including variations in atmospheric pressure; extraction or injection of water by pumping etc. (Todd, 1989). Careful examination of the cause of water level variations is necessary to correctly identify water level rises that can be attributed to recharge events.
(b) Accurate estimation of the value of specific yield is a problem. There is always an uncertainty issue (Rasmussen & Andreasen, 1959) with estimating Sy either in the laboratory or in the field or even empirical methods.
(c) When rainfall events occur for a long duration with low intensity, then recharge estimation by this method will normally be underestimated.
(d) When rainfall events occur with high-intensity, then air can be trapped in the unsaturated zone increasing the rise in water levels and thus overestimating aquifer recharge (Heliotis & DeWitt, 1987).

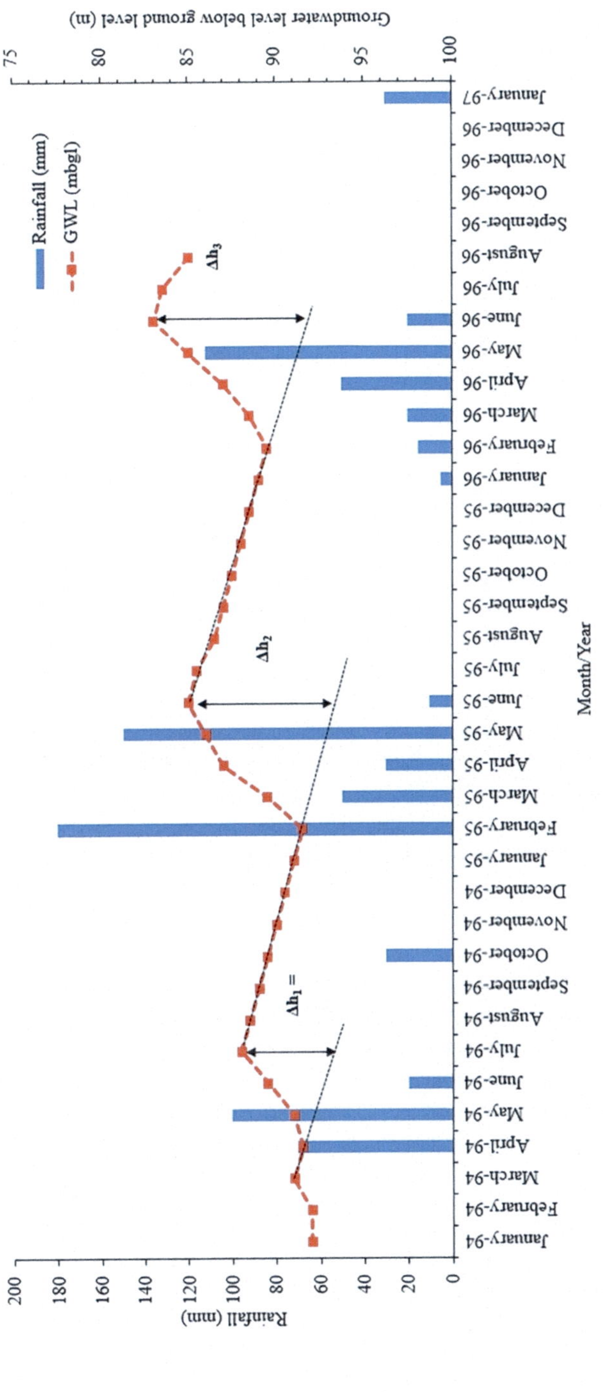

Fig. 15.1 The concept fluctuation in water tables as a result of rainfall events to estimate aquifer recharge

15.2.3 *Statistical Methods Relating Recharge as Percentage of Rainfall*

15.2.3.1 Sedimentary Aquifers

This method utilizes rainfall data in a number of rainfall stations all-over the concerned catchment to calculate the long-term annual average for the desired available time series period (Al-Qallaf et al., 2020; Hussary et al., 1995). The following statistical methodology is developed in this study to calculate recharge to aquifers on monthly basis rather than annual events to avoid providing recharge to aquifers in dry months. The methodology also adopts accumulation of rainfall for 3 months only. This methodology assumes recharge to be a single correlation with rainfall on monthly basis. The concept of Eq. 15.3 is adopted:

$$R_a = \sum_{i}^{12} b_i P_{ci} \tag{15.3}$$

R_a is the annual recharge(mm)for that year that includes months from $i = 1$ *to* $i = 12$.
i is the month of the year from i – 1 (January) to i = 12 (December).
b_i is the recharge coefficient for the month i.
P_{ci} is the total contribution of all rainfall events into month i.

The recharge cofficient (b_i) for each month is taken as the average of rainfall of each month over the period (take Kuwait as an example form 1962 to 2022) for all available rainfall stations divided by the average annual rainfall of the same period for all rainfall stations in Kuwait. If the airport rainfall station of Kuwait was taken, then the percentage of monthly rainfall contribution to the annual average rainfall at the airport looks as shown in Fig. 15.2. As an example the month December is contributing almost 20% of the long term annual average rainfall in the airport of Kuwait. The months June, July, August and September contribute nothing to the annual rainfall over Kuwait airport.

Then the monthly contribution was taken for all available rainfall stations and produced avarge coefficents as seen in Fig. 15.3.

The contribution of each month for all rinfall station is assumed initially to be recharge coeffcient for that month and is summarized in Table 15.1.

The P_{ci} values are taken as the average of 3 months: the current month and previous 2 months as in Eq. 15.4.

$$P_{ci} = \frac{(P_{ci}) + (P_{ci,i-1}) + (P_{c,i-2})}{3} \tag{15.4}$$

The approach developed in this paper assumes that the underlying aquifer stores water as recharge from rainfall of the 3 consecutive months: the current month and two

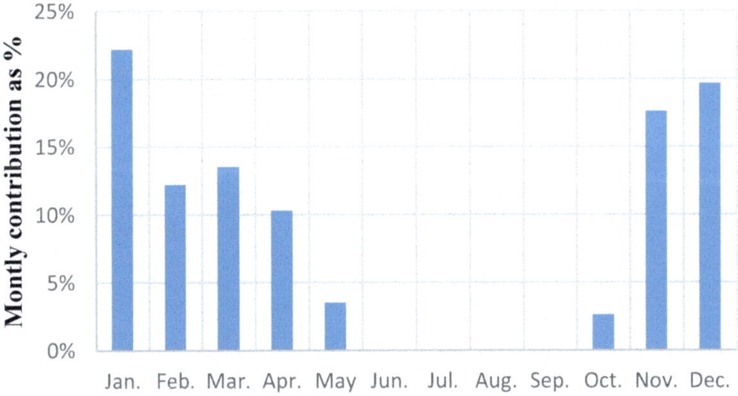

Fig. 15.2 Monthly rainfall contribution of long term annual rainfall over the period (1962–2022) in Kuwait

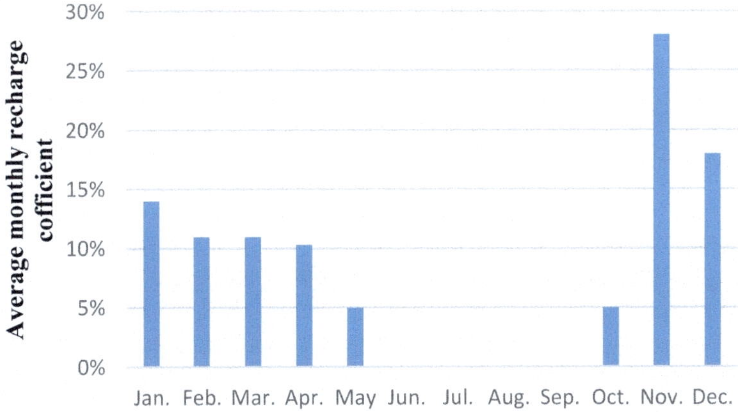

Fig. 15.3 Monthly recharge coffiecien (bi) from average long term annual rainfall over the period (1962–2022) for all available rainfall stations

previous months. The recharge coefficient of each month is calculated as presented in Table 15.1. This means that the effect of rainfall (to appear as recharge in the aquifer) is accumulated over a period of 3 consecutive months.

15.2.3.2 Carbonate Aquifers

This study follows the approach developed by Abu Sada et al. It is observed for a catchment underlined by carbonate (karst) aquifers that recharge to the aquifer will increase gradually till after 4 months when the maximum of this event is reflected in the water level rise of the aquifer. After that the contribution of this event to aquifer

Table 15.1 Long term monthly recharge coefficients for Kuwait upper Aquifer

Month	Recharge coefficient
January	0.14
February	0.11
March	0.11
April	0.10
May	0.05
June	0
July	0
August	0
September	0
October	0.05
November	0.28
December	0.18
Total	1.02 (error 2%)

recharge reduces gradually till the end of a period of about a few more months from the peak. It was clearly observed that the month of April exhibits the peak of rise in the aquifer's water level which indicates the delay and the accumulation in the rainfall events over the rainy season is 6 to 7 months.

15.2.4 Chloride Mass-Balance (CMB)

The Cl^- can be readily utilized for recharge estimation, particularly in arid and semiarid regions. However, careful consideration is warranted regarding the nature of the receiving aquifer, whether it is freshwater (with TDS < 1000 mg/l) or brackish (with 10,000 > TDS > 1000 mg/l). The long-term concentration of Cl^- in the aquifer from other sources, known as background concentration, must be accounted for to ensure the successful application of CMB especially in brackish aquifers. The classic method for estimating aquifer recharge as a percentage of rainfall, using the ratio of Cl^- concentration in rainfall to that in the groundwater of the underlying aquifer, is inaccurate for several reasons: (a) it assumes that chloride's sole source is rainfall (McKenzie, 2001), whereas it is well-known that Cl^- originates from other sources such as mineral dissolution, raw wastewater, and industrial and urban wastes. (b) The classic CMB disregards the vertical extent in salinity profile necessary to identify the effect of recharging water in the aquifer, crucial for determining the recharge rate from an external source. Building upon the aforementioned discussion, Aliewi et al. (2022) developed new Eq. 15.5 to estimate aquifer recharge (for both freshwater and brackish) in sedimentary and carbonate aquifers as follows (Eq. 15.5).

$$R = A.b.\overline{S}.\left[\frac{Cl_0 - Cl_m}{Cl_m - Cl_P}\right] \tag{15.5}$$

R: Recharging flux (Mm^3 per unit time); $A.b.\overline{S}.$: volume of that portion of aquifer under recharge stresses (Mm^3 per unit time); A: the surface area of that portion of aquifer under recharge stresses (m^2); b: saturated thickness of the aquifer only affected by infiltrating water (m); $\overline{S}$: specific yield (Sy) of the unconfined aquifer and Storage coefficient (S) of the confined aquifer under recharge stresses; Cl_0: Background chloride concentration (mg/l) that represents all other sources; Cl_P: Chloride (mg/l) in rainfall; Cl_m: Measured chloride (mg/l) in water samples taken from the aquifer after rain events.

15.2.5 Method of Contour Maps

The method is based on Darcy Equation

$$Q = -K.A.\frac{\Delta H}{\Delta l} \tag{15.6}$$

When Darcy's Equation is applied to the Fig. 15.6, it reads as:

$Q = -K.(Thickness.Width).\frac{\Delta H}{\Delta l}$, for the lateral flow of Fig. 15.4 coming from the left side to the right side provided that (*K. Thickness*) for unconfined aquifers and (*K.b*) for confined aquifers is Transmissivity (T), thus Eq. 15.6 can be written as

$$Q = -(T.width).\frac{\Delta H}{\Delta l} \tag{15.7}$$

Equation 15.7 can be used to estimate both lateral recharge alongside trans-boundary with width y and submarine groundwater discharge as shown on Fig. 15.4.

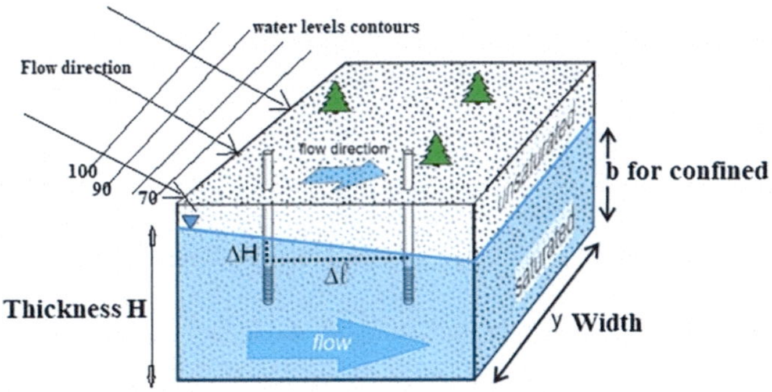

Fig. 15.4 Schematic of Darcy's law for lateral flows

This concept will be utilized to estimate transboundary fluxes and SGD to both utilized aquifers in Kuwait which are Kuwait Group and Dammam Aquifers.

15.3 Results and Discussion

15.3.1 Sedimentary Aquifers

15.3.1.1 Statistical Methods Application

The monthly average of rainfall for each month from 1962 to 2022 were taken for all available rainfall stations in Kuwait to apply Eq. 15.3 and 15.4. The results were presented in Table 15.2 which clearly shows that the total annual long term aquifer recharge is 15 mm which is 14% of the average annual long term (1962–2022) rainfall in Kuwait. Figure 15.5 shows that the monthly rainfall is dry in June to September for period (1962–2022) but when the average accumulation of the current month and previous two months were taken, the rainfall contribution to recharge events was dry only in the months of August and September. The red bar depicted in Fig. 15.5 illustrates aquifer recharge, constituting a relatively small fraction (14%) of the total rainfall. In practical scenarios, monthly rainfall data offers a more realistic approach compared to annual rainfall figures. SubsEquationuently, Eq. 15.3 and 15.4 were employed to estimate monthly recharge to the aquifer (Table 15.2). Figure 15.5 shows that for the long term average year, the effect of average accumulated rainfall is greater than monthly rainfall for the current month from January to September while it is less from Octobr to December. This is because this contribution value is affected by rainfall of the previous two months.

15.3.1.2 Contour Maps Application

It has always been very important to determine the lateral influx (recharge) and the submarine groundwater discharge for the Kuwait Group Aquifer (Upper Aquifer). To do so, Eq. 15.7 as $Q = -(T.width).\frac{\Delta H}{\Delta l}$ was applied as follows:

(i) Lateral transboundary influx

The input data is as follows:

- Southern boundary of Kuwait through which the transboundary influx takes place = 222,000 m.
- Transmissivity (T) value near the southern boundary of Kuwait: Calibrated hydraulic conductivity 10 m/d (Aliewi et al., 2021a), effective thickness that allows of transboundary influx = 125 m. Then, T = 1250 m^2/day.
- Referring to Fig. 15.6:

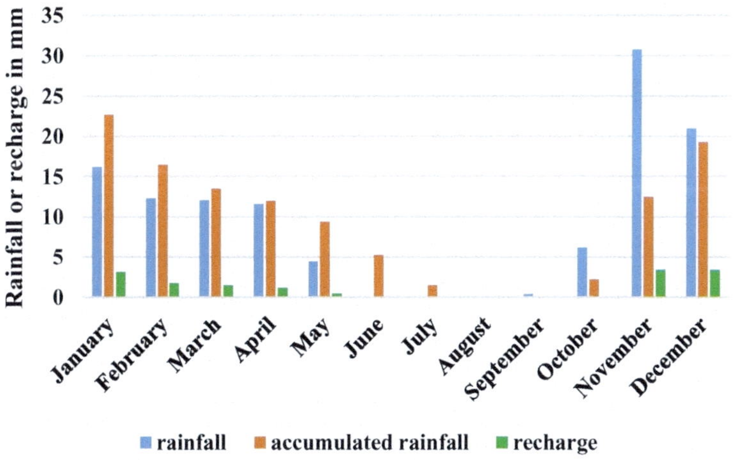

Fig. 15.5. Application of Eqs. 15.3 and 15.4 to estimate monthly aquifer recharge in Kuwait

Table 15.2 Average long term (1962–2022) monthly rainfall and recharge over Kuwait compiled from several rainfall stations

Month	Recharge coeffiecint (bi)	Average long term Rainfall (mm)	Application of Eq. 15.4 (Pci) (average values of accumulated rainfall for each month) (mm)	Application of Eq. 15.3 (bi * Pci) values Recharge (mm)
January	0.14	16.2	22.7	3.2
February	0.11	12.3	16.5	1.8
March	0.11	12.1	13.5	1.5
April	0.10	11.6	12.0	1.2
May	0.05	4.5	9.4	0.5
June	0	0.0	5.3	0.0
July	0	0.0	1.5	0.0
August	0	0.0	0.0	0.0
September	0	0.4	0.1	0.0
October	0.05	6.2	2.2	0.1
November	0.28	30.8	12.5	3.5
December	0.18	21.0	19.3	3.5
Total	1.02 (error = 2%)	Annual = 113.1	Annual = 115 (error = 1.7%)	Annual = 15.2

- $\Delta H = 110 - 110 = 10$ m.
- Average distance between contour 110 m asl and 100 m asl = Δl = 15,000 m.
- Width of southern boundary = 222,000 m.

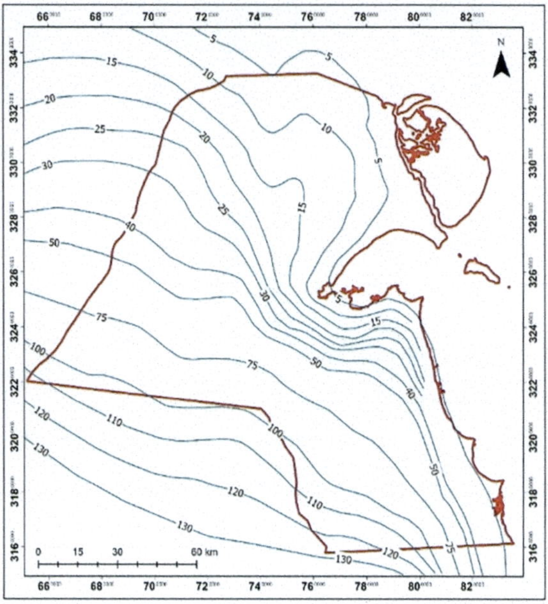

Fig. 15.6 Water levels contours of Kuwait Group aquifer in Kuwait, pre-1960 (Mukhopadhyay et al., 1996)

Then applying Eq. 15.7 (1250 * 222,000 *[10/15,000]) yields: lateral influx or recharge = 185,000 m^3/day = 67.5 Mm^3/year.

(ii) Submarine Groundwater Discharge (SGD)

The input data is as follows:

- Sea boundary with Kuwait through which the SGD takes place = 420,000 m.
- Transmissivity (T) value the western boundary of Kuwait with Arabian Gulf: Calibrated hydraulic conductivity = 25 m/day (Aliewi et al., 2021a), Effective thickness that allows of SGD to take place = 25 m. Then, T = 625 m^2/day.
- Referring to Fig. 15.6:
  - ΔH-northern section = 5 − 0 = 5 m.
  - ΔH-southern section = 15 − 0 = 15 m.
  - Average distance between contour 5 m asl and sea level = Δl-northern section = 12,000 m.
  - Average distance between contour 15 m asl and sea level = Δl-southern section = 5000 m.
  - Width of northern part of sea boundary = 210,000 m.
  - Width of southern part of sea boundary = 210,000 m.

SGD-northern part of sea boundary: (625 * 210,000 * [5/12,000]) yields = 54,688 m^3/day = 20 Mm^3/year.

SGD-southern part of sea boundary: (625 * 210,000 * [15/5000]) yields = 393,750 m^3/day = 144 Mm^3/year.

Total SGD = 164 Mm^3/year.

15.3.1.3 Modified CMB Method Application

In Eq. 15.5, Cl^- in rainfall dilutes the brackish water in the upper zone of the brackish aquifer. Thus, it possible to determine the thickness of the zone that was affected by recharging water with less Cl concentration (Aliewi et al., 2020; Aliewi et al., 2021b). The background concentration Cl_0 is the long-term concentration of Cl from other sources than rainfall. Equation 15.5 was applied to Kuwait aquifers utilizing the data of extreme rainfall events that took place in the first two weeks of November 2018 (Table 15.3). The results are shown in Fig. 15.7 which illustrates that the average aquifer recharge is about 9.3% from rainfall in November 2018. Figure 15.7 also shows that wherever the water level is closer to the ground surface, the aquifer recharge from rainfall is high and for deep water levels, the aquifer recharge from rainfall is less than 1%.

15.3.2 Carbonate Aquifers

15.3.2.1 Statistical Method Application

Wadi Natuf Catchment in Palestine (Fig. 15.8) is taken to estimate recharge utilizing the statistical method of monthly recharge explained earlier.

The Natuf catchment (in yellow) in Fig. 15.8 is located in the Ramallah Area. The recharge zone (in green) is shown with contour lines ranging 400–600 mm/year. The Natuf catchment extends over an area of 95 km^2. A rainfall map (Fig. 15.9) was superimposed over the entire surface of the Natuf Catchment using rainfall data of 12 stations. Then the zones of long-term rainfall over each geological outcrop (Fig. 15.10) were determined, then the spatial rainfall distributions over the outcrop formations were superimposed and finally the recharge coefficients were estimated from based on statistical methods (Table 15.4).

In previous studies, annual recharge has been calculated as a percentage of the long-term annual average rainfall over the basin as a whole. Scarpa (1994) came out with 20%, while Blake and Goldschmidt (1947) with 22–25%, Rofe and Raffety (1963, 1965) with 20–55% and Guttman and Zukerman (1995) with 25–60%. The problem with the above studies is that they use a percentage of recharge from the annual rainfall (except for Guttman & Zukerman, 1995) who takes percentages for 3 ranges of rainfall and that they neglect the effect of geology and hydrogeology when they estimate the recharge on aquifer outcrops.

Table 15.4 shows that the total rainfall over the entire Natuf catchment is 49.70 Mm^3/year while Table 15.5 shows that the total recharge is 12.75 Mm^3/year which effectively means that aquifer recharge is 25.7% form rainfall. This result goes in line

Table 15.3 Application of Eq. 15.5 to estimate recharge over entire State of Kuwait (After Aliewi et al., 2022)

Rainfall station →	A	B	C	D[a]	E	F	G	H	I-1[a]	I-2	Total
A (km^2)	1529	3105	2960	2087	763	2501	457	485	1458	1099	16,444
Rainfall (Mm^3)	263	199	299	292[a]	159	540	74	88	204[a]	154	1776
Cl_p (mg/l)	16	16	16	16	16	16	16	16	16	16	
Cl_m (mg/l)	3960	5000	1910	939[a]	3081	5180	18,900	4110	30[a]	2315	
Cl_p/Cl_m	0.004	0.003	0.008	0.017[a]	0.021	0.003	0.0008	0.0133	0.53[a]	0.02	
b (m)[b]	3	3	3	1	2	3	1.5	2	3	3	
Sy (%)	1.7	1.7	1.7	1	1	1	1	1	1.7	1.7	
Groundwater (Mm^3)	76	155	148	21	15	75	7	10	73	55	635
Cl_0 (mg/l)	5325	5001	2250	1007	10,271	6300	23,640	13,700	4148	3830	
R_{CC} (Mm^3)	26.5	0	26.6	1.5[a]	35.8	16.3	1.7	22.7	172.4[a]	36.2	165.1
% of recharge	10.1	0.0	8.9	0.3[a]	22.5	3.0	2.3	25.8	85.0[a]	23.5	9.3 (avg)
$(Cl_0–Cl_m)/(Cl_m–Cl_p)$	0.35	0	0.18	0.074[a]	2.35	0.22	0.25	2.34	2.37[a]	0.66	

[a] rainfall stations excluded from analysis because of a human error in measuring Cl_m

[b] the saturated thickness affected by direct rainfall recharge in the vertical salinity profiles

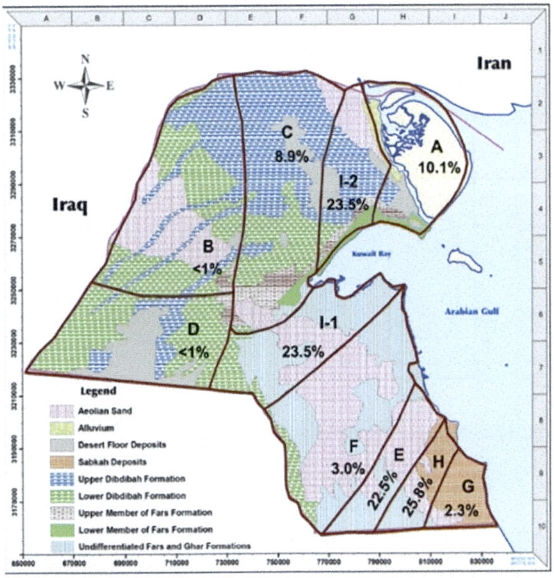

Fig. 15.7 Aquifer recharge as percentage of rainfall over Kuwait during 1–15 November 2018 (After Aliewi et al., 2022)

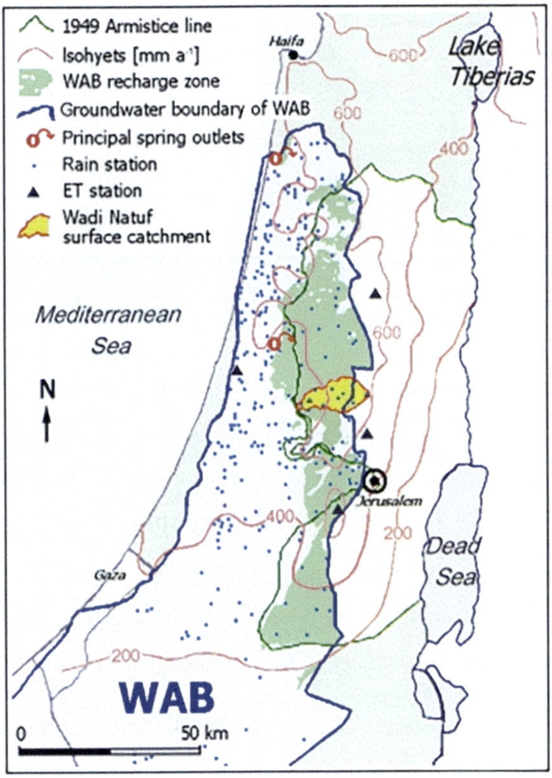

Fig. 15.8 Natuf surface catchment in Palestine (After Messerschmid et al., 2020)

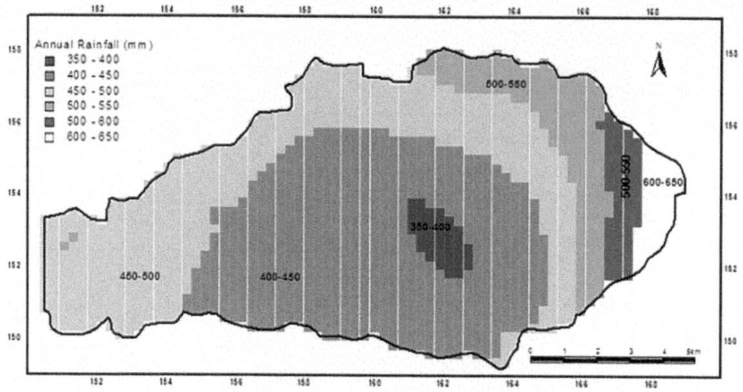

Fig. 15.9 Annual average rainfall map for Natuf Catchment

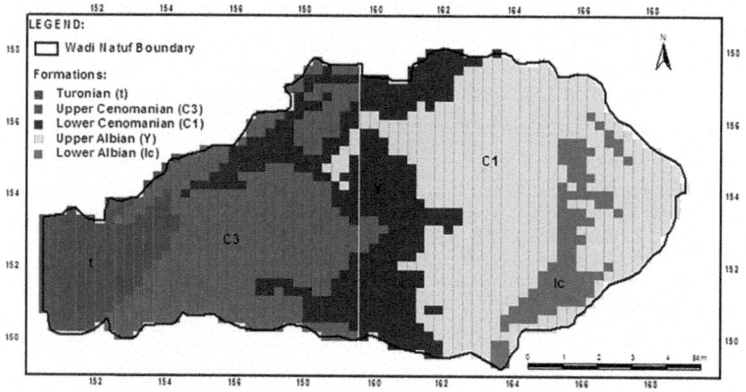

Fig. 15.10 Outcrop formations of the Natuf Catchment

with the results of Blake and Goldschmidt (1947), Rofe and Raffety (1963, 1965) and Guttman and Zukerman (1995).

15.3.2.2 Contour Maps Method Application

This method is applied for the Dammam aquifer in Kuwait. The lateral transboundary influx through Dammam aquifer can be estimated from the following input data:

- Southern boundary of Kuwait through which the transboundary influx takes place = 222,000 m.
- Transmissivity (T) value near (far left) of the southern boundary of Kuwait = 300 m^2/day.

Table 15.4 Observed rainfall over Natuf outcrops

	Observed rainfall (in 1000 m^3)												Total Mm^3
	Oct	Nov	Dec	Jan	Feb	Mar	Apr	May	June	July	Aug	Sep	
Tur	124.1	618.8	1350.5	1333.8	1131.4	381.2	74.5	10.0	0	0	0	0	5.02
Up.Cen	361.4	1643.1	3100.7	3033.0	2556.7	915.0	222.6	29.5	0	0	0	0	11.86
Lo.Cen	311.0	1379.5	2420.0	2433.6	2116.5	868.7	202.2	30.3	0	0	0	0	9.76
Up.Alb	650.5	2680.6	4673.2	4920.4	4254.9	2160.2	468.8	62.5	0	0	0	0	19.87
L.Cret	100.0	418.6	742.8	793.5	684.7	358.7	76.0	7.7	0	0	0	0	3.18
Total	1547.0	6740.5	12,287.3	12,514.3	10,744.1	4683.8	1044.1	139.9	0	0	0	0	49.70

Table 15.5 Recharge estimation in Natuf catchment

	Calculated monthly recharge (in 1000 m^3)												Total Mm^3
	Oct	Nov	Dec	Jan	Feb	Mar	Apr	May	Jun	Jul	Aug	Sep	
Tur	3.2	12.9	45.5	107.1	198.0	277.7	297.3	253.0	164.7	82.9	33.3	8.4	1.48
Up.Cen	8.7	32.3	105.7	242.6	436.5	595.1	628.7	533.2	349.3	179.3	73.8	20.0	3.21
Lo.Cen	6.6	23.2	73.4	167.4	299.1	405.5	431.8	372.7	250.4	131.8	55.7	16.1	2.23
Up.Alb	18.4	59.3	185.0	423.0	753.4	1031.3	1114.8	981.8	684.3	370.7	160.0	49.7	5.83
L.Cret	0.0	0.0	0.0	0.0	0.0	0.0	0.0	0.0	0.0	0.0	0.0	0.0	0.00
Total	36.9	127.7	409.5	940.1	1686.9	2309.6	2472.5	2140.7	1448.7	764.7	322.7	94.3	12.75

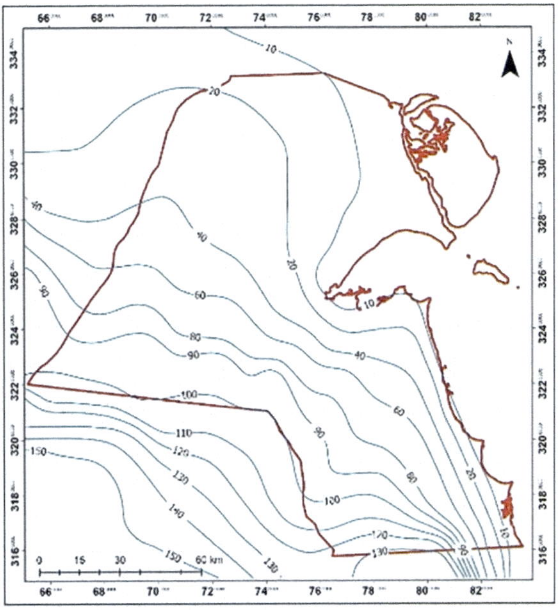

Fig. 15.11 Water levels contours of Dammam Aquifer in Kuwait

- Transmissivity (T) value near (far right) of the southern boundary of Kuwait = 750 m^2/day.
- Referring to Fig. 15.11:
- Far left part
- ΔH = 110 − 110 = 10 m.
- Average distance between contour 110 m asl and 100 m asl = Δl = 10,000 m.
- Far left width = 152,000 m.
- Then applying Eq. 15.7 (300 * 152,000 * [10/10,000]) yields: lateral influx or recharge = 45,600 m^3/day = 16.7 Mm^3/year.
- Far right part
- ΔH = 130 -120 = 10 m.
- Average distance between contour 130 m asl and 120 m asl = Δl = 5,000 m.
- Far right width = 70,000 m.
- Then applying Eq. 15.7 (750 * 70,000 * [10/5,000]) yields: lateral influx or recharge = 105,000 m^3/day = 38.3 Mm^3/year.
- Total transboundary recharge to Kuwait Dammam Aquifer = 55 Mm^3/year.

15.4 Conclusions

This paper offers a comprehensive review of freshwater and brackish water aquifer recharge mechanisms in arid and semi-arid environments. The paper highlights challenges associated with applying the water budget method in arid settings, stemming from the necessity of accurately estimating spatial and temporal water budget inputs and outputs. Additionally, it discusses the limitations of relying solely on fluctuations in groundwater levels for estimating aquifer recharge, particularly over short periods, as this method assumes a clear correlation between groundwater level rise and recharge, disregarding other potential causes. Moreover, accurate estimation of specific yield poses a significant challenge to this method. The study suggests that monthly recharge estimation, based on long-term records of rainfall, provides a more precise reflection of aquifer recharge in arid and semi-arid environments compared to yearly estimates, considering the variability in rainfall throughout the year, especially during dry months. To validate this approach, the monthly recharge method was applied to the Natuf catchment in Palestine, demonstrating a more accurate estimation of aquifer recharge, totaling 25.7% from rainfall. Furthermore, the study reveals discrepancies in classic chloride mass balance methods, which tend to underestimate aquifer recharge to brackish aquifers in arid environments. This underestimation arises from the neglect of chloride sources other than rainfall and the failure to account for chloride accumulation due to evaporation. The research findings suggest that the optimal percentage of aquifer recharge from rainfall for brackish aquifers falls between 10 and 20%, contrasting with the classic method's estimation in brackish aquifers in Kuwait, which ranges from 0.5% to 3% due to rainfall. Moreover, employing the water level contour maps method unveils significant transboundary recharge from Saudi Arabia to Kuwait Group Aquifer and Kuwait Dammam Aquifer, estimated at 67.5 Mm^3/year and 55 Mm^3/year, respectively. Additionally, the method highlights substantial submarine groundwater discharge (SGD) from Kuwait's eastern boundary to the Arabian Gulf, reaching 164 Mm^3/year.

Acknowledgements The authors express sincere gratitude to the administration of the Kuwait Institute for Scientific Research and Kuwait University for their invaluable support throughout the research process and also for their financial support, which enabled the completion of this study.

References

Aliewi, A., Bhandary, H., Sabarathinam, C., & Al-Qallaf, H. (2022). A new modified chloride mass balance approach based on aquifer hydraulic properties and other sources of chloride to assess rainfall recharge in brackish aquifers. *Hydrological Processes Journal, 36*(3). https://doi.org/10.1002/hyp.14513

Aliewi, A. S., Bhandary, H., Al-Qallaf, H., Sabarathinam, C., & Al-Kandari, J. (2021a). Assessment of the groundwater yield and sustainability of the transboundary Dibdibba aquifer using numerical modelling approach. *Groundwater for Sustainable Development Journal, 15*(2021), 100678. https://doi.org/10.1016/j.gsd.2021.100678

Aliewi, A. S., Al-Kandari, J., Al-Khalid, A., Bhandary, H., & Al-Qallaf, H. (2021b). Modelling the effect of high level of total dissolved solids (TDS) for the sustainable utilization of brackish groundwater from saline aquifers in Kuwait. *Environment Development and Sustainability, 23*, 2204–2223. https://doi.org/10.1007/s10668-020-00670-9.

Aliewi, A., Al-Enezi, H., Al-Maheimid, I., Kandari, J. Al-Haddad, A., Al-Qallaf, H, Rashid, T., & Sadequi, D. (2020). Sustainability of brackish groundwater utilization from the Eocene Aquifer for oil exploration operations in central Kuwait. *Environment, Development and Sustainability, 22*, 4639–4653. https://doi.org/10.1007/s10668-019-00401-9.

Al-Qallaf, H., Aliewi, A., & Abdulhadi, A. (2020). Assessment of the effect of extreme rainfall events on temporal rainfall variability in Kuwait. *Arabian Journal of Geosciences, 13*, 1129. https://doi.org/10.1007/s12517-020-06086-z

Al Rashed, M., Safelnasr, A., Sherif, M., Murad, A., Alshamsi, D., Aliewi, A., & Ebraheem, A. (2023). Novel concept for water security quantification considering nonconventional and virtual water resources in arid countries: application in Gulf Cooperation Council Countries. *Science of the Total Environment*, 163473, https://doi.org/10.1016/j.scitotenv.2023.163473. https://www.sciencedirect.com/science/article/pii/S0048969723020922)

Al-Rashed, M., & Aliewi, A. (2018). Water resources sustainability in Kuwait against United Nations sustainable development goals. In E. Azar & M. Raouf (Eds.), *Sustainability in the Gulf: Challenges and opportunities* (312 pp, pp. 37–56). Published by Routledge Explorations in Environmental Studies. ISBN 9781138040687 (hb).

Blake, G. S., & Goldschmidt, M. J. (1947). *Geology and water resources of Palestine, Government of Palestine, department of land settlement and water commissioner, Jerusalem.*

Bredehoeft, J. D., Papadopoulos, S. S., & Cooper H. H. Jr. (1982). *Groundwater –the water-budget myth.* Scientific Basis of Water Resource Management, National Academy Press, pp. 51–57.

Delin, G. N., Healy, R. W., Landon, M. K., & Bohlke, J. K. (2000). Effects of topography and soil properties on recharge at two sites in an agricultural field. *American Water Resources Association, 36*, 1401–1416.

Gburek, W. J., & Folmar, G. J. (1999). A ground water recharge field study—Site characterization and initial results. *Hydrological Processes, 13*, 2813–2831.

Guttman, J., & Zukerman, C. (1995). Setup and calibration of the flow and salinity models. Technical report # TAHAL 01/95/72.

Heliotis, F. D., & DeWitt, C. B. (1987). Rapid water table responses to rainfall in a northern peatland ecosystem. *Water Resources Bulletin, 23*, 1011–1016.

Healy, R. W., & Cook, P. G. (2002). Using groundwater levels to estimate recharge. *Hydrogeology Journal, 10*, 91–109.

Huet, M., Chesnaux, R., Boucher, M. A., & Poirier, C. (2016). Comparing various approaches for assessing groundwater recharge at a regional scale in the Canadian shield. *Hydrological Sciences Journal, 61*(12), 2267–2283.

Hogan, J. F., Phillips, F. M., & Scanlon, B. R. (2004). *Groundwater recharge in a desert environment, the southwestern United States* (Vol. 9). Book Series: Water Science and Application. American Geophysical Union. https://doi.org/10.1029/WS009.

Hussary, S., Aliewi, A., & Najjar, T. (1995). *Analysis of secondary source rainfall data from the northern West Bank.* Publications of Newcastle University (UK) and Palestinian Hydrology Group.

Izbicki, J. A., Radyk, J., & Michel, R. L. (2002). Movement of water through the thick unsaturated zone underlying Oro Grande and Sheep Creek Washes in the western Mojave Desert, USA. *Hydrogeology Journal, 10*, 409–427.

Li, M., Xie, Y., & Dong, Y. (2023). Review: Recent progress on groundwater recharge research in arid and semiarid areas of China. *Hydrogeology Journal.* https://doi.org/10.1007/s10040-023-02656-z

Levin, S. B., Briggs, M. A., Foks, S. S., Goodling, P. J., Raffensperger, J. P., Rosenberry, D. O., Scholl, M. A., Tiedeman, C. R., & Webb, R. M. (2023). Uncertainties in measuring and

estimating water-budget components: Current state of the science. *Wires Water, 10*(4), e1646. https://doi.org/10.1002/wat2.1646
Maurer, D. K., & Thodal, C. E. (2000). *Quantity and chemical quality of recharge, and updated water budgets, for the BasinFill Aquifer in Eagle Valley, Western Nevada.* Water Resources Investigations Report 99–428, U.S. Geological Survey.
McKenzie, A. A. (2001). *West Bank aquifers-conceptual recharge estimation.* Technical report no. SUSMAQ-REC#04V0.1. Sustainable Management for the West Bank and Gaza Aquifers Project (SUSMAQ). Palestinian Water Authority (Palestine) and Newcastle University (UK).
Messerschmid, C., & Aliewi, A. S. (2022). Spatial distribution of groundwater recharge, based on regionalised soil moisture models in Wadi Natuf karst aquifers, Palestine. *Hydrology and Earth System Sciences, 26,* 1043–1061. https://doi.org/10.5194/hess-26-1043-2022 © Author(s) 2022. This work is distributed under the Creative Commons Attribution 4.0 License. https://doi.org/10.5194/hess-26-1043-2022.
Messerschmid, C., Sauter, M., & Lange, J. (2020). Field-based estimation and modelling of distributed groundwater recharge in a Mediterranean karst catchment, Wadi Natuf, West Bank. *Hydrol Earth System Science, 24*(887–917), 2020. https://doi.org/10.5194/hess-24-887-2020
Mukhopadhyay, A., Al-Sulaimi, J., & Al-Ruwaih, F. (1996). An overview of the Tertiary geology and hydrogeology of the northern part of the Arabian Gulf region with special reference to Kuwait. *Earth-Science Reviews, 40*(3–4), 259–295. https://doi.org/10.1016/0012-8252(95)00068-2
Rasmussen, W. C., & Andreasen, G. E. (1959). *Hydrologic budget of the Beaverdam Creek Basin, Maryland.* Water-Supply Paper 106, U.S. Geological Survey.
Rofe and Raffety. (1963). *Jerusalem and District water supply, geological and hydrological report.* Hashemite Kingdom of Jordan. Central Water Authority. Rofe & Raffety Consulting Engineers. Westminster, London.
Rofe and Raffety. (1965). *Nablus District water resources survey, geological and hydrological report.* Hashemite Kingdom of Jordan. Central Water Authority. Rofe & Raffety Consulting Engineers. Westminster, London.
Scarpa, D. (1994): Eastward groundwater flow from the mountain aquifer. In Isaac, J. & Shuval, H. (Ed.), *Water and Peace in the Middle East. Proceedings of the First Israeli-Palestinian International Academic Conference on Water.* Zürich, Switzerland. 10–13 Dec 1992 (Vol. 58, pp. 193–204). Studies in Environmental Science, Elsevier, Amsterdam.
Todd, D. K. (1989). *Groundwater Hydrology.* Wiley.
Walker, D., Parkin, G., Schmitter, P., Gowing, J., Tilahun, S. A., Haile, A. T., & Abdu Yimam, Y. (2019). Insights from a multi-method recharge estimation comparison study. *Ground Water, 57*(2), 245–258. https://doi.org/10.1111/gwat.12801

Chapter 16
An Applied Methodology for the Quantification of Socio-Economic Valuation of Groundwater Resources in Malta

Nicholas Ellul, Jonathan Spiteri, and Daniel Gravino

Abstract Groundwater resources provide a myriad of economic benefits to businesses, consumers and society at large, from supporting agricultural holdings in the cultivation of crops to ensure food security to the provision of clean water to households and tourists alike. Nonetheless, there are significant lacunae in the valuation of groundwater as an economic resource, particularly from a policy perspective, given the significant market failures that are typically associated with such common pool resources. This paper seeks to develop a tractable approach for the valuation of groundwater resources in Malta, a small island EU member state. Malta provides a compelling context for the undertaking of such an analysis. As a small island located in the middle of the Mediterranean with a semi-arid climate and a fast-growing population, Malta's natural water supplies are sufficiently under stress, even if used sustainably, with the onset of climate change further exacerbating these issues further. The focus of this research is on the private use of groundwater in the cultivation of crops within the agricultural sector, given that it constitutes the principal private abstractor of groundwater in Malta (beyond government for public water supply). This study proposes a paradigm for the identification, characterization and—where possible—quantification of social demand for groundwater based on its use within crop cultivation. Such paradigm follows the social demand model in mainstream economics that captures both private financial benefits resulting from the sector's use of these resources (e.g., enhanced agricultural productivity) as well as the external

N. Ellul (✉)
Malta Food Agency, Pitkali Road, Ta' Qali ATD 4000, Malta
e-mail: nicholas.ellul.2@gov.mt

J. Spiteri
Department of Insurance and Risk Management, Faculty of Economics, Management & Accountancy, University of Malta, Msida MSD 2080, Malta
e-mail: jonathan.v.spiteri@um.edu.mt

D. Gravino
Department of Economics, Faculty of Economics, Management & Accountancy, University of Malta, Msida MSD 2080, Malta
e-mail: daniel.gravino@um.edu.mt

A. Sefelnasr et al. (eds.), *Water Resources Management and Sustainability*, Water Science and Technology Library 114, https://doi.org/10.1007/978-3-031-80520-2_16

benefits emanating from water use (e.g., recreational value and carbon sequestration). These values are estimated using a variety of recognized approaches ranging from the residual value approach to values derived using the UN's System of Environmental-Economic Accounting—Ecosystem Accounting which identify the economic impact of groundwater use across various sectors, including agriculture. The framework's inherent simplicity and flexibility allows for its application across a wide variety of economic sectors, and its pragmatic approach to environmental valuation lends itself to adaptation in different countries, territories and indeed natural resources. The results underscore the significant economic contribution of groundwater resources to crop cultivation in Malta, while also highlighting the need to conduct systematic economic valuation to inform policy and thus the effective management of these resources.

Keywords Environmental valuation · Social demand · Groundwater use

16.1 Introduction

Economics' fascination with the value of water dates back to the writings of its modern founding father. In his seminal work, Adam Smith (1776) pondered upon the irony that a necessity, such as water, fetches a lower price than a luxury good, such as diamonds, and argued that such goods have "value in exchange", while the former have "value in use"—the so-called diamond-water paradox that is staple in any introductory economics textbook. Smith's famous "invisible hand" (1776) would theoretically have contributed towards an efficient allocation of resources and hence solved this dilemma. However, for that to happen, various assumptions would have to be respected—as if there would need to be near-perfect conditions for the market to operate—something completely removed from reality in the case of water and groundwater in particular. Pigou (1920) elaborated on Marshall's externalities concepts (1890) with these being the 'missing component' that explain why certain goods, like education and healthcare, are under-provided by the market. The corollary of this thinking is that the opposite also applies: negative externalities where third parties suffer welfare losses from the transaction of others also exists, and hence such products are supplied in more quantities than optimally desirable from society's point of view. The policy response to this was what was later coined as the Polluter Pays Principles, with its official adoption by the OECD in 1992. There subsequently were other valid contributions to this area of study, perhaps most notably the *Coase theorem* (1960); Hardin's *Tragedy of the Commons* implication (1968); and more recently Elinor Ostrom's (1990) rebuttal of the latter view, as her research showed several case studies where under certain settings the commons are well preserved and thus their exploitation is not inevitable.

This brief overview of the economic dilemmas surrounding natural resources—and hence groundwater—serves more than just an intellectual exercise; it shapes some of the concepts and frameworks that will be explored to in greater detail throughout the rest of the paper. The objective of this paper is to formulate a theoretical framework—a *paradigm* if you will—that can serve as a pragmatic guideline for applied researchers when taking on the challenging task of economic valuation of groundwater. As shall be explained, this paradigm is (only) semi-prescriptive, otherwise it would directly contradict one of the cardinal reasons for its raison d'être—the fact that public policy researchers and applied economists are often faced with myriad studies (on the topic of environmental valuation) which offer limited external validity as each is particularly case and context specific. The development of a pragmatic paradigm offers a conceptual tool the application of which does not depend on the economic use of the groundwater body, and/or the location in which it is applied. Besides, in understanding it, it empowers the researcher to develop knowledge about the economic use (and valuation) of that groundwater body and hence make informed decisions in its application for economic valuation. Moreover, the development of a tool that draws upon concepts beyond the strict confines of economic literature allows: a better understanding of the intellectual contributions from other fields that have been developed (even if only indirectly related) to this area; unlocks 'scope economies' in terms of knowledge; facilitates the cross-pollination of knowledge in multi-disciplinary teams while avoiding unnecessary misinterpretations of jargon and ideas (given that it builds on previous contributions); and finally, it is also more congruent with a progressive contemporary ecological economics perspective as opposed to the more conventional environmental economics approach (e.g., Gujarati, 2004; Daly & Farley, 2011).

In the last part of this paper, an initial attempt at applying this paradigm to arrive at an economic valuation of groundwater used for crop agriculture in Malta is made. The green paper issued by the Government of Malta (Environment & Resources Authority, 2023) on the Regulation of Groundwater Abstraction in the Maltese Islands in November 2023 showed that agriculture is the single most important user in terms of groundwater abstracted– with 76% of the amount of this resource that is abstracted for private purposes (i.e. excluding public water utilities) going towards this sector. Besides, local authorities also estimate that a minimal amount of the latter proportion (i.e., circa around 6%) is directed towards animal husbandry. In this light, the special focus of the application of the methods described in this paper were directed towards the crop-production sector.

16.2 Methods

Economics is concerned with the optimal allocation of resources (Koopmans, 1951) and Fig. 16.1 illustrates the powerhouse of mainstream economics. This can be explained as the market allocation referenced in the earlier section which happens at the intersection of the MPB (Marginal Private Benefit or Demand) and MPC (Marginal Private Costs or Supply) schedules in a free market—and augmented by the externalities concepts, thus incorporating the contributions described in the earlier section by Smith, Marshall and Pigou succinctly. Put differently, in the presence of externalities, the *social equilibrium*—i.e., the most desirable allocation from the point of view of society- would happen at the intersection of the MSC (Marginal Social Costs) and MSB (Marginal Social Benefit), where theoretically society as a whole (as opposed to a private firm or individual) values the last unit supplied equally to the last unit demanded. However, while this framework is a suitable starting point in elucidating why certain goods cannot be left to the unfettered free market to have a desirable allocation from the point of view of society, it offers less in terms of the applied economic valuation of natural resources.

This paper takes a first principles approach towards building a paradigm that allows the valuation of groundwater that can be possibly applied to different circumstances and characteristics as may be required by the applied economist or policy analyst in question. Inspiration was directly taken from Michela Miletto and Richard Connor's preface to the UN World Water Development Report for 2021, who eloquently point out that that "there is no one 'true' value of water", and that this is dependent on a number of variables including "its level of abundance or scarcity, its quality, and its availability", as well as its "use" and "benefits".

Environmental valuation is the process of assigning a monetary value to the various goods and services provided by the environment and its wide array of natural resources (Adamowicz, 2004). The purpose is to ensure that the myriad benefits

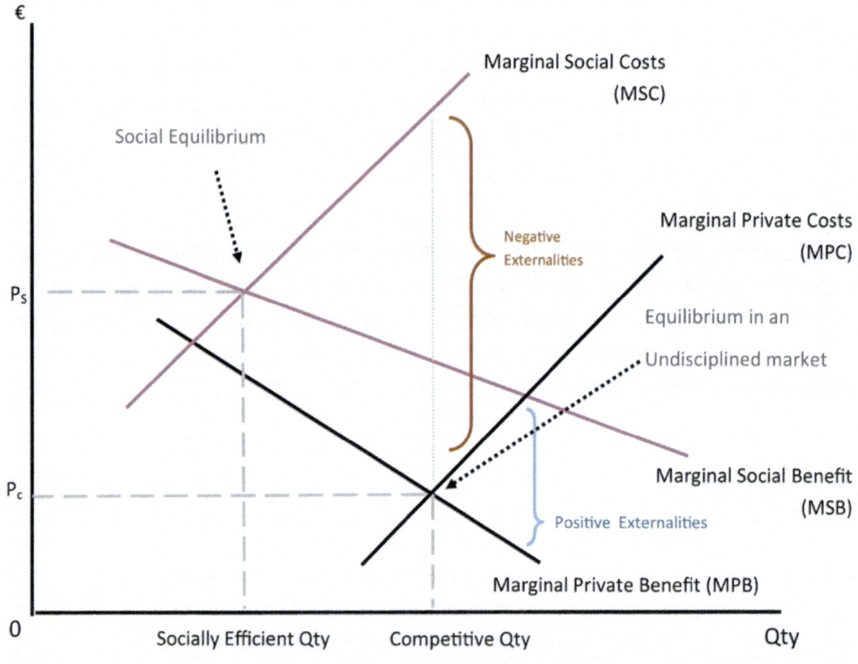

Fig. 16.1 Competitive versus socially efficient outcome. Adapted from Tietenberg and Lewis (2018)

provided by the environment to society are duly valued and accounted for in a systematic and holistic manner, particularly since there is often no market price or value attached to these benefits. Therefore, assigning a monetary value enables citizens, policymakers, and stakeholders to fully grasp the extent of these ecological, social and economic benefits, using a metric (money) that is immediately understandable to all, irrespective of background.

A key tenet of environmental valuation is the concept of Total Economic Value (TEV), which encompasses all of the different benefits and values provided by natural resources to society (Pearce & Turner, 1990). Total economic value can broadly be divided into two categories, namely *use* and *non-use values* as per Fig. 16.2, adopted from Plottu and Plottu (2007). Use values capture all of those benefits which accrue to society as a result of utilising the natural resource in some manner (Dietz et al., 2005). These can further be subdivided into three groupings, specifically *direct use values*, which are the benefits resulting from direct usage of the resource (e.g., the use of groundwater for irrigation purposes); *indirect use values*, which accrue from the incidental use of the resource (e.g., cropland serving as a carbon sink); and *option values*, which reflect the desire to potentially benefit from resource use in the future (e.g., future abstraction). Non-use values in turn are the benefits that are not generated from using the natural resource per se (Brouwer, 2000), and can also be subcategorised into three groups, namely *existence values* (the benefit of merely knowing that a natural resource exists); *altruistic values* (the warm-glow benefit resulting from knowledge that others are using the resource); and *bequest values* (the benefits accruing from knowing that future generations will benefit from the resource).

Somewhat similarly, albeit from a slightly different angle, the Millennium Ecosystem Assessment (2005) brought into the mainstream the idea of ecosystem services, that is the concept that humankind obtains benefits from the use of the natural environment and functioning ecological systems. Innovatively, it grouped them into four overarching categories, also in aid of explaining the benefits they provide to humankind, as shown in Table 16.1.

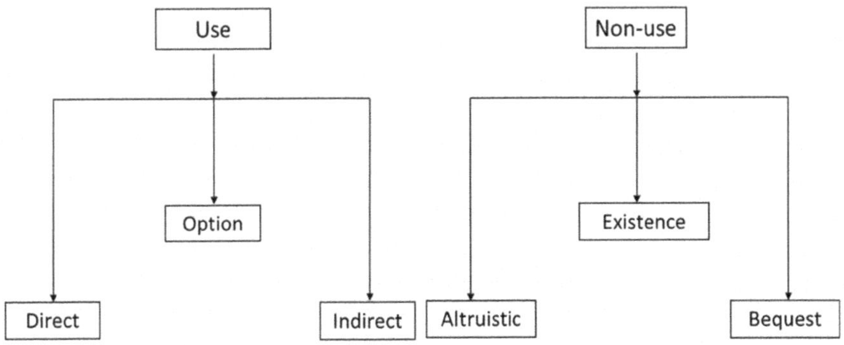

Fig. 16.2 Dimensions of total economic value (Plottu & Plottu, 2007)

Table 16.1 Ecosystem services by category

Type	Examples
Provisioning	Food, water, fibre and fuel
Regulating	Climate regulation, water purification, disease regulation
Cultural	Spiritual, aesthetic, recreation and education
Supporting	Soil formation, Habitat Provision, Nutrient Recycling, Eco-Tourism

Source Millennium Ecosystem Assessment (2005)

A particularly important contribution in elucidating the similarities between terminologies is that put forward by the National Research Council (NRC) (1997), in what it terms as "Taxonomy of Ground Water Valuation Terminology" as demonstrated by Table 16.2.

The exploration and mapping of different frameworks and terminology is not a self-indulgent intellectual exercise but helps to clarify ideas such that as NRC (1997) point out, different terminologies might obscure understanding between flow and stock concepts, and subsequently such confusion might lead to "double counting or other errors".

The *Resource: Sector-Use-Services* or more succinctly **R-SUS paradigm** was constructed by combining -partially or wholly- these various frameworks into a single template. It is intended to aid mapping the parts where they overlap and to provide a logical thinking sequence for the applied researcher to appraise the natural resource in question in a clear manner as opposed to more tradition approaches

Table 16.2 Taxonomy of ground water valuation terminology

Physical state terminology	Economic terminology	Accounting terminology	
		Stocks	Flows
Extractive values			
Municipal use values	Use value		*
Industrial use values	Use value		*
Agricultural use values	Use value		*
Other extractive use values	Use value		*
In situ *values*			
Ecological values	Use value	*	*
Buffer values	Use value	*	*
Subsidence avoidance values	Use value	*	*
Recreational values	Use value		*
Sea water intrusion values	Use value	*	*
Existence values	Non-use value		*
Bequest values	Non-use value		*

Adapted from NRC (1997)

that tackle ecosystem services of a particular resource without necessarily distinguishing between its various components—a more detailed analysis is deliberated in the discussion section below.

Figure 16.3 shows the R-SUS paradigm with the application congruent to the theme (and subsequent practical application) of this paper whilst following the important aforementioned axiom that there is "no one 'true' value of water". The first step was to identify the natural resource in question that warrants valuation: groundwater in Malta. The second step was to identify the economic sector which is making use of this natural resource: *Crop-agricultural production sector* i.e. the main users of this natural resource. However, it is important to point out that there exists a bi-directional relationship between the sector in question and the natural resource, since as Sakuyama and Stringer (2006) contend, "agriculture is essentially a human-made ecosystem" however, as opposed to natural ecosystems it provides both positive as well as "negative externalities for society" (Hassan et al., 2005).

The next stage of the paradigm discernment process involved the dimensions of economic use (as noted in Fig. 16.2) that warrant investigation for the assignment at hand. Here exists a trade-off: the more dimensions of economic use considered, the more exhaustive the valuation at hand. However, it is also true that non-use values are notoriously more costly and time-consuming to obtain, not to mention they increase the level of uncertainty. For the purposes of this study, a conservative approach was adopted, where the direct and indirect dimensions of TEV as per Fig. 16.2 were considered. The fourth level (or final step) of the paradigm involves identifying the ecosystem services to be valued within the parameters localised in the previous steps. For the purposes of this study, it stands to reason that Crop Provision (CP) was considered since it is an intrinsic function of the crop-agriculture sector. The second

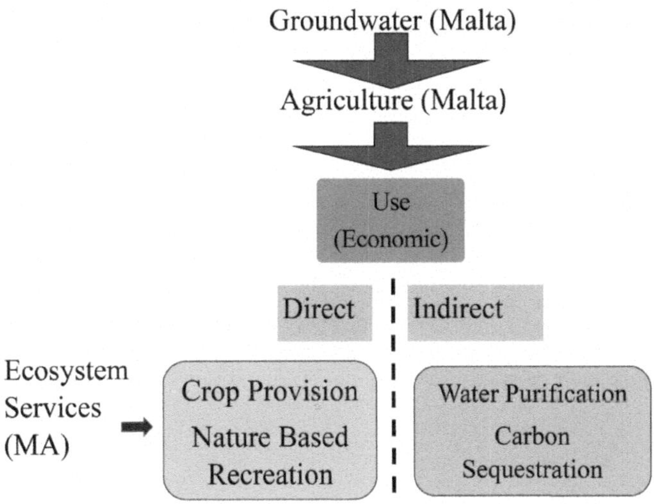

Fig. 16.3 An application of the R-SUS paradigm to groundwater use for the crop-agricultural sector in Malta

Direct Use value that was considered was Nature Based Recreation (NBR) given the importance of tourism in the local context (Sacco & Cassar, 2019). With regards to Indirect Use values, this paper considered ecosystems services with the potential to act as sinks for pollution—Water Purification (WP) and Carbon Sequestration (CS).

After having populated the R-SUS paradigm, the next step entailed the estimation of the ecosystem services in the identified context. By virtue of the economic sector identified as described above, Crop Provision was the first to be estimated, especially since it impinges on the estimations of the other ecosystem services. The gold standard for estimating the value of groundwater as an input in agriculture is the *Productivity Change* method. In essence, the productivity change method considers environmental resources as inputs in the production of a particular (marketable) good or service, and thus seeks to value these resources on the basis of their contribution to the production of the good or service in question (Barbier, 2007). For example, groundwater is a critical input in the production of various crops within the agricultural sector; thus, the value of groundwater can be estimated by understanding its contribution to agricultural crop productivity and thus turnover, while controlling for other inputs like land and labour (Nunez et al., 2006).

The downside of the productivity change method is that it is highly data intensive, thus a second-best solution was chosen: the residual value method. This approach also considers the value of an environmental resource in terms of its productive capabilities in relation to a particular good or service. In this instance, the starting point is the market value of the good or service in question, which represents the total revenue generated from the sale of this product (Moller-Gulland et al., 2020). The costs of each input used in the production process are then subsequently deducted from this total revenue, with the exception of the environmental input. The remaining or residual value would thus represent the monetary value of the environmental resource in question. Thus, a key proviso in the use of the residual value approach is that the environmental resource under consideration constitutes a key input in the production of the output in question.

For the purposes of this study, the "residual" environmental resource in question is groundwater abstracted for crop provision for the entire Maltese sector. Figure 16.4, shows the importance of groundwater for crop production, constituting 83% of its irrigation needs.

An adaptation of the residual approach taken by Moller-Gulland et al. (2020)—who estimated of "the value of water to agriculture in the Barind Tract" in Bangladesh was adopted. These authors' approach is shown in Eq. 16.1, where the water net benefits to agriculture (N_i) was approximated by deducting a number of costs from the benefit of the (crop) output R_i (per hectare). The main inputs costs considered were materials and equipment (M_i,) labour (L_i,) and Land (D_i,), and Capital (K_i,)

$$N_i = R_i - M_i - M_i - L_i - K_i - D_i \qquad (16.1)$$

For the purposes of this paper, considering the appropriate data limitations, Agriculture Sector Economic Accounts data published by Malta's National Statistics Office (NSO) were used were in a similar fashion but using national aggregates (given the

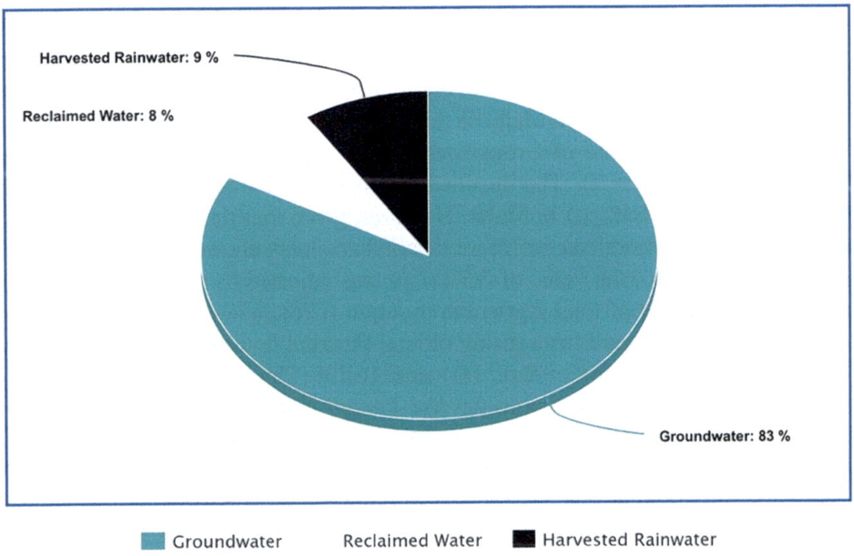

Fig. 16.4 Agricultural water supply in Malta. *Source* Environmental and Resources Authority (2023)

geographical scope of the study, rather than per hectare) to find N—the Net Benefit of Water for the Agricultural sector or NBW_{agri} as denoted in Eq. 16.1a below i.e. this study's adaptation of Eq. 16.1 used by Moller-Gulland et al. (2020). In the local context this involved deducting, Intermediate Consumption (M) of raw materials and other inputs in the agriculture production process, Consumption of Fixed Capital (K), Compensation for employees (L) and operating surplus (or profits) π from the Output figure (R) of the Production account of the entire agricultural sector (only the aggregate sector was available from NSO).

$$\overline{NBW_{agri}} = \overline{R} - \overline{M} - \overline{L} - \overline{K} - \overline{\pi} \tag{16.1a}$$

Given that the reported operating surplus (a standard measure for profit) from the aggregated agricultural accounts inevitably also includes the residual component, an imputed figure was estimated instead. Thus, the proxy for π that was used was an application of the economy-wide ratio of Operating Surplus-to-GVA—which stood at 54%—to the reported GVA of the agricultural sector as shown in Eq. 16.1b where π_{Agri} is the imputed profit figure to be estimated; GVA_{agri} is the average Gross Value Added for the Maltese Agricultural sector; and $\pi_{Economy}$ and $GVA_{Economy}$ are the economy-wide averages for operating surplus while Gross Value Added respectively

$$\overline{\pi}_{agri} = \overline{GVA}_{agri} \times \overline{\pi}_{Economy} / \overline{GVA}_{Economy} \tag{16.1b}$$

Chapter 17
Delineation of Groundwater Recharge Potential Zones Using GIS: A Case Study for Yalamlam Watershed in Saudi Arabia

Ibrahim H. Elsebaie, Atef Q. Kawara, and Ali O. Alnahit

Abstract In arid and semi-arid areas, it is crucial to identify suitable areas for effective groundwater recharge to ensure the long-term sustainability of aquifers. This study used a Geographic Information Systems (GIS) approach to select potential groundwater recharge areas in the Yalamlam watershed. Several data were used, including hydrogeological, land use, precipitation, elevation, and soil data. It was found that areas with significant recharge potential typically have lithology, moderate slopes, minimal land use disturbances, and higher average precipitation. On the other hand, areas with poor geological features and intense land use practices were found to have low recharge potential zones. These findings are crucial for creating sustainable groundwater management strategies and informing the development of policies for specific interventions to recharge groundwater. The findings will also contribute to developing strategies to promote the sustainable use and extraction of groundwater in the study area.

Keywords GIS · Groundwater recharge · Yalamlam · Saudi Arabia

17.1 Introduction

Groundwater, an underground aquifer, is considered one of Earth's most important natural resources (Naghibi et al., 2015). The effect of groundwater has significant implications for the well-being of humans, the expansion of the economy, and the maintenance of environmental equilibrium (Prentice et al., 2001). Surface water,

I. H. Elsebaie (✉) · A. Q. Kawara · A. O. Alnahit
Department of Civil Engineering, King Saud University, P.O. Box 800, Riyadh 11421, Saudi Arabia
e-mail: elsebaie@ksu.edu.sa

A. Q. Kawara
e-mail: 439106883@student.ksu.edu.sa

A. O. Alnahit
e-mail: alialnaheet@ksu.edu.sa

A. Sefelnasr et al. (eds.), *Water Resources Management and Sustainability*, Water Science and Technology Library 114, https://doi.org/10.1007/978-3-031-80520-2_17

such as lakes, marshes, reservoirs, and rivers, makes up a mere 0.3% of the world's freshwater resources, but groundwater accounts for approximately 30% (Senanayake et al., 2016). Precipitation and snow melting infiltrate the soil and move through the aquifer using interconnected soil pores, providing the main groundwater source. Groundwater is considered a preferable and potable alternative to surface water due to its greater accessibility and lower susceptibility to contamination (Mallick et al., 2019).

To determine the geographical distribution of the Groundwater Recharge Potential Zone (GRPZ), it is necessary to map and identify certain criteria accurately. GIS technologies may be used to analyze spatial by integrating several datasets (Mishra & Sinha, 2020). The Analytical Hierarchy Process (AHP) is a widely used technique for assessing several alternatives (Rajasekhar et al., 2019; Vahidnia et al., 2009).

Groundwater recharge is influenced by several factors, including lineament density, slope, rainfall, drainage density, soil composition, elevation, Topographic Wetness Index (TWI), and Land Use Land Cover (LULC) (Khan et al., 2023). To calculate the Area of Potential Zone (APZ), weights were assigned to the stated factors based on their respective contributions to groundwater potentiality. Techniques for assessing the potential of an aquifer included the use of geographic information systems (GIS) and remote sensing (RS). GIS facilitated the distribution of hydrogeological, geological, and geographic data. Remote sensing (RS) methods are used to identify land use and land cover (LULC), soil type, and the presence of cracks. Subsequently, they began using GIS for the purpose of analysis and data overlays. The definition of the GRPZ was established by the use of the Analytic Hierarchy Process (AHP), as indicated by the sources (Adiat et al., 2012; Echogdali et al., 2022; Machiwal et al., 2011), The system effectively handles the outputs and condenses intricate viewpoints into pairwise comparisons of series. In recent years, the practice of assigning weights to thematic layers has become increasingly prevalent by combining GIS methodologies with Multi-Criteria Decision-Making (MCDM) technology, such as the analytic hierarchy process (AHP) (Saranya & Saravanan, 2020). Incorporating methods such as structure, appropriateness, and accuracy may enhance the decision-making process (Kumar & Krishna, 2018). The use of analytical hierarchy process (AHP) methods has increased in environmental problem-solving and management. The Analytic Hierarchy Process (AHP) has been used by several researchers in various fields, such as ecological impact evaluations (Memarbashi et al., 2017), flood risk mapping (Elsebaie et al., 2023; Wang et al., 2019), and solid waste management (Khoshand et al., 2018). Mapping potential groundwater resources and identifying suitable sites for recharging is a widely used and effective AHP technology (Arumugam et al., 2023; Dar et al., 2021).

Saudi Arabia uses several water sources, such as desalination, processing, and renewable resources, to fulfill its water requirements (Arumugam et al., 2023; Dar et al., 2021). The replenishment of the shallow aquifer, enhanced by precipitation and sudden floods, is a vital factor in augmenting the total water supply. Similarly, precipitation plays a significant role in the country's water storage, representing around 2.04 billion cubic meters (Masoud et al., 2018). Approximately 84% of the groundwater in this nation is allocated for agricultural use (Chowdhury & Al-Zahrani, 2015). The depletion of shallow aquifers has occurred as a result of excessive extraction of groundwater (Zaidi et al., 2015). The shallow aquifer's groundwater storage has decreased due to excessive abstraction (Zaidi et al., 2015).

The main objective of this study is to use GIS technology to detect GRPZ in the Yalamlam watershed. To achieve this, various data were collected, including hydrogeological data, land use, precipitation, slope, and soil properties. The results of this study will provide valuable information to decision-makers about the location of groundwater recharging zones, improve sustainable groundwater management strategies, and contribute to the development of policies for recharging groundwater in the study area.

17.2 Study Area

The distance between Wadi Yalamlam and Al Lith is 90 km to the north and 100 km to the south of Mecca. With longitudes ranging from 39° 50′ 00″ to 40° 36′ 00″, the precise coordinates of its position are 20° 33′ 00″ and 21° 07′ 00″. The watershed of the Yalamlam Wadi is around 1665 km^2 in size. The watershed boundary in the downstream region encompasses almost the whole lower-level area. Wadi Yalamlam begins in the Al Shafa and winds down from the Hijaz Mountains near Taif. Annual precipitation averages 140 mm. The wadi's elevations vary greatly, from 2600 to 25 m. Figure 17.1 shows a digital elevation model (DEM) of Wadi Yalamlam and a generic location map. The top and lower portions of the watershed are mostly covered with sharp and distinct natural vegetation. In contrast, the lower part of the wadi consists of slopes made of granitoid and metamorphosed basalt, which have undergone extensive alteration due to deposits, sand dunes, and dispersed debris.

17.3 Database

The data processing method is a crucial step in producing a map that indicates possible groundwater recharge zones. The data for the research area comes from several sources, as shown in Fig. 17.2. Twelve-meter resolution data for the Digital Elevation Model (DEM) was used. The several layers that were produced as a consequence of the DEM analysis using ArcGIS software are shown in Fig. 17.2 and include lineament density, slope, rainfall, drainage density, LULC, soil type, elevation, and TWI. The Ministry of Environment, Water, and Agriculture's, database provided daily precipitation statistics. The Inverse Distance Weighting (IDW) approach was used to interpolate the mean daily maximum rainfall data for the research region. The digital soil map of the globe were obtained from the following link (https://data.apps.fao.org/map/catalog/srv/eng/catalog.search#/metadata/446ed430-8383-11db-b9b2-000d939bc5d8). The US Geological Survey's

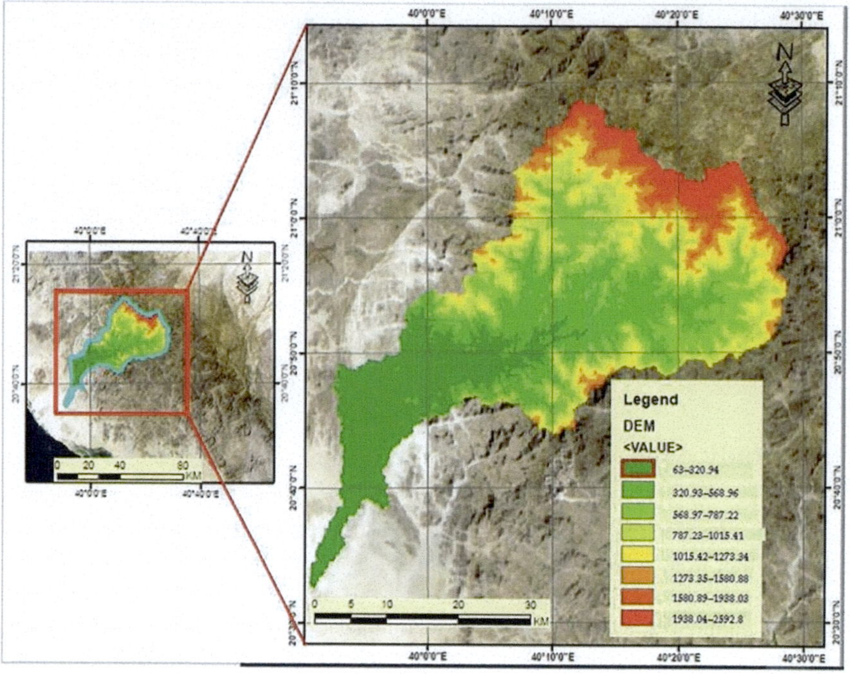

Fig. 17.1 Wadi Yalamlam coordinates and DEM (Kawara & Elsebaie, 2022)

website (https://earthexplorer.usgs.gov/) provided Earth Explorer satellite images, which were then obtained and analyzed for LULC data. The LULC map was created in ArcGIS 10.8 using satellite photos, image recognition methods, and a maximum likelihood algorithm. A single database was created by resampling the geographic layers into a raster format.

17.4 Methodology

Figure 17.2 shows the methodology used to identify the GRPZ in the study area. Seven factors were selected to create the GRPZ. The first step involved gathering data from various sources, including satellite imagery, meteorological data, and the Digital Elevation Model (DEM), to locate areas with potential for groundwater recharge. The DEM provides information on the density of lineaments, slope, drainage density, and elevation. Meteorological data was used to obtain the rainfall information for the study area, and satellite imagery provided information on soil and land uses for the study area. The factors selected in the study area were lineament density, slope, drainage density, elevation, soil type, land use, and rainfall.

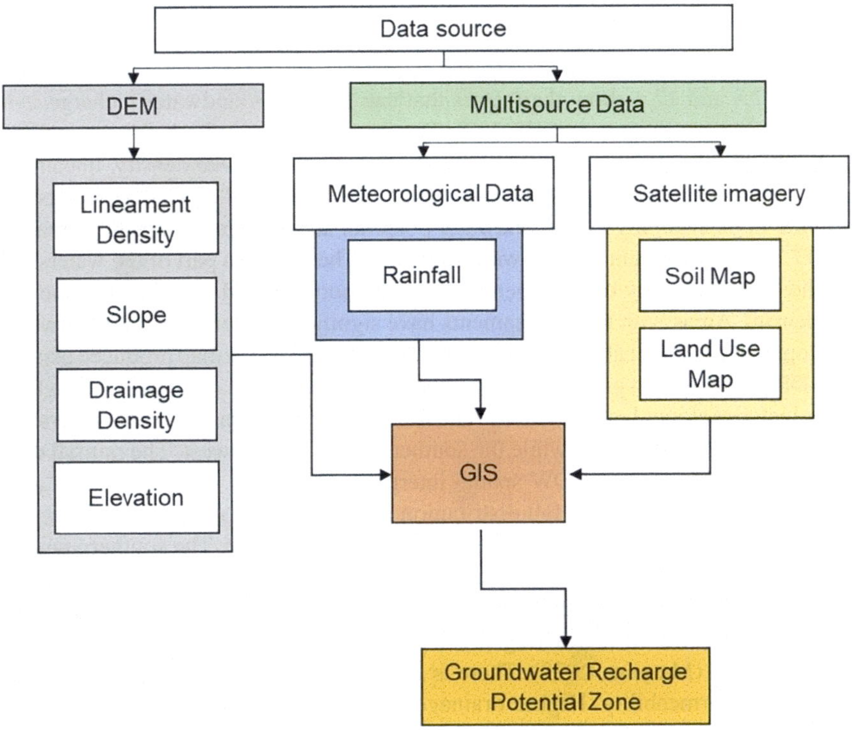

Fig. 17.2 Flow chart illustrating the approach used in the current research

After collecting the data, ArcGIS was used for processing and analysis. The DEM analysis included the extraction of lineaments, calculation of their density, determination of the slope, extraction of drainage networks, and acquisition of elevation data. Meteorological data was collected from past rainfall records, while satellite photography was used to categorize various soil types and comprehend the pattern of land usage. In ArcGIS, all data layers were assigned geographic coordinates and aligned in location.

By incorporating the data into the GIS, an extensive analysis of the variables impacting groundwater recharge can be conducted, guaranteeing a reliable and precise selection of GRPZ. The resulting maps provide valuable information for the management and development of sustainable water resources.

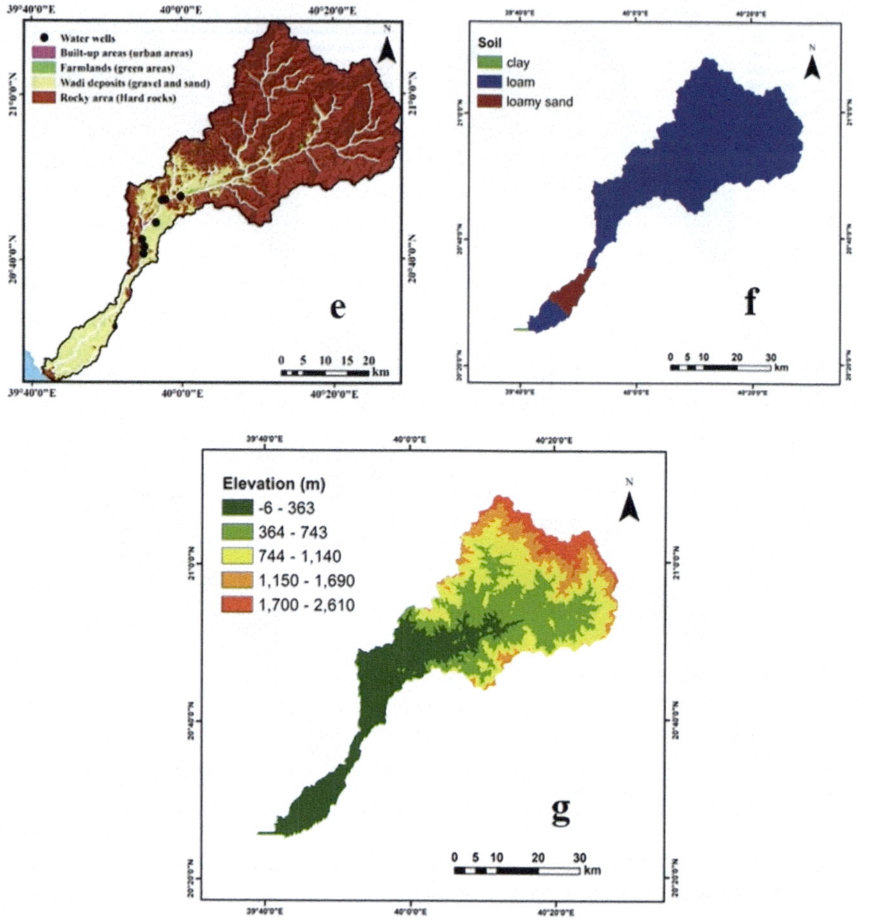

Fig. 17.4 The hydrologic parameters which affect the groundwater in Wadi Yalamlam basin: **e** land use/land cover, **f** soil, **g** elevation

17.6 Conclusions

In the last few decades, the expansion of urban areas has led to a decrease in natural water resources, resulting in excessive water usage by large city populations. The study identified several factors associated with groundwater potential and storage, including lineament density, slope, rainfall, drainage density, land use/land cover (LU/LC), soil composition, and elevation. Loose deposits in certain areas have high infiltration rates and significant groundwater potential, while areas dominated by large hard-rock formations experience significant surface runoff and limited infiltration rates. Structural elements, such as lineaments and shear zones, impact groundwater recharge by providing pathways for groundwater flow. Areas with high drainage density exhibit elevated surface runoff and less infiltration. Areas with a small slope, where the slope degrees are less than 5°, facilitate water penetration. The best places for groundwater occurrences and recharging in the Wadi Yalamlam watershed are the areas with loose soil types, lineament junctions, and moderate slopes, which can be classified as zones with high and moderate potential.

The findings could be valuable for water resources managers in ensuring that the potential groundwater recharge area in the study region is utilized and monitored effectively to ensure its longevity. Additionally, a more comprehensive understanding of the region's groundwater recharge capacity may be achieved through future research that incorporates a broader range of factors and includes economic and social data.

Acknowledgements The author extends their sincere appreciations to the National Plan for Science Technology and Innovation (NPST)-King Saud University for funding the Research Project (13-WAT1027-02). The author thanks the staff of the NPST for great support and cooperation.

References

Adiat, K. A. N., Nawawi, M. N. M., & Abdullah, K. (2012). Assessing the accuracy of GIS-based elementary multi criteria decision analysis as a spatial prediction tool—A case of predicting potential zones of sustainable groundwater resources. *Journal of Hydrology, 440–441*, 75–89. https://doi.org/10.1016/j.jhydrol.2012.03.028

Arumugam, M., Kulandaisamy, P., Karthikeyan, S., Thangaraj, K., Senapathi, V., Chung, S. Y., Muthuramalingam, S., Rajendran, M., Sugumaran, S., & Manimuthu, S. (2023). An assessment of geospatial analysis combined with AHP techniques to identify groundwater potential zones in the Pudukkottai District, Tamil Nadu, India. *Water, 15*(6). Article 6. https://doi.org/10.3390/w15061101

Azad, S., Al-Manmi, D. A. M., & Aziz, B. Q. (2023). Integrated GIS, remote sensing, and electrical resistivity tomography methods for the delineation of groundwater potential zones in Sangaw Sub-Basin, Sulaymaniyah, KRG-Iraq. *Water*. https://doi.org/10.3390/w15061055

Chowdhury, S., & Al-Zahrani, M. (2015). Characterizing water resources and trends of sector wise water consumptions in Saudi Arabia. *Journal of King Saud University—Engineering Sciences, 27*(1), 68–82. https://doi.org/10.1016/j.jksues.2013.02.002

Dar, T., Rai, N., & Bhat, A. (2021). Delineation of potential groundwater recharge zones using analytical hierarchy process (AHP). *Geology, Ecology, and Landscapes, 5*(4), 292–307. https://doi.org/10.1080/24749508.2020.1726562

Echogdali, F. Z., Boutaleb, S., Bendarma, A., Saidi, M. E., Aadraoui, M., Abioui, M., Ouchchen, M., Abdelrahman, K., Fnais, M. S., & Sajinkumar, K. S. (2022). Application of analytical hierarchy process and geophysical method for groundwater potential mapping in the Tata Basin, Morocco. *Water, 14*(15). Article 15. https://doi.org/10.3390/w14152393

Elsebaie, I. H., Kawara, A. Q., & Alnahit, A. O. (2023). Mapping and assessment of flood risk in the Wadi Al-Lith Basin, Saudi Arabia. *Water, 15*(5). Article 5. https://doi.org/10.3390/w15050902

Ike, S. A., Sahoo, K., Milewski, A., & Mishra, D. R. (2019). Impact of land use land cover (LULC) change on surface runoff in an increasingly urbanized tropical watershed. *Water Resources Management*. https://doi.org/10.1007/S11269-019-02320-W

Ilugbo, S. O., Aigbedion, I., & Ozegin, K. O. (2023). Structural mapping for groundwater occurrence using remote sensing and geophysical data in Ilesha Schist Belt, Southwestern Nigeria. *Geology, Ecology, and Landscapes*. https://doi.org/10.1080/24749508.2023.2182063

James, P. R. S., Finch, J., Ireson, A., & Jackson, C. R. (2014). Comparison of varied complexity models simulating recharge at the field scale. *Hydrological Processes*. https://doi.org/10.1002/HYP.9752

Kanak, N. M., Pande, C. B., Gautam, V., & Singh, S. K. (2023). Integration of hydrogeological data, GIS and AHP techniques applied to delineate groundwater potential zones in sandstone, limestone and shales rocks of the Damoh district, (MP) central India. *Environmental Research*. https://doi.org/10.1016/j.envres.2023.115832

Kawara, A. Q., & Elsebaie, I. H. (2022). Development of rainfall intensity, duration and frequency relationship on a daily and sub-daily basis (case study: Yalamlam Area, Saudi Arabia). *Water, 14*(6). Article 6. https://doi.org/10.3390/w14060897

Khan, M. Y. A., ElKashouty, M., Zaidi, F. K., & Egbueri, J. C. (2023). Mapping aquifer recharge potential zones (ARPZ) using integrated geospatial and analytic hierarchy process (AHP) in an Arid Region of Saudi Arabia. *Remote Sensing, 15*(10). Article 10. https://doi.org/10.3390/rs15102567

Khoshand, A., Kamalan, H., & Rezaei, H. (2018). Application of analytical hierarchy process (AHP) to assess options of energy recovery from municipal solid waste: A case study in Tehran, Iran. *Journal of Material Cycles and Waste Management, 20*(3), 1689–1700. https://doi.org/10.1007/s10163-018-0736-3

Kumar, A., & Krishna, A. P. (2018). Assessment of groundwater potential zones in coal mining impacted hard-rock terrain of India by integrating geospatial and analytic hierarchy process (AHP) approach. *Geocarto International, 33*(2), 105–129. https://doi.org/10.1080/10106049.2016.1232314

Machiwal, D., Jha, M. K., & Mal, B. C. (2011). Assessment of groundwater potential in a semi-arid region of India using remote sensing, GIS and MCDM techniques. *Water Resources Management, 25*(5), 1359–1386. https://doi.org/10.1007/s11269-010-9749-y

Mallick, J., Khan, R. A., Ahmed, M., Alqadhi, S. D., Alsubih, M., Falqi, I., & Hasan, M. A. (2019). Modeling groundwater potential zone in a semi-arid region of Aseer using fuzzy-AHP and geoinformation techniques. *Water, 11*(12). Article 12. https://doi.org/10.3390/w11122656

Maryam. (2022). Transmissivity and groundwater flow exert a strong influence on drainage density. *Earth Surface Dynamics*. https://doi.org/10.5194/esurf-10-1-2022

Masoud, M. H. Z., Basahi, J. M., & Rajmohan, N. (2018). Impact of flash flood recharge on groundwater quality and its suitability in the Wadi Baysh Basin, Western Saudi Arabia: An integrated approach. *Environmental Earth Sciences, 77*(10), 395. https://doi.org/10.1007/s12665-018-7578-0

Memarbashi, E., Azadi, H., Barati, A. A., Mohajeri, F., Passel, S. V., & Witlox, F. (2017). Land-use suitability in Northeast Iran: Application of AHP-GIS hybrid model. *ISPRS International Journal of Geo-Information, 6*(12). Article 12. https://doi.org/10.3390/ijgi6120396

Michael, W. S., Döring, A. O., Möck, M., Sedrez, M., Mumm, O., Schneider, A.-K., Weber, S., & Schröder, B. (2019). The "hidden urbanization": trends of impervious surface in low-density housing developments and resulting impacts on the water balance. *Frontiers in Environmental Science*. https://doi.org/10.3389/FENVS.2019.00029

Mishra, K., & Sinha, R. (2020). Flood risk assessment in the Kosi megafan using multi-criteria decision analysis: A hydro-geomorphic approach. *Geomorphology, 350*, 106861. https://doi.org/10.1016/j.geomorph.2019.106861

Naghibi, S. A., Pourghasemi, H. R., Pourtaghi, Z. S., & Rezaei, A. (2015). Groundwater qanat potential mapping using frequency ratio and Shannon's entropy models in the Moghan watershed, Iran. *Earth Science Informatics, 8*(1), 171–186. https://doi.org/10.1007/s12145-014-0145-7

Prentice, I. C., Heimann, M., & Sitch, S. (2001). Contribution of working group I to the third assessment report of the Intergovernmental Panel on Climate Change. In *Chapter climate change* (pp. 183–237).

Rajasekhar, M., Sudarsana Raju, G., Sreenivasulu, Y., & Siddi Raju, R. (2019). Delineation of groundwater potential zones in semi-arid region of Jilledubanderu river basin, Anantapur

District, Andhra Pradesh, India using fuzzy logic, AHP and integrated fuzzy-AHP approaches. *HydroResearch, 2*, 97–108. https://doi.org/10.1016/j.hydres.2019.11.006

Saranya, T., & Saravanan, S. (2020). Groundwater potential zone mapping using analytical hierarchy process (AHP) and GIS for Kancheepuram District, Tamilnadu, India. *Modeling Earth Systems and Environment, 6*(2), 1105–1122. https://doi.org/10.1007/s40808-020-00744-7

Senanayake, I. P., Dissanayake, D. M. D. O. K., Mayadunna, B. B., & Weerasekera, W. L. (2016). An approach to delineate groundwater recharge potential sites in Ambalantota, Sri Lanka using GIS techniques. *Geoscience Frontiers, 7*(1), 115–124. https://doi.org/10.1016/j.gsf.2015.03.002

Vahidnia, M. H., Alesheikh, A. A., & Alimohammadi, A. (2009). Hospital site selection using fuzzy AHP and its derivatives. *Journal of Environmental Management, 90*(10), 3048–3056. https://doi.org/10.1016/j.jenvman.2009.04.010

Vesna, Z., Bračič Železnik, B., Pintar, M., & Cencur Curk, B. (2020). Assessment of groundwater recharge for a coarse-gravel porous aquifer in Slovenia. *Hydrogeology Journal*. https://doi.org/10.1007/S10040-020-02152-8

Wang, Y., Hong, H., Chen, W., Li, S., Pamučar, D., Gigović, L., Drobnjak, S., Tien Bui, D., & Duan, H. (2019). A hybrid GIS multi-criteria decision-making method for flood susceptibility mapping at Shangyou, China. *Remote Sensing, 11*(1). Article 1. https://doi.org/10.3390/rs11010062

Zaidi, F. K., Nazzal, Y., Jafri, M. K., Naeem, M., & Ahmed, I. (2015). Reverse ion exchange as a major process controlling the groundwater chemistry in an arid environment: A case study from northwestern Saudi Arabia. *Environmental Monitoring and Assessment, 187*(10), 607. https://doi.org/10.1007/s10661-015-4828-4

Chapter 18
Risk Assessment Analysis Related to Managed Aquifer Recharge (Rye Island Pilot Area, Slovakia)

Andrej Šoltész, Michaela Červeňanská, Dana Baroková, Andrea Vranovská, and Jakub Mydla

Abstract Coming from the literature sources Managed aquifer recharge (MAR) is defined as the intentional infiltration of water into aquifers with the purpose of either later recovering that water for different purposes like agricultural, industrial, urban or obtaining an environmental benefit (Dillon et al. in Managed aquifer recharge: an introduction. National Water Commission, Canberra, 2009). It has replaced the term "artificial recharge" which maybe evoked something what seemed to be unnatural in the environment (Dillon in Hydrogeol J 13:313–316, 2005). There are different types of MAR methodologies and examples. Some of them were proposed or even already realized in frame of the DEEPWATER-Central Europe (CE) project. To increase long-term retention of water in aquifers and its subsequent use in drier or increased demand periods, MAR techniques were studied in 4 pilot areas in the DEEPWATER-CE project funded by the Interreg CENTRAL EUROPE program. Originating from transnational toolbox for designating potential MAR locations in Central Europe, the MAR pilot sites in the project participating countries were identified, namely in Hungary, Poland, Croatia and Slovakia. Slovak partner was designated to utilize the MAR methodology for agricultural purposes using as a source the surface water in channel system of the greatest lowland in Slovakia—the Rye Island (Žitný ostrov

A. Šoltész (✉) · M. Červeňanská · D. Baroková · J. Mydla
Faculty of Civil Engineering, Slovak University of Technology in Bratislava, Radlinskeho 11, 810 05 Bratislava, Slovak Republic
e-mail: andrej.soltesz@stuba.sk

M. Červeňanská
e-mail: michaela.cervenanska@stuba.sk

D. Baroková
e-mail: dana.barokova@stuba.sk

J. Mydla
e-mail: jakub.mydla@stuba.sk

A. Vranovská
Water Research Institute, Nábr. arm. gen. L. Svobodu 5, 812 49 Bratislava, Slovak Republic
e-mail: andrea.vranovska@vuvh.sk

A. Sefelnasr et al. (eds.), *Water Resources Management and Sustainability*, Water Science and Technology Library 114, https://doi.org/10.1007/978-3-031-80520-2_18

in Slovak language or Schüttinsel in German). The territory of the Rye Island is bordered by the Danube River from the West, the Váh River form the South-East and the Little Danube River from the North-East. The whole territory of the Rye Island is protected against flooding by flood protection dikes. To have the possibility to collect and dewater the surplus water from agricultural land after heavy rain or snowmelt the drainage channel system was built in the last century. The existing form was achieved in the second half of the century and it was completed after putting Gabčíkovo hydropower plant (GHPP) into operation in 1992 (Šoltész et al. in STU in Bratislava, p. 29, 2021). The main objective for the Slovak side in frame of the project was to induce infiltration from surface water body created by surface water in drainage channel system (sometimes ground water on the surface) into aquifer. The groundwater recharge using MAR techniques was realized using one of the former drainage system (SVII) in chosen territory of the upper Rye Island. More detailed explanation and description of the water management is presented in the introduction.

Keywords Aquifer · Channel system · Infiltration · Managed aquifer recharge · Risk assessment · Surface and ground water interaction

18.1 Introduction

The risk assessment report introduces the recharge dam MAR system located in Rye Island pilot area in Slovakia (Fig. 18.1) where the MAR system was utilized for agricultural purposes. In this regard, there were recognized possible risks for the MAR system coming from environmental and human health, technical, socio-economic, governance and legislative risks as well as risks related to the sensitivity of MAR to climate-induced extreme situations.

Risks were assessed by combination of two methods; i.e. quantitative risks were identified according to the methodology developed by MARSOL EU project and qualitative risk assessment was done in accordance with Australian guidelines (the likelihood and the severity of a risk is interpreted by risk factor matrix). According to MARSOL methodology, there were quantitative technical and non-technical risks evaluated (social, economic, governance and legislation) during design, construction and operational phases.

Lack of private/public funding was identified as very high risk as a non-technical risk during design and construction phase. Possible treatment is to disseminate and publicize the MAR schemes benefits to be able to involve as many investors as possible. Within this group, two other risks were evaluated as high: low price of water and high installation cost. These risks can be overcome by additional support for the use of MAR facilities (state support, private financial sources) in order to promote its financial viability.

Evaluation of technical risks revealed several high probabilities of occurrence during design and construction phase: construction technical difficulties, risk of low water storage, hydrogeological setting. These risks can be treated by preparation of

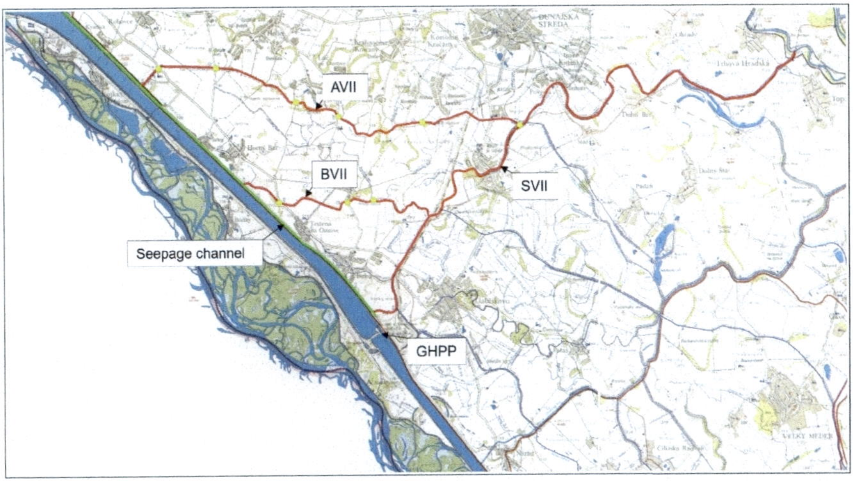

Fig. 18.1 Location of pilot area for MAR purposes in Slovakia—upper Rye Island

specific technical project, and proper and detailed geological and hydrogeological investigation. During operational phase, high risks are represented by swelling clays, nutrients, droughts and rainfall events periodicity, changes in water demand and supply.

To ensure effective functioning of the MAR system and to be informed about all risks mentioned above in time, the thorough monitoring system of risks should be put into operation.

18.2 Selected Risk Assessment Methods

The main objective for the Slovak side in frame of the project was to induce infiltration from surface water body created by surface water in drainage channel system (sometimes groundwater on the surface) into aquifer. The groundwater recharge using MAR techniques was realised via one of the former drainage system (SVII) in chosen territory of the upper Rye Island. This region is more than 100 years intensively cultivated. The area is comprehensively investigated; there is lot of hydrological, morphological and hydrogeological data available. Of course, lot of conditions have changed since then, but land farmers are still active, channel systems exist, there are tools for water flow control (sluices in operation last 50 years) and in addition, the channel system can be supplied by water.

After putting GHPP into operation in 1992, this system has been supplied by water from the seepage channel of the GHPP (Fig. 18.1). This process enabled another very important function of the channel system—to supply the aquifer by control of water flow on sluices as well as supply the unsaturated zone by percolation from several irrigation systems. There are 4 channels located in the investigated area—SVII (primary channel, length 28.825 km, whole drainage area 32,024 ha) and secondary channels AVII (length 18.200 km), BVII (length 9900 km) and CVII (length 2.469 km). All mentioned channels are hydraulically connected to seepage channel from which they are supplied by water. The source of the water in the seepage channel is the water infiltrated from the Hrušov reservoir (part of the GHPP project), i.e. in fact groundwater from the quality viewpoint.

The surface water supply into drainage channel system started to work from May 1993 when all intake structures were completed. The operation can be divided into two seasonal parts.

The first one is the winter (non–vegetation) period (i.e. late October until spring time, mostly April). In this period, all sluices are open and the natural discharge and water level regime is stabilized, although water is supplied from seepage channel. This operation is performed by the administrator of the channel system—Slovak Water Management Enterprise (SWME) due to safety reasons, i.e. ice cover formation (possible damage).

The second one is the summer (vegetation) period when sluices built on this drainage channels are closed, the surface water level is increased and conditions for the MAR system are created in full extent (Šoltész et al., 2021).

The operation of MAR system—creation of recharge dam MAR type on the channels network—pose some risks on building, financing, operation, users viewpoint and ecosystems. Therefore, based on selected methodology, there were identified the risks related to environment and human health, technique, socio-economic issues, governance, legislation, and risks related to the sensitivity of MAR to climate-induced extreme situations. Consequently, the risk treatment measures were proposed in order to tackle the above-mentioned risks effectively. The risk identification and proposal of risk treatment is summarised in the form of tables. Within risk analysis, also the monitoring of identified risks was proposed.

Referring to results of the MARSOL EU FP7 project managed aquifer recharge can be affected by many risks (Rodríguez-Escales et al., 2018). Those risks are related to different technical and non-technical aspects of recharge. Very detailed risk assessment analysis and description of events that can lead to MAR failure were compiled in above mentioned MARSOL project and presented for different facilities according to infiltration typology all over the world (Rodríguez-Escales et al., 2018), e.g. infiltration using deep wells, surface infiltration (several types) or vadose infiltration.

Following the methodology for risk assessment of Rodríguez-Escales et al. (2018) based on Tartakovsky (2007) and Bedford and Cooke (2003) it consists mostly of four steps:

- definition of the concept of the system failure and the identification of the basic events (factors) that would potentially result in such failure,
- construction of the fault tree depicting the combination of all possible events,
- developing a probabilistic representation of the fault tree and

- calculating the individual probability of event occurrence (using expert opinions) to calculate the global risk of the MAR facility.

The critical operational component of any MAR scheme is the interaction between the infiltrated water and the aquifer (Nandha et al., 2015). Mentioned events (problems) can be divided according to Rodríguez-Escales et al. (2018) by categories to technical and non-technical problems. The main technical problem at infiltration process from surface ponding system but from personal system analysis, as well, is clogging (Alam et al., 2021). Due to almost 30-year operation on the channel system, this process was observed with increased attention when realizing the field measurements and surface and ground water modelling investigation.

The Australian MAR guidelines suggest a quantitative risk assessment method specific for MAR systems. In this method the likelihood and the severity of a risk is assessed by risk factor matrix (Fig. 18.2). The likelihood of a hazard to occur is identified by the expected recurrence of the hazard (indicated in years), and this likelihood is expressed by using a five-step scale. A hazard recurrence interval of 100 years is defined as 'rare' (lowest scale) and a recurrence of several times per year is defined as 'almost certain' (highest scale), although such exact timely ranking cannot be determined for every case.

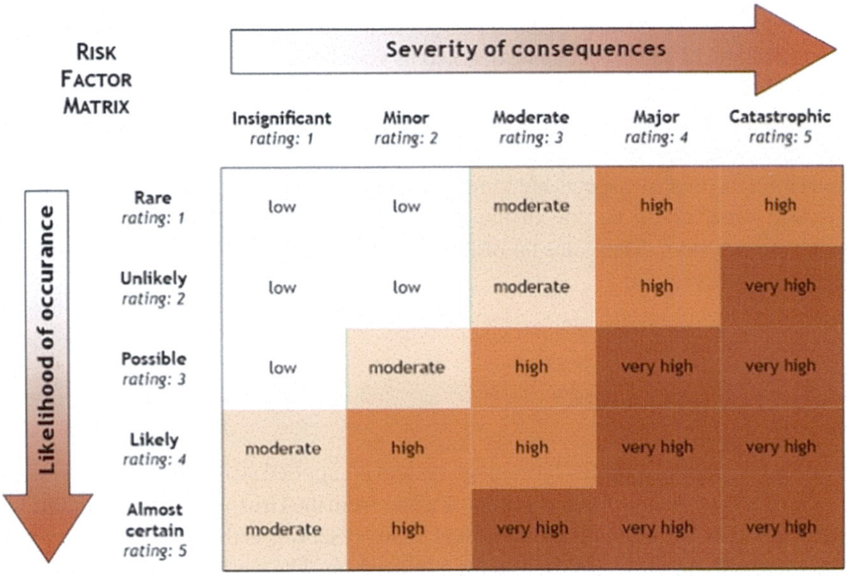

Fig. 18.2 Risk factor score matrix for qualitative risk assessment, relating the likelihood of hazards to the severity of consequences (after Swierc et al., 2005)

18.3 Risk Identification and Risk Treatment

According to the agricultural purpose of the MAR system dedicated in the DEEP-WATER CE project for the Slovak partner, source of the water for the proposed MAR system and operation on the SVII drainage system which is very flexible and farmer—friendly (good cooperation between the SWME and farmers) we have recognized following risks. Due to the fact that the drainage system is already operating, it has no sense to mention the planning process. At the end we will just recommend possible (or necessary) improvement of the system.

18.3.1 Environmental and Human Health Risks

The source of the MAR system in the selected Rye Island region can be divided into two parts. Secondary channels (AVII, BVII, CVII) are supplied—as already mentioned—from the seepage channel of the GHPP reservoir in Hrušov (i.e. groundwater). The primary and longest channel SVII is supplied from the power supply canal of the GHPP (Fig. 18.1), i.e. surface water. The power canal is nowadays a component of the international waterway on the Danube River and possible accident of a boat on the navigation way could represent a risk from water quality point of view. The likelihood of such risk can be denoted as very low due to the fact that the power channel is fully insulated and has no contact to the groundwater of the adjacent area.

The whole SVII drainage system is situated in agriculturally exploited lowland area, but some parts of the system are flowing through villages and municipalities. Therefore, the quality aspects are of increased importance. The possibility of release of waste water could be one of the possible polluter of the transported water in the channel system. There are no official groundwater sources (wells) for drinking purposes in this region.

18.3.2 Technical Risks

As already mentioned the SVII drainage system is supplied artificially and the system is dependent on the infiltrated amounts of water from the Hrušov reservoir. Therefore, the clogging of banks and bottom of the Hrušov reservoir is the most important risk of the functioning of the MAR system in this region. The evidence of this statement were decreased groundwater levels in observation wells of the Slovak Hydrometeorological Institute (SHMI) from 1998 when the clogging firstly appeared after 6 years of operation of the GHPP (Šoltész, 2002) as well as decreased measured discharge in the seepage channel (down to 2 $m^3\ s^{-1}$). That was the reason for improving the infiltration rate by dredging of sediments in the Hrušov reservoir. Nowadays the dredging of sediments is almost everyday activity of the reservoir administrator (SWME) and the system can be specified as stable.

The clogging of the Hrušov reservoir is not the only risk of the MAR system in the selected region. The clogging was observed in the river bed of channels, as well. It is quite complicated to distinguish the sort of clogging (physical, chemical or biological). According to operation on the channel system (mentioned seasonal change—vegetation and non-vegetation period) when the gates on channels are elevated for the winter season, the sudden change in velocity of the released surface flow is taking certain fine soil material from the bottom of channels with.

Next risk which was recognized for the MAR system is connected with the climate change. It is apparent that there will be still more water needed in the vegetation period for the agricultural plants because of greater differences between precipitation on one side and evapotranspiration on the other, especially in vegetation period. The MAR system in that case is not sufficient to cover the water demand for the plants. Due to operation through inlet structures and gates on the drainage channels (adequate flow rate and water depth) there is a possibility to supply by irrigation the unsaturated (root) zone of the soil profile. It is understandable that this is not a real managed aquifer recharge but it stores the water for the nature. Otherwise, the water from channels would flow via Little Danube and the Danube River into the Black Sea.

From technical (operation) point of view another identified risk for the MAR system is the damage of the inlet structure as well as damage of existing gates. It happened more times already and then the problem is going to be an economic risk because of lack of funding. Flood appearance is possible mostly in the spring time after snowmelt (mostly for weeks) and after heavy rain in the summer (mostly for days). The duration of flooding depends on infiltration rate of soil cover layers. In the spring time it is often connected with the air temperature (frozen soil) and the consequence is the delay in agricultural works.

18.3.3 Socio-economic Risks

Socio-economic risks are all the time connected with humankind. Therefore, when identifying socio-economic risks, the most important is the relation between the water supply administrator of the channel system (SWME) and the farmers cultivating the agricultural land. Up to now, the presented MAR system using SVII drainage system in the area of interest is running upon unwritten common agreement between mentioned partners. The administrator is storing the water to maximum possible extent for increasing the infiltration rate into subsoil. There are still possibilities to improve the system (Šoltész et al., 2021), but that is again dependent on funding and the problem is connecting with governance and legislation risks (EU funding, public and private funding, etc.).

The consequence of no operation on the drainage channel systems on Rye Island would lead to collapse of the agricultural production. Other water sources for the agriculture would be necessary (groundwater wells) and this issue is connected with additional unplanned costs for the farmers, as well.

Groundwater flooding is in this region very rare. The drainage system intersects in some cases villages in the area of interest, but the flooding of basements of family houses is mostly insignificant, the excessive amount of water can be dewatered by drainage channels.

18.3.4 Governance and Legislation Risks

The whole Rye Island is a huge groundwater reservoir. The capacity of it was approximately estimated to be more than 18 $m^3 s^{-1}$ (Šoltész et al., 2021). Therefore, very strict legislation regulations were introduced for this region in last 30 years. They are covering the private and public water use as well as the use of fertilizers what was not so strictly controlled in the past. Another point which has to be taken into consideration are changes in property rights and with it connected land use. There were farmer cooperatives in the past which were transferred during last 30 years into smaller private farmer units.

Waterworks which are supplying the population of the region for drinking purposes are situated close to Hrušov reservoir northerly of the SVII drainage area. Due to very convenient hydrogeological conditions the drinking water is pumped from wells in lower horizons (60–80 m below the surface) with a very high quality where almost no treatment is necessary (low content of oxygen).

The greatest legislation risk for farmers—from my point of view—is the introduction of charging for water used for irrigation of agricultural land which was prepared by the Ministry of the Environment. Introduced fee is next of the costs that farmers bear due to the necessary irrigation of agricultural crops. In the case of the irrigation system, which belongs to the state-owned company Hydromeliorations, farmers have to reckon with additional costs. These include rents for the irrigation system, energy, investment in system repairs and its protection.

Increase in the cost of irrigating agricultural land is closely related to the increase in the prices of Slovak food and the decline in their competitiveness. The increased costs will be reflected in the price of commodities and there is a risk that production will not be competitive, which may lead to the cessation of cultivation of some crops.

18.3.5 Risks Related to the Sensitivity of MAR to Climate-Induced Extreme Situations

Climate change is expected to have a significant impact on the hydrologic cycle, creating changes in freshwater resources. The Intergovernmental Panel on Climate Change (IPCC) predicts that, as a result, floods and prolonged droughts will take place at increasingly frequent periods (IPCC, 2012). The Mediterranean has been described as one of the main climate change "hot-spots", with recent simulations showing a collective picture of substantial drying and warming (Tsanis et al., 2011).

The reality is that the problem of drying and warming occurs more frequently in the Central and Eastern Europe, as well. Developed sensitivity analysis for extreme situation cases (Vranovská et al., 2020) has identified trigger parameters causing natural hazards in wet and dry periods. In dry periods due to extremely low amount of precipitation and extremely high evapotranspiration the MAR system in investigated SVII drainage channel system can respond to mentioned extreme situation. It should work to full extent, i.e. maximum water supply through inlet structures and closed gate structures increasing the water level to increase the aquifer recharge and minimize the consequences of drought.

In opposite climate—induced extreme situation (water surplus) in the lowland the flash flood is very rare and improbable. Mostly, the excessive inland water (ponding after snowmelt in the spring time) or groundwater flood (after long—lasting precipitation period) can appear. According to data from boreholes of the SHMI observation system (Červeňanská and Baroková, 2021) and due to high permeable aquifer the MAR system is not really vulnerable and changes of groundwater table for investigated precipitation rich years (2010, 2013) are in limited figures.

18.3.6 Definition of Risk Monitoring Milestones/Risk Monitoring Scheme

Risk assessment determines risks that might affect the MAR process. There are several methodologies for risk identification, but the risk can be treated effectively only in case when it is thoroughly identified.

Risk monitoring and control is the process of keeping track of the identified risks, monitoring residual risks and identifying new risks, ensuring and evaluating the effectiveness in reducing risk. Good risk monitoring and control processes provide information that assists with making effective decisions in advance of risk occurrence. The purpose of risk monitoring is to determine if:

risks were identified from all points of view (health, environmental, technical, socio-economic, governance and legislation, etc.),

- risk responses have been implemented and are measurable,
- risk trigger has occurred, evaluation of its extent is necessary,
- proper operational procedures are followed.

The SVII channel system is operating for more than 28 years. There was never requirement for risk monitoring system, neither connected with MAR system. There was an attempt in the past to observe the impact of the water level regime in SVII channel system on groundwater level during the drainage as well as during the recharge period. The observation showed the interval of groundwater level changes and the extent of the effect.

There are no special national regulations in place concerning operational monitoring of MAR systems. Anyway, existing rules of surface water and groundwater monitoring can be effectively applied for monitoring of MAR system operation.

Responsible for risk assessment of the MAR site will be the company/institution managing the MAR facility (for instance in Slovak pilot site it is Slovak Water Management Enterprise). Surface water and groundwater monitoring is done under EC Directives, particularly the Water Framework Directive (2000/60/EC), Groundwater Directive (2006/118/EC) and Nitrates Directive (91/676/EEC).

Surface water monitoring is performed in accordance with the Act No 364/2004 Coll. (Water Act); Act No 201/2009 Coll. (on state hydrological service and state meteorological service); Act No 7/2010 Coll., (on flood protection) and Government Regulation No 269/2010 Coll. (on requirements to achieve good status of waters), Government Regulation No 167/2015 Coll. (on environmental quality standards in the field of water policy), Government Regulation No 201/2011 Coll. (on technical specifications concerning chemical analyses and monitoring of water), Government Regulation No 354/2006 Coll. (on drinking water standards) in accordance with Decree No 418/2010 Coll. of the Ministry of Agriculture, Environment and Regional Development of the Slovak Republic (on occurrence, monitoring and assessment of quantity and quality of surface water and groundwater).

Groundwater monitoring relates to item 4 of Collection of Slovak Republic, Act No 364/2004 Coll. (Water Act), Act No 201/2009 Coll. (on state hydrological service and state meteorological service), Act No 569/2007 Coll. (Geological Act), Act No 7/2010 Coll. (on flood protection), Government Regulation No 416/2011 Coll. (on the assessment of chemical status of groundwater body) and Decree No 418/2010 Coll. of the Ministry of Agriculture, Environment and Regional Development of the Slovak Republic.

Surface water and groundwater monitoring data are stored in the Slovak Hydrometeorological Institute. Surface water monitoring is partially performed by the Slovak Water Management Enterprise and Water Research Institute; other monitoring activities are covered by the Slovak Hydro-meteorological Institute. Within the River Basin Plan of Slovakia, framework programs of water monitoring covering 5 years are prepared.

The SVII channel system is operated by the Slovak Water Management Enterprise (SWME), for more than 28 years. In the past after putting Gabčíkovo Water Structure into operation, there was observed the impact of the water level regime in SVII channel system on groundwater level during the drainage and recharge periods. The observation showed the interval of groundwater level changes and the extent of this effect.

Risk monitoring milestones will be specified in detail based on specific MAR case. In general, the periodically will be monitored identified non-technical risks during design and construction phase (i.e. the lack of private/public funding; low price of water and high installation cost); and during operational phase (i.e. low price of water); as well as technical risks during design and construction phase (i.e. construction technical difficulties, risk of low water storage, hydrogeological setting) and operational phase (i.e. swelling clays, nutrients, droughts and rainfall events periodicity, changes in water demand and supply).

According to the experience in operating the sluices during the year, there are still possibilities how to improve the recharge the water from channel system into aquifer. As already mentioned nowadays it is dependent on beginning of the freezing period, but sometimes the administrator of the channel system (SWME) opens the sluices too early. Improved operation plan would help to replenish the groundwater sources in enhanced rate as well as drilling the observation boreholes for monitoring of groundwater level regime in adjacent area. That would be the best way for risk identification and for following operational interference. The surface water and groundwater quality should be monitored.

18.4 Conclusions

The analysis of climate change models resulting data indicates that present extreme events will intensify, i.e. precipitation will be probably less frequent but more intense and droughts are likely to be more frequent and serious in some regions (Tsanis et al., 2011). Because of expected intensive climate change impacts, the MAR systems are the solutions how to tackle droughts. Water retention in the landscape and consequently its infiltration into groundwater should ensure sustainable usage of water sources.

Within the potential location of MAR system in Rye Island area, there was performed risk assessment in pilot area. Risks were assessed by combination of two methods; i.e. quantitative risks were identified according to the methodology developed by MARSOL project and qualitative risk assessment was done in accordance with Australian guidelines (the likelihood and the severity of a risk are interpreted by risk factor matrix). According to MARSOL methodology, there were evaluated quantitative technical and non-technical risks (social, economic, governance and legislation) during design, construction and operational phases. However, taking into account all specific aspects of MAR system in pilot area, we identified other special risks related to pilot site; on the other hand, many of risks from MARSOL project were not applicable to Rye Island.

As non-technical risks during design and construction phase, the lack of private/public funding was identified as very high risk. Possible treatment is to disseminate and publicize the MAR schemes benefits to be able to involve as many investors as possible. Within this group, two other risks were evaluated as high: low price of water and high installation costs. Low price of water risk was evaluated in operational phase as well. These risks can be overcome by additional support for the use of MAR facilities (state support, private financial sources) in order to promote its financial viability.

Evaluation of technical risks revealed several high probabilities of occurrence during design and construction phase: construction technical difficulties, risk of low water storage, hydrogeological setting. These risks can be treated by preparation of specific technical project, and proper and detailed geological and hydrogeological investigation. During operational phase, high risks are represented by swelling clays, nutrients, droughts and rainfall events periodicity, changes in water demand and supply. To treat this group of risks there considered following issues accordingly: (i) detailed study on geological/rock mechanic conditions of the channels banks; (ii) agricultural pollution can be avoided by applying of Good Agricultural Practice and waste water pollution by centralised sewage system and WWTP; (iii) efficient manipulation with water in channels; (iv) proper regulation of water flow in channels and monitoring of water quantity/quality.

Acknowledgements The contribution was developed within framework and based on the financial support of the APVV-19-0383 project as well as VEGA project 1/0728/21.

References

Alam, S., et al. (2021). Managed aquifer recharge implementation criteria to achieve water sustainability. *Science of the Total Environment 768*, https://doi.org/10.1016/jscitotenv.2021.144992

Bedford, T., & Cooke, R. (2003). *Probabilistic risk analysis: Foundations and methods*. Cambridge University Press.

Červeňanská, M., & Baroková, D. (2021). Evaluation of the flood risk due to increased groundwater level. In *STU Bratislava*, 111 p.

Dillon, P. (2005). Future management of aquifer recharge. *Hydrogeology Journal, 13*, 313–316. https://doi.org/10.1007/s10040-004-0413-6

Dillon, P., et al. (2009). *Managed aquifer recharge: An introduction* (Technical report, 13). National water Commission, Canberra.

IPCC. (2012). In C. B. Field, V. Barros, T. F. Stocker, D. Qin, D. J. Dokken, K. L. Ebi, M. D. Mastrandrea, K. J. Mach, G.-K. Plattner, S. K. Allen, M. Tignor, & P. M. Midgley (Eds.), *Managing the risks of extreme events and disasters to advance climate change adaptation. A special report of working groups I and II of the Intergovernmental Panel on Climate Change* (p. 582). Cambridge University Press.

Nandha, M., Berry, M., Jefferson, B., & Jeffrey, P. (2015). Risk assessment frameworks for MAR schemes in the UK. *Environmental Earth Sciences 73*, 7747–7757. https://doi.org/10.1007/s12665-014-3399-y.

Rodríguez-Escales, P., et al. (2018). A risk assessment methodology to evaluate the risk failure of managed aquifer recharge in the Mediterranean Basin. *Hydrology and Earth System Sciences, 22*(3213–3227), 2018. https://doi.org/10.5194/hess-22-3213-2018

Swierc J., Page D., & Van Leeuwen J. (2005). Preliminary hazard analysis and critical control points plan (HACCP)—Salisbury stormwater to drinking water aquifer storage transfer and recovery (ASTR) project of Montana. *Water*.

Šoltész, A. (2002). Water management in regions affected by water structure construction and operation. In *Advances in Hydro-Science and—Engineering. Abstracts. Proceedings of the 5th International Conference on Hydro-Science and—Engineering* (pp. 127–136). University of Technology, Warsaw.

Šoltész, A., et al. (2021). Interreg Central Europe: DEEPWATER-CE final report on cooperation on activities implementation of the project. In *STU in Bratislava* (p. 29).

Tartakovsky, D. M. (2007). Probabilistic risk analysis in subsurface hydrology. *Geophysical Research Letters, 34*, L05404. https://doi.org/10.1029/2007GL029245

Tsanis, I. K. et al. (2011). Severe climate-induced water shortage and extremes in Crete. *Climatic Change*, 667–677. https://doi.org/10.1007/s10584-011-0048-2

Vranovská, A., et al. (2020). *DEEPWATER-CE, workpackage T2, activity T2.3 commonly developed sensitivity analysis for extreme situation cases* (p. 15). Water Research Institute.

Chapter 19
Computation of Vertical Flows for Groundwater Modelling in Multi-aquifer Setup

Hema Vanar, Pradeep Kumar Majumdar, and Niraj Shah

Abstract Hydrological processes creating storage of groundwater beneath the earth surface is complex and functionally nonlinear. Hence more often required to be assessed using advanced numerical solution techniques. Accuracy in physics is attributed to the assumptions considered in correlating the point logs, whereas, mathematical precision depends upon, how close the discretized function approaches linearity. One of the major uncertainties in numerical modelling of the multi-aquifer domain, is the appreciation of the leakage that occurs into the interfaces between two layers. In the present paper, to address the problem, we take the opportunity of cluster wise lithological correlations and fitting of the exact Theim, Thies and Hantush solutions of pump test, to assess the vertical flows across the interfaces. We suggest that this technology be practiced for each layer of a multi-aquifer domain explicitly, and estimated vertical flows be used as an input to model application, in order to gain more certainty in the model results.

Keywords Aquifer · Aquitard · Correlations · Litho-logs · Vertical flow · Well function

---

H. Vanar (✉)
CHARUSAT University, Changa, Gujrat, India
e-mail: hemavanar@gmail.com

P. K. Majumdar
C. U. Shah University, Surendranagar, India
e-mail: patmajum@gmail.com

N. Shah
School of Engineering, P. P. Savani University, Surat, India
e-mail: niraj.shah@ppsu.ac.in

A. Sefelnasr et al. (eds.), *Water Resources Management and Sustainability*, Water Science and Technology Library 114, https://doi.org/10.1007/978-3-031-80520-2_19

19.1 Introduction

Computational numerical models are one of the most powerful tools to advance theoretical and practical knowledge in the characterization of water flow through underground media, which must always be compared and calibrated with experimental measurements. In practice, it is common to carry out these kinds of studies with a wide combination of geophysical methods and isotope techniques. Physical models are generally expressed mathematically, in terms of partial deferential equations (PDE). Groundwater dynamics, when pose the PDE analytically unsolvable, numerical models are best suited.

It is clear that these models' have extensions that predict objective functions for upcoming regimes. The models' parameters and predictions are likely to be speculative in many cases for the same study domain because they are location-specifically calibrated for a hydrologic phenomenon under different stresses. It is a well-known fact that even black boxes can be solved with the help of GIS's enormous capacities, the laws of numerical integration, and calibration optimization techniques. Perfectly matchable trials with the observed points, however, do not ensure unique hydrologic realization, and it is doubtful that such calibrated models can be used for purposes other than providing answers to what-if questions. More site-specific stress values that can be explicitly computed and input for the model run can lead to more computationally error-free solutions in such practices.

The classic quasi-three-dimensional approach, which assumes that groundwater flows vertically in the weakly permeable layer (aquitard) and horizontally in the highly permeable layer (aquifer), has a high computational efficiency when it comes to the numerical simulation of groundwater flow in multi-layer formations. The plane two-dimensional flow control equation is derived in each soil layer, taking into account the infiltration of the upper and lower interfaces and using the average water head as the control variable. The two-dimensional calculation theory is extended to apply to the general multi-layer situation. At the interface, the flow velocity is constant. Perform quadratic interpolation on the average water head and the water head at the upper and lower interfaces in each soil layer, conditionally coupling the governing equations of each layer.

Multi-aquifer wells are encountered in applications involving relatively deep water-supply wells in the entire world. Multi-aquifer wells are open across two or more water bearing strata that have different hydraulic properties and water levels. Wells that are open across multiple flow zones in fractured rock settings should also be conceived as multi-aquifer wells. These wells can have a profound effect on ground water flow, regardless of pumping. They change the physical system and the equations that describe it by establishing direct hydraulic links between units that are otherwise isolated. The simulation techniques developed for multi-aquifer wells can also be applied to represent wells that span multiple model layers, whether the layers involved represent distinct aquifers or arbitrary subdivisions of a single aquifer. Although the properties and hydraulic head distributions may differ among the units penetrated by a multi-aquifer well, a single water level is established within the wellbore. Multi-aquifer system drawn by pumping well is describe in (Fig. 19.1).

All the stresses on an aquifer are periodic and influence the water levels. The stress created by the draft through pumping wells are crucial due to non-uniform manual intervention beyond specific yield criteria and thereby disturb the equilibrium. The steady state condition, once disturbed, takes longer duration to regain, even if not influenced by the granular deformations due to overburden and pore water pressure. With uninterrupted pumping throughout the year, water table depletes continuously. Church and Granato (1996), Hutchins and Acree (2000) and Puls and Barcelona (1996) have described the groundwater flow in multi-aquifer system, to be proportional to the aquifer permeability, when abstracted by a well. However, these wells serve as an artificial conduit allowing another kind of vertical flows between the layers, otherwise may remain an unconnected leaky geological unit.

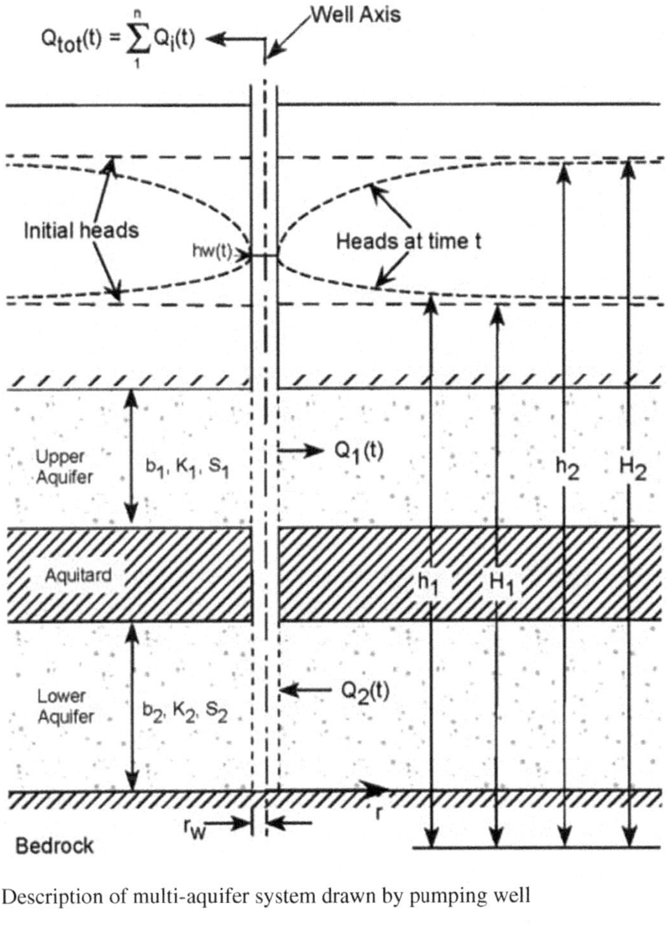

Fig. 19.1 Description of multi-aquifer system drawn by pumping well

Neuman and Witherspoon (1972) outlined the process of determining the hydraulic characteristics of leaky multiple aquifer systems in the field. The hydraulic properties of the aquitard can be assessed by comparing the drawdown in the aquitard to that recorded in the aquifer at the same time and radial distance from the pumping well. In a more recent study, Ha et al. (2019) examined the hybrid algorithm's performance in conjunction with the ratio method and Neuman and Witherspoon model using multiple pumping tests. They came to the conclusion that, in other hydrogeology conditions, the hybrid algorithm's performance warrants more investigation.

Mehdizadeh et al. (2014) highlighted approaches to freshwater leakage through aquitards in sharp-interface models, comparing interface position and vertical fluxes for sharp-interface and dispersive models, and sand-tank experiments. Shen et al. (2015) conducted four groups of single well pumping tests in each aquifer to obtain the hydrogeological parameters of the aquifers and investigate the hydraulic connection among the aquifers. The results of the tests indicated that the upper three aquifers are hydraulically connected. Furthermore, the hydrogeological parameters were analysed using both analytical and numerical methods. The phreatic aquifer was solved analytically through the application of Dupuit's equation, while the confined aquifers were solved using the Cooper and Jacob method. A finite element method was used to carry out the numerical simulation (FEM). However, it was doubtful that either the wellbore storage or the aquitard's leakage effect would have an impact on the analytical solution.

The ambient vertical flows in wells are as ubiquitous as vertical flows through aquifers and will occur at least to some degree in all aquifers (Elci et al., 2001). In addition to that, later arrival of groundwater entering the distant ends of the screen farthest from the pump intake (Martin-Hayden, 2000a, 2000b; Martin-Hayden et al., 2014; Reilly & Gibs, 1993), well casing storage (Barber & Davis, 1987), well screen and sand pack design (Kozuskanich et al., 2012), partial mixing of inflowing water with water within the well screen during pumping (Martin-Hayden & Wolfe, 2000) and purging methods (Robbins & Martin-Hayden, 1991) do not allow such stabilized permeability weighted drafts for quite appreciable durations, since the beginning of the pumping, depending mainly upon the screen lengths.

Craig and Gracie (2011) used XFEM to the problem of transient leakage from abandoned or free-flowing artesian wells in perforated aquifer–aquitard systems. The impact of aquifer geometric heterogeneity on leakage rates is assessed and seen to be an important factor in determining total leakage. McMillan et al. (2014) through numerical modelling demonstrated that greater anisotropy between vertical to horizontal hydraulic conductivity including discrete presence of low permeability layers affect the chemical composition of groundwater samples drafted using pumps.

Practically speaking, hydrogeological conditions in the arid alluviums do not remain that stable, as depicted in the theories, including the engineering geology. In addition, there is continuous subsidence of the layers as a result of compaction of the granular materials due to excessive pumping. In such situation, vertical flows in multi-aquifer systems play more dominant role in groundwater hydrology. Many multi layered aquifers even could approach single layer behaviors with gradual increase in the number of bore wells explored and also leakage through the interfaces. Hence it is pertinent for modelling studies to cater for the vertical flows in the multi aquifer systems, like never before. Present paper explores possibility of explicitly computing the vertical flows through interfaces using lithology and type curves in well hydraulics.

19.2 Problem Definition

As per the Neville and Tonkin (2004) study, the standard Well Package in MODFLOW is strictly correct only if flow is steady and purely radial in each stratum, with no vertical flow along the wellbore. Several corrections for converging flow have been developed; however, there is no simple method to estimate the transient composite water level in a multi-aquifer well based on application of the standard Well Package. The second approach for modeling multi-aquifer wells consists of representing the connection along the wellbore of the multi-aquifer well by assigning high vertical hydraulic conductivities to the cells it intersects. To simulate active pumping with this approach, the MODFLOW Well Package is used, and the total discharge is applied to the uppermost or the lowermost aquifer penetrated by the multi-aquifer well. For non-pumping conditions, the Well Package is not required; it is sufficient to incorporate the connection along the well with the high vertical conductivities.

The High vertical hydraulic conductivities in the Well Block approach can accommodate vertical flow along the wellbore. Furthermore, this approach does not require assumptions regarding the pumping level in the well or the allocation of the total discharge among the strata penetrated. Both the pumping level and allocation of flow are determined as part of the solution. However, it is not immediately obvious what constitutes a high vertical hydraulic conductivity in this context.

The appropriate vertical hydraulic conductivity can be estimated when the size of the well block approaches the physical dimensions of the wellbore, using Hagen-Poiseuille pipe-flow theory (Reilly et al., 1989): Eq. (19.1)

$$K_{pipe} = \rho\, g/\mu\, (r_w^2)/8 \tag{19.1}$$

where ρ and μ designate the density and dynamic viscosity of water, g is the acceleration due to gravity, and r_w is the radius of the well screen.

For a 0.3 m diameter well, the effective conductivity of the well is ~ 28,000 m/s. Generally speaking, it is not practical to specify a vertical hydraulic conductivity this high because it could cause convergence issues if the properties of neighbouring cells differ too much. To guarantee that the head is almost uniform throughout the wellbore, it is sufficient to specify a vertical conductivity that is sufficiently high. The theory of the MAW1 Package is formulated for a single well located at the center of a grid block, in a model with uniform grid spacing where radial symmetry is not disturbed by heterogeneity or hydrologic boundaries. Although the assumptions of the conceptual model are rarely satisfied in practice, deviations from them tend not to limit the applicability of the method. indicates that although complex cases may present difficulties, "ignoring the effects of multi-aquifer wells in a simulation will inevitably produce erroneous results." The approach implemented in the MAW1 Package provides an approximate means to account for converging flow in relatively large grid blocks. The approach implemented in the MNW Package is more general and offers flexibility in representing situations that are more complex.

The main drawback of Visual MODFLOW is its numerical formulation, which makes it impossible to "interpolate" between cell values in the absence of measured or fed data. Instead, the software dries out the cell by giving it a zero head value, resulting in a computational error. Other programs, like SWIFT, SUTRA, MIKESHE, FEFLOW, etc., encounter additional difficulties when applying wells that draw groundwater from multiple layers that are interfaced by impervious layers.

Worth mentioning here is that similar computational difficulty arises with analytical solutions for multi aquifer setup also, and remedies prescribed there is; allowing leakance through wells or interfaces. Single aquifer type curve solutions therefore could not be extended to multi-aquifer system, unless a continuity equation was coupled (Majumdar et al., 2009).

The above-mentioned computational difficulties and laborious solution techniques, if can be replaced by some exact solution results, smoothness in model run and certainty in calibration can improve. Lithological correlations are generally drawn using the litho-logs or bore logs to conceive a geometry that includes the multi-aquifers and impermeable layers. Analytical solutions available in Well hydraulics can be applied, in a cluster of wells, situated in a way that number of well logs surround a central well at the center. The aquifers and aquitards thus available after the lithology correlation analysis among central and each surrounding well can be used for the calculation of the vertical flow. We demonstrate the solution technique using an example of multi-aquifer system defined in Table1 for Sabarmati River basin in Gujarat, India.

19.3 Multi-aquifer System

Bradley and Phadtare (1989) highlighted two distinct aquifer systems in Sabarmati River basin in western parts of India, viz. a phreatic aquifer varying in thickness from 9 to 35 m overlying a series of aquifers and aquitards and isolated perched aquifers within the phreatic aquifer. They found that the total thickness of the aquifer system is around 250 m. The complete aquifer-aquitard system that was generated out of the lithological analysis is tabulated in Table 19.1. The system includes 5 confined aquifers and 4 aquitards derived from the approximation of the space between point litho-logs.

Multi-aquifer head fluctuations with increasing depths were examined during the periods between months May to October, in the years 2000, 2005, 2010 and 2013. The seasonal recharge trends show uniformity, irrespective of the piezometric depth.

Table 19.1 Emerged layers of the aquifer-aquitard system from lithological study

Name	Easting	Northing	Surface soil	Aquifer 1	Aquitard 1	Aquifer 2	Aquitard 2	Aquifer 3	Aquitard 3	Aquifer 4	Aquitard 4	Aquifer 5
MSW-223	235,708.0	724,815.0	184.8	181.8	172.8	170.8						
MSW-224	235,308.0	723,714.0	165	162	152	152	140	133	121	118	104	99
MSW-225	235,056.0	724,518.0	159.4	156.4	147.4	145.4	132.4	131.4				
MSW-227	232,600.0	722,513.0	87.28	84.28	39.28	36.28	– 1.72	– 6.72	– 27.72			
MSW-230	232,921.0	721,228.0	71.24	68.24	29.24	20.24	– 10.76	– 20.76	– 53.76	– 62.76	– 82.76	– 93.76
MSW-236	233,328.0	723,252.0	107.33	104.33	68.33	63.33	32.33	28.33	0.33	– 4.67		
MSW-241	234,443.0	724,720.0	150.5	147.5	115.5	113.5	99.5	97.5				
MSW-242	232,933.0	723,140.0	105.73	102.73	55.73	49.73	13.73	3.73	– 22.27	– 27.27	– 38.27	
MSW-243	232,825.0	722,104.0	83.13	80.13	14.13	11.13	– 16.87	– 20.87	– 36.87	– 41.87	– 66.87	
MSW-247	234,605.0	722,656.0	128.81	125.81	100.81	92.81	81.81	76.81	38.81	35.81	16.81	
MSW-250	234,651.0	723,322.0	139.18	136.18	118.18	114.18	101.18	94.18	88.18	80.18	60.18	56.18
MSW-252	234,707.0	723,830.0	155.61	152.61	140.61	140.61	121.61	115.61	87.61	85.61	60.61	55.61
MSW-254	234,926.0	723,655.0	149.17	146.17	124.17	120.17	109.17	107.17	95.17	92.17	68.17	52.17
MSW-257	234,049.0	724,255.0	137.64	134.64	88.64	82.64	69.64	60.64	56.64	45.64	16.64	13.64
MSW-258	233,505.0	724,044.0	124.85	121.85	73.85	69.85	46.85	42.85	25.85	19.85	– 12.15	– 20.15
MSW-272	234,948.0	722,157.0	115.56	112.56	59.56	$^{9}55^{9}.56$	30.56	27.56	– 2.44	– 9.44	– 29.44	

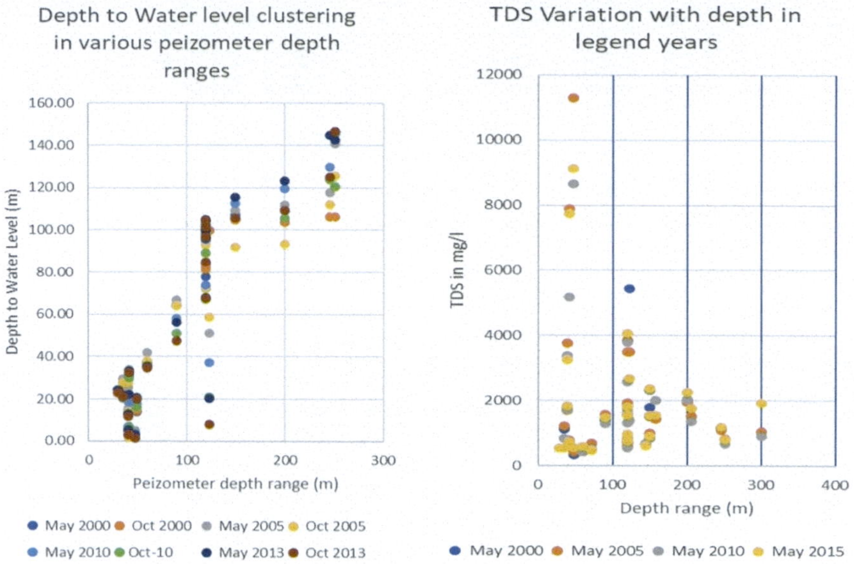

Fig. 19.2 Conceptualizing multi-aquifer domains

Almost all the piezometers are found, more or less gaining hydraulic head, in a quite similar manner, as if it is a single aquifer behavior. Such conditions in multi-aquifer system are possible in cases of some continuity of vertical flow through interfaces and/or aquifer interactions through wells, provided some piezometers draw water from perched aquifers. Analysis of seasonal groundwater level and TDS data of 29 piezometer setup conceptualize the flow and tracer movement in the multi aquifer system. On the left of Fig. 19.2, pre and post monsoon water levels in the years 2000, 2005, 2010 and 2015 are zoned in respect of the respective depths of the piezometers. The TDS values for May on right are presuming in steady non-reactive state.

Figure 19.2 shows that the vertical movement of water is finally stopped by a continuous barrier between 50 and 100 m depth zone. Critically saline zones are 0–50 m and 100–150 m, rest of the zones are gradually catching up to become critical. Source of the concentration could be leakage from the shallow unconfined aquifer, spreading downward through aquitards and it's discontinuities at different locations.

19.3.1 *Theoretical Background*

Bore hole lithology and characteristics are the foundation of studying the formation of aquifers and confining layers in between them. Different lithology forms the groundwater reservoir of different reservoir space and physical property. Hence, a significant difference exists in the formation mechanism. Fence diagrams are the outcome of corelating two adjacent lithologs. In general software like ROCKWORKS is used to generate aquifer proximities in a shape of panel diagram, which is nothing but a typical type of Fence diagram only. We selected the fence diagrams in clusters with a central well surrounded by numbers of adjacent wells. In a way profiles of formations allowing flow towards and outwards of a central well is incorporated. Radial flow type curve analysis is carried out in each of the profiles of a selected cluster using Thiem (1906), Theis (1935) and Hantush (1956) well functions for pumping test data analysis.

Thiem well function is derived under steady or equilibrium condition, can be used to assess the parameters of an aquifer layer, with well data of a period with negligible manual interventions. The non-equilibrium method as presented by Theis (1935) can be then used for the well function computation, for a non-leaky condition. It has been verified and modified by the leading scientists in groundwater hydrology. When the field conditions approximate the assumptions made in the development of the theory, the results are strikingly reliable. The non-equilibrium method as presented by Theis is the general equations which define the flow toward a pumped well penetrating the entire thickness of the aquifer are as follows: Eq. (19.2)

$$s = \frac{Q}{4\pi T} \int_{u}^{\propto} \left(\frac{e^{-u}}{u} \right) \cdot du \tag{19.2}$$

where the integral is called Theis well function and denoted as W(u) may be expanded as; Eq. (19.3)

$$W(u) = 0.577216 - \ln(u) + u - \frac{u^2}{2 \cdot 2!} + \frac{u^3}{3 \cdot 3!} - \frac{u^4}{4 \cdot 4!} + \cdots + \frac{(-1)^{n+} \cdot u^n}{n \cdot n!} \tag{19.3}$$

and

$$u = \frac{r^2 S}{4Tt}$$

s = drawdown at any point in the aquifer (L), Q—discharge of pumped well (L^3/T), T = coefficient of transmissibility of the aquifer (L^2/T), t = time since pumping started in days (T), r = distance from the discharging well (L), S = coefficient of storage of the aquifer (dimensionless).

The solution of W(u) is too tedious for frequent use. Wenzel (1942) has provided a simplified solution through a table of values of W(u) for a wide range of values of u. Knowing the values of u, W(u) can be obtained from the table with sufficient accuracy for most practical purposes.

Hantush (1956) developed the first rigorous mathematical model of transient flow of water to a pumping well in a leaky confined aquifer. Their mathematical equation for flow to a fully penetrating line sink discharging at a constant rate in a homogeneous, isotropic and leaky confined aquifer of infinite extent is as follows:

Eq. (19.4)

$$s = \frac{Q}{4\pi T} \int_{u}^{\propto} \frac{e^{-u^{-r^2/4B^2 u}}}{u} du \tag{19.4}$$

where $u = \frac{r^2 S}{4Tt}$ and $B = \sqrt{\frac{Tb'}{K'}}$, with notations as below.

Groundwater hydrologists commonly refer to the integral in the Hantush and Jacob solution as the Hantush and Jacob well function for leaky confined aquifers, abbreviated as w(u, r/B). Therefore, we may write the Hantush and Jacob equation in compact notation as follows: Eq. (19.5)

$$s = \frac{Q}{4\pi T} W(u, r/b) \tag{19.5}$$

Hantush and Jacob well function for leaky confined aquifers, w (u, r/B).

Note that Kruseman and de Ridder (1994) refer to the Hantush and Jacob curve-fitting procedure as Walton's method (Walton, 1960) and use the symbol w(u, r/L) to represent the Hantush and Jacob well function for leaky confined aquifers.

For each confining layer of a section between central and surrounding well, vertical flow (v) can be computed as; Eqs. (19.6) and (19.7)

$$v = W(u) - W(u, r/b) - W(u, r/b) \tag{19.6}$$

$$v = \int_{u}^{\propto} \left(\frac{e^{-u}}{u} \right) \cdot du - \int_{u}^{\propto} \frac{e^{-u^{-r^2/4B^2 u}}}{u} du \tag{19.7}$$

where, v is vertical flow.

It is not difficult to assess v as dimensionless in the previous equation, hence considered as a ratio or percentage.

19.4 Analysis of Vertical Flows

Figures 19.3, 19.4 and 19.5 are specially developed cluster fence diagrams resulted out of setting co relations between central and surrounding wells, shown for 3 selected clusters. Hydraulic gradients of the flow through various aquifers in above three selected central wells surrounded by 4, 5 and 6 numbers of wells respectively are computed using Darcy equation. Positive values of hydraulic gradients represent flow towards the well and vice-versa.

The average regional hydraulic gradient with which the flow is occurring through the aquifers of Table 19.1 are: 0.0294, 0.0347, 0.0409, 0.0401 and 0.0238 in aquifer

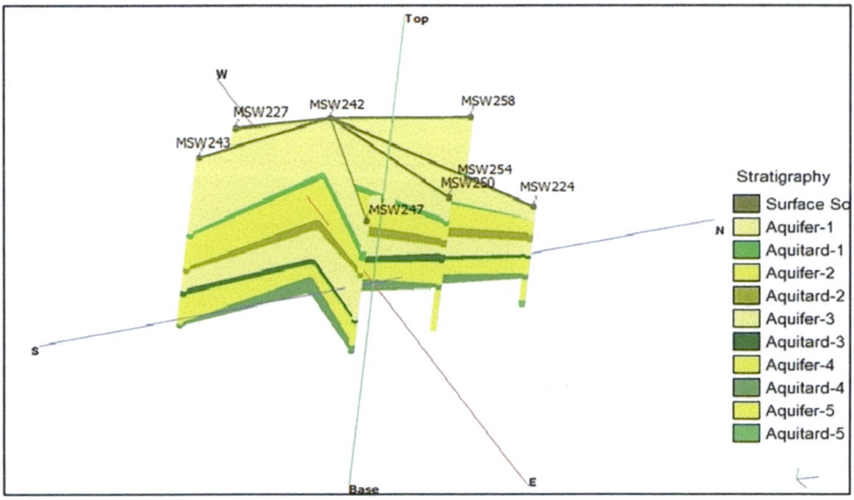

Fig. 19.3 Cross sectional view of correlation between MSW242 with surrounding MSWs (224, 227, 243, 247, 254 and 258)

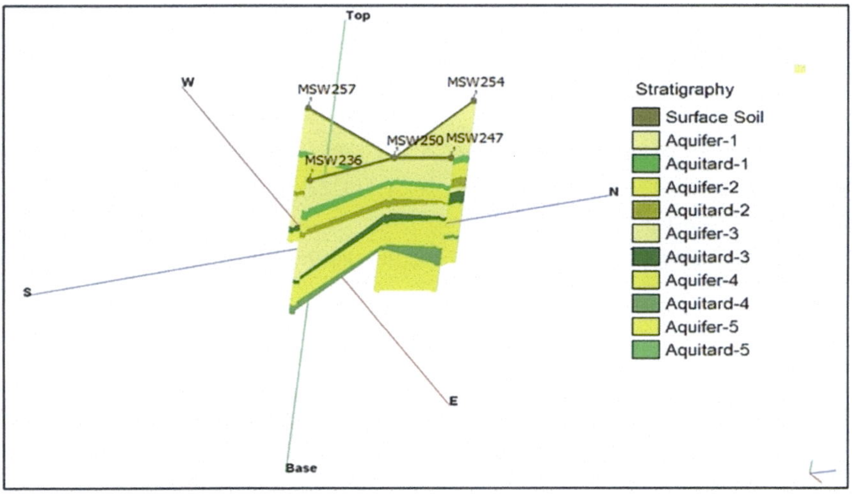

Fig. 19.4 Cross sectional view of correlation between MSW250 with surrounding MSWs (236, 247, 254 and 257)

1–5 respectively. In comparison, the local clusters of the wells around MSW242, MSW250 and MSW224 are flowing with hydraulic gradient as shown in Table 19.2.

Following the local setup of the wells as in Table 19.2, further study of the well hydraulics behaviors of the cluster of wells around MSW242, MSW250 and MSW224 were carried out. Considering the well data of pre-monsoon (PR) periods under unsteady state condition due to maximum well drafts and post monsoon (PO)

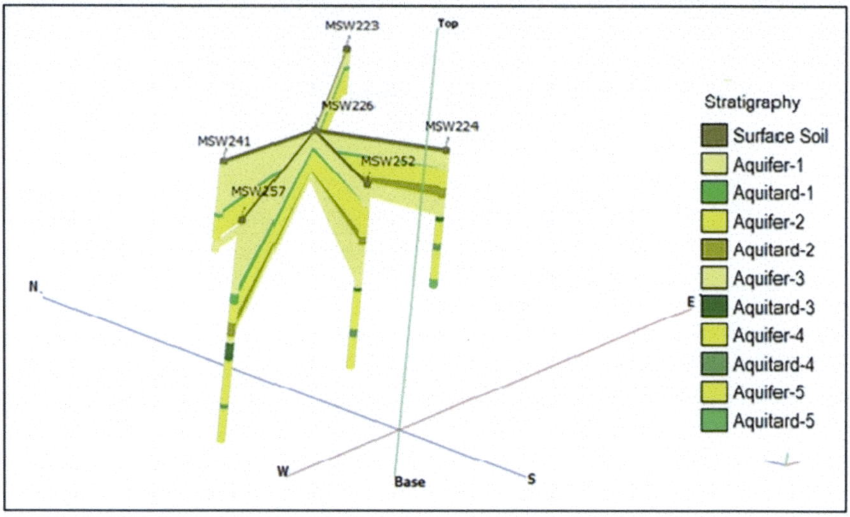

Fig. 19.5 Cross sectional view of correlation between MSW225 with surrounding MSWs (223, 224, 241, 252 and 257)

Table 19.2 Comparison of regional and local aquifer hydraulic gradients

Scale/cluster	Average Aquifer Hydraulic Gradient of groundwater flow				
	Aquifer1	Aquifer2	Aquifer3	Aquifer4	Aquifer5
MSW242	0.030249025	0.037251536	0.037388964	0.039746054	
MSW250	0.026029965	0.031972295	0.046213227	0.040892926	0.023798571
MSW225	0.03111916	0.03357046			
Regional	0.029413987	0.034691445	0.040918669	0.040176131	0.023798571

data to be under steady state condition with negligible draft because of sufficiency in canal supply and soil moisture confronted by the data in Table 19.3, with less coefficient of variation in the periodical data for pre-monsoon time and more in post monsoon period as shown in Table 19.3. The reason may be the groundwater is no more the safe resources for irrigation, people rely more on rainfed and canal irrigation. However, the merit in the numbers in Table 19.3 are processed for adoptability of well hydraulics theories in a particular cluster by including surface and bottom level elevations of respective wells.

The fitness of the well observations with theoretical solutions of Thiem (1906) and Theis (1935) for steady and unsteady conditions respectively are also shown in Table 19.4. Steady state pressure head is the ratio of pumping rate to the transmissivity of the aquifer. Calculations were carried out using the equilibrium equation of Thiem (1906). The error is the difference of calculated value from the average observed value for unit observed average value. In Table 19.4 the error in the fitness comes out to be between 0.26 and 0.48 irrespective of the depth of the cluster. This indicates a

Table 19.3 Transformed data for steady and unsteady well hydraulics behaviors in clustered multi-aquifers

CW-SW cluster	Bottom level CW-SW (m)	Radial distance (m)	Radial well head difference in the times (m)							
			2000		2005		2010		2013	
			PR	PO	PR	PO	PR	PO	PR	PO
MSW242-224	– 275	2443	–	141.37	170.57	157.42	184.47	159.17	180.17	184.77
MSW242-227	– 113	710	–	6.45	25.47	11.20	24.85	1.95	24.55	23.55
MSW242-243	18	1042	–	– 22.52	0.55	– 9.15	– 6.90	– 25.70	24.70	– 1.10
MSW242-247	– 154	1741	–	–	59.93	49.23	64.38	42.78	60.78	67.98
MSW242-250	– 164	1728	–	–	75.90	61.40	76.25	55.75	76.15	82.95
MSW242-258	– 120	1070	–	–	–	–	60.22	39.32	61.42	66.62
MSW250-236	32	1325	–	–	– 27.85	– 27.05	– 24.65	– 22.45	– 27.95	– 19.65
MSW250-247	10	668	–	–	– 15.97	– 12.17	– 11.87	– 12.97	– 15.37	– 14.97
MSW250-254	– 10	432	–	–	24.39	37.09	46.69	54.89	45.29	56.59
MSW250-257	27	1110			25.47	30.92	50.82	52.06	49.52	51.52
MSW225-223	– 31	716	48.90						22.70	23.40
MSW225-224	49	843	– 35.62	– 34.25	– 31.65	– 31.00	– 32.30	– 24.70	– 28.90	– 20.80
MSW225-241	34	645	– 22.57	– 21.70	– 21.15	– 19.10	– 24.40	– 23.40	– 22.40	– 22.40
MSW225-252	89	771	19.71	20.51	– 46.64	– 48.79	– 57.29	– 49.99	– 50.09	– 50.29
MSW225-257	132	1041	1.74	2.54	– 63.60	– 59.50	– 51.80	– 45.76	– 48.90	– 44.70

uniform impact of the well pumping in the multi aquifers very similar to a well drawing water from an equivalent single aquifer.

Computations for unsteady well functions for each cluster is carried out based upon the average pre and post monsoon piezometric water level fluctuations and using pumping rate-transmissivity ratio in steady state analysis. Amid anticipation of vertical flows in the multi aquifers these well functions are considered as that of Hantush (1956) and it's deviations from Theis (1935) allowed us to estimate the extent of vertical flows through the multi-aquifers. Results of the analysis is shown in Table 19.5. It varies cluster to cluster between 1 and 12% of the horizontal well flows.

Table 19.4 Fitness of the Thiem (1906) functions to the well observations

CW-SW cluster	Datum head (m)	Average (m)		Coefficient of variation		Steady state potential head	
		PR	PO	PR	PO	Head (m)	Error
MSW242-224	– 275	178.40	160.68	0.056	0.151	224.48	0.26
MSW242-227	– 113	24.96	10.78	0.026	1.439	35.19	0.41
MSW242-243	18	– 10.35	– 14.62	– 1.772	– 1.253	14.07	0.36
MSW242-247	– 154	61.70	53.33	0.054	0.347	80.01	0.29
MSW242-250	– 164	76.10	66.7	0.003	0.304	98.75	0.29
MSW242-258	– 120	60.82	52.97	0.014	0.364	82.47	0.35
MSW250-236	32	– 26.82	– 23.05	– 0.099	– 0.229	35.65	0.33
MSW250-247	10	– 14.40	– 13.37	– 0.217	– 0.152	20.43	0.42
MSW250-254	– 10	38.79	49.52	0.455	0.308	57.49	0.48
MSW250-257	27	41.94	44.83	0.481	0.380	56.67	0.35
MSW225-223	– 31	35.80	23.4	0.366	0	50.44	0.41
MSW225-224	49	– 27.69	– 27.69	– 0.224	– 0.296	38.40	0.38
MSW225-241	34	– 21.65	– 21.65	– 0.134	– 0.147	30.82	0.42
MSW225-252	89	– 32.14	– 32.14	– 1.062	– 0.946	44.96	0.40
MSW225-257	132	– 36.86	– 36.85	– 0.893	– 0.694	50.10	0.36

Table 19.5 Calculation for the vertical flows

CW-SW cluster	Datum head (m)	Unsteady well function		
		W (u, r/b)	W (u)	Vertical flows (%)
MSW242-224	– 275	0.9915	1.06	6.46
MSW242-227	– 113	5.0567	4.64	8.98
MSW242-243	18	3.8096	3.7	2.96
MSW242-247	– 154	1.3134	1.22	7.66
MSW242-250	– 164	1.1955	1.11	7.71
MSW242-258	– 120	1.1955	1.11	7.70
MSW250-236	32	1.3271	1.22	8.78
MSW250-247	10	0.6351	0.63	0.82
MSW250-254	– 10	2.3450	2.38	1.47
MSW250-257	27	0.6420	0.63	1.91
MSW225-223	– 31	0.2748	0.31	11.32
MSW225-224	49	1.2490	1.16	7.67
MSW225-241	34	0.3821	0.41	6.80
MSW225-252	89	0.3843	0.41	6.24
MSW225-257	132	0.8604	0.91	5.44

19.5 Conclusion

The present paper is an attempt to solve the implicit mathematical difficulty in the numerical modelling of multi-aquifer groundwater domains by explicit computation of vertical flows through interfaces of a multi-aquifer system using the lithological information and the fundamental type curves of well functions of Theis (1935) and Hantush (1956). Satisfying observed steady state well head condition with Thiem (1906) well function, is a precondition for using present method Although the procedure is demonstrated for the top layer, however, the same can be repeated for each of the layers. These values could be an input to the numerical modelling of groundwater flow in layered aquifers, so as to approach the activity with more certainty in the model application to a conceptual multi-aquifer domain generated out of lithological correlation. Investigated the impact of sea tides on mean water levels in a coastal aquifer system that consists of a semi-permeable layer, an unconfined aquifer, and a confined aquifer Li and Jiao (2003).

Acknowledgements I take this opportunity to express my special thanks to Dr. P. K. Majumdar, Emeritus Professor, C. U. Shah University, for helping me understand the nitty-gritty of the work. Due to his personal attention, I express my gratitude for sharing his insights on text processing software that has been very helpful in preparing the reports efficiently within the stipulated time. I express my sincere gratitude to him for his constant support during the research work for my thesis. Also, I thank, Dr. Niraj D. Shah. Director/Dean, School of Engineering, P. P. Savani University, Surat. His observations and guidance helped me to establish the overall direction of the research and move forward with the investigation in depth. His vital support at each juncture culminated in the successful completion of my research work.

References

Barber, C., & Davis, G. B. (1987). Representative sampling of ground water from short-screened boreholes. *Groundwater, 25*(5), 581–587.

Bradley, E., & Phadtare, P. N. (1989). Paleohydrology affecting recharge to overexploited semi-confined aquifers in the Mehsana area, Gujarat State, India. *Journal of Hydrology, 108*, 309–322.

Church, P. E., & Granato, G. E. (1996). Bias in ground-water data caused by wellbore flow in long-screen wells. *Groundwater, 3*(2), 262–273.

Craig, J. R., & Gracie, R. (2011). Using the extended finite element method for simulation of transient well leakage in multilayer aquifers. *Advances in Water Resources, 34*(9), 1207–1214.

Elci, A., Molz, F. J. I., & Waldrop, W. R. (2001). Implications of observed and simulated ambient flow in monitoring wells. *Groundwater, 39*(6), 853–862.

Ha, D., Zheng, G., Zhou, H., Zeng, C., & Zhang, H. (2019). Estimation of hydraulic parameters from pumping tests in a multi-aquifer system. *Underground Space*. https://doi.org/10.1016/j.undsp.2019.03.006

Hantush, M. S. (1956). Analysis of data from pumping test in leaky aquifers. *Transactions American Geophysical Union, 37*, 702–714.

Hutchins, S. R., & Acree, S. D. (2000). Ground water sampling bias observed in shallow, conventional wells. *Ground Water Monitoring and Remediation, 18*, 86–93.

Kozuskanich, J., Novakowski, K. S., & Anderson, B. C. (2012). Influence of piezometer construction of groundwater sampling in fractured rock. *Groundwater, 50*(2), 266–278.

Kruseman, G. P., & de Ridder, N. A. (1994). *Analysis and evaluation of pumping test data*. International Institute for Land Reclamation and Improvement.

Li, H., & Jiao, J. J. (2003). Tide-induced seawater–groundwater circulation in a multi-layered coastal leaky aquifer system. *Journal of Hydrology, 274*(1–4), 211–224.

Majumdar, P. K., Mishra, G. C., Sekhar, M., & Sridharan, K. (2009). Coupled analytical solutions for forced well recharge in confined aquifers. *ASCE Journal of Hydrological Engineering, 14*(12), 1351–1358.

Martin-Hayden, J. M. (2000a). Sample concentration response to laminar wellbore flow: Implications to ground water data variability. *Groundwater, 89*(1), 12–19.

Martin-Hayden, J. M. (2000b). Controlled laboratory investigations of wellbore concentration response to pumping. *Groundwater, 38*(1), 121–128.

Martin-Hayden, J. M., & Wolfe, N. (2000). A novel view of wellbore flow and partial mixing: Digital image analyses. *Groundwater Monitoring and Remediation, 20*(4), 96–103.

Martin-Hayden, J. M., Plummer, M., & Britt, S. L. (2014). Controls of wellbore flow regimes on pump effluent composition. *Groundwater Monitoring and Remediation, 52*(1), 96–104.

McMillan, L. A., Rivett, M. O., Tellam, J. H., Dumble, P., & Sharp, H. (2014). Influence of vertical flows in wells on groundwater sampling. *Journal of Contaminant Hydrology, 169*, 50–61.

Mehdizadeh, S. S., Werner, A. D., Vafaie, F., & Badaruddin, S. (2014). Vertical leakage in sharp-interface seawater intrusion models of layered coastal aquifers. *Journal of Hydrology, 519*(A), 1097–1107.

Neuman, S. P., & Witherspoon, P. A. (1972). Field determination of the hydraulic properties of leaky multiple aquifer system. *Water Resources Research, 8*, 1284–1298.

Neville, C. J., & Tonkin, M. J. (2004). Modelling multiaquifer wells with MODFLOW. *Ground Water, 42*(6), 910–919.

Puls, R. W., & Barcelona, M. J. (1996). *Low-flow (minimal drawdown) ground water sampling procedures*. EPA/540/S-95/504.

Reilly, T. E., Franke, L. O., & Prince, R. K. (1989). Analysis of the shallow groundwater flow system near Connetqout Brook, Long Island, New York. *Journal of Hydrology, 107*(1–4), 223–250.

Reilly, T. E., & Gibs, J. (1993). Effects of physical and chemical heterogeneity on water-quality samples obtained from wells. *Groundwater, 31*(5), 805–813.

Robbins, G. A., & Martin-Hayden, J. M. (1991). Mass balance evaluation of monitoring well purging: Part I. Theoretical models and implications for representative sampling. *Journal of Contaminant Hydrology, 8*, 203–224.

Shen, S., Wu, Y., Xu, Y., Hino, T., & Wu, H. (2015). Evaluation of hydraulic parameters from pumping tests in multi-aquifers with vertical leakage in Tianjin. *Computers and Geotechnics, 68*, 196–207.

Theis, C. V. (1935). The relation between the lowering of the piezometric surface and the rate and duration of discharge of a well using ground water storage. *Transactions American Geophysical Union*, 519–524.

Thiem, G. (1906). *Hydrolische Methoden*. J. M. Gebhardt, Leipzig.

Walton, W. C. (1960). Leaky artesian aquifer contamination in Illinois. In *Report of investigation* (Vol. 39, p. 27). State Water Survey, Urbana.

Wenzel, L. K. (1942). Methods for determining permeability of water-bearing materials. *U.S. Geological Survey Water-Supply Paper, 887*, 83–87.

Part IV
Water Quality

Chapter 20
Ceramic Membrane Filtration of Carwash Wastewater Sample from Al-Ain City, United Arab Emirates

Hakim Saibi, Nouf Alblooshi, Amira Alnaqbi, Shamma Alzaabi, Rauda Almazrouei, Fidha Ismail, Nakamura Arata, and Kawamura Kazuyuki

Abstract This study introduces a pioneering ceramic membrane-based produced water treatment technology (CMPWT) and validates its efficacy by treating a carwash water sample sourced from Al-Ain, situated in the Abu Dhabi Emirate, United Arab Emirates (UAE). A carwash sample procured in May 2023 from a local car washing facility in Al-Ain City served as the basis for this investigation. Utilizing the dead-end filtration mode with a filtration rate of 3 $m^3/m^2/d$, the transmembrane pressure peaked at approximately 30 kPa (temperature normalized to 25 °C). Air blowing was employed to counteract this pressure buildup, resulting in a substantial reduction of approximately 5 kPa. Within a span of 1 h, the total suspended solids (TSS) in the water decreased remarkably from an initial concentration of 489 mg/L to a final concentration of 2 mg/L. Comprehensive water chemistry analysis revealed a composition primarily comprising Cl–HCO_3–Na–Ca in the carwash wastewater. This study underscores the remarkable capability of CMPWT to efficiently reduce the TSS in carwash wastewater. The treated water stands are poised for safe discharge into the environment without causing harm or, alternatively, for potential reuse within the same facility. This technological advancement addresses the pressing need for effective wastewater treatment solutions and underscores its potential for broader environmental and industrial applications. The successful deployment of CMPWT in this context marks a significant step forward in sustainable water management practices, particularly in arid regions such as the UAE, where water conservation and quality are paramount. By providing tangible evidence of its effectiveness in real-world scenarios, this study provides valuable insights into the ongoing discourse surrounding wastewater treatment technologies and their implications for environmental sustainability and resource management.

H. Saibi (✉) · N. Alblooshi · A. Alnaqbi · S. Alzaabi · R. Almazrouei · F. Ismail
Geosciences Department, College of Science, United Arab Emirates University, P.O. Box 15551, Al-Ain, United Arab Emirates
e-mail: hakim.saibi@uaeu.ac.ae

N. Arata · K. Kazuyuki
Japan Organization for Metals and Energy Security, Tokyo, Japan

A. Sefelnasr et al. (eds.), *Water Resources Management and Sustainability*, Water Science and Technology Library 114, https://doi.org/10.1007/978-3-031-80520-2_20

Keywords Al-Ain · Carwash wastewater · Ceramic membrane · Filtration

20.1 Introduction

Carwash wastewater (CWW) is a global environmental concern because of its potential contaminants, including oils, suspended solids, and other pollutants. To address this issue, several techniques have been employed worldwide to mitigate and filter CWW effectively. Advanced filtration systems, such as ceramic membrane-based produced water treatment technology, including the Collaborative Membrane Purification Water Treatment (CMPWT) developed in Japan, offer a precise and efficient means of removing suspended solids and oil while maintaining dissolved ions. These technologies have advantages such as a sharp pore size distribution, higher hydrophilicity to reduce fouling risks, mechanical robustness, integrity, and extended service life. Other methods involve the use of environmentally friendly biofiltration systems, oil–water separators, and chemical coagulation processes to remove contaminants from CWW. These mitigation techniques are crucial in safeguarding local water resources, mitigating water pollution, and contributing to sustainable water management practices in the automotive industry on a global scale.

Over the last two decades, Abu Dhabi, United Arab Emirates (UAE), has experienced a nearly threefold increase in population and water consumption. This surge presents a significant challenge in ensuring an adequate supply of clean water to meet the diverse needs of various sectors, including industry, agriculture, and the population. Water pollution further exacerbates this challenge by introducing substances that render water unsuitable for essential activities such as drinking, cooking, cleaning, and recreational use. These pollutants include chemicals, waste, bacteria, and parasites (Sun et al., 2022; Xie et al., 2022; Yin et al., 2022). It is imperative to address all forms of pollution, because they inevitably enter water sources. This challenge is further underscored by the prevalence of persistent organic pollutants emanating from anthropogenic industrial, agricultural, and urban activities, which significantly complicate wastewater treatment systems (Kumari et al., 2022).

Various techniques have been employed to filter and treat CWW to mitigate its environmental impact and comply with water quality regulations. Some commonly applied techniques include: (1) Oil–water separators, which are designed to separate oils and greases from water, preventing them from entering drainage systems. They function by allowing the oils to float to the surface, where they can be skimmed off and properly disposed. (2) Sedimentation tanks, which help settle suspended solids from the carwash water. By allowing the particles to settle at the bottom, cleaner water can be decanted from the top, reducing the load on the downstream treatment processes. (3) Biological filtration, which uses natural processes involving microorganisms to degrade pollutants. Constructed wetlands or biofiltration basins can be implemented to promote the biological degradation of contaminants. (4) Chemical coagulation involves the addition of chemicals to facilitate particle aggregation, making them easier to remove. This method is effective for reducing turbidity and removing suspended solids from carwash effluent. (5) Ceramic membrane filtration: Advanced ceramic membrane technologies, such as CMPWT, provide precise filtration by selectively allowing water and dissolved ions to pass through while capturing suspended solids and oils. These systems are known for their efficiency, longevity, and low environmental impact. (6) Recycling systems: Water recycling systems have become increasingly popular in car washing facilities. These systems collect, treat, and reuse water for subsequent wash cycles, thereby minimizing water consumption and reducing the volume of wastewater generated. (7) Activated carbon filtration: Activated carbon is effective at adsorbing organic contaminants, including certain chemicals and odors present in carwash water. It can be used as part of a filtration system to improve the water quality. The choice of the filtration technique often depends on factors such as the volume and characteristics of the wastewater, local environmental regulations, and the level of investment that a carwash facility is willing to make in the water treatment infrastructure. Implementing a combination of these techniques can provide a comprehensive treatment and contribute to more sustainable water management in the carwash industry.

The CMPWT system, a product of a Japanese joint research initiative involving JOGMEC, INPEX Corporation, CHIYODA Corporation, and METAWATER Co., Ltd., is a membrane technology designed to effectively eliminate suspended solids and oil from water while maintaining dissolved ions during filtration (Saibi et al., 2023). JOGMEC emphasizes several advantages of CMPWT over other existing water technologies, including a highly precise pore size distribution, enhanced hydrophilicity to minimize fouling risks from oil and pollutants, superior mechanical robustness, heightened integrity, and extended service life.

This study focused on applying a CMPWT to treat wastewater from a car washing facility in Al-Ain (Abu Dhabi, UAE). Subsequent advanced chemical analyses were conducted to discern the compositional differences in the water before and after filtration. These findings have significant implications for addressing local water pollution in Al-Ain and safeguarding the environment, encompassing the hydrosphere and the ecosphere. Therefore, the primary objective of this study was to assess the efficacy of ceramic-membrane filtration technology in treating car washing water samples collected from Al-Ain. The study outcomes provide valuable insights into the filtration capabilities of suspended materials in water, aiding environmental protection and preventing groundwater aquifer pollution. Additionally, this study offers potential support to agricultural and industrial sectors by proposing a filtration technology suitable for purifying water for use in these domains. The study area was located in the city of Al-Ain (Abu Dhabi Emirate, UAE) (Fig. 20.1).

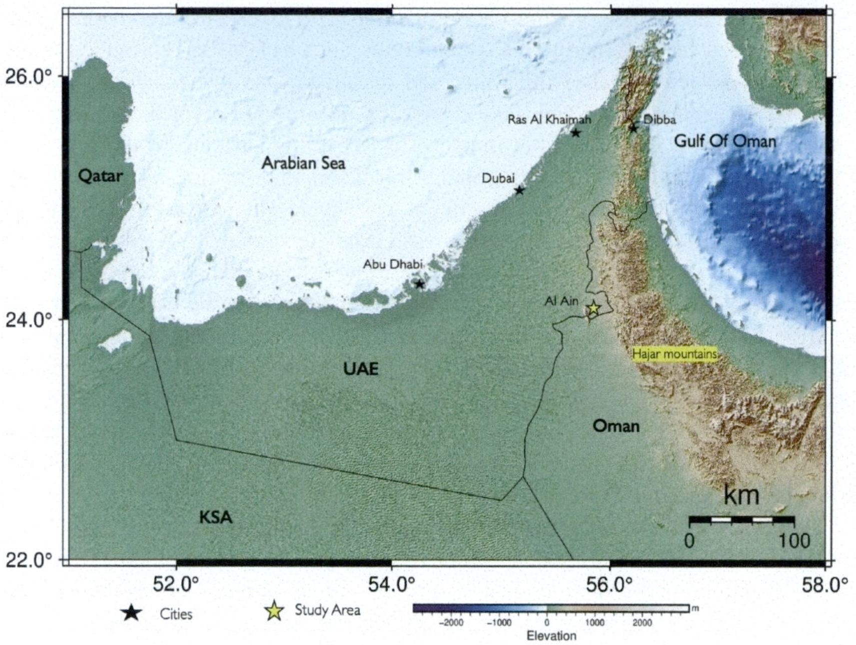

Fig. 20.1 Location of the study area within the UAE. The background shows the elevation relative to sea level

20.2 Methods

20.2.1 Specification of the CMPWT Unit

In this study, a Portable CMPWT Test Unit was used (Fig. 20.2). This equipment uses a ceramic membrane (Fig. 20.3) with φ30/100 mm L, an effective filtration surface area of 0.04 m^2 and nominal pore size of 0.1 μm. The maximum filtration capacity was approximately 110 mL/min (corresponding to a filtration flux of 4 $m^3/m^2/d$). This equipment can control the flow rate from 0 to 110 mL/min at the control panel and record and monitor the flow rate, temperature, inlet water pressure, and outlet water pressure every 5 s. It selects both dead-end and cross-flow filtration modes; however, the dead-end filtration mode was used in this study. This equipment has a backwash function that cleans the ceramic membrane.

20.2.2 Water Sampling

One water sample (ID: CWW) was collected from CWW in Al-Ain City (Abu Dhabi Emirate, UAE). CWW was sampled from a car washing facility in Al-Ain city (UAE),

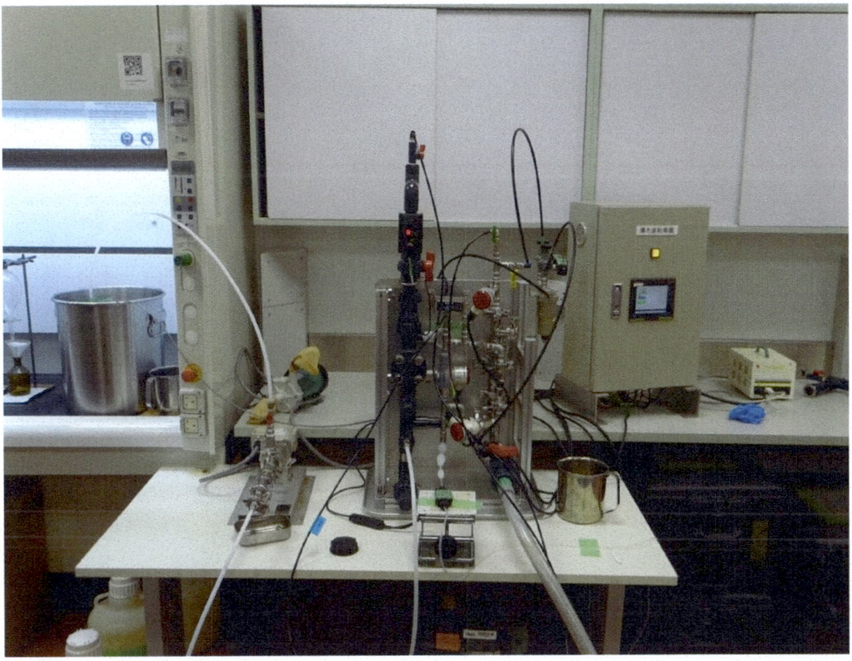

Fig. 20.2 Portable CMPWT Test Unit in UAE University (Building E4, Lab. 0008). Dead-end B Set with pressure sensor and flow meter

and the experimental conditions are summarized in Table 20.1. The water sample was transported to the laboratory (UAE University, Al-Ain, UAE) in 20 L plastic bottles and stored in a refrigerator until filtration testing.

Water samples were filtered without coagulation pretreatment. We selected the sole filtration approach because the samples were taken from a surface source and groundwater and both samples had almost no visible suspended matter.

The flow configuration was set to best suit the type of water for filtration; normally, cross flow was selected to avoid oil plugging if water contained oil. Because oil was only slightly visible on the water, dead-end filtration was used for the study. According to Van der Bruggen (2018), dead-end filtration can be applied at a constant flux and increasing pressure to compensate for flux reduction over time or at a constant pressure with decreasing flux over time. In this system, the backwash and blowdown processes were deployed if the pressure increased at a constant flux and reached 2 bar (200 kPa). Based on this rule of thumb, the filtration flux was initially set to a value that was predicted to be appropriate, that is 3 $m^3/m^2/d$.

The following parameters were evaluated: filtration stability [transmembrane pressure (TMP)], water quality, suspended solids, ions (sodium, calcium, potassium, magnesium, barium, strontium, total iron, dissolved iron, chloride, sulfate, carbonate, and bicarbonate), and pH. Analyses were performed at Core Laboratories International B.V. (Abu Dhabi, UAE).

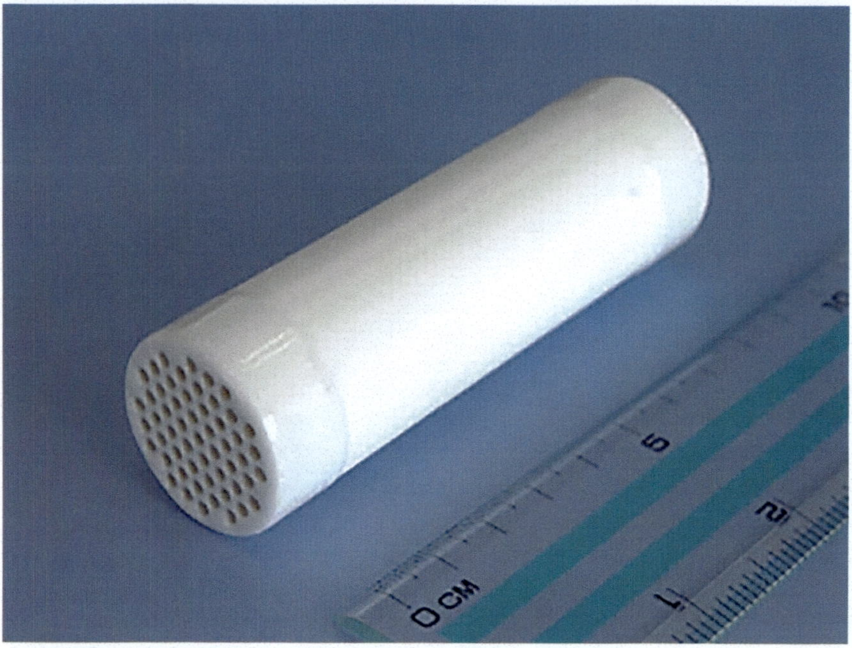

Fig. 20.3 A ceramic membrane used for the portable CMPWT test unit. *Source* NGK

Table 20.1 Experimental conditions

Test No.	Water	Filtration mode	Filtration flux ($m^3/m^2/d$)
CWW	Car wash wastewater	Dead-end	3.0

20.3 Results and Discussion

20.3.1 Water Chemistry

The water quality analysis results for the studied samples are summarized in Table 20.2. The raw carwash water samples contained low concentrations of sodium (59 mg/L) and chloride (91 mg/L) ions. These samples correspond to the Na–Cl water type.

20.3.2 TMP

The TMP over the filtration time for sample CWW is shown in Fig. 20.4. A stable filtration performance was observed during the test period for the CWW. Because a

Table 20.2 Water analysis data results of raw and filtrate CCW samples

Components		Unit	Carwash water	Carwash water
			Raw water	Filtrate water
Oil and grease (gravimetric, pH adjusted)		mg/L	26	< 5
Appearance			Turbid	Clear
TSS (0.45 μm)	No rinse	mg/L	**489**	**2**
	After rinse	mg/L	–	–
Dissolved oxygen		ppm		
Determined resistivity (Ohm m at 60 °F)			24.893	27.156
Ion balance		%	104.6	98.7
Hydrogen sulfide			ND	ND
Specified gravity at 60 °F			1.000	1.000
pH at 25 °C			6.81	7.36
Cations	Sodium	mg/L	59	57
	Potassium	mg/L	1.8	1.3
	Calcium	mg/L	34	28
	Magnesium	mg/L	3.8	3.5
	Barium	mg/L	0.18	0.13
	Strontium	mg/L	0.40	0.30
	Total iron	mg/L	ND	ND
	Dissolved iron	mg/L	< 0.01	< 0.01
Anions	Chloride	mg/L	91	90
	Sulfate	mg/L	31	29
	Bicarbonate	mg/L	74	66
	Carbonate	mg/L	NIL	NIL
	Hydroxide	mg/L	NIL	NIL

ND not determined (Extreme values are given in bold)

constant flux (3 m/d) was maintained throughout the experiment, TMP represents an indicator of the progression of membrane fouling. The TMP gradually increased to approximately 30 kPa in 60 min (Fig. 20.4), which is not a high value. The increase in the TMP was caused by the small size of the raw water tank. The apparent particle density increases as the water level in the raw water tank decreases. Figure 20.5 shows photographs of the raw and filtered CWW water samples. Table 20.2 lists the chemical concentrations of the sampled CWW water (before and after filtration).

During filtration, the fluids flow vertically to the filter surface in dead-end mode and the retained particles instantly solidify on the filter surface to create a filter cake. This filter cake then clogs the filter pores, dramatically reducing the filtration effectiveness in a short period. However, air blowing helped remove these clogs. The operation of the dead-end mode is straightforward, and thus ideal for small-scale situations with high product recovery.

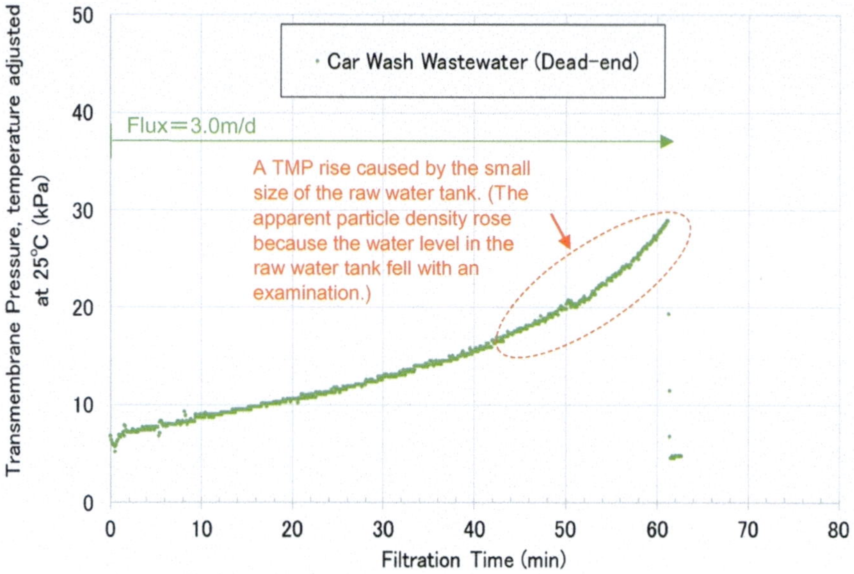

Fig. 20.4 TMP during the filtration of the CCW sample

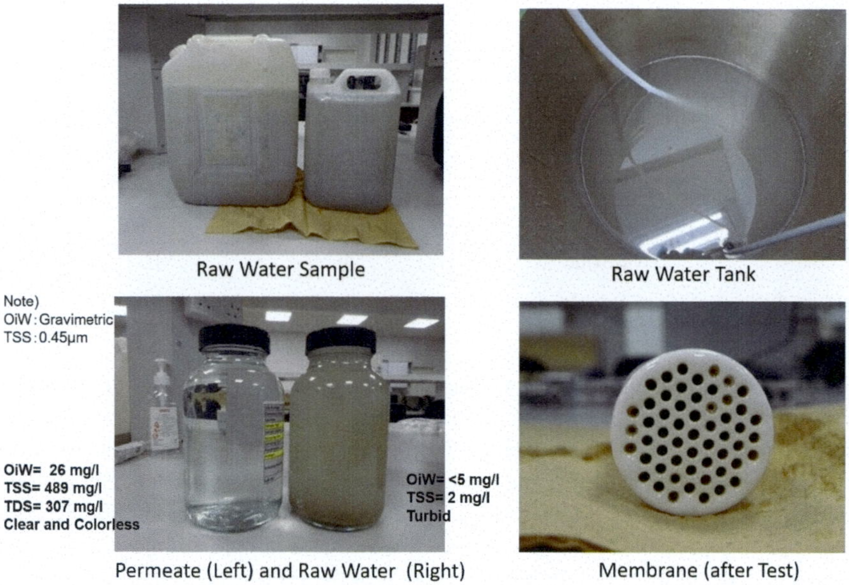

Fig. 20.5 Photographs of the raw and permeate water

Table 20.3 TSS results of raw water and permeate (CMPWT filtrate)

Components		Unit	Carwash water	
			Raw water	CMPWT filtrate
TSS (0.45 μm)	No rinse	mg/L	**489**	**2**

(Extreme values are given in bold)

Table 20.3 lists the total suspended solids (TSS results) for the raw and permeate CWW samples. The TSS concentrations of the raw and filtrate samples were 489 mg/L and 2 mg/L, respectively.

20.4 Conclusions

The conclusions of this study are as follows:

- CMPWT demonstrated the effective filtration of a carwash water sample from Al-Ain, UAE.
- Utilizing dead-end filtration at a rate of 3 $m^3/m^2/d$, the CMPWT successfully reduced the TSS from 489 to 2 mg/L within 1 h. This TSS level of the filtrate was enough to reuse it for carwash water.
- During a 1-h filtration test, the TMP increased from 5 to 30 kPa. After air blowing, the TMP was efficiently reduced from 30 kPa to approximately 5 kPa, which was almost the same as the initial TMP. This indicated that the suspended solids remaining on the filtration surface of the membrane were successfully removed by air blowing and the membrane returned to its initial state.
- Water chemistry analysis revealed a Cl–HOC_3–Na–Ca composition in the CWW.
- The filtrate water quality was compatible with the Water Quality Regulations 2021 (Department of Energy, UAE). Although a more detailed analysis is needed, this is drinking water quality.
- This study emphasized the potential of CMPWT for treating CWW, enabling environmentally safe discharge or reuse within the same facility.
- Overall, CMPWT offers a sustainable solution for TSS reduction in CWW, contributing to environmental conservation and water resource management.

Acknowledgements The authors thank the UAE University Research Office for its financial support (SURE PLUS Project 2023–2024).

References

Kumari, P., Bahadur, N., Conlan, X. A., et al. (2022). Atomically-thin schottky-like photo-electrocatalytic cross-flow membrane reactors for ultrafast remediation of persistent organic pollutants. *Water Research*, 218. https://doi.org/10.1016/j.watres.2022.118519

Saibi, H., Nakamura, A., Kawamura, K., et al. (2023). Ceramic membrane filtration of water samples from Ain Fayda Lake and cement factory (Al-Ain), United Arab Emirates. *Environmental Progress and Sustainable Energy*, e14268. https://doi.org/10.1002/ep.14268

Sun, Z., Zhang, Y., Guo, S., et al. (2022). Confining FeNi nanoparticles in biomass-derived carbon for effectively photo-Fenton catalytic reaction for polluted water treatment. *Advanced Composites and Hybrid Materials, 5*, 1566–1581. https://doi.org/10.1007/s42114-022-00477-4

Van der Bruggen, B. (2018). *Fundamental modeling of membrane systems: Membrane and process performance* (pp. 25–70).

Xie, K., Wei, S., Alhadhrami, A., et al. (2022). Synthesis of $CsPbBr_3/CsPb_2Br_5$@silica yolk-shell composite microspheres: Precisely controllable structure and improved catalytic activity for dye degradation. *Advanced Composites and Hybrid Materials, 5*, 1423–1432. https://doi.org/10.1007/s42114-022-00520-4

Yin, H., Zhong, W., Yin, M., et al. (2022). Carboxyl-functionalized poly(arylene ether nitrile)-based rare earth coordination polymer nanofibrous membrane for highly sensitive and selective sensing of Fe^{3+} ions. *Advanced Composites and Hybrid Materials, 5*, 2031–2041. https://doi.org/10.1007/s42114-022-00547-7

Chapter 21
Water Conservation in Jordan: Connecting the Dots for a Sustainable Future

Majd Aideh and Wael Sheta

Abstract Due to Jordan's paucity of natural resources, there are various environmental challenges that make it difficult to maintain population increase. It is anticipated that Jordan, one of the nations with the fewest natural resources, would slip below the poverty line. This study addresses the most pressing environmental concerns confronting Jordan, including water constraint, pollution, and desertification. The topic of water shortage is then discussed, along with the environmental repercussions of current practices and proposals for improvement. This study describes a "sustainability package" that may help avert these problems and establish a sustainable future. Governmental and non-government organizations are closely engaged in environmental concerns to build a solid foundation for environmental protection. Citizens and entities should collaborate to discover realistic answers to the nation's concerns. The primary goal of this research is to establish a framework for environmental water preservation in Jordan, ensuring that water resources are managed and utilized in a sustainable, egalitarian, and health-protective manner. This includes balancing the demands of various water users, safeguarding water quality, and preserving water supplies for future generations. It is feasible to encourage water conservation, enhance water management practices, and safeguard water resources for future generations by establishing such a framework for environmental water preservation in Jordan. This may help guarantee that there is enough water to fulfil people's and the environment's requirements, and that aquatic ecosystems stay healthy and resilient in the face of increasing demands from human activities.

Keywords Desertification · Environmental challenges · Pollution · Sustainability package · Water conservation · Water quality

---

M. Aideh · W. Sheta (✉)
Faculty of Engineering and IT, The British University in Dubai, Dubai, UAE
e-mail: wael.sheta@buid.ac.ae

M. Aideh
e-mail: 22000446@student.buid.ac.ae

W. Sheta
Architecture Department, Faculty of Engineering, Al-Azhar University, Cairo, Egypt

A. Sefelnasr et al. (eds.), *Water Resources Management and Sustainability*, Water Science and Technology Library 114, https://doi.org/10.1007/978-3-031-80520-2_21

21.1 Overview of Environmental Problems in Jordan

Jordan's climate is quite diverse, situated between the Mediterranean and the arid deserts. Until the 1970s, environmental challenges in Jordan were uncommon. Both population density and industrialization contribute to environmental issues. Jordan's economy is one of the worst in the Middle East as a result of a lack of natural resources and usually ineffective use of these resources (Hadadin & Tarawneh, n.d.). To bridge the budget shortfall and foster development, the government mostly depends on foreign help. Poverty, a lack of employment opportunities, and government borrowing are continuing economic concerns that must be addressed by inventive ways administered by relevant governments. Air pollution is a major concern on a worldwide scale owing to its negative effects on human health and well-being. Jordan is coping with a major air pollution issue. Transportation, electricity production, and industry account for the majority of air emissions. High-density areas like as Amman and Irbid are becoming more worried about air pollution. Predicted carbon dioxide (CO_2) emissions in the country were 5.18 million tons, with the power sector accounting for 2.3% of total emissions (Al-Jaafreh & Nagy, 2018; Hadadin, 2015). According to a Greenpeace MENA (Middle East/North Africa) research, nitrogen dioxide (NO_2) air pollution levels in Amman, Jordan's capital, have climbed by 81% by 2021. The report's recommendations said that "there is an urgent need to reduce nitrogen dioxide air pollution in populated areas" in order to protect people's health (Mustafa, 2021). According to field measurements in Amman and Zarqa (Fig. 21.1), an average of 70 g/m^3 nitrogen dioxide concentrations were observed in the city center of Amman, which equates to 100 g/m^3 NO_2, i.e., the permissible national limit (Ababsa, 2013). According to Zarqa Municipality, 52% of Jordan's industrial activity takes place in Zarqa, and the governorate's biggest environmental pollution issues include emissions and phosphate buildup from nearby distilleries, additionally, industrial wastewater discharges, brick and stone quarry dust, and pollutants from the neighboring thermal power station have an impact on the neighborhood (Ababsa, 2013; Abu Qdais et al., 2019). Amounts of hydrocarbons HC and carbon monoxide (CO) from gasoline-powered cars were discovered in rates of non-compliance with Traffic Law restrictions (Al-Jaafreh & Nagy, 2018). New rules were enacted in 2006 to reduce greenhouse gas emissions while also limiting industrial pollutants in the air and soil. The introduction of two kinds of unleaded petrol (90 and 95 octanes) and government-mandated limits on pollutant gas emissions from cars and industries resulted in significant reductions in lead pollution (Ababsa, 2013). Within this context, this study will shed light on three of Jordan's key environmental concerns, with a focus on the water environmental problem, which presents the greatest threat to future generations in the region and in Jordan in particular. The fundamental aim of this study is to provide a framework for environmental water preservation in Jordan to ensure that water resources are managed and used in a way that is sustainable, equitable, and protects the health of aquatic ecosystems. This involves balancing the needs of different water users, protecting water quality, and conserving water resources for future generations. By providing such framework for environmental

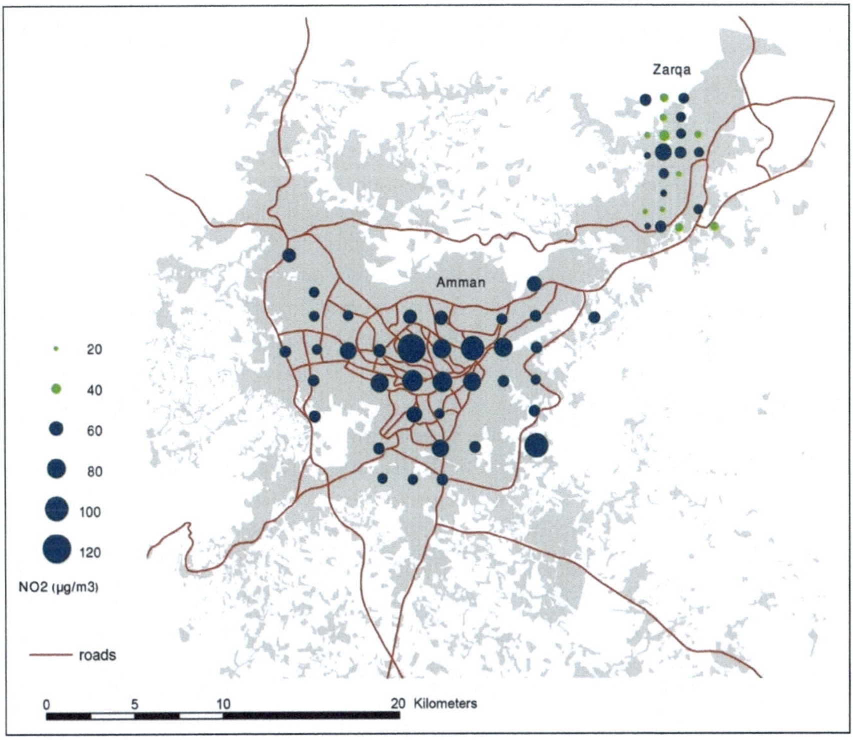

Fig. 21.1 Nitrogen Dioxide (NO_2) levels in different locations in Amman and Zarqa in 2007

water preservation in Jordan, it is possible to promote water conservation, improve water management practices, and protect water resources for future generations. This can help ensure that water is available to meet the needs of people and the environment, and that aquatic ecosystems remain healthy and resilient in the face of growing pressures from human activities.

21.1.1 Agriculture and Land Pollution

Several heavy metals are naturally found in agricultural soil in trace concentrations. However, when their concentrations reach a certain level due to pollutants imported from the outside, soil contamination and contaminated agricultural products develop. Al-Khashman and Shawabkeh (2006) revealed in an experiment conducted by collecting several samples from an area near a factory in the south of Jordan, that the tested samples included considerably high levels of lead, zinc, and cadmium, affecting the quality of the agricultural soil. The wastewater from olive mills is another form of pollution in Jordan, affecting the soil surface and producing serious environmental difficulties since the waste is released straight into the surrounding media. In most instances, mills do not have treatment plants because of cost constraints (Al-Kharabsheh, 2020; Khdair et al., 2019). Excessive use of chemical fertilizers and pesticides is a serious problem in Jordan, causing soil deterioration. It changes the properties of the soil and increases the amount of harmful substances in the environment (Jaber et al., 2004). Only 6% of Jordan's land is utilized for agricultural, which is dependent on rainfall and irrigation availability. Urbanization, soil erosion, and pesticide contamination all contribute to the continuous loss of agricultural areas (Jaber et al., 1997). Over the second part of the twentieth century, an estimated 25% of all agricultural areas were lost (Jaber et al., 2004). The government has started implementing many steps to avoid transferring agricultural land to other incompatible uses that do not restrict green space. Towns and houses will be relocated to rocky terrain or other unsuitable agricultural locations. Several measures have already been done in this direction, such as switching to organic fertilizers to preserve soil structure through the use of natural resources and encouraging ecological activity, reduce vegetable-producing soils' concern with nitrogen, lessen the impact on agricultural land and the use of chemical pesticides and take care of domesticated animals, including shelter, feeding, health, rearing, and reproducing (Al-Karadsheh et al., 2012).

21.1.2 Desertification

Jordan's tree cover declined by 826 ha (10%) between 2000 and 2020. Desertification removes the fertile layer of soil, which includes organic and metallic components. Living things, such as plants and animals, have a substantial impact on the ecosystem's balance. Overgrazing and human activities deplete the plants quickly, hastening erosion. The key contributors to desertification in Jordan include a lack of a comprehensive strategy for planning and distributing land uses on a sustainable basis, as well as inadequate land management (Al-Karadsheh et al., 2012). Many leading institutions have carried out proper planning and execution of operations, such as the Ministry of Agriculture, which has completed a variety of connected development projects, including the one for Sustainable Rangeland Management, as Jordan has four Rangeland Management strategies in use in arid and semi-arid areas (Al-Karablieh et al., 2010). Furthermore, the Ministry of Agriculture implemented development initiatives in certain areas, such as the Hamad basin, with specific goals of range development and management, animal productivity and health, and livestock water development (Al-Karadsheh et al., 2012).

21.1.3 Water Pollution and Scarcity

Jordan is regarded as one of the world's driest nations. There are no regional or local rivers in the nation, and the Jordan River is filthy and salty, running practically dry most of the year. 80% of Jordan is considered desert since it is in an arid-semi-arid climate zone and receives fewer than 100 mm of precipitation per year on average. Only 1.3% receive more than 500 mm per year, whereas 12.5% receive 100–200 mm/year, 3.8% receive 200–300 mm/year, 1.8% receive 300–500 mm/year, and 3.8% receive 200–300 mm/year (Hadadin, 2015; Hadadin et al., 2010). The average annual precipitation on Jordan's territory is less than 300 mm over just 3% of the country's area. Jordan gets little rainfall, and that rainfall is unevenly distributed, thus there are little groundwater and surface resources accessible for home use, as well as agriculture and industry. Jordan's water resources are used by four industries: agriculture, municipal services, industry, and tourism. Agriculture is the largest consumer of the nation's water resources, accounting for around 64% of the total water supply. In contrast, 30% is utilized for urban uses, 5% for industrial reasons, and 1% for tourism (Altz-Stamm, 2012). The Jordan River is Jordan's sole native river. The yearly availability of renewable water resources is just around 100 cubic meters per person, much below the 500-cubic-meter threshold that is considered "absolute water shortage." 1 More than 90% of low-income families in Jordan are critically vulnerable to water scarcity if water stress levels continue to rise at the projected yearly pace of 1%–1.5% until 2100 (UNICEF & Economist Impact, 2022). Jordan is therefore the world's second most water-scarce country (Al-Karadsheh et al., 2012). There are two main components to analyzing water quality that must be separated so that the quality of a water source may be appropriately measured, the first is natural water quality is an environmental effect on water quality, while the second is human activities such as urbanization, industrialization, agriculture, and building have all resulted in changes in the original features of water (Salameh et al., 2018).

Temperature and climate change have a significant influence on natural water resources. Jordan's climate ranges from the Mediterranean in the west to the desert in the east and south; winters in certain areas may be very cold, while summers are often hot and dry (Lelieveld et al., 2012). Jordan is situated in a region where temperature increases caused by climate change are predicted to be 20% greater than in other areas (Ghazal, 2023). Even a 20% increase in rainfall would not be enough to compensate for a two-degree Celsius increase in evaporation, which would diminish surface runoff. The Köppen climate classification divides the world into five basic climatic categories, with subgroups within each defined by seasonal temperature and precipitation trends. As shown in Figs. 21.2 and 21.3, between 1980 and 2016, Jordan has six distinct climatic subgroups along the area. However, in the desert environment type, these subgroups were reduced to four (UNICEF & Economist Impact, 2022). Jordan has one of the world's lowest per-capita water levels, making climate change vulnerable to water resources that are fast depleting. Climate change is predicted to exacerbate Jordan's water scarcity, leading to harsher consequences such as a drop in water quality, a fall in reservoir water volume, and a decrease in groundwater levels (Ratcliffe, 2022; United Nations, 2022). Another factor contributing to Jordan's water crisis is the country's population. Jordan has a population of around 11 million people as of 2021. In 2016, Jordan had the most refugees per person in the world; they mostly came from Syria, Iraq, and the West Bank of Palestine. Furthermore, numerous South Asian, Indonesian, and Egyptian workers work and reside in Jordan (United Nations, 2022). Jordan has a population of 10,439,344 people, accounts for 0.13% of the world total. Despite this, the fertility rate fell by 0.63% between 2015 and 2020; the population density grew by 11 persons per square kilometer in 2020 compared to 2015. Jordan has received around 622,000 Syrian refugees since February 2015, according to the UNHCR (United Nations, 2022). Politics have substantially impacted Jordan's water situation. Due to a history of instability and warfare, Jordan has taken in numerous refugees from many counties: Syria during the last ten years, Iraq after the 2003 war, Palestine after 1948, and the War in 1967.

Figure 21.4 illustrates Jordan's population density and dispersion. The total number of people living in regions designated as urban by the government is referred to as the urban population. In 2020, cities, towns, and the country's capital housed 9,332,603 of the country's 10,203,134 residents. The pace of population expansion shown in this data places a significant pressure on the water demand sector, affecting both the quality and availability of drinking water. The absence of a comprehensive regulatory framework, recent changes in agricultural, commercial, and industrial activity, and the population's fast expansion (about 4.5% per year) have all had detrimental environmental implications (Batarseh, 2021; Odeh, 2019). Rapid population growth, increasing urbanization, industry, and exploitation of groundwater have all contributed to the environmental concerns associated with dwindling water resources.

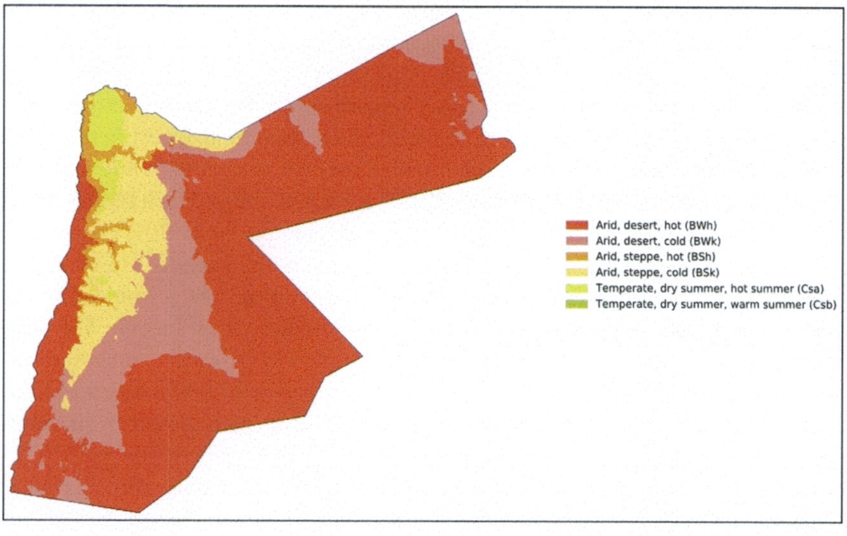

Fig. 21.2 Climate change map for Jordan for 1980–2016 as per The Köppen climate classification

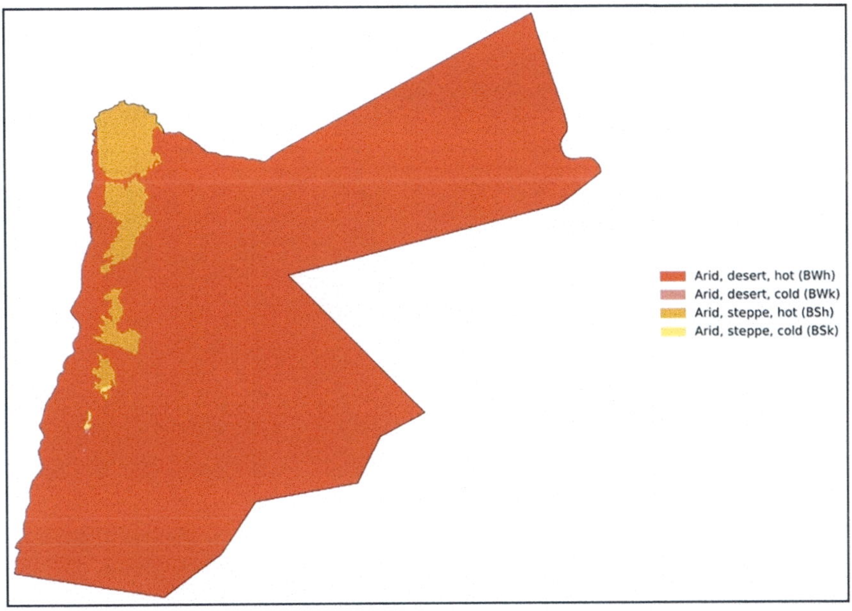

Fig. 21.3 Climate change map for Jordan for 2071–2100 as per The Köppen climate classification

This is due to industrial operations being located near or immediately upstream of potable water sources. Furthermore, excessive use of herbicides, insecticides, fungicides, and fertilizers is common, contaminating both groundwater and surface water (Hadadin & Tarawneh, n.d.).

21.2 Water Resources in Jordan

Water resources in Jordan are limited, with the country facing severe water scarcity due to its arid climate, high population growth, and rapid urbanization. The main sources of water in Jordan are surface water, groundwater, and desalination. Figure 21.5 shows that surface water accounts for 36% of Jordan's water resources, while groundwater accounts for 53% and treated wastewater accounts for 11%. Surface water is naturally lost via evaporation, evapotranspiration, ocean discharge, and groundwater recharge. The major surface water sources include rivers, streams, valleys, and winter floodwater. The north and west of Jordan are rich in surface water, but the south and east are deficient (Ababsa, 2013). Water may be found under the Earth's surface in soil, rock, and rock formation gaps. Jordan has twelve groundwater basins (Fig. 21.6); groundwater is recognized as the major water source and, in some locations, the sole supply. The majority of Jordan's public water use is derived from groundwater aquifers. Nonetheless, those aquifers are being severely

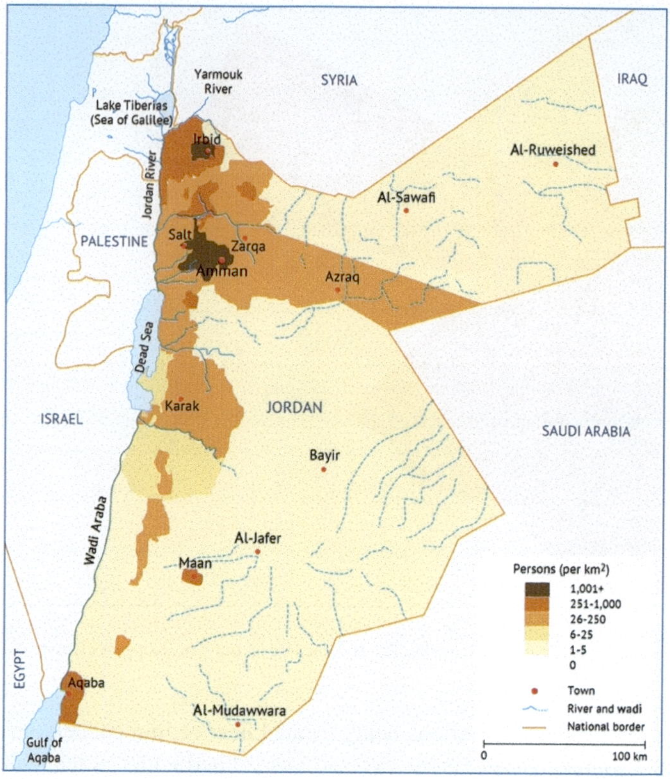

Fig. 21.4 Population density in Jordan

strained by the agricultural sector, which consumes 70% of available resources, with the remaining 20% being used for urban and industrial purposes. Wastewater treatment is the process of converting sewage from industrial waste or a municipality into water that may be reused in a variety of ways. All of the water used in a house that is flushed down the toilet or collected in sewage systems is referred to as wastewater. Jordan has twelve groundwater basins; groundwater is recognized as the principal source of water and, in some locations, the sole supply (Mohsen, 2007). Jordan is making a strong push to transition to a green economy, which refers to a kind of economic growth that aims to achieve sustainable development while minimizing water environmental concerns and scarcity, as well as eliminating pollution. Water resource management includes planning, improving, allocating, and managing the optimal use of water resources. It is a component of water cycle management.

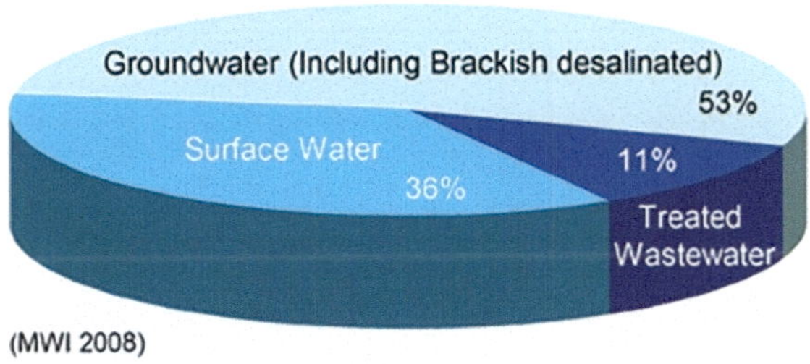

Fig. 21.5 Jordan water resources

21.3 Environmental Framework for Water Preservation in Jordan

There are several steps that can be taken to provide a framework for environmental water preservation. The study proposed a framework should be that guided by the principles of integrated water resource management, which recognizes that water is a finite and interdependent resource that is essential for human well-being, economic development, and environmental health.

21.3.1 Develop Water Conservation Policies

The future of Jordan will depend on how well the government can find solutions and how well the country can deal with problems. Governments and organizations should develop policies to conserve water resources and regulate the use of water. These policies should be based on sound scientific and economic principles and should be implemented in a way that is fair and equitable. "Water for Life: Jordan's Water Strategy 2008–2022" is the current plan for the whole country to keep an eye on the water sector and make sure the best standards are met. There is strong evidence that Jordan has met the Millennium Development Goals (MDGs) for access to clean water and sanitation and has shown best practices in this area. The Ministry of Water and Irrigation (MWI) has proposed an updated National Water Strategy for 2016–2030 to make sure that national goals are in line with the country's changing needs and the new Global Sustainable Development Goals (SDGs) for the water sector (Ibrahim, 2016; Salameh et al., 2018). Jordan should enhance its water supply by employing new resources while decreasing pressure on current over-extracted groundwater resources in order to meet its growing demand (Fanack's Water editorial team, 2022a, 2022b). In this context, it is proposed that innovative water resource management strategies be used to handle water development in the nation in the next years. One option is to include IWRM (Integrated Water Resources Management) into Jordan's National Strategy. (IWRM) is an approach that promotes the integrated use of water, land, and associated resources in order to maximize social and economic advantages while preserving the sustainability of critical ecosystems. Furthermore,

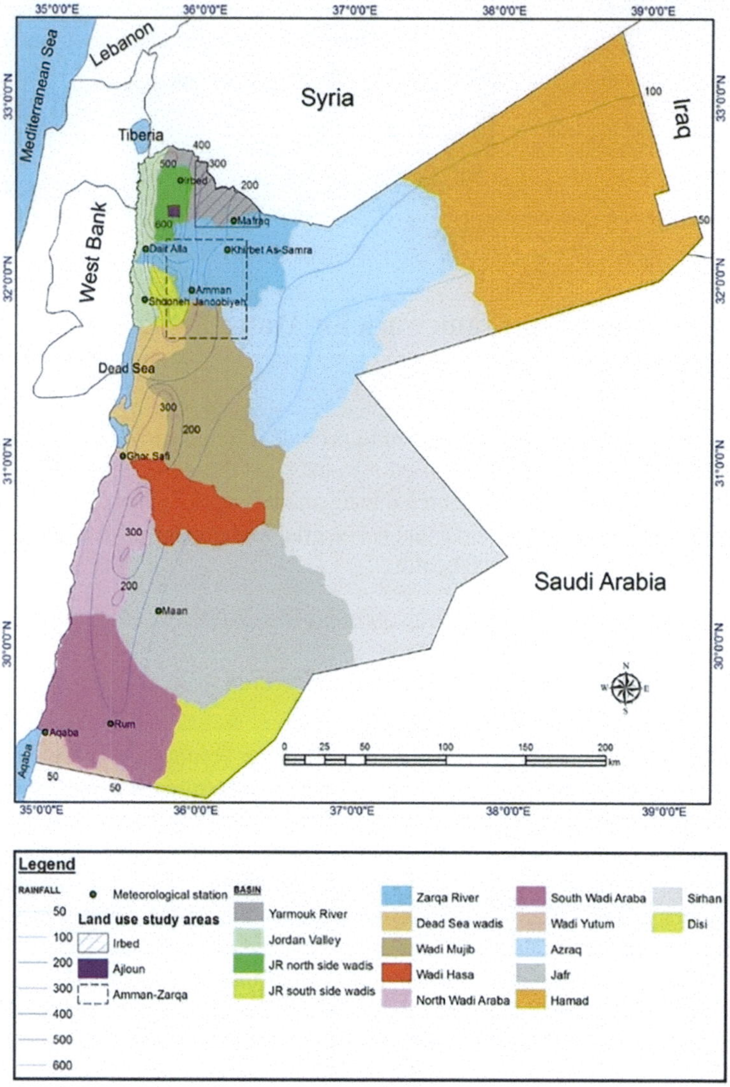

Fig. 21.6 The main groundwater basins of Jordan (MWI, 2017)

the new WASH agenda may be entrenched as a strategy for equitable access to clean drinking water, sanitation, and hygiene for all people globally. National WASH goals and strategic goals may be included into the National Water Strategy framework and executed in its plan. Collaboration with other sectors such as agriculture, energy, industry, and the environment will also be considered. This may assist to maintain a coordinated approach for planning and providing sustainable water supply for the demands of other sectors' national economic progress. Related ministries should develop standards and procedures to increase efficiency, service delivery balance, and resource use. Implementing water efficiency measures, such as reducing water waste, improving irrigation practices, and promoting water-saving technologies, can help reduce water consumption and preserve water resources. Furthermore, to meet the present economic circumstances, the government must create adequate, cost-effective water and wastewater rates for urban, agricultural, and industrial water usage (Fanack's Water editorial team, 2022b; Salameh et al., 2018).

21.3.2 Protect and Restore Aquatic Ecosystems

Protecting and restoring aquatic ecosystems, such as rivers, lakes, and wetlands, is essential for preserving water resources. This can be done by reducing pollution, restoring damaged habitats, and protecting sensitive species. To enhance the development, management, and exploitation of water resources, the government has devised several plans, initiatives, and projects. Jordan's Water for Life Policy, which runs from 2008 to 2022, focuses on future challenges and how to overcome them (Al-Kharabsheh, 2020; Almuhtady et al., 2019). Because rainfall is one of Jordan's most vital water suppliers, the Jordanian government has made major attempts to absorb every drop of rain as much as possible by constructing massive dams and various small-scale water-harvesting activities. Jordan has various dam projects to capture water, including the Wadi Arab, Zeqlab, King Talal, and Karak Dams. As shown in Fig. 21.7, the Desert Dams (Constructed type) are instances of micro-scale water collecting activities in Jordan, with 61 counts scattered over the area as of 2017 (Al-Kharabsheh, 2020). The Ministry of Water and Irrigation announced a plan to construct 11 dams in 2019 in order to enhance its collection of water resources. According to the Jordan Times, this takes the total number of dams in the country to 25. Dams, according to Jordan, have a high potential for recharging groundwater supplies and decreasing over-abstraction. Desalination is now used primarily to address domestic, industrial, and agricultural water supply demands. However, it is unrealistic to expect desalination to alleviate water supply difficulties in the future, particularly for agriculture, unless additional treatment facilities with bigger capacity are built (Ibrahim, 2016). Jordan does have brackish water reserves, and a few brackish water desalination plants have been built. More plants with the required increased capacity have been built around the area (Mohsen, 2007). The Red Sea-Dead Sea Water Conveyance Project, an ambitious project, provides the cornerstone for developing sustainable water solutions in Jordan. The Read-Dead project, together with desalination and access to deep aquifers, is intended to provide a permanent solution to Jordan's water supply. The effort attempts to connect the Red Sea with the Dead Sea, a landlocked body of water, in order to find a long-term solution to Jordan's water scarcity problem (Ibrahim, 2016; Fanack's Water editorial team, 2022a; United Nations, 2022).

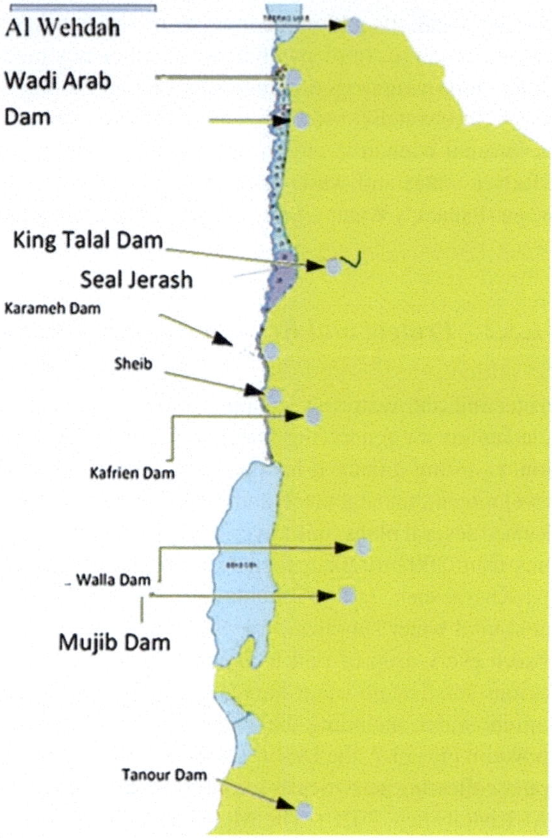

Fig. 21.7 Major dams in Jordan

21.3.3 *Monitor and Enforce Water Quality*

(Al-Karadsheh et al., 2012; Ibrahim, 2016; Mohsen, 2007) Reusing wastewater for agricultural purposes is becoming increasingly popular throughout the globe as more countries do it. Jordan is one of the worst-affected nations. In accordance with research published in the ICE Virtual Library, employing treated wastewater as a climate change adaptation method may minimize the imbalance between demand and capacity in Jordan by 48% by 2025 (Fanack's Water editorial team, 2022b; Odeh, 2019). Jordan has at least 26 wastewater treatment facilities for treating and recycling untreated wastewater, which must be expanded to satisfy the ongoing demand. Jordan is pushing forward with plans to acquire desperately needed water and is investigating if water imports may benefit the country. A study was done in 1983 for delivering 160 MCM/y from the Euphrates River in Iraq to serve the country's north. Two important potential projects have been considered. The first project is to construct the "Red Sea-Dead Sea" canal and associated wastewater treatment facilities, which may increase Jordan's water supply by up to 850 million cubic meters per year. The major purpose of this proposal, in addition to providing hydroelectric power, is to transport water from the Red Sea to the Dead Sea. Another major project is the Turkish Peace pipeline, which will bring water from Turkey. Jordan and other nations will benefit from water from the Ceyhan and Seyhan rivers in southern Turkey as a result of this project (Ibrahim, 2016; Mohsen, 2007). However, they might be large-scale undertakings that need a huge financial investment and take several years to execute. Jordan would get 200 million cubic meters of desalinated water from Israel each year as part of a contract between the United States and the United Arab Emirates (Al-Karadsheh et al., 2012; Ibrahim, 2016; Mohsen, 2007). The Disi-Mudawwara pipeline will run 325 km from the Disi aquifer on Jordan's border with Saudi Arabia to Amman. The projected necessary construction cost is $1 billion, and the city will get around 100 million cubic meters of water per year as a result. In terms of water quality, Jordan's Royal Scientific Society, Ministry of the Environment, and Water Authority Spring water worked together to monitor the country's water quality and has produced data stating that biological contamination in spring water is about 70% in most situations. Jordan's water resources contain high quantities of pollutants. Surface water has high amounts of fecal coliforms due to pollution sources such as overloaded wastewater treatment facilities. Industrial waste is treated ineffectively or not at all. Nitrates and phosphate levels in water resources have grown due to available fertilizer and pest control. Excessive groundwater extraction for agricultural purposes has reduced the water table by 5 m in certain aquifers and quadrupled salinity.

The MWI published a Water Demand Management Policy in 2016. In 2008, the government enacted a water-demand management policy with the following primary goals: to maximize the use of available water, reduce water losses, preserve water resources, support better water use efficiency, and adapt to the challenge of water scarcity in order to close the demand–supply gap (i.e., increase water use efficiency to meet this sector's water needs across the nation). This policy also addresses structure, standards, and specifications, as well as a reduction in non-revenue water consumption, a reduction in water losses, non-traditional water resources, water-conserving gardening, irrigation replacement and recycling, efficient irrigation, and farm-scale water harvesting. Jordan also views dam development as a tremendous chance to supply water for agriculture, industry, and urban animal usage. Some suggestions may be made to improve the operating mechanisms of existing dams. This involves erecting a checking dam to monitor water quality by letting sediments settle out while reducing the flow velocity generated by flash floods. They are also reducing the likelihood of floods by enhancing the capacity to move water from one side of the road to the other. Dam post-production maintenance (operation and safety) is critical; therefore, additional technical human resources are required to carry out duties and tasks. Furthermore, the gap between water supply and demand might be narrowed by constructing additional dams of excellent quality in various regions. More plants with increased capacity have been built around the area. It should be emphasized that the population center is located at high altitudes, making pumping water to these locations exceedingly costly and difficult owing to the difficulties of monitoring the quality of water available in the areas. Therefore, more money must be spent on building treatment facilities with high-quality criteria to meet the ongoing demand. Jordan likewise intends to desalinate its water by 2027. The government has invited five groups to submit bids for the construction of a 300 million cubic meter per year plant in Aqaba.

The amount of treated wastewater should be raised due to fast population growth and growing sewer regions. Even while over 80 million cubic meters of wastewater are being reused each year, only a small percentage is institutionalized with licenses and dedication to reuse standards, thus all treatment facilities and buildings should be inspected on a regular basis. It is advised to create wetlands and surface water tanks to store and utilize water. They are water storage structures that may collect and store rainwater for later use during dry periods for irrigation or even fish farming; it is a low-cost and ecologically beneficial option for water recycling. Regular monitoring and enforcement of water conservation policies and practices is necessary to ensure their effectiveness and prevent water waste. This can be done through regular audits, inspections, and enforcement actions. The rising water sector shortfall, which damages all other sectors, poses a severe danger to Jordanians' well-being, security, and economic future. The environmental consequences of water pollution control may be used to assess effectiveness. Appropriate management is evidenced by effective water usage, wastewater treatment and reuse, and the avoidance of negative environmental consequences (Al-Khashman & Shawabkeh, 2006). Water resource management is ineffective, as indicated by water and environmental contamination, misuse of water resources, failure of sustainability standards, and disregard for generation equality. For example, a large portion of Jordan's water stability is made up of amounts lost due to evaporation from soils and open water bodies. Rainwater from the vast territory, when gathered properly, has the potential to significantly increase the nation's water supplies.

21.3.4 Promote Public Awareness

Raising public awareness about the importance of water preservation and the impact of water use on the environment is critical for promoting water conservation. This can be done through educational programs, public outreach efforts, and media campaigns. Jordan is giving much-needed assistance to awareness-raising activities, which have substantially enhanced public knowledge of the country's serious water problems. Conservation education may help to better manage natural resources and reduce environmental impact. It attempts to raise public awareness of the importance of ecological processes and natural resources, demonstrate how people may improve environmental management by exposing them to environmental dangers, and motivate people to take action. Identifying, attaining, and assessing these three objectives distinguish conservation education from other types of learning. Jordan is giving much-needed assistance to awareness-raising activities, which have substantially enhanced public knowledge of the country's serious water problems. Jordan organizes educational projects and activities like the Sustainable Schools Programs. Jordan's water sector includes the following ministries and organizations: Ministry of Water and Irrigation (MOWI), which operates the Water Authority of Jordan (WAJ) and the Jordan Valley Authority (JVA), the Ministry of Municipal Affairs, the National Center for Agricultural Research and Technology Transfer, the Ministry of Agriculture, and the Ministry of Environment. University students' environmental awareness improved linearly because of following planned strategies such as events, volunteer work, and workshops to understand the impacts of human behaviors on the surrounding environment and the importance of its protection during their studying years. It is suggested that the relevant government enhance sustainability awareness by adopting suitable initiatives in close partnership with both public and private university administrations. One idea is to provide complementary curriculums and university majors in sustainability. A mandatory curriculum for all school grades may be supplemented with hands-on projects to gain previous knowledge. Moreover, collaborating with stakeholders, such as water users, industry, and community groups, can help ensure that water conservation efforts are effective and sustainable. This can be done through partnerships, stakeholder engagement, and collaborative decision-making processes.

21.4 Conclusion

Jordan's water industry might become highly dependent on energy in the future decades due to the rising need for desalination and the anticipated longer pumping distances. Strategically, it is essential for the water sector to convert as much of its energy consumption as feasible to renewable sources in order to preserve the cost of water and depend on locally available sources of energy to produce and pump fresh water, rather than foreign sources of energy. Jordan has begun to produce "new water" by collecting rainfall, reusing, inventing, changing, and using new sustainable technologies reliant on renewable energy sources such as wind and solar power. The National Water Strategy (2016–2025) and the Energy Efficiency and Renewable Energy Policy are two promising efforts implemented by the Jordanian government to promote energy efficiency and the use of renewable energy in the water sector. According to Jordan's energy minister, the country has developed plans to increase the proportion of renewable energies in its overall energy mix to 50 percent by 2030. The expected development of the use of renewable energy sources should include a large portion for the water sector, which is a vital strategic sector with a high energy demand. Due to the substantial positive effects of sustainable water projects on reducing greenhouse gas emissions and, consequently, global warming and climate change, the introduction of sustainable energy in the water sector can make water investments more sustainable and attractive to donor countries, international institutions, and the Green Climate Fund. Climate change may be mitigated by increasing the availability of inexpensive water as one of the most important instruments for adaptation and by achieving the targeted, sustainable development goals of renewable energy. Increasing the use of renewable energy and energy efficiency in Jordan's water sector should be a top priority on the country's agenda.

The Hashemite Kingdom of Jordan's principal environmental concerns are water reservation and pollution, as discussed in this study's overview of the country's present environmental difficulties. There are various sustainability initiatives and strategies in Jordan. They are, however, mostly concentrated in Amman, Jordan's capital. The government should prioritize integrating and improving national-level projects that can be implemented in all cities. There are strong links between social and physical systems and their effect on national water policy formation, water consumption, and sustainable livelihoods in the water sector. Jordan must find effective, long-term solutions to its water concerns, which necessitates a paradigm shift in water administration and management. It is feasible to develop an effective, sustainable, and fair framework for environmental water preservation by implementing the procedures outlined and described in this study. A comprehensive environmental framework for water preservation in Jordan should consider the water-energy nexus, greywater reuse, water environment restoration and conservation, public awareness, and attitudes towards water consumption. Jordan may strive towards sustainable water management and reduce the challenges created by water shortage by tackling the aforementioned problems.

References

Ababsa, M. (Ed.). (2013). *Atlas of Jordan: History, territories and society*. Presses de l'Ifpo. https://doi.org/10.4000/books.ifpo.4560.

Abu Qdais, H., Abdulla, F., & Kurbatova, A. (2019). Wastewater reuse in Jordan and its potential as an adaptation measure to climate change. *Journal of Environmental Engineering and Science, 14*(4), 203–211. https://doi.org/10.1680/jenes.19.00029

Al-Jaafreh, O., & Nagy, I. (2018). The environmental challenges, problems, and management: A case study of Jordan. *Zbornik Radova Departmana Za Geografiju, Turizam i Hotelijerstvo, 47–2*, 53–70. https://doi.org/10.5937/ZbDght1801053A

Al-Karablieh, E. K., Jabarin, A. S., Al-Karablieh, E. K., & Jabarin, A. S. (2010). *Different rangeland management systems to reduce livestock feeding costs in arid and semi-arid areas in Jordan*. https://doi.org/10.22004/AG.ECON.155543.

Al-Karadsheh, E., Akroush, S., & Mazahreh, S. (2012). *Land degradati on in Jordan—Review of knowledge resources*. International Center for Agricultural Research in the Dry Areas.

Al-Kharabsheh, A. (2020). Challenges to sustainable water management in Jordan. *Jordan Journal of Earth & Environmental Sciences, 11*(1).

Al-Khashman, O. A., & Shawabkeh, R. A. (2006). Metals distribution in soils around the cement factory in southern Jordan. *Environmental Pollution, 140*(3), 387–394. https://doi.org/10.1016/j.envpol.2005.08.023

Almuhtady, A., Alshwawra, A., Alfaouri, M., Al-Kouz, W., & Al-Hinti, I. (2019). Investigation of the trends of electricity demands in Jordan and its susceptibility to the ambient air temperature towards sustainable electricity generation. *Energy, Sustainability and Society, 9*(1), 39. https://doi.org/10.1186/s13705-019-0224-1

Altz-Stamm, A. (2012). *Jordan's water resource challenges and the prospects for sustainability*. GIS for Water Resources.

Batarseh, E. (2021). *Energy efficiency and renewable energy policy for the water sector*. Deutsche Gesellschaft für Internationale Zusammenarbeit. https://www.mwi.gov.jo/ebv4.0/root_storage/ar/eb_list_page/energy_efficiency_and_renewable_-_energy_policy_for_the_water_sector.pdf.

Fanack's Water editorial team. (2022a). *Jordan water report*. Water Fanack. https://water.fanack.com/jordan/ (Fanack's Water editorial team, 2022a, 2022b).

Fanack's Water editorial team. (2022b). *What does the future hold for water in Jordan?* Water Fanack. https://water.fanack.com/jordan/.

Ghazal, M. (2023). *Climate change to hit impoverished Jordanians the hardest, streamlining funding process vital to build resilience*. The Jordan Times. https://jordantimes.com/news/local/climate-change-hit-impoverished-jordanians-hardest-streamlining-funding-process-vital

Hadadin, N. (2015). Dams in Jordan current and future perspective. *Canadian Journal of Pure and Applied Sciences*. https://www.cjpas.net/wp-content/uploads/pdfs/9/1/12-Nidal-7040%20POK%20Dec%2031%20F.pdf.

Hadadin, N., Qaqish, M., Akawwi, E., & Bdour, A. (2010). Water shortage in Jordan—Sustainable solutions. *Desalination, 250*(1), 197–202. https://doi.org/10.1016/j.desal.2009.01.026

Hadadin, N., & Tarawneh, Z. (n.d.). Environmental issues in Jordan, solutions and recommendations. *American Journal of Environmental Sciences*.

Ibrahim, E. (2016). *A summary of Jordan national water strategy 2016–2025*. The Water Network. https://thewaternetwork.com/_/water-in-the-middle-east/article-FfV/a-summary-of-jordan-national-water-strategy-2016-2025--NtvJ5wUAZVqd1oPOgsorg.

Jaber, J. O., Badran, O. O., & Abu-Shikhah, N. (2004). Sustainable energy and environmental impact: Role of renewables as clean and secure source of energy for the 21st century in Jordan. *Clean Technologies and Environmental Policy, 6*(3). https://doi.org/10.1007/s10098-003-0232-9.

Jaber, J. O., Probert, S. D., & Badr, O. (1997). Energy and environmental issues for Jordan. *Applied Energy, 57*(1), 45–101. https://doi.org/10.1016/S0306-2619(97)00026-3

Khdair, A. I., Abu-Rumman, G., & Khdair, S. I. (2019). Pollution estimation from olive mills wastewater in Jordan. *Heliyon, 5*(8), e02386. https://doi.org/10.1016/j.heliyon.2019.e02386

Kippert, K., Ali, M., Hamaideh, A., & Fricke, A. (2020). *Waste-derived products and secondary raw materials as a chance to reduce MSW management costs in developing countries—Case study for Jordan*. 5th EurAsia Waste Management Symposium, İstanbul, Türkiye.

Lelieveld, J., Hadjinicolaou, P., Kostopoulou, E., Chenoweth, J., El Maayar, M., Giannakopoulos, C., Hannides, C., Lange, M. A., Tanarhte, M., Tyrlis, E., & Xoplaki, E. (2012). Climate change and impacts in the Eastern Mediterranean and the Middle East. *Climatic Change, 114*(3–4), 667–687. https://doi.org/10.1007/s10584-012-0418-4

Mohsen, M. S. (2007). Water strategies and potential of desalination in Jordan. *Desalination, 203*(1–3), 27–46. https://doi.org/10.1016/j.desal.2006.03.524

Mustafa, M. I. (2021). *Amman ranks fourth in NO2 air pollution in MENA—Report there's a strong relationship between strict lockdowns and NO2 pollution rates*. The Jordan Times. https://jordantimes.com/news/local/amman-ranks-fourth-NO2-air-pollution-mena-%E2%80%94-report.

Odeh, O. (2019). Water shortage in Jordan. *International Journal of Engineering and Management Sciences, 4*(4), 277–286. https://doi.org/10.21791/IJEMS.2019.4.31.

Ratcliffe, V. (2022). *Running low on water, Jordan looks to imports and desalination*. Bloomberg. https://www.bloomberg.com/news/articles/2022-01-27/running-low-on-water-jordan-looks-to-imports-and-desalination#xj4y7vzkg.

Salameh, E., Shteiwi, M., & Al Raggad, M. (2018). *Water resources of Jordan* (Vol. 1). Springer International Publishing. https://doi.org/10.1007/978-3-319-77748-1.

UNICEF and Economist Impact. (2022). *Water stress in Jordan report: The costs of the water crisis in Jordan*. UNICEF. https://www.unicef.org/jordan/media/11356/file/water%20stress%20in%20Jordan%20report.pdf.

United Nations. (2022). *UN: Recycled wastewater provides a window for Jordan to address water scarcity*. United Nations. https://jordan.un.org/en/175501-un-recycled-wastewater-provides-window-jordan-address-water-scarcity.

Chapter 22
Irrigation with Brackish Water: Impacts on Tomato Photosynthesis, Mineral Nutrition, and Root Water Uptake

Hui Yang and Manoj K. Shukla

Abstract With increasing surface water shortage for irrigation, brackish water is increasingly used for irrigating a variety of crops in arid and semi-arid agroecosystems. This study evaluated the impacts of brackish water irrigation on gas exchange and water uptake in tomato plants and proposed a new model to predict responses of tomato evapotranspiration and yield to irrigation water salinity. A greenhouse experiment was conducted under five salinity treatments: tap water of EC 0.6, brackish water of EC 2 and 3, and reserve osmosis (RO) concentrate of EC 4 and 6 dS m^{-1}. Results indicated that increasing water salinity reduced plant evapotranspiration (ET) and consequently increased deep percolation (DP) and leaching fraction (LF). Tomato yield decreased with increasing salinity, but moderate salinity (2 dS m^{-1}) stimulated leaf photosynthesis rate (Pn), transpiration rate (Tr), and stomatal conductance (Gs). The increasing salinity of irrigation water increased Na and Mg but decreased K and Ca accumulations in plants. The accumulations of micronutrients (Fe, Mn, Cu, B, and Al) except Zn in the whole plant decreased with increasing salinity. Tomato yield was significantly correlated to accumulations of Na, Ca, P and N in whole plants. The proposed new model provided a good estimation of both yield and ET under saline conditions. The calculated soil salinity threshold values of EC_e* from the new model were 1.73 dS/m for yield and 2.52 dS/m for ET, respectively, indicating that the reduction of tomato yield was more sensitive to salinity than the decrease of ET. To achieve an optimal comprehensive plant growth and nutrition, this study showed that tomato plants can be irrigated up to a water salinity of 2 dS m^{-1} without causing stress.

Keywords Evapotranspiration · Mineral nutrition · Root water uptake model · Salinity irrigation · Yield

---

H. Yang · M. K. Shukla (✉)
Department of Plant and Environmental Sciences, New Mexico State University, MSC 3Q, P.O. Box 30003, Las Cruces, NM, USA
e-mail: manoj.shukla@oregonstate.edu

M. K. Shukla
Department of Crop and Soil Science, Oregon State University, Corvallis, OR, USA

A. Sefelnasr et al. (eds.), *Water Resources Management and Sustainability*, Water Science and Technology Library 114, https://doi.org/10.1007/978-3-031-80520-2_22

22.1 Introduction

The groundwater in the United States is utilized by big industrial farms and sprawling cities, and reports indicate that more water is being pumped out than nature can replenish over the past 40 years (The New York Times, 2023). Every year since 1940, more and more wells have reported falling water levels than rising levels (Fig. 22.1). The shrinking of the snowpack that feeds rivers and more surface water evaporation is contributing to enhanced reliance on groundwater to sustain domestic, industrial, and agricultural water needs (Wu et al., 2020). Owing to the shortages of fresh water, producers are forced to apply deficit irrigation using fresh water, irrigation with available saline water or a combination of fresh water and saline water, alternate fresh water and brackish water application as well as leaving the land uncultivated (Albaladejo et al., 2017; Baath et al., 2017; Sarah et al., 2023). Thus, crops are subjected to salt stress, water stress or both, and a consequence yield and growth are adversely affected by the two abiotic stresses and associated nutritional imbalances induced in plants (Ore & Suarez, 2017; Yang et al., 2023).

To clarify the mechanisms of plant responses to salt stress and provide solutions for improving crop acclimation to abiotic stresses, we tested tomato, one of the most important vegetable crops worldwide, under different gradients of salt stresses in both northwest China and southwest United States. Both regions are arid to semiarid with low rainfall and high evapotranspiration. In the experiments conducted in China, three salts of sodium chloride (NaCl), magnesium sulfate ($MgSO_4$) and calcium sulfate

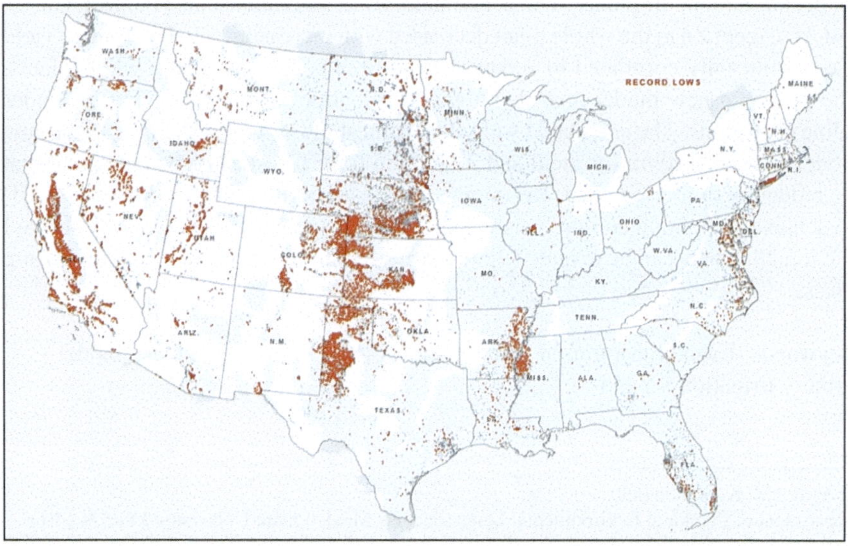

Fig. 22.1 Record low annual water levels in the past decade, four of every 10 sites hit all-time lows. *Source* America is draining its groundwater like there's no tomorrow—the New York Times (nytimes.com)

($CaSO_4$) were mixed with soil to create four different salt treatments. Increasing soil salinity decreased plant sap flow, dry biomass, and yield, leading to a greater decrease in water use efficiency of tomato (Yang et al., 2019a). However, fruit quality parameters including TSS (total soluble solids), TSSC (total soluble sugar content) and CI (color index) improved significantly with increasing soil salt content (Yang et al., 2019b). Nutrient imbalances and increases in leaf Na^+ and Cl^- concentrations resulted in plants being stunted or dead (De Pascale et al., 2007).

Tomato yield was reported linearly (Ben-Gal et al., 2003) or nonlinearly (Patanè et al., 2011) correlated to evapotranspiration. Although tomato plant is adversely affected by increasing salinity, yield and evapotranspiration might not respond to salinity in the same way. Thus, determining the specific threshold values for yield and evapotranspiration gives producers and irrigation engineers insights into how much salinity they can use for irrigation before it negatively impacts the crops. Plant response to salinity is mathematically expressed in terms of relative yield or evapotranspiration as a function of the electrical conductivity of the saturated soil paste extract (EC_e). There are several macroscopic water uptake functions proposed by different researchers, the linear model of Mass and Hoffman (1977), the nonlinear models of van Genuchten and Hoffman (1984), the Dirksen et al. (1993) and Homaee et al., (2002a, 2002b). However, the plant's real response to stress has two break points: first, a plant can tolerate a certain amount of salinity before it shows a response; eventually under higher salinity, there is another break point after which the slope of yield reduction to salinity is reduced (Maggio et al., 2007). Therefore, the quantized responses of tomato yield and evapotranspiration to increasing salinity should be further explored and a model that better reflects the plant realistic response to salinity is needed.

Apart from planting crops on salt-accumulated agricultural lands, brackish groundwater is voluntarily used for irrigation in southern New Mexico of United States. Using brackish groundwater for irrigation is more realistic for testing plant responses to salinity because of the diversity of ions present in the natural brackish groundwater. The aims of this study were to (a) investigate the effects of irrigation with brackish waters on photosynthesis, evapotranspiration, and yield of tomato; (b) determine ion accumulations in tomato plants under different gradients of brackish waters; and (c) propose a model that relates tomato evapotranspiration and yield to salinity.

22.2 Methods

22.2.1 Experimental Descriptions

The experiment was conducted in a solar greenhouse at the Fabian Garcia Science Center of New Mexico State University, Las Cruces (32.28°N, 106.77°W, elevation 1186 m asl) from January 24 to May 18 2018. Pots, 17 cm in diameter and 33 cm

Table 22.1 Ion Concentrations of Na^+, K^+, Mg^{2+}, Ca^{2+}, Zn^{2+}, and Cl^{-1}, pH, electrical conductivity (EC) and cation exchange capacity (CEC) in the experimental soil packed into the pots

Parameter	Value	Unit
Na	115.3 ± 14.4	mg/kg
K	163.7 ± 17.7	mg/kg
Ca	2233.0 ± 132.9	mg/kg
Mg	123.0 ± 13.0	mg/kg
Zn	1.03 ± 0.1	mg/kg
Cl	105.3 ± 1.5	mg/kg
pH	8.10 ± 0.0	dimensionless
EC	1.06 ± 0.1	dS/m
CEC	13.1 ± 0.9	dimensionless

The results are mean values and standard deviation (SD)

depth, were filled using 9 kg of air-dried soil with a bulk density of 1.36 ± 0.5 g cm^{-3}. The soil was collected from the top 30 cm at the research Center, dried, and sieved through a 4-mm sieve before packing. The soil texture was sandy loam. The chemicals of the soil prior to packing are given in Table 22.1. 14 g fertilizer of Miracle-Gro was homogeneously mixed with soil prior to packing into the experimental pots. Two tomato seeds (cv. 'Sow easy') were sown within a soil depth of 1–2 cm in each pot. After the germination and seedling stages, only one healthy plant was kept from February 28.

The treatments were set as a non-saline water treatment of EC 0.6 dS m^{-1} (S0) and four salinity treatments of EC 2 (S1), 3 (S2), 4 (S3), and 6 dS m^{-1} (S4). Each treatment had four replicates. The tap water was used as the control (S0). The brackish groundwater with EC 4 dS m^{-1} and reverse osmosis (RO) concentrate with EC 8 dS m^{-1} were provided by the Brackish Groundwater National Desalination and Research Facility (BGNDRF) in Alamogordo, NM. The brackish water and control water were appropriately mixed to create S1 and S2, and RO concentrates were mixed to create S3 and S4. The diluted brackish groundwater and RO concentrate were applied to plants from March 17 up to harvest, plants were continuously irrigated with the same volume of water every 3–4 days based on the weight change of reference pots in control. The chemical compositions of irrigation water used for each treatment are given in Table 22.2.

22.2.2 *Measurements*

During the growth stages, pots were weighed before and after irrigation in each irrigation event, and the evapotranspiration for each pot was calculated based on the water balance method. The volume and EC of leachate collected after each

Table 22.2 The chemical compositions of irrigation water used for each treatment

Treatment	EC	Na	K	Mg	Ca	Cl
	dS/m	mg/kg	mg/kg	mg/kg	mg/kg	mg/kg
Control	0.62	77.8 ± 1.8e	12.06 ± 0.1b	12.52 ± 1.7e	55.0 ± 2.0e	65.7 ± 3.1e
S1	2	166.3 ± 7.1d	14.79 ± 0.5a	73.26 ± 2.6d	173.0 ± 17.8d	300.5 ± 6.5d
S2	3	270.2 ± 44.2c	15.25 ± 0.7a	135.20 ± 20.0c	307.6 ± 20.7c	514.5 ± 16.5c
S3	4	332.9 ± 13.2b	14.82 ± 0.6a	172.00 ± 6.4b	366.5 ± 20.7b	620.5 ± 14.5b
S4	6	510.3 ± 16.5a	15.29 ± 0.3a	283.60 ± 9.0a	606.7 ± 10.4a	1031.0 ± 49.0a

The results are mean values and standard deviation (SD). Different letters within a column correspond to Duncan's multiple range test at $P \leq 0.05$

irrigation were continuously recorded. The leaching fraction (LF) was decided by the ratio of leachate volume and irrigation water volume (Ayers & Wescot, 1985). The leaf photosynthesis rate (Pn), transpiration rate (Tr), and stomatal conductance (Gs) of tomato were determined at 11:00 a.m. on days 72, 83, 93, 105, and 115 after sowing with a Licor LI-6400XT portable system; leaf chlorophyll content was determined using a SPAD meter right after the gas exchange measurements. At the end of the experiment, tomato plants were harvested by leaves, fruits, roots, and stems separately. Fruit number (FN) per plant was recorded and fresh fruit weight was measured as yield. Then the plant biomass were dried at 75 °C to a stable weight and weighed using an electronic balance with 0.1 g accuracy, fruit dry biomass (FDM) was then determined Afterward, the dry plant biomass were ground to determine the ion concentrations including sodium (Na), magnesium (Mg), potassium (K), calcium (Ca), sulfur (S), phosphorus (P), zinc (Zn), copper (Cu), iron (Fe), manganese (Mn), boron (B), as well as aluminum (Al), more details of measurements were reported in Yang et al., (2023).

22.2.3 Data Analysis

The data of evapotranspiration, yield, photosynthesis and plant ion accumulations were analyzed by one-way ANOVA using SPSS version 23.0 (IBM, Inc., Armonk, NY, USA). To test the statistical significance among five treatments, Duncan's test was used at 0.05 level. Ion accumulation for each organ (mg) was determined by ion concentration for each organ (mg/kg) multiplying organ dry weight (kg). The ion accumulation of whole plant was the sum of ion accumulations in roots, stems, leaves and fruits.

The statistical indexes, including the coefficient of determination (r^2), the root mean square error (RMSE), Willmott index of agreement (d), normalized RMSE (NRMSE), coefficient of residual mass (CRM), as well as modeling efficiency

(EF), were used to assess the difference between simulated and measured values (Kobayashi & Salam, 2000; Nash & Sutcliffe, 1970; Willmott & Matsuura, 2005).

22.3 Results

22.3.1 *Yield, Evapotranspiration, and Photosynthesis of Tomato Under Different Salinity Treatments*

The water use and gas change parameters under different salinity are given in Fig. 22.2. Decreases in total evapotranspiration (ET) were observed with increasing salinity (Fig. 22.2a), therefore, increases in deep percolation (DP) and leaching fraction (LF) were observed (Fig. 22.2b and c). Tomato yield for treatments S1, S2, S3, and S4 decreased by 12.2%, 26.5%, 32.5%, and 45.5%, respectively compared to S0 (Fig. 22.2d). Average Pn, Gs, and Tr showed an increase of 3.9%, 23.9%, and 7.9%, respectively for S1, after which average Pn, Gs, and Tr decreased with increasing salinity, but the differences among the five treatments were not significant (Fig. 22.2g, h and i). Fruit number and foliar chlorophyll content did not show significant differences among the five treatments; fruit number for treatment S1 was 1.2% higher than that for S0 (Fig. 22.2e), the leaf foliar chlorophyll content for S1, S2, S3, and S4 treatments reduced by 4.1%, 5.5%, 2.0%, and 5.2%, respectively (Fig. 22.2f).

22.3.2 *Plant Mineral Nutrition Accumulations of Tomato Under Different Salinity Treatments*

Na accumulation in tomato plants increased with increasing salinity, the differences became significant under S3 and S4 treatments (Table 22.3). Compared to S0, Ca and K accumulations for treatments S1, S2, S3, S4 decreased by 11.6%, 20.9%, 23.4%, and 41.5%, and 5.14%, 14.8%, 6.96%, 25.7%, respectively. Mg accumulation under S4 slightly increased, while S and Cl decreased in comparison with S0, however, the differences were not significant among five treatments for these three ions. When salinity was up to 6 dS m^{-1}, P accumulation showed a significant decrease but the accumulation of N significantly decreased at a salinity level of 2 dS m^{-1}. Moreover, irrigation water salinity had significant effects on the accumulations of micronutrients in tomato plant except for Zn (Table 22.4). The accumulations of Cu, Mn, Fe, Al, and B decreased with increasing salinity, the difference between S0 and S4 treatments was significant for each micronutrient.

The result of Pearson correlation analysis is showed in Fig. 22.3, tomato yield was significantly correlated with ion accumulations of Na, Ca, N and P in the whole plant of tomato. While fruit water content (FWC) and fruit dry weight (FDM) were negatively correlated with Na accumulation, the fruit number (FN) didn't show significant

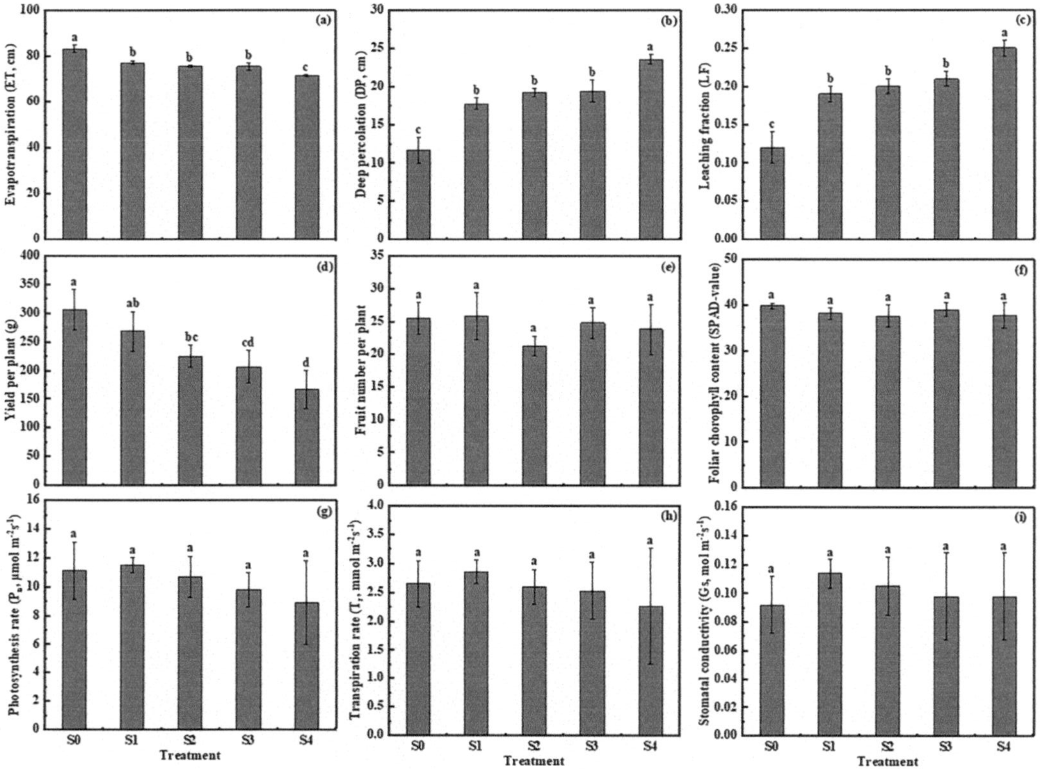

Fig. 22.2 Tomato ET, DP, LF, yield, and average Pn, Tr, Gs, foliar chlorophyll content under different salinity treatments. Results are mean values and SD of four replicates. Different letters within one figure correspond to Duncan's multiple range tests at $P < 0.05$. Total irrigation amount for each treatment is 94.9 cm

Table 22.3 The total uptake of sodium (Na), calcium (Ca), magnesium (Mg), potassium (K), chloride (Cl), nitrogen (N), sulfur (S), and phosphorus (P), zinc (Zn), manganese (Mn), copper (Cu), iron (Fe), boron (B), aluminum (Al) in tomato plants under different salinity treatments

Ions	Treatment				
	S0	S1	S2	S3	S4
Na, mg	73.0 ± 6.8c	79.1 ± 6.9c	82.6 ± 6.5c	111.2 ± 5.8b	136.5 ± 19.5a
Ca, mg	1168.8 ± 48.5a	1033.8 ± 119.7ab	924.4 ± 141.2b	894.8 ± 66.7b	684.0 ± 37.1c
Mg, mg	214.7 ± 9.8a	228.7 ± 16.9a	226.2 ± 15.0a	244.0 ± 17.1a	236.5 ± 21.6a
K, mg	1236.7 ± 56.7a	1173.1 ± 110.1a	1053.4 ± 16.9b	1150.6 ± 20.8ab	918.3 ± 21.9c
Cl, mg	661.4 ± 29.2a	560.4 ± 123.0a	569.9 ± 103.1a	620.8 ± 87.3a	572.2 ± 44.5a
N, mg	706.8 ± 24.4a	570.7 ± 40.5b	545.6 ± 66.7b	570.6 ± 6.95b	501.6 ± 40.4b
S, mg	343.3 ± 11.1a	345.5 ± 66.1a	321.2 ± 55.8a	344.7 ± 29.4a	326.4 ± 36.2a
P, mg	115.2 ± 11.0a	109.4 ± 8.1a	101.8 ± 8.5ab	105.3 ± 6.0ab	93.2 ± 3.5b

Different letters within a row correspond to Duncan's multiple range tests at 0.05 P level

Table 22.4 The total uptake of zinc (Zn), copper (Cu), manganese (Mn), iron (Fe), aluminum (Al), and boron (B) in tomato plants under different salinity treatments

Ions	Treatment				
	S0	S1	S2	S3	S4
Zn, mg	1.00 ± 0.06a	1.02 ± 0.07a	0.91 ± 0.10a	1.09 ± 0.05a	1.01 ± 0.20a
Cu, mg	0.98 ± 0.03a	0.83 ± 0.05b	0.72 ± 0.03 cd	0.66 ± 0.06d	0.80 ± 0.06bc
Mn, mg	1.74 ± 0.12a	1.45 ± 0.06b	1.24 ± 0.02c	1.28 ± 0.02c	0.99 ± 0.13d
Fe, mg	12.3 ± 1.3a	11.6 ± 1.3ab	9.7 ± 0.4bc	8.3 ± 1.3 cd	7.3 ± 0.9d
Al, mg	11.5 ± 1.0a	9.8 ± 1.7ab	9.4 ± 0.3abc	8.2 ± 1.1bc	7.4 ± 1.1c
B, mg	2.22 ± 0.02a	1.96 ± 0.18b	1.69 ± 0.19c	1.79 ± 0.02bc	1.42 ± 0.10d

Different letters within a row correspond to Duncan's multiple range tests at 0.05 P level

correlation with Na accumulation. The gas exchange parameter of Pn was significantly correlated with ion accumulations of Na and Ca, while the correlations between Gs and ions were not significant.

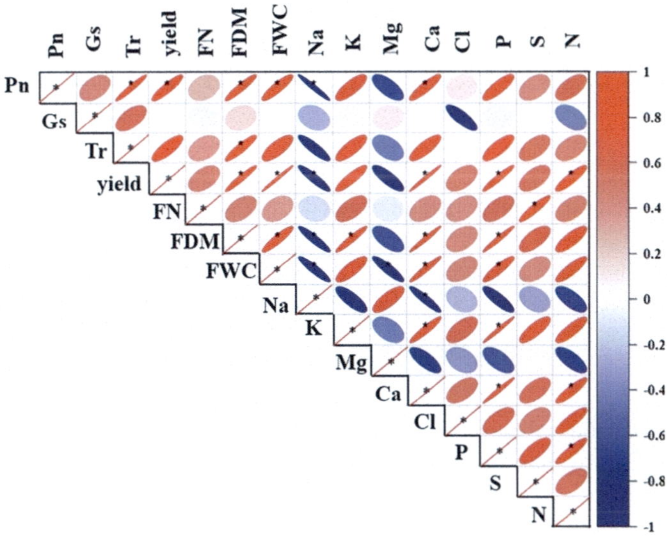

Fig. 22.3 The triangular matrix of Pearson correlation analysis for different ion accumulations in tomato plants and leaf gas exchange parameters, yield as well as yield-relative parameters. FN, fruit number per plant; FWC, fruit water content; FDM, fruit dry biomass; The direction of the ellipse in upper triangular matrix indicates positive and negative, the color of ellipse represents the range of Pearson correlation coefficients. *Significant differences at 0.05 level

22.3.3 Proposed New Model for Yield Responses to Salinity

The existed water uptake models of Maas and Hoffman (1977), Dirksen et al. (1993), and Homaee et al., (2002a, 2002b) demonstrated a threshold after which the yield started to decrease. However, Maggio et al. (2007) reported that there was a second break point after which the slope of yield decline function to salinity was decreased. In order to better describe the plant adaptation to salinity higher than the first break point and avoid the limitation of determining the complex parameters, a new model based on an inverse function was proposed as following:

$$\alpha(EC_e) = \frac{1}{A(EC_e - B)} + C \tag{22.1}$$

where EC_e is the electrical conductivity of soil saturated paste extract, A is a shape parameter, B (dS m^{-1}) is the asymptote which parallels to the y axis, C is the asymptote parallels to x axis, indicating the potential lowest value for relative yield and relative ET. When $\alpha(EC_e) = 1$, the calculated EC_e from Eq. (22.1) is the salinity threshold value (EC_e^*).

This new model was proposed based on the relationships between EC_e and plant relative yield as well as relative ET, however, translating the model to unit of water salinity could make it more useful and convenient for irrigation engineers and producers. It was reported that electrical conductivity of irrigation water (EC_w) was linearly correlated to EC_e (Maggio et al., 2004; Shalhevet & Yaron, 1973; Yang et al., 2020):

$$EC_w = m \times EC_e + n \tag{22.2}$$

where EC_w is the salinity of irrigation water, dS m^{-1}; *m* and *n* are slope and intercept, respectively.

According to Eq. (22.1) and Eq. (22.2), the model can be written as:

$$\alpha(EC_w) = \frac{1}{\frac{A}{m}(EC_w - n - m \times B)} + C \tag{22.3}$$

When used under saline water irrigation conditions, assuming $A/m = A'$, $n + mB = B'$, Eq. (22.3) can be expressed as

$$\alpha(EC_w) = \frac{1}{A\prime(EC_w - B\prime)} + C \tag{22.4}$$

where EC_w is the EC of irrigation water (dS m^{-1}), A' is the shape parameter, B' is the asymptote (parallel to the y-axis, dS m^{-1}), and C is the potential lowest value of relative yield or relative ET.

The existing models including Maas and Hoffman, van Genuchten, Dirksen et al., and Homaee et al. as well as the new model were fitted to the measured relative yield and relative ET. The statistical indexes for the five models showed that the new model performed the best with the highest r^2, d and EF values and lowest RMSE values for both relative yield and relative ET (Table 22.5). The CRM value for the new model were closer to 0 (0.002) than that of the four existing models for relative yield, which indicated that the estimated relative yield of the new model was closer to the measured ones.

To further identify the applicability of the new model, the relationships between EC_w and relative yield, and relative ET were fitted using our experimental data and data collected from literature (Fig. 22.4). The responses of observed tomato yield and ET to salinity were categorized into four salt tolerance varieties, denoted as I, II, III and IV (Table 22.6). The parameter of A' indicated the curve shape and the C represented the lowest relative yield or relative ET under salinity, respectively. They reflected the salt tolerance of various tomato cultivars. A' and C decreased from varieties I to IV (Table 22.6).

Table 22.5 The statistical indexes for the comparison between measured relative yield, relative ET and those simulated by different models

Model	r^2	RMSE	d	CRM	EF
Relative yield					
Mass and Hoffman	0.590	0.175	0.739	−0.187	0.957
van Genuchten and Hoffman	0.688	0.096	0.889	0.011	0.838
Dirksen et al.	0.863	0.063	0.962	0.012	0.962
Homaee et al.	0.864	0.063	0.961	0.006	0.962
New model	0.894	0.055	0.971	0.002	0.994
Relative ET					
Mass and Hoffman	0.644	0.052	0.770	−0.044	0.778
van Genuchten and Hoffman	0.638	0.035	0.892	0.000	0.994
Dirksen et al.	0.689	0.041	0.889	−0.007	0.793
Homaee et al.	0.834	0.022	0.952	0.001	0.793
New model	0.855	0.021	0.959	−0.002	0.999

Note the coefficient of determination (r^2), the root-mean-square error (RMSE), Willmott index of agreement (d), coefficient of residual mass (CRM), and modeling efficiency (EF)

22.4 Discussion

Sodium accumulation in the whole plants was significantly correlated to leaf Pn, fruit water content (FWC), fruit dry weight (FDM), and yield except fruit number (FN) (Fig. 22.3), suggesting that yield decrease due to Na toxicity is caused by reduction in and FWC and FDM rather than FN. In addition, the accumulation of Na decreased FDM by suppressing leaf photosynthesis, ultimately affecting tomato yield. All results were consistent with the source-sink relationships of tomatoes reported by Liu et al. (2017). Compared to the S0, Pn and Gs of tomato plants under S1 were 3.9 and 23.9% higher, respectively (Fig. 22.2), which was consistent with the report that moderate saline conditions stimulated photosynthesis (Maggio et al., 2004). The differences of leaf chlorophyll content among the five salinity treatments were not significant, we even found a slight increase of chlorophyll content under S3 treatment (Fig. 22.3), given that the salt effect on chlorophyll synthesis was not associated with the decline of photosynthesis. However, a negative effect of saline stress (EC of 3 dS/m) on chlorophyll contents was reported on passion fruit (da Silva Lima et al., 2020).

Due to the surface water scarcity in arid and semiarid areas, high salinity groundwater is widely used for irrigation (Flores et al., 2017; Lucker et al., 2023). Thus, the direct descriptions of plant yield and ET as a function of irrigation water salinity (EC_w) could achieve timely feedback for irrigation scheduling and provide growth modelers highly available relationships between plant and salinity (Ben-Gal et al., 2003). Our proposed new model provided the salinity threshold values for yield and ET under saline conditions, and all the parameters of A', B', and C had biological

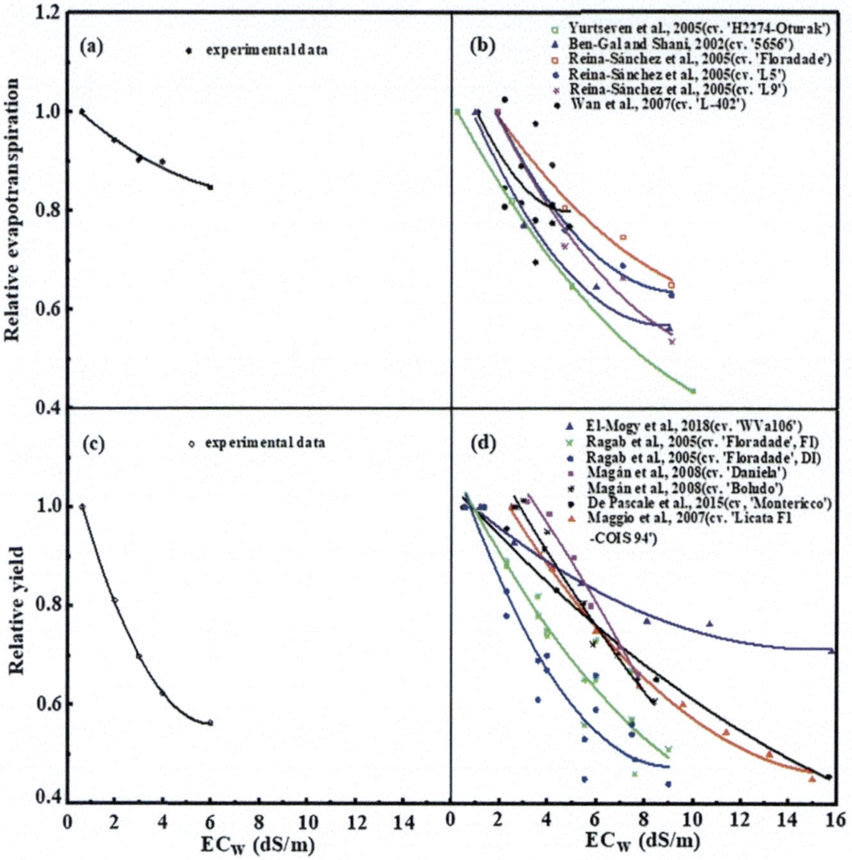

Fig. 22.4 Relationships between electrical conductivity of irrigation water (EC_W), and relative evapotranspiration (ET) as well as relative yield, for observed data from the experiment (**a** and **c**) and collected data from the literature (**b** and **d**)

Table 22.6 The parameters A′, B′, and C of the proposed new model when simulating relative yield and relative ET for four different salt-tolerant varieties of tomato

Four salt-tolerance categories	Relative ET			Relative yield		
	A′	B′	C	A′	B′	C
I	1.337	−1.159	0.669	1.06	−0.358	0.411
II	0.437	−6.155	0.662	0.365	−5.279	0.581
III	0.246	−4.626	0.302	0.334	−2.817	0.215
IV	0.152	−6.118	0.176	0.226	−2.23	0.207

Note A′ represents shape parameter, B′ represents the asymptote parallel to the y-axis, and C is the lowest value that of relative yield and relative ET under salinity could reach. I, II, III and IV are four salt-tolerant levels of tomato cultivars significance. The parameter A′ reflected the sensitivity of plant response to irrigation water salinity, the greater the value A′, the better the plant tolerance would be. The value of C represented that the potential minimal relative yield under different EC_e was 0.45 while the minimal relative ET was 0.82. The calculated salinity threshold values of EC_e* calculated from Eq. (22.1) were 1.73 dS m^{-1} and 2.52 dS m^{-1} for relative yield and relative ET respectively when $\alpha(EC_e) = 1$, given that the reduction of tomato yield was more sensitive than the decrease of ET under saline conditions. This new model can be easily used to determine plant ET and yield responses to brackish water irrigation.

22.5 Conclusion

In this study, tomato evapotranspiration, yield, photosynthesis and plant ion accumulation were analyzed under different salinity treatments. The results showed that increasing irrigation water salinity decreased tomato yield and evapotranspiration. Moderate salinity up to 2 dS m^{-1} stimulated leaf photosynthesis rate, transpiration rate and stomatal conductivity. Na accumulation in plants increased with increasing salinity while K and Ca decreased. The uptakes of micronutrients including Mn, Cu, Fe, B and Al decreased with increasing salinity except for Zn. A new model was proposed to describe plant ET and yield as a function of electrical conductivity of irrigation water. The model can be easily used to predict responses of plant yield and water uptake under brackish water irrigation.

Acknowledgements We would like to thank the Nakayama Professorship Endowment, NIFA, New Mexico State University Agricultural Experiment Station, and Brackish Groundwater National Desalination Research Facility for all the supports and assistances.

References

Albaladejo, I., Meco, V., Plasencia, F., Flores, F. B., Bolarin, M. C., & Egea, I. (2017). Unravelling the strategies used by the wild tomato species *Solanum pennellii* to confront salt stress: From leaf anatomical adaptations to molecular responses. *Environmental and Experimental Botany, 135*, 1–12.

Ayers, R. S., & Wescot, D. W. (1985). *Water quality for agriculture*. FAO. Retrieved from http://www.fao.org/docrep/003/T0234E/T0234E02.htm.

Baath, G. S., Shukla, M. K., Bosland, P. W., Steiner, R. L., & Walker, S. J. (2017). Irrigation water salinity influences at various growth stages of Capsicum annuum. *Agricultural Water Management, 179*, 246–253.

Ben-Gal, A., Karlberg, L., Jansson, P. E., & Shani, U. (2003). Temporal robustness of linear relationships between production and transpiration. *Plant and Soil, 251*, 211–218.

De Pascale, S., Martino, A., Raimondi, G., & Maggio, A. (2007). Comparative analysis of water and salt stress-induced modifications of quality parameters in cherry tomatoes. *Journal Horticultural Science Biotechnology, 82*(4), 283–289.

Dirksen, C., Kool, J. B., Koorevaar, P., & van Genuchten, M. T. (1993). HYSWASOR simulation model of hysteretic water and solute transport in the root zone. In D. Russo & G. Dagan (Eds.), *Water flow and solute transport in soils* (pp. 99–122). Springer.

Flores, A. M., Shukla, M. K., Schutte, B. J., Picchioni, G., & Daniel, D. (2017). Physiologic response of six plant species grown in two contrasting soils and irrigated with brackish groundwater and RO concentrate. *Arid Land Research and Management, 31*, 182–203.

Homaee, M., Feddes, R. A., & Dirksen, C. (2002a). A macroscopic water extraction model for nonuniform transient salinity and water stress. *Soil Science Society of America Journal, 66*, 1764–1772.

Homaee, M., Feddes, R. A., & Dirksen, C. (2002b). Simulation of root water uptake: III. Non-uniform transient combined salinity and water stress. *Agricultural Water Management, 57*, 127–144.

Kobayashi, K., & Salam, M. U. (2000). Comparing simulated and measured values using mean squared deviation and its components. *Agronomy Journal, 92*, 345–352.

Liu, Y., Li, P., Xu, G. C., Xiao, L., Ren, Z. P., & Li, Z. B. (2017). Growth, morphological, and physiological responses to drought stress in *Bothriochloa ischaemum*. *Frontiers in Plant Science, 8*, 230.

Lucker, A. O. S., Picchioni, G. A., Consford, J. D., Steiner, R. L., Guzman, I., Schutte, B. S., Shukla, M. K., & Young, R. B. (2023). Gypsiferous groundwater and its desalination brine concentrate: Biomass, water use, and salt mining of three Southwestern USA native halophytes. *Agricultural Water Management, 289*, 108553.

Maas, E. V., & Hoffman, G. J. (1977). Crop salt tolerance-current assessment. *Journal of the Irrigation and Drainage Division, 103*, 115–134.

Maggio, A., De Pascale, S., Angelino, G., Ruggiero, C., & Barbieri, G. (2004). Physiological response of tomato to saline irrigation in long-term salinized soils. *European Journal of Agronomy, 21*, 149–159.

Maggio, A., Raimondi, G., Martino, A., & De Pascale, S. (2007). Salt stress response in tomato beyond the salinity tolerance threshold. *Environmental and Experimental Botany, 59*, 276–282.

Nash, J. E., & Sutcliffe, J. V. (1970). River flow forecasting through conceptual models part I—A discussion of principles. *Journal of Hydrology, 10*, 282–290.

Ors, S., & Suarez, D. L. (2017). Spinach biomass yield and physiological response to interactive salinity and water stress. *Agricultral Water Management, 190*, 31–41.

Patanè, C., Tringali, S., & Sortino, O. (2011). Effects of deficit irrigation on biomass, yield, water productivity and fruit quality of processing tomato under semi-arid Mediterranean climate conditions. *Scientia Horticulturae, 129*, 562–571.

Rojanasakul, M., lavelle, C., Migliozzi, B., & Murray, E. (2023). *America is draining its groundwater like there's no tomorrow*. The New York Times.

Sarah, M. C., Shukla, M. K., Jeon, S., & O'Meara, S. (2023). The effects of Brackish irrigation on soil ion accumulation and growth of *Atriplex* species. *Soil Systems, 7*(4), 84.

Shalhevet, J., & Yaron, B. (1973). Effect of soil and water salinity on tomato growth. *Plant and Soil*, (pp. 39–285–293).

da Silva Lima, L. K., de Jesus, O. N., Soares, T. L., dos Santos, I. S., de Oliveira, E. J., & Coelho Filho, M. A., (2020). Growth, physiological, anatomical and nutritional responses of two phenotypically distinct passion fruit species (Passiflora L.) and their hybrid under saline conditions. *Scientia Horticulturae, 263*.

van Genuchten, M. T., & Hoffman, G. J. (1984). Analysis of crop production. In I. Shainberg & J. Shalhevet (Eds.), *Soil salinity under irrigation* (pp. 258–271). Springer.

Willmott, C. J., & Matsuura, K. (2005). Advantages of the mean absolute error (MAE) over the root mean square error (RMSE) in assessing average model performance. *Climate Research, 30*, 79–82.

Wu, W. Y., Lo, M. H., Wada, Y., Famiglietti, J. S., Reager, J. T., Yeh, P. J. F., Ducharne, A., & Yang, Z. L. (2020). Divergent effects of climate change on future groundwater availability in key mid-latitude aquifers. *Nature Communications, 11*, 3710.

Yang, H., Shukla, M. K., Mao, X. M., Kang, S. Z., & Du, T. S. (2019a). Interactive regimes of reduced irrigation and salt stress depressed tomato water use efficiency at leaf and plant scales by affecting leaf physiology and stem sap flow. *Frontiers in Plant Science, 10.*

Yang, H., Du, T. S., Mao, X. M., Ding, R. S., & Shukla, M. K. (2019b). A comprehensive method of evaluating the impact of drought and salt stress on tomato growth and fruit quality based on EPIC growth model. *Agricultural Water Management, 213*, 116–127.

Yang, H., Du, T. S., Mao, X. M., & Shukla, M. K. (2020). Modeling tomato evapotranspiration and yield responses to salinity using different macroscopic reduction functions. *Vadose Zone Journal, 19*, e20074.

Yang, H., Shukla, M. K., & Du, T. S. (2023). Assessment of plant mineral nutrition concentrations of tomato irrigated with brackish water and RO concentrate. *Journal of Plant Nutrition, 46*(19), 4639–4656.

Part V
Water Resources Management

Chapter 23
Impact Projections of Aggregate Food Production System and Water Stress on Water Productivity in Gulf Cooperation Countries (GCCs): A Review-Based Analysis

Mohamed Farig, Mohamed Dawoud, and Waleed Hassan Abou El Hassan

Abstract The scarcity of water resources is one of the main barriers to further food production. On the other hand, the historical evaluation of food production systems regarding crop water productivity is a key factor in understanding the country's food security. Thus, this manuscript describes the historical development of water productivity components, along with water stress and imported food, in the Gulf Cooperation Countries (GCCs) of Bahrain, Kuwait, Oman, Qatar, Kingdom of Saudi Arabia (KSA), and the United Arab Emirates (UAE) from 2000 to 2020. This is to highlight the vulnerability of the GCCs food supply to overseas water use in imported food. Water productivity is calculated based on Gross Domestic Product (GDP) per cubic meter of total freshwater withdrawal. The annual freshwater withdrawal refers to the withdrawal from storage basins, desalination plants, and non-renewable aquifers, and also includes the withdrawal for agriculture and livestock. The review analysis is based on the Food and Agriculture Organization (FAO), AQUASTAT data, the World Bank, the Organisation for Economic Co-operation and Development (OECD) for

---

M. Farig · M. Dawoud · W. H. Abou El Hassan (✉)
National Water Research Center, Water Management Research Institute, P.O. Box: 13621/5, Cairo, Egypt
e-mail: Waleed.Abouelhassan@fao.org

M. Farig
e-mail: Mohamed.Farig@fao.org

M. Dawoud
e-mail: mdawoud@ead.gov.ae

M. Dawoud
Environment Agency Abu Dhabi, P.O. Box: 45553, Abu Dhabi, United Arab Emirates

W. H. Abou El Hassan
Food and Agriculture Organization, United Nations, Sub Regional Office for the GCC States and Yemen, 62072 Abu Dhabi, United Arab Emirates

M. Farig
Food and Agriculture Organization United Nations, Egypt Office, P.O. Box 100, Cairo, Egypt

A. Sefelnasr et al. (eds.), *Water Resources Management and Sustainability*, Water Science and Technology Library 114, https://doi.org/10.1007/978-3-031-80520-2_23

GDP estimates, and the World Research Institute-Aqueduct. The results revealed that between 2000 and 2020, the annual freshwater withdrawals for livestock and crops dropped by 20.8%. The cereal production harvested for dry grain also dropped by about 36% between 2000 and 2020. The decrease in water withdrawals and cereal production has not affected water productivity, where it has increased, growing at an average rate of 121.8% from 92.8 in 2000 to 205.9 in 2020. The increase in livestock production index (meat, milk, cheese, eggs, honey, raw silk, wool, hides, and skins) by about 189% between 2000 and 2020 could be the main reason for improving water productivity. Therefore, livestock production is the most crucial component of agriculture in GCCs, contributing the biggest percentage of water productivity. As a result, the food imports were 11.3% as a share percent of total merchandise imports. This percentage is equivalent to 90% of the food demand in the GCCs, meaning an increase in water imported in agricultural products. This research highlighted that despite the improvements in water productivity, the GCCs are not self-sufficient in food, as about 90% of GCCs food demand is imported, and consequently, the impact on water use and productivity is overseas. Hence, it is important to understand, and account for, the impact of imported food on water productivity overseas. In the future, funders are recommended to make greater engagements with international organizations concerned with agriculture and water use to promote sustainable production chains between both GCCs and the countries to which food is imported; water security perspectives.

Keywords Arid lands · Food production · Gulf cooperation countries · Virtual water · Water productivity · Water resources scarcity

23.1 Introduction

Water shortages are anticipated to increase in more than 80% of croplands globally (UNDRR, 2022). The significant worldwide water shortages will worsen due to climate change, where every region of the world is influenced by climate change, and over 40% of people worldwide suffer from water scarcity (World Bank, 2022). Water productivity is directly impacted when unforeseen changes in the weather, such as droughts or floods, occur that affect water availability (Blanc & Noy, 2023; Pun et al., 2023). Recent challenges to food policy have come from emerging stresses, such as water scarcity (Christoforidou et al., 2023; Hanjra & Qureshi, 2010), energy crises (Buko et al., 2021; WFP, 2022), global financial crises (Braun, 2009; FAO, 2009), and climate change (Borgomeo et al., 2020; Fitton et al., 2019). Consequently, there is a severe impact on global food security, with 60% of irrigated agriculture experiencing severe water stress worldwide (WRI, 2023). The situation is further exacerbated in the Gulf Cooperation Countries (GCCs), where out of 6 GCCs (United Arab Emirates, Kingdom of Saudi Arabia, Oman, Kuwait, Bahrain, and Qatar), 4 countries (Bahrain, Kuwait, Oman, and Qatar) have been classified among the five most water-stressed countries in terms of low supply in conjunction with residential, agricultural, and industrial demand (WRI, 2023). Hence, increasing physical water productivity in agriculture is a critical solution to the growing water scarcity because it reduces the need for additional products in response to the increasing water shortage (Brauman et al., 2013; Khorchani et al., 2024).

Increasing water productivity can be achieved through water and soil conservation measures (Niacsu et al., 2022; Reicosky et al., 2023), processing food to raise its value (IFAD, 2014; Salminen et al., 2022), and integrating animals into agricultural systems (Kijne et al., 2006; FAO, 2014). Assessing agricultural output per unit of water utilized and incremental welfare per unit of water applied in agriculture are two of the objectives of investigating water productivity (Molden et al., 2001). Furthermore, the concept of water productivity elucidates the wider framework within which water utilization is employed to generate a range of goods, services, and advantages, such as agroforestry, mixed systems, crops, livestock, and fisheries (Bastiaanssen & Steduto, 2017). Increasing production per unit of resource is necessary to ensure food security given the scarcity of land and water resources; For example, this can be achieved by combining the production per unit of water used with the production per unit of land (Cai & Rosegrant, 2003). However, if crop water consumption remains constant, decreasing irrigation water withdrawals might not increase water productivity (USAD, 2014). Also, when virtual water is taken into account, increasing water productivity does not represent an optimal use of water per unit area, especially in countries where there are severe water shortages. Thus, considering virtual water when analyzing water productivity, better decisions could be made to assist sustainable development, resource efficiency, and food and water security.

Virtual water analysis can improve water-efficient farming practices and sustainably promote food production systems (SSWG, 2023). Given the variation in water productivity, virtual water trade results in real water-savings for importing countries and the whole globe (FAO, 2002). Two of the most significant indicators in water accounting are virtual water and water productivity, both of which can help with policy decision-making (Elfkih et al., 2023; Letseku & Grové, 2022). The Food and Agriculture Organization of the United Nations (FAO) and other International Organizations have launched programs to increase crop water productivity in response to food and water security concerns. Thus, this research aims to show how much water productivity and its components increased in Gulf Cooperation Countries under the conditions of severe water scarcity between 2000 and 2020. The analysis is based on the Food and Agriculture Organization (FAO), AQUASTAT data, World Bank, Organisation for Economic Co-operation and Development (OECD) for GDP estimates, and World Research Institute-Aqueduct. This is to provide a framework for developing and testing hypotheses regarding how aggregate food production and water stress react to overseas water productivity (imported food).

23.2 Methods

23.2.1 *Gulf Countries' General Profile*

Generally, the term gulf has been used for the large, indented, coastal navigational flats. The Gulf Cooperation Countries (GCCs) are Bahrain, the Kingdom of Saudi Arabia (KSA), the United Arab Emirates (UAE), Kuwait, Oman, and Qatar (hereinafter; GCCs). The GCCs have been named because they share the border with the Arabian Gulf. The GCCs are bordered by Iran on the North, Northeast, and East, from the southeast and south GCCs are bordered by the UAE and part of Oman. Qatar, Bahrain, and KSA bordered the GCCs from the southwest and west, while Iraq and Kuwait bordered the GCCs on the northwest, as shown in Fig. 23.1.

The biggest country among the GCCs is the Kingdom of Saudi Arabia (KSA), which covers around 80% of the Arabian Peninsula (UNFCCC, 2024). Bahrain is the smallest Gulf country with a total area of 695 km^2. The GCCs are important, providing 40% of the world's crude oil and 15% of oil product exports (Din, 2007). At the same time, the GCCs provide a significant market for agricultural exports from the European Union (EU), including food preparation and cereal. Also, the European Union is an important market for the GCCs, where the GCCs export animal, vegetable oil, and live animals (European Union Official Website).

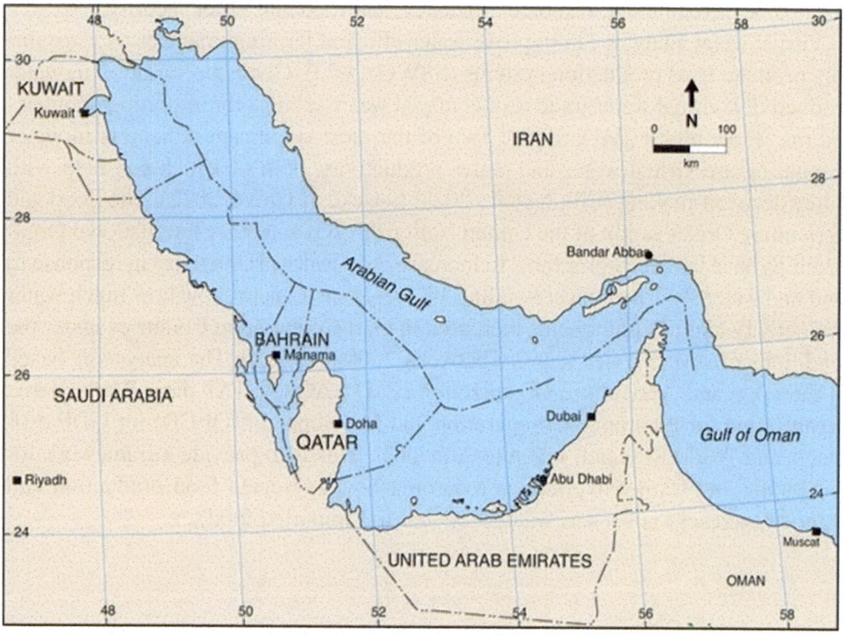

Fig. 23.1 The Arabian Gulf Map (*Source* Rehren, 2011)

23.2.2 Description of the Assessed Indicators

Water stress, groundwater decline, and interannual variability are based on data from World Water Atlas (Aqueduct 4.0). The global hydrological model of PCR-GLOBWB 2 was used to obtain data (Kuzma et al., 2023; Sutanudjaja et al., 2018; Wada et al., 2014). For water stress and interannual variability, the spatial units were sub-basins (Gassert et al., 2013), while for groundwater table decline, the spatial units were aquifers (Gleeson et al., 2012). Agricultural land, freshwater withdrawals, cereal production, livestock production index, food imports, and water productivity are based on data from the Food and Agriculture Organization, AQUASTAT data, and World Bank and OECD GDP estimates (World Bank Data Website). Data are for the years from 2000–2020.

Agricultural land (%): The portion of land that is arable, planted with permanent crops, and used for permanent pastures is referred to as agricultural land. The FAO defines arable land as land that is temporarily planted with crops (double-cropped areas are counted once), briefly covered with pasture or meadows for mowing and temporarily planted with market or kitchen gardens, and land temporarily fallow. Land abandoned as a result of shifting cultivation is excluded as well as land under trees grown for wood or timber.

Water stress: Water stress was measured as the ratio of total water demand (domestic, industrial, agricultural, and livestock uses) to the renewable surface and groundwater sources that were available. The ratio is classified into five categories when the value ranges from 0 to 1, it means low water stress. A value between 1 and 2 means low to medium water stress. A value from 2 and 3 means medium to high water stress. When the range is between 3 and 4, the water stress is considered high, while between 4 and 5 means extremely high-water stress. The classification also includes arid and low water use which means a high-water stress. Higher values indicate increased competition among users.

Groundwater table decline (cm): The average groundwater drop from 1990 to 2014 is measured by the groundwater table decline. PCR-GLOBWB 2 and MODFLOW are used in the calculation of groundwater decline to account for lateral groundwater flow mechanisms. Higher values indicate higher levels of unsustainable groundwater withdrawals, while the lower values mean a low and insignificant decline trend. A value between 0 and 1 means a low to medium decline. A value from 1 and 4 means medium to high decline. When the range is between 4 and 6, the groundwater decline is considered high, while higher than 6 means an extremely high decline.

Interannual variability: The amount of water that varies between years, including groundwater and renewable surface water supplies, is measured as interannual variability. Higher values imply wider fluctuations in the water supply that are accessible from year to year. Water supply is calculated using the total blue water of each catchment according to the National Aeronautics and Space Administration (NASA,

2012). The classification has five categories, low (<0.25), low-medium (0.25–0.50), medium–high (0.50–0.75), high (0.75–1.00), and extremely high (>1.00).

Freshwater withdrawals (%): Total water withdrawals for agriculture and livestock production are referred to as annual freshwater withdrawals; evaporation losses from storage basins are not included in this calculation. It also includes withdrawal from desalination plants in countries where they are a significant source.

Cereal production (tons): Data on cereal production only include crops that are harvested for dry grain, while cereal crops used for grazing and those harvested for hay or green for food, feed, or silage are excluded.

Livestock production index: The livestock production index includes meat and milk from all sources, dairy products such as cheese, and eggs, honey, raw silk, wool, and hides and skins.

Water productivity: GDP (Gross Domestic Product) at constant prices is divided by the country's annual total water withdrawal to determine water productivity.

Food imports (%): It is counted as a percent of merchandise imports, and includes food and live animals, beverages and tobacco, vegetables, fats, and oil (oil seeds, oil nuts, and oil kernels).

23.3 Results and Discussions

23.3.1 Agricultural Lands

The proportion of agricultural land to the total land of each country is shown in Fig. 23.2. Land designated for agriculture is any area that can grow both temporary and permanent crops, such as pastures and crops, excluding land covered with trees planted for timber or wood (World Bank, 2015). Notably, the agricultural lands were the same from 2000 to 2020 for all GCCs without any significant increase, as shown in Fig. 23.2. This indicates the land shortages faced by the GCCs, and emphasizes the importance of increasing water productivity per unit area as a vital solution to mitigate land shortages. Water shortages, decline of groundwater, deficient soils, arid climate, and anticipated climate change impacts in the Gulf all contribute to the limitation of increasing agricultural lands (Shahid & Ahmed, 2014).

23.3.2 Overall Water Risk in GCCs

The overall baselines for water-stress risks in GCCs are shown in Table 23.1. In geographic terms, the Arabian Peninsula encompasses Bahrain, Kuwait, Oman, Qatar, Saudi Arabia, the United Arab Emirates (UAE), Yemen, southern Iraq, and

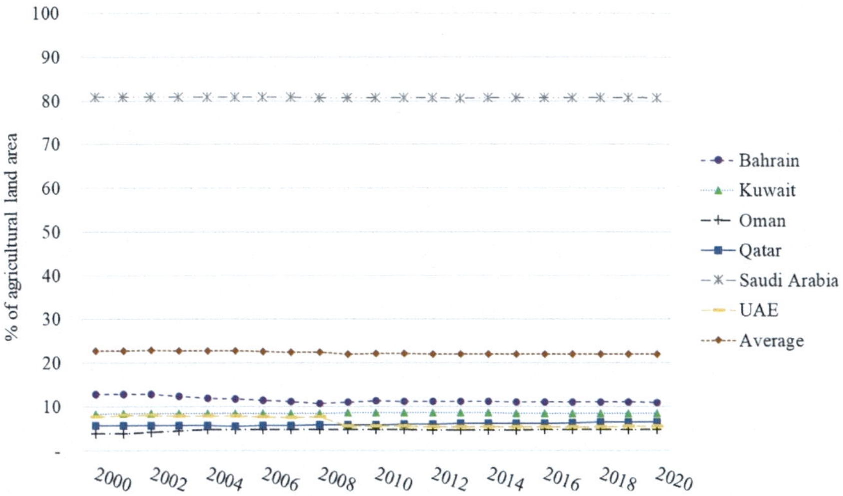

Fig. 23.2 The proportion of agricultural land to the total land of each country (*Source* FAO-AQUASTAT)

Jordan (Britannica, 2005). In this study, Yemen, Iraq, and Jordan are excluded to maintain the same water environment, as they share different basins. The main basins of Yemen, Iraq, and Jordan are the eastern coast of the Red Sea; Tigris Euphrates, and Dead Sea, respectively (WRI, 2023). The Arabian aquifer system is preliminary located in Saudi Arabia, and also in the other GCCs, as shown in Table 23.1. Generally, the Arabian aquifer system is recognized as a vital supply of water for the GCCs, however, it is under stress and is the world's most overstressed groundwater system (Richey et al., 2015). The water stress is characterized by extremely high (>80%) in the GCCs, except Kuwait and Oman, the water stress is characterized by arid and low water use (high water stress), as shown in Table 23.1.

The ratio of groundwater withdrawal to aquifer recharge rate is known as groundwater decline (Gleeson et al., 2012). The groundwater decline varied from an insignificant trend in the UAE and KSA to an extremely high (>8 cm/year) in Bahrain. While, the decline was Low-Medium (0–2 cm/year) in Kuwait, Oman, and Qatar. In countries where the value is over one, there is unsustainable groundwater consumption, which has an impact on both the ecology and sustainability of groundwater.

The interannual variability is the average year-to-year variation in the amount of water that is available, encompassing both renewable surface and groundwater supplies. Overall, the interannual variability of water supply ranges from medium–high to extremely high in the GCCs. In Qatar and KSA, it was moderately high, high in Bahrain and Kuwait, and extremely high in Oman and the UAE. The higher variability means that yearly rainfall is more irregularly distributed around the annual mean, which presents problems for agricultural water management and ultimately impacts food production (Sloat et al., 2018; Harp & Horton, 2023).

Table 23.1 The overall baselines for water-stress risks in GCC countries

Country	Major Basin	Aquifer Systems	Water stress	Groundwater decline	Interannual variability
Bahrain	Arabian Peninsula	Arabian aquifer system	Extremely high (>80%)	Extremely high (>8 cm/year)	High (0.75–1.00)
Kuwait	Arabian Peninsula	Arabian aquifer system	Arid and low-water use	Low–medium (0–2 cm/year)	High (0.75–1.00)
Oman	Arabian Peninsula	Arabian aquifer system	Arid and low-water use	Low–medium (0–2 cm/year)	Extremely high (>1.00)
Qatar	Arabian Peninsula	Arabian aquifer system	Extremely high (>80%)	Low–medium (0–2 cm/year)	Medium–high (0.50–0.75)
Saudi Arabia	Arabian Peninsula	Arabian aquifer system	Extremely high (>80%)	Insignificant trend	Medium–high (0.50–0.75)
UAE	Arabian Peninsula	Arabian aquifer system	Extremely high (>80%)	Insignificant trend	Extremely high (>1.00)

Source Aqueduct 4.0; WRI (2023)

23.3.3 Freshwater Withdrawals, Cereal Production, and Livestock Production Index in GCCs

This section shows the main components of water productivity in the GCCs. The annual freshwater withdrawal (%) for agriculture is shown in (Fig. 23.3), and cereal production and livestock production index are shown in Figs. 23.4 and 23.5, respectively. Generally, the Gulf region's conventional water resources include groundwater (shallow and deep aquifers) and surface water (MEI, 2009). The Tigris, Euphrates, and Krn Rivers provide most of the minor freshwater intake into the Gulf (Graham, 2023). Overall, there is a reduction in annual freshwater withdrawal for all GCCs by about 20.8%, except Kuwait has an increase of about 16.1% in the total water withdrawal from 2000 to 2020. In Qatar and UAE, a drastic decrease in freshwater withdrawal of about 49% and 40.8% was observed, respectively. The same reduction was observed in Bahrain (32.1%), Saudi Arabia (7.0%), and Oman (10.7%) as well, as shown in Fig. 23.3. The main reason is that the GCCs are in the "extremely high-water stress" range (Table 23.1). This is due to the average rainfall being almost less than 100 mm, besides, heat and high evaporation also prevent the natural recharge of groundwater.

The total volume of all cereals produced in the GCCs is 1.7 million tons in 2020, as shown in Fig. 23.4. While, in the year 2000, the total production was 2.3 million tons. The total reduction in cereal production in GCCs is 36% between 2000 and 2020. In Kuwait, the cereal production is increased by about 118% between 2000 and 2020. Also in Kuwait, the total freshwater withdrawal (mentioned in Fig. 23.3) is increased by about 16.1%. As of Bahrain, cereal production was zero throughout the assessed years. Contrary to cereal production which decreased by about 36% between 2000 and 2020, the livestock production index for GCCs is 135.4 in 2020,

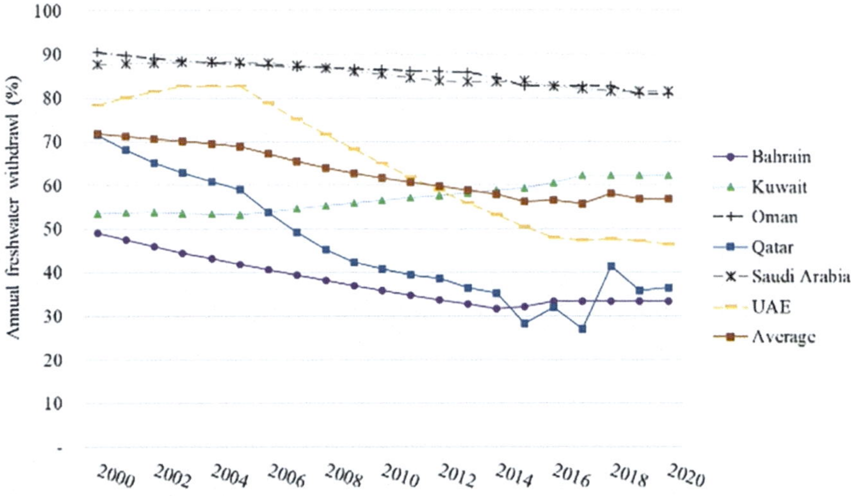

Fig. 23.3 Annual freshwater withdrawal (%) for agriculture and livestock (*Source* FAO-AQUASTAT)

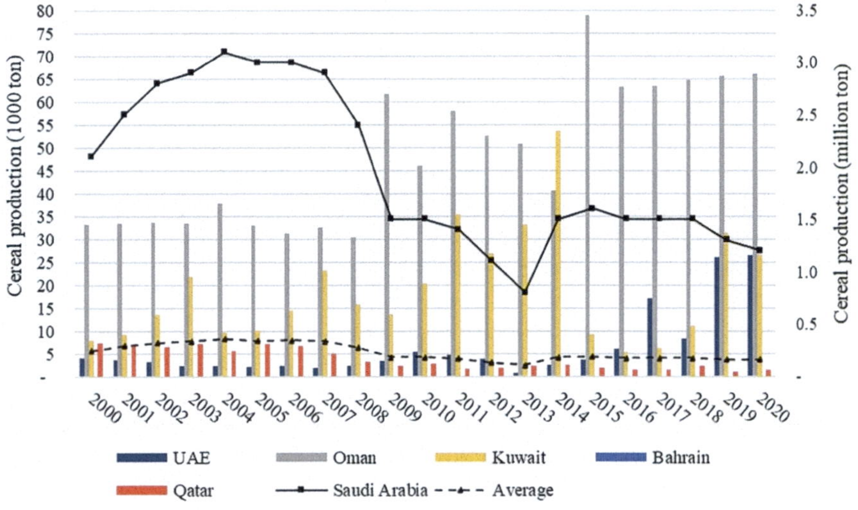

Fig. 23.4 Cereal production (ton) (*Source* FAO-AQUASTAT)

and 46.7 in 2000, growing at an average of 120%, as shown in Fig. 23.5. Overall, the livestock production index has drastically increased in each country of the GCCs, rather than the drastic decrease in cereal production. Therefore, livestock production is the most important component of agriculture, contributing the biggest percentage of agriculture and water productivity.

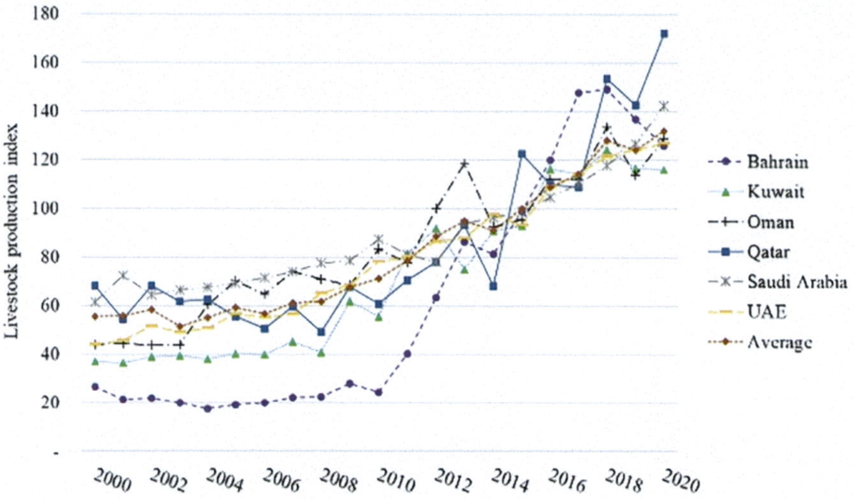

Fig. 23.5 Livestock production index (*Source* FAO-AQUASTAT)

23.3.4 Water Productivity and Food Imports

This section shows water productivity (Fig. 23.6) versus food imported (Fig. 23.7). This is for a better understanding the relationship between food production system and water scarcity as well as the role that imported food/virtual water plays in compensating water shortages and improving water productivity. Overall, water productivity increased from 92.8 in 2000 to 205.9 in 2020, growing at an average of 121.8%. Water productivity increased between the years 2000 and 2020 in all GCCs, except for Kuwait water productivity decreased by about 15.4%. In Qatar, a drastic increase in water productivity with an average of 214% was observed throughout the assessed period. Despite the improvement in water productivity, the GCCs food imports recorded 11.3% as a share of merchandise imports throughout the assessed years. In 2020, food imports for GCCs tended to increase and recorded 13.6%. However, imported food fluctuated substantially between 8.7% in 2007 (lowest record) to 13.6% (highest record) in 2020. Kuwait recorded the highest imported food at an annual rate of 20.1%, followed by Oman and KSA, at an annual rate of 15.5% in 2020. The situation is further exacerbated when those percentages (food imports as a share of merchandise imports) are considered 90% of the GCCs food demand (Writer, 2016).

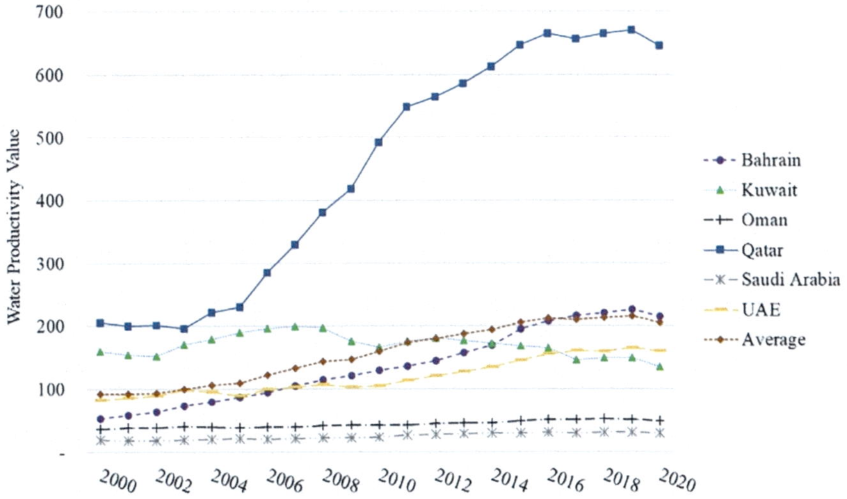

Fig. 23.6 Water productivity (*Source* FAO-AQUASTAT)

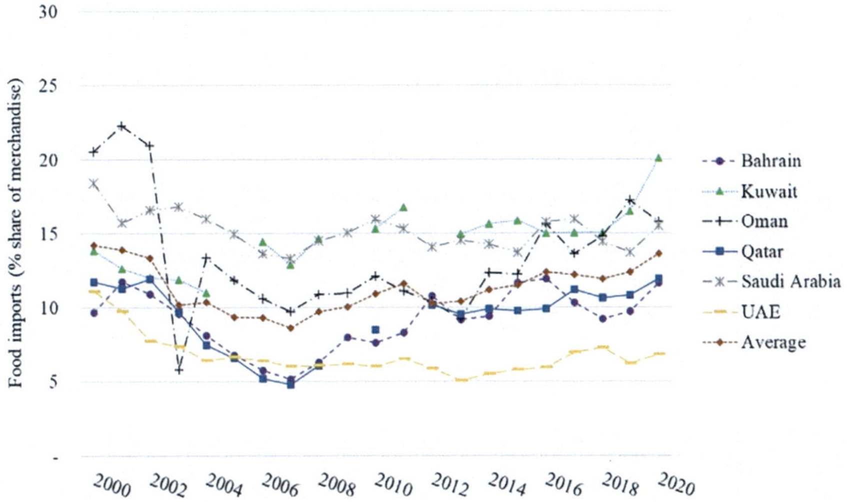

Fig. 23.7 Food imports as a share of merchandise imports (*Source* FAO-AQUASTAT)

23.3.5 Sustainable Development of Land and Water Productivity Strategies

This study reviews the key issues of food production systems in GCCs, concerning water productivity components and virtual water in imported food. This is to develop a framework for effectively using development resources and promoting sustainable agricultural and water practices. Overall, severe water scarcity and excessive usage of vulnerable groundwater aquifers are the main challenges facing the GCCs, revealing the challenges of intensive agriculture. Therefore, creating small rural communities that can sustain a subsistence farming system is crucial and will contribute to sustainable development (IFAD, 2020). In such communities, when water is limited, it is therefore suggested to establish small-scale farms of 50 ha to 100 ha (Freeg et al., 2023). The establishment of more sustainable production methods for emerging technologies is then required; These include the development and establishment of (I) Water systems equipped with telecommunications systems, such as Supervisory Control and Data Acquisition systems (SCADA) (Brad et al., 2021), (II) Watershed management systems and rainwater harvesting techniques, such as diversion drains, canals, and farm ponds (Ranasingha & Wattage, 2024), (III) Artificial recharge of groundwater, such as water traps, infiltration basins, and diversion of excess flows from irrigation canals (Zhai & Wu, 2024), (IV) Modern irrigation techniques, such as drip irrigation, sprinkler irrigation, subsurface irrigation, and center pivot irrigation system (Abde et al., 2021), (V) Crop diversity and bio-diversity, such as intercropping systems, multiple cropping, and growing crops in rotation (Arenas-Salazar et al., 2024), (VI) Climate-Smart Agriculture (CSA), such as sensors to monitor temperature, light, humidity, and soil conditions, hardware and software to support robotics, automation and Internet of Things-based solutions (World Bank, 2021), (VII) Nexus between energy (wind and solar), water, and land, such as using non-traditional energy sources of solar, wind, geothermal, hydropower, ocean, and bioenergy and utilizing renewable energy-powered desalination equipment (Climate Council, 2022), (VIII) Use of non-conventional water sources, such as agricultural drainage water, brackish groundwater and/or seawater, and treated municipal wastewater (Qadir et al., 2022), (IX) Circular economy, which helps create a gender-sensitive economy and is also a suitable strategy for labor-intensive smallholder farmer (Barros et al., 2020), and eventually (X) Policies, institutions, and capacity building, such as crafting rules, incentives instruments, appropriating of finance, establishing water community-based, and training and capacity building (Ferrero et al., 2019).

23.4 Conclusions

This research aims to investigate the aggregate food production systems and water stress on water productivity in the Gulf Cooperation Countries (hereinafter; GCCs) based on data obtained from different sources. This is for a closer and more detailed analysis of the controlled water environment agriculture (arid conditions) in the GCCs.

The results showed that the GCCs face significant challenges that must be considered to fully realize the potential to develop sustainable food production systems under water stress conditions. These include; (I) The most reliable challenge is the GCCs are facing extremely high-water stress which means almost water is used for their renewable supply, and limits the possibilities of expanding the agricultural lands that can be reclaimed, (II) GCCs have limited available options to meet the rising food demand when they lack import food finance, and any future increases in food demand will need to be compensating by food imports, which are becoming more vulnerable to price increases and volatility as a result of political issues and climatic change, (III) The livestock production system is the most crucial component of agriculture in GCCs, contributing the biggest percentage of water productivity, and more promising future strategies should be given to that sector, (IV) Unlikely, redistributing agricultural water to other economic sectors is anticipated to occur as a result of the development of other economic sectors (it is recommended to cover the expected future increase in food imports), which will raise demand for the extremely scarce water resource base, (V) The establishment of more sustainable production methods for emerging technologies, and eventually (VI) This study opens up a wealth of opportunities to conduct further research on the combination of virtual water and water productivity throughout different regions in GCCs to understand the diverse production system in each region.

References

Abde, Y. A., Allam, A. M., & Darwisch, W. M. B. (2021). Effect of irrigation systems and agriculture methods on the productivity and quality of snap bean crop under sandy soil conditions. *Journal of Environmental Studies and Researches, 11*(2), 332–340. https://doi.org/10.21608/jesr.2021.244298.

Arenas-Salazar, A. P., Schoor, M., Parra-Pacheco, B., García-Trejo, J. F., Torres-Pacheco, I., & Feregrino-Pérez, A. A. (2024). Intercropping systems to modify bioactive compounds and nutrient profiles in plants: Do we have enough information to take this as a strategy to improve food quality? A review. *Plants, 13*, 194. https://doi.org/10.3390/plants13020194.

Barros, M. V., Salvador, R., Francisco, A. C., & Piekarski, C. M. (2020). Mapping of research lines on circular economy practices in agriculture: from waste to energy. *Renewable and Sustainable Energy Reviews, 131*, 109958. https://doi.org/10.1016/j.rser.2020.109958

Bastiaanssen, W. G. M., & Steduto, P. (2017). The water productivity score (WPS) at global and regional level: Methodology and first results from remote sensing measurements of wheat, rice and maize. *Science of the Total Environment, 575*, 595–611. https://doi.org/10.1016/j.scitotenv.2016.09.032

Blanc, E. & Noy, I. (2023). Impacts of droughts and floods on agricultural productivity in New Zealand as measured from space. Environment Research Climate. https://doi.org/10.1088/2752-5295/acd4da.

Borgomeo, E., Khan, H. F., Heino, M., Zaveri, E., Kummu, M., Brown, C., & Jägerskog, A. (2020). Impact of green water anomalies on global rainfed crop yields. *Environmental Research Letters, 15*(12), 124030. https://doi.org/10.1088/1748-9326/abc587

Brad, S., Murar, M., Vlad, G., Brad, E., & Popanton, M. (2021). Lifecycle design of disruptive SCADA systems for waste-water treatment installations. *Sustainability, 13*, 4950. https://doi.org/10.3390/su13094950

Brauman, K. A., Siebert, S., & Foley, J. A. (2013). Improvements in crop water productivity increase water sustainability and food security—A global analysis. *Environmental Research Letters, 8*, 024030. https://doi.org/10.1088/1748-9326/8/2/024030

Britannica. (2005). *Arabian Basin*. Encyclopedia Britannica. https://www.britannica.com/place/Arabian-Basin. Accessed September 20, 2023.

Bruan, J. V. (2009). Food and financial crises implications for agriculture and the poor. *Food Policy Report*. https://doi.org/10.2499/0896295346. Accessed May 24, 2024.

Buko, J., Duda, J., & Makowski, A. (2021). Food production security in times of a long-term energy shortage crisis: The example of Poland. *Energies, 14*, 4725. https://doi.org/10.3390/en14164725

Cai, X., & Rosegrant, M. W. (Eds.) (2003). Water productivity in agriculture: Limits and opportunities for improvement. In J. W. Kijne, R. Barker, & D. Molden (Eds). https://www.iwmi.cgiar.org/Publications/CABI_Publications/CA_CABI_Series/Water_Productivity/Unprotected/0851996698ch10.pdf. Accessed April 24, 2024.

Christoforidou, M., Borghuis, G., Seijger, C., Halsema, G. E., & Hellegers, P. (2023). Food security under water scarcity: A comparative analysis of Egypt and Jordan. *Food Security, 15*, 171–185. https://doi.org/10.1007/s12571-022-01310-y

Climate Council. (2022). *11 Countries leading the charge on renewable energy*. https://www.climatecouncil.org.au/11-countries-leading-the-charge-on-renewable-energy/. Accessed June 2, 2024.

Din, M. A. E. (2007). The transport importance of the Arabian (Persian) Gulf. *Transport Reviews, 10*, 127–148. https://doi.org/10.1080/01441649008716749

Elfkih, S., Hadiji, O., Ben Abdallah, S., & Boussadia, O. (2023). Water accounting for food security: Virtual water and water productivity in the case of Tunisian olive oil value chain. *Agriculture, 13*, 1205. https://doi.org/10.3390/agriculture13061205

European Union official website. *Agriculture and rural development*. https://agriculture.ec.europa.eu/international/agricultural-trade/bilateral-agreements/middle-east-gulf-countries_en#overview. Accessed May 26, 2024.

FAO (Food and Agriculture Organization). (2002). *Value of virtual water in food: Principles and virtues*. Paper presented at the UNESCO-IHE workshop on virtual water trade 12–13 December 2002, Delft, The Netherlands. Accessed May 23, 2024.

FAO (Food and Agriculture Organization). (2009). *Food security and the financial crisis*. World summit on food security Rome 16–18 November 2009. https://www.fao.org/4/k6360e/k6360e.pdf. Accessed May 15, 2024.

FAO (Food and Agriculture Organization). (2014). Family poultry development—Issues, opportunities and constraints. In: *Animal production and health working paper*. No. 12. Accessed May 24, 2024.

Ferrero, G., Setty, K., Rickert, B., George, S., Rinehold, A., DeFrance, J., & Bartram, J. (2019). Capacity building and training approaches for water safety plans: A comprehensive literature review. *International Journal of Hygiene and Environmental Health, 222*(4), 615–627. https://doi.org/10.1016/j.ijheh.2019.01.011

Fitton, N., Alexander, P., Arnell, N., Bajzelj, B., Calvin, K., Doelman, Gerber, J. S., Havlik, P., Hasegawa, T., Herrero, M., Krisztin, T., Meijl, H., Powell, T., Sands, R., Stehfest, E., West, P. C., & Smith, P. (2019). The vulnerabilities of agricultural land and food production to future water scarcity. *Global Environment Change, 58*, 101944. https://doi.org/10.1016/j.gloenvcha.2019.101944.

Freeg, A., Riad, P., Hassan, N. A., & Nassar, A. A. (2023). Appraisal of multilayer aquifer system for sustainable groundwater management in Wadi Araba, Eastern Desert, Egypt. *Groundwater for Sustainable Development, 21*, 100942. https://doi.org/10.1016/j.gsd.2023.100942

Gassert, F., Landis, M., Luck, M., Reig, P., & Shiao, T. (2013). *Aqueduct global maps 2.0. Working paper*. World Resources Institute. Available online at http://www.wri.org/publication/aqueduct-metadata-global. Accessed May 27, 2024.

Gleeson, T., Wada, Y., Bierkens, M. F. P., & van Beek, L. P. H. (2012). Water balance of global aquifers revealed by groundwater footprint. *Nature, 488*(7410), 197–200. https://doi.org/10.1038/nature11295

Graham, E. (2023). Persian Gulf. *Encyclopedia Britannica*, September 26, 2023. https://www.britannica.com/place/Persian-Gulf. Accessed April 27, 2024.

Hanjra, M. A., & Qureshi, M. E. (2010). Global water crisis and future food security in an era of climate change. *Food Policy, 35*, 365–377. https://doi.org/10.1016/j.foodpol.2010.05.006

Harp, R. D., & Horton, D. E. (2023). Observed changes in interannual precipitation variability in the United States. *Geophysical Research Letters, 50*, e2023GL104533. https://doi.org/10.1029/2023GL104533.

IFAD (International Fund for Agricultural Development). (2014). Proceedings of an E-conference 16 January–3 February 2012. In: *Family poultry interactions with other production systems (Forestry, tree crops, annual crops, large animals, fisheries, etc.): Nutritional opportunities and constraints*. Accessed April 15, 2024.

IFAD (International Fund for Agricultural Development). (2020). *Why small farms are key to the future of food—And how we can support them*. https://www.ifad.org/zh-TW/web/latest/-/blog/why-small-farms-are-key-to-the-future-of-food-and-how-we-can-support-them. Accessed May 30, 2024.

Khorchani, M., Awada, T., Schmer, M., Jin, V., Birru, G., & Dangal, S. R. S. (2024). Long-term croplands water productivity in response to management and climate in the western US corn belt. *Agricultural Water Management, 291*, 108640. https://doi.org/10.1016/j.agwat.2023.108640

Kijne, J. W., Barker, R., Molden, D. (Eds). (2006). Water productivity in agriculture: Limits and opportunities for improvement. *CAB International, Wallingford, UK, in association with the International Water Management Institute (IWMI), Sri Lanka, 2004*, (pp. 354). US$110 (hardback), ISBN 0-85199-669-8. Accessed May 29, 2024.

Kuzma, S., Bierkens, M. F. P., Lakshman, S., Luo, T., Saccoccia, L., Sutanudjaja, E. H., & Van Beek, R. (2023). *Aqueduct 4.0: Updated decision-relevant global water risk indicators*. Technical Note. World Resources Institute. https://doi.org/10.46830/writn.23.00061. Accessed April 25, 2024.

Letseku, V., & Grové, B. (2022). Crop water productivity, applied water productivity and economic decision making. *Water, 14*, 1598. https://doi.org/10.3390/w14101598

MEI (Middle East Institute). (2009). *Water issues in the Gulf: Time for action*. January, 2009. https://www.mei.edu/. Accessed May 22, 2024.

Molden, D. J., Sakthivadivel, R., & Habib, Z. (2001). *Basin-level use and productivity of water: Examples from South Asia*. Research Report 49, International Water Management Institute, Colombo, Sri Lanka.

National Aeronautics and Space Administration (NASA). (2012). *Global land data assimilation system version 2 (GLDAS-2), Goddard earth sciences data information services center*. http://disc.sci.gsfc.nasa.gov/hydrology/data-holdings.

Niacsu, L., Bucur, D., Ionita, I., & Codru, I. C. (2022). Soil conservation measures on degraded land in the hilly region of eastern Romania: a case study from Puriceni-Bahnari catchment. *Water, 14*, 525. https://doi.org/10.3390/w14040525

Pun, M., Nirman, S., Petra, S., & Zemadim, B. B. (2023). *A framework to monitor crop-specific drought and flood impacts using remote sensing datasets*. Colombo, Sri Lanka: International water management institute (IWMI). CGIAR initiative on excellence in agronomy (p. 29).

Qadir, M., Smakhtin, V., Koo-Oshima, S., & Guenther, E. (2022). *Unconventional water resources*. ISBN 978-3-030-90146-2 (eBook). https://doi.org/10.1007/978-3-030-90146-2

Ranasingha, C. M., & Wattage, P. (2024). Rainwater harvesting (RWH) systems: Is the conservation of water in Colombo Urban areas worth it? *Conservation, 4*(1), 23–35. https://doi.org/10.3390/conservation4010002

Rehren, T. (2011). The UCL institute of archaeology and Qatar. *Archaeology International, 13*, 28–29. https://doi.org/10.5334/ai.1308

Reicosky, D., Brandt, D., Reeder, R., Lal, R., & Montgomery, D. R. (2023). Plowing: Dust storms, conservation agriculture, and need for a soil health act. *Journal of Soil and Water Conservation, 78*, 5. https://doi.org/10.2489/jswc.2023.0619A

Richey, A. S., Thomas, B. F., Lo, M. H., Reager, J. T., Famiglietti, J. S., Voss, K., Swenson, S., Rodell, M. (2015). Quantifying renewable groundwater stress with GRACE. *Water Resource Research, 51*, 5217–5238. https://doi.org/10.1002/2015WR017349.

Salminen, J., Maatta, K., Haimi, H., Maidell, M., Karjalainen, A., Noro, K., Koskiaho, J., Tikkanen, S., & Pohjola, J. (2022). Water-smart circular economy—conceptualization, transitional policy

instruments and stakeholder perception. *Journal of Cleaner Production, 334*, 130065. https://doi.org/10.1016/j.jclepro.2021.130065
Shahid, S. A., & Ahmed, M. (Eds). (2014). Environmental cost and face of agriculture in the Gulf cooperation council countries: Fostering agriculture in the context of climate change. *Gulf Research Centre Cambridge*. https://doi.org/10.1007/978-3-319-05768-2_1. Accessed May 18, 2024.
Sloat, L. L., Gerber, J. S., Samberg, L. H., Smith, W. K., Herrero, M., & Ferreira, L. G. (2018). Increasing importance of precipitation variability on global livestock grazing lands. *Nature Climate Change, 8*(3), 21.
SSWG (Seven Seas Water Group). (2023). *The significance of virtual water*. 14400 Carlson circle Tampa, Florida 33626. https://sevenseaswater.com/significance-of-virtual-water/. Accessed May 13, 2024.
Sutanudjaja, E. H., Van Beek, R., Wanders, N., Wada, Y., Bosmans, J. H. C., Drost, N., & Van der Ent, R. J. (2018). PCR-GLOBWB 2: A 5 arcmin global hydrological and water resources model. *Geoscientific Model Development, 11*(6), 2429–53. https://doi.org/10.5194/gmd-11-2429-2018.
UNDRR (United Nations Office for Disaster Risk Prevention). (2022). *Water scarcity predicted to worsen in more than 80% of croplands globally this century*. https://www.preventionweb.net/quick/71368. Accessed April 28, 2024.
UNFCCC (United Nations Framework Convention on Climate Change). (2024). *Kingdom of Saudi Arabia*. Report submitted to the second biennial update report (BUR 2). https://unfccc.int/sites/default/files/resource/Kingdom%20of%20Saudi%20Arabia%C2%A0Bur2.pdf. Accessed May 29, 2024.
USDA (United States Department of Agriculture, Economic Research Services). (2014). *Water productivity indicators: What do they tell us?*. International workshop on going beyond agricultural water productivity, December 8–9, 2014. World Bank, Washington D. Accessed September 29, 2023.
Wada, Y., Wisser, D., & Bierkens, M. F. B. (2014). Global modeling of withdrawal, allocation and consumptive use of surface water and groundwater resources. *Earth System Dynamics, 5*(1), 15–40. https://doi.org/10.5194/esd-5-15-2014
WFP (World Food Program). (2022). Understanding the energy crisis and its impact on food security. Via Cesare Giulio Viola, 68, 00148 Rome RM, Italy. 19 August 2022. https://www.wfp.org/publications/understanding-energy-crisis-and-its-impact-food-security. Accessed May 27, 2024.
World Bank. (2015). *Agricultural land* (% of land area). http://data.worldbank.org/indicator/AG.LND.AGRI.ZS. Accessed May 18, 2024.
World Bank. (2021). *Climate-smart agriculture*. Last updated: April 05, 2021. https://www.worldbank.org/en/topic/climate-smart-agriculture. Accessed May 30, 2024.
World Bank. (2022). *World water day, 2022*. https://www.worldbank.org/en/country/india/brief/world-water-day-2022-how-india-is-addressing-its-water-needs. Accessed May 26, 2024.
WRI (World Resources Institute). (2023). *25 Countries, housing one-quarter of the population, face extremely high-water stress*. https://www.wri.org/. Accessed May 23, 2024.
Writer, S. (2016). *Food security is a growing issue in the Gulf region*. Gulf Business. September 19, 2016. https://gulfbusiness.com/author/staff-writer/. Accessed May 22, 2024.
Zhai, Y., & Wu, J. (2024). River ecological restoration and groundwater artificial recharge II. *Water, 16*(10), 1328. https://doi.org/10.3390/w16101328

Chapter 24
Water Productivity and Water Balance Assessment in Furfuro Small-Scale Irrigation Scheme Using Agrohydrological Model

Kedrala Wabela, Ali Hammani, Sirak Tekleab, and Abdelilah Taky

Abstract Irrigation is by far the largest consumer of freshwater worldwide. Thus, effective water management in this sector is crucial to enhancing water productivity (WP). In this study, a physically based agrohydrological model, the Soil Water Atmosphere Plant (SWAP), was used to investigate the WP of the main irrigated crops and water balance in the Furfuro irrigation scheme, Ethiopian Rift Valley. The SWAP model requires meteorological, soil, crop, and water data to simulate soil moisture and crop growth. The meteorological data (1991–2020) was collected from the Ethiopian National Meteorology Agency. The physical properties of the soil were determined in the laboratories. The basic crop data for selected crops was collected from Food and Agricultural Organization (FAO) documents. Field experiments were conducted in the scheme command area in 2022–23 to collect data for the calibration and validation of the SWAP model. The field experiments were conducted on six experimental plots using the main irrigated crops (wheat, onion, and tomato) in the study area. The six experimental plots were categorized into two groups: Researcher plots and Farmer plots. Each group contains three plots, one for each crop type. Crop water requirements and irrigation scheduling were determined for researcher plots based on climatic, crop, and soil data of the area. All irrigation practices (amount and timing of application) and field management for farmer experimental plots were carried out by farmers based on their own experiences. However, the depth of irrigation water was measured during each irrigation at all six experimental plots using a 5.08*90 cm Cutthroat flume. Soil moisture content after each irrigation (for the wheat plot) was determined using the gravimetric method. Crop data such as leaf area, growth stage-based dry matter, and yield were collected from each experimental

K. Wabela (✉) · A. Hammani · A. Taky
Institute of Agronomy and Veterinary (IAV) Hassan II, Rabat, Morocco
e-mail: h.kedrala@iav.ac.ma

K. Wabela
Worabe Agricultural Research Center, Worabe, Ethiopia

S. Tekleab
Ethiopian Institute of Water Resources (EIWR), Addis Ababa University, Addis Ababa, Ethiopia

A. Sefelnasr et al. (eds.), *Water Resources Management and Sustainability*, Water Science and Technology Library 114, https://doi.org/10.1007/978-3-031-80520-2_24

plot. The SWAP model was then calibrated and validated using soil moisture and crop growth data collected from experimental plots. The water balance components (rainfall, irrigation, transpiration, evaporation, and percolation) of the experimental plots were simulated using the calibrated SWAP model. The WP of the main crops grown in the study area was then determined using the simulated water balance components and measured yield in the experimental plots. The results indicated that the simulated irrigation depth at both researcher and farmer plots accurately represented the field condition. There was no significant difference between the researcher and the farmer plots in the evapotranspiration of the same crop. The percolated depth was 210.9 and 308.5 mm for wheat, 167.1 and 252.6 mm for onion, and 169.8 and 253.9 mm for tomato at the researcher and farmer plots, respectively. The physical and economic WP of researcher plots was greater than that of farmer plots at all water balance components. The physical and economic WP of wheat was lower than that of onion and tomato in both researcher and farmer plots, and tomato demonstrated higher physical and economic WP in both researcher and farmer plots. The findings in this study can provide technical assistance for effective irrigation water management to save irrigation water and enhance WP.

Keywords Evapotranspiration · Experiment · Transpiration · Water balance · Water productivity

24.1 Introduction

In many regions of the world, the need for freshwater by industries, households, agriculture, and nature reserves is higher than the available freshwater, thus signifying the need for improved freshwater management (Van Dam et al., 2006). Along with the growing demand, drought, climate change (Ronco et al., 2017), and rapid population growth (Kaur et al., 2010) are causing freshwater availability to decline, which affects the supply. In particular, in the Ethiopian Rift Valley, agricultural production has been severely hampered by frequent droughts and declining annual rainfall (Girma et al., 2007; Kassie et al., 2014). The agroecology in the Central Ethiopian Rift Valley has experienced longer dry spells and low precipitation more frequently, which has affected crop production (Belay et al., 2017). Due to climate change, the stream flow of the Bilate River, which is a source of water for several irrigation schemes in the Ethiopian Rift Valley, has been projected to decline from 8.9% to 26.6% in the near future (Orke & Li, 2022). The main components of the water balance have been projected to decline due to climate change, which is expected to occur in the Ethiopian Central Rift Valley in the coming decades (Muluneh, 2020). Therefore, effective management of the available agricultural water is essential to sustaining agricultural production. Agriculture output needs to be measured in terms of how much water is consumed rather than how much land is used in areas where water is the most limiting factor (Molden et al., 2007). This will aid in improving the output per unit of water consumed, WP. WP is a term used to describe the value or benefit attained from the use of water, which encompasses many different facets of water management and is crucial, particularly in arid and semi-arid regions (Droogers & Bastiaanssen, 2002; Kijne et al., 2003; Molden & Sakthivadivel, 1999).

Due to the growing water shortage in agriculture, water allocation and management have gained attention from researchers across the globe. Irrigated agriculture is by far the biggest freshwater consumer worldwide. Thus, effective irrigation water management in this sector is vital to saving water and enhancing WP. WP in agriculture can be improved in several ways under field conditions. The application of deficit irrigation, which aims to reduce the amount of irrigation water, is the most popular and widely applied method (Molden et al., 2010; Nagaz et al., 2012; Schneider & Howell, 2001). To reduce soil evaporation and improve water productivity, soil management techniques like film plastic mulching and crop straw mulching can make a significant difference (Chakraborty et al., 2008; Zhao et al., 2012). Agricultural water productivity indicators can provide a clear picture of where and when water can be saved. They are also helpful for examining the potential rise in crop yield brought on by increased access to water (Singh et al., 2006). Therefore, measurable data on water productivity indicators is essential for the planning of effective irrigation water management practices (Vazifedoust et al., 2008).

Understanding the relationships between soil, crop, water, and atmosphere is important to investigate which farming practices help to produce "more crops per drop". Taking into account the spatial variability in soil, irrigation water application techniques, crop growth and development, and field management conditions, knowledge of water balance components at the field level will provide information on the WP of individual crops grown in the irrigation schemes. Therefore, it is essential to figure out the associations among water hydrological components such as transpiration, evaporation, and percolation under various ecohydrological situations to enhance water management and productivity (Singh et al., 2006). However, it is difficult, time-consuming, and expensive to measure all hydrological components under field conditions. In the past few years, several agrohydrological models have been developed to understand various processes occurring in space and time. Soil–Water–Atmosphere–Plant (SWAP) is a one-dimensional vertically directed model, and it can simulate the transport of water, solutes, and heat in topsoil in interaction with vegetation development (Kroes et al., 2017). The SWAP model describes a domain from the top of the canopy into the groundwater, which may be in interaction with the surface water system. The model gives a chance to learn in-depth details about how the system behaves in space and time, combined with field experiments (Singh et al., 2006), and is used to simulate field-level water balance components such as rainfall, irrigation, transpiration, evaporation, evapotranspiration, and percolation. SWAP (version 4.2.0) incorporates three crop growth modules (one simple module for static crops and two detailed modules for dynamic crops) to simulate irrigation practices, crop growth, and yield (Kroes et al., 2017).

The SWAP model has been extensively applied for irrigation water management, forecasting crop yield, and water productivity for various crops in different parts of the world (Jiang et al., 2015; Singh et al., 2006; Van Dam et al., 2006; Vazifedoust et al., 2008; Xue & Ren, 2017). Using the SWAP model, the impacts of irrigation water management on groundwater dynamics have also been studied (Ma et al., 2015; Qureshi et al., 2013; Xu et al., 2015). This study aimed to determine the water balance components and analyze the WP of the main crops of the Furfuro irrigation scheme in the Ethiopian Rift Valley using the SWAP agrohydrological model. The WP of the main crops grown in the scheme was analyzed with respect to different water balance components.

24.2 Materials and Methods

24.2.1 Description of the Study Area

The Furfuro irrigation scheme is located in the Wulbareg district of southern Ethiopia (Fig. 24.1), in the ranges of 7°39′21″ to 7°40′24″ latitude and 38°10′47″ to 38°11′53″ longitude. The scheme is situated about 25 km southeast of Worabe City, the capital of Siltie Zone, in the western escarpment of the Ethiopian Rift Valley. The scheme was constructed in 2007 with a command area of 200 ha. Its water source is the Furfuro River, which flows from the northern edge of the upper Dijo catchment to Shala Lake. The study area has humid and sub-humid climatic conditions with a bimodal rainfall pattern (short and main rainy seasons). The short rainy season known as *Belg* typically begins in March and lasts until May, which is used to supplement irrigation. The main cropping season, called *Kiremt*, lasts from June until the end of September. The study area usually experiences dry condition from November to March, which is the typical irrigation season. Onion, tomato, wheat, and other small vegetable crops are most popularly grown in the scheme. According to the meteorological data, the area's annual mean rainfall varies between 560 and 1300 mm (Fig. 24.2). The mean minimum and maximum temperatures in the study area are 9 °C and 26 °C, respectively (Fig. 24.3). The dominant soil types in the area are sandy loam and sandy clay loam (Abrar et al., 2023), with an intensively cultivated agriculture system.

24.2.2 Data Collection

The SWAP model requires meteorological, soil, crop, and water data to simulate water balance and crop dry matter. The average long-term (1991–2020) meteorological data was used, which was collected from the Ethiopian National Meteorology Agency (ENMA). The physical properties of the soil in the experimental area were determined in the laboratories. Field experiments were conducted at the scheme command area in 2022–23 to collect data for the calibration and validation of the SWAP model (Table 24.1). The experiments were conducted on six experimental plots using the main irrigated crops (wheat, onion, and tomato) in the study area. The six experimental plots were categorized into two groups: Researcher plots and

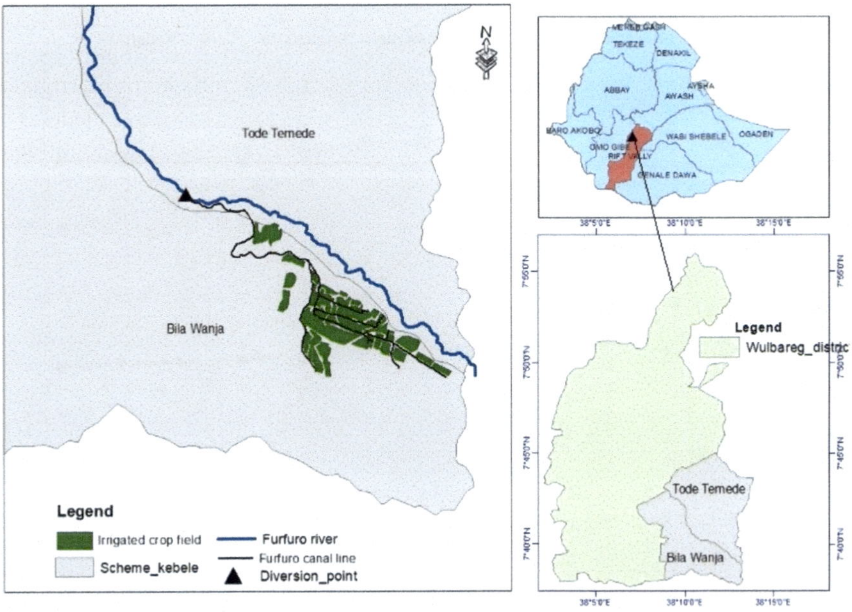

Fig. 24.1 Location map of the study area

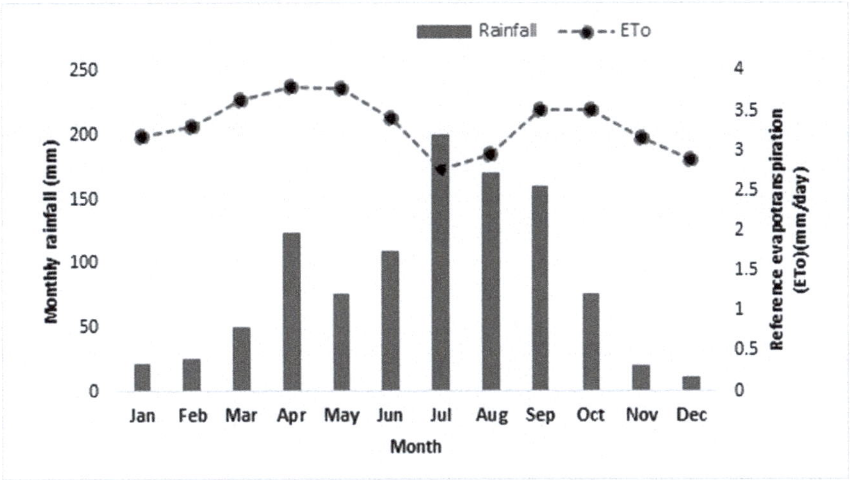

Fig. 24.2 Rainfall and reference evapotranspiration of the study area

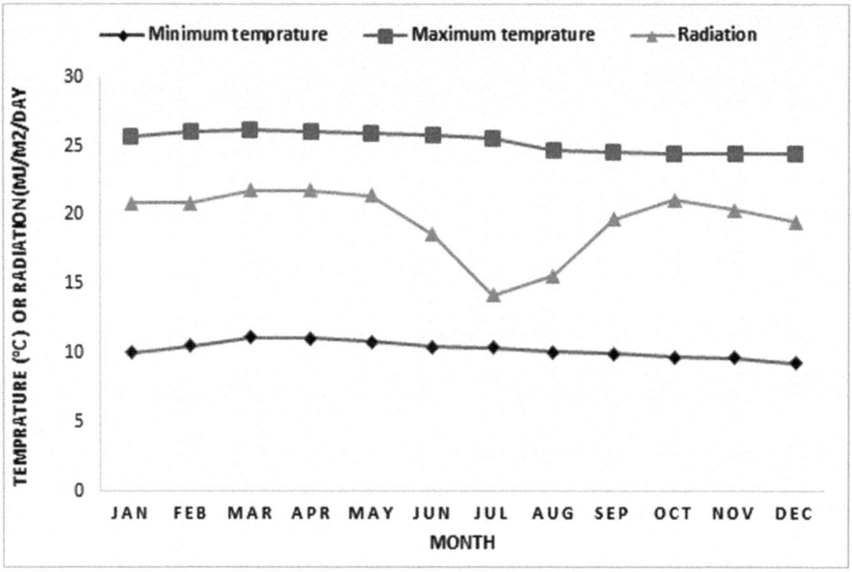

Fig. 24.3 Temperature and radiation of the study area

Farmer plots. Each experimental group contains three plots, one plot for each crop type. Researcher plots were irrigated based on predetermined crop water requirements (CWR) and irrigation schedules. The CWR was determined using CROPWAT computer model based on the climatic, soil and crop data of the study area. All irrigation practices (amount and timing of water application) and field management for farmer experimental plots were carried out by farmers based on their own experiences. However, the depth of irrigation water was measured during each irrigation event at all six experimental plots using a 5.08*90 cm Cutthroat flume. Before each irrigation, the soil moisture level of researcher plots was checked using Watermark. The soil moisture content before and after each irrigation (for the wheat plot) was determined at different soil depths using the gravimetric method to calibrate and validate the SWAP model:

$$\theta_m = \frac{\text{Weight of wet soil} - \text{Weight of oven dry soil}}{\text{Weight of oven dry soil}} * 100 \quad (24.1)$$

where, θ_m is the gravimetric soil water content

Gravimetric water content is then converted to volumetric water content (θ_v) using the soil bulk density and density of water.

$$\theta_v = \theta_m * \frac{\text{Soil bulk density}}{\text{Water density}} \quad (24.2)$$

Table 24.1 Collected data for input and calibration and validation of the SWAP model

Data type	Method of collection	Purpose
Climatic data	Meteorological station	Input
Irrigation depth	Filed measurement using Cutthroat flume	Input
Soil water content	Field soil sampling and gravimetric method	Calibration and validation
Soil properties	Laboratory analysis	Input
Crop development stage	Field observation	Input
Plant height	Field measurement	Input
Leaf area	Field measurement	Calibration and validation
Dry matter partitioning	Field measurement	Calibration and validation
Rooting depth	Field measurement	Input
Crop yield	Field measurement	Calibration and validation

Crop data such as leaf area, growth stage-based dry matter, plant height, and yield were collected from each experimental plot. The leaf area index (LAI) was calculated from the collected leaf area (Kang et al., 2003) by:

$$LAI = 0.759{*}P * \sum_{i=1}^{m}\sum_{i=1}^{n}\left(\frac{L_{ij} * W_{ij}}{m}\right) \quad (24.3)$$

where, P is plant density per area, n is the number of leaves of the nth sample plant, m is the number of measured plants, L_{ij} is leaf length, W_{ij} is the maximum leaf width

24.2.3 SWAP Model

The SWAP model determines the field-scale transport process in a physical and deterministic manner. The soil water transport and root water extraction are described by Richards' equation:

$$C_w(h)\frac{\partial h}{\partial t} = \frac{\partial}{\partial z}\left[K(h)\left(\frac{\partial h}{\partial z} + 1\right)\right] - S_a(z) \quad (24.4)$$

where C_w is the differential water capacity (cm^{-1}), h is the soil water pressure head (cm), K is the hydraulic conductivity ($cm\ d^{-1}$), S_a is root water extraction rate ($cm^3\ cm^{-3}\ s^{-1}$) and z is soil depth (cm)

To solve Eq. (24.4), relationships among the soil moisture content θ, soil water pressure head h, and hydraulic conductivity K are required. The soil hydraulic functions of Mualem (1976) and van Genuchten (1980) have been applied to describe the soil moisture retention curve:

$$\theta(h) = \theta_{res} + \frac{\theta_{sat} - \theta_{res}}{\left[1 + |\alpha h|^{n}\right]^{\frac{(n-1)}{n}}} \tag{24.5}$$

where, θ_{res} is residual water content (cm^3 cm^{-3}), θ_{sat} saturated water content (cm^3 cm^{-3}), and α (cm^{-1}) and n (−) are empirical shape factors.

The hydraulic conductivity curve is given by:

$$K(\theta) = K_{sat}\, S_e^{\lambda}\left[1 - \left(1 - S_e^{\frac{n}{n-1}}\right)^{\frac{(n-1)}{n}}\right]^2 \tag{24.6}$$

where, K_{sat} is the saturated hydraulic conductivity (cm d^{-1}), $S_e = (\theta - \theta_{res})/(\theta_{sat} - \theta_{res})$ relative saturation (−), and λ an empirical coefficient (−).

The precipitation, potential evapotranspiration (ET_P), and irrigation fluxes determine the upper boundary condition. SWAP applies the Penman–Monteith equation to calculate ET_P based on daily weather data (temperature, solar radiation, wind speed, and humidity) and canopy characteristics such as surface albedo, minimum crop resistance, and crop height (Allen et al., 1998). The lower boundary can extend up to the top of the groundwater flow system. Solute transport is modelled by the equation of convection–dispersion. The detailed crop growth module of the SWAP model is based on the WOrld FOod STudies (WOFOST) model. WOFOST simulates photosynthesis, crop growth, and crop production, including potential and actual dry matter yields, and accounts for salt and water stress conditions (Spitters et al., 1989; Supit et al., 1994). The byproduct of photosynthesis helps crops grow their leaves, stems, roots, and reproductive organs and maintain their respiratory systems. The potential gross assimilation of a crop under ideal conditions can be calculated using the effective incoming radiation to simulate the dry matter growth rate. The produced dry matter is split among roots, stems, leaves, and storage organs using partitioning factors depending on the stage of crop development (Kroes et al., 2017). The 4.2.0 version of SWAP does not account for the effects of nutrient deficiency, pests, weeds, and diseases on crop growth and production. However, the WOFOST module provides feedback for crop growth conditions under different salt and water stress levels.

24.2.4 Calibration and Validation of the Model

Calibration and validation of the model were conducted using the collected soil moisture contents from the field experimental plot at different depths. The one-at-a-time (OAT) technique was used to identify sensitive parameters and calibrate the model manually. The sensitive parameters were adjusted until the difference between observed and simulated soil moisture content met acceptable values. Irrigation was scheduled in the main input data of the SWAP module for the six experimental plots based on field-obtained data. The seasonal irrigation water consumptions of the six experimental plots were divided into ten-day irrigation intervals based on crop growth stages and simulated individually. There is no shallow groundwater in the study area, and the water source for irrigation is always canal water; thus, the free drainage option was selected in the bottom boundary condition. Since there is no risk of shallow groundwater in the study area, lateral drainage was not simulated. Similarly, solute transport was not simulated, assuming that the quality of the canal water was good (<2 dS m^{-1}). Root mean square error (RMSE) and Nash–Sutcliffe model efficiency (NSE) were used to evaluate the level of difference between observed and simulated values.

$$\mathrm{RMSE} = \sqrt{\frac{\sum_{i=1}^{n} (Q_o - Q_s)^2}{n}} \tag{24.7}$$

$$\mathrm{NSE} = 1 - \left[\frac{\sum_{i=1}^{n} (Q_o - Q_s)^2}{\sum_{i=1}^{n} (Q_o - Q_{savr})^2} \right] \tag{24.8}$$

where, n represents the number of observed data, Q_o denotes observed soil moisture ($cm^3\ cm^{-3}$), Q_s represents simulated soil moisture ($cm^3\ cm^{-3}$) and Q_{savr} represents the average simulated soil moisture ($cm^3\ cm^{-3}$).

The detailed crop growth module (WOFOST) was used to simulate irrigation and crop growth parameters. After calibrating the model with the soil hydraulic properties and irrigation depth, the crop growth parameters in the WOFOST sub-module were manually adjusted by running the simulation several times to increase the precision of the output values. An initial value of 70 s m^{-1} was used for the minimum canopy resistance of all crops (Singh et al., 2006; Verma et al., 2012). A green vegetation cover has an albedo of about 0.20–0.25 (Allen et al., 1998); thus, 0.22 was used for onion and tomato, but 0.19 was used for wheat to be accurate with the calibrated model. Plant height, LAI, and dry matter partitioning were adjusted with data collected directly from experimental plots.

24.2.5 Water Productivity

WP can be defined in various ways, which offers valuable indicators for assessing water use and determining when and where water savings can be achieved (Singh et al., 2006). The water balance components (rainfall (R), irrigation (I), transpiration (T), evaporation (E), evapotranspiration (ET), and percolation (Q)) of the experimental plots were simulated using the calibrated SWAP model. The WP of the main crops grown in the study area was then calculated using the simulated water balance components and actually measured yield in the experimental plots (Singh et al., 2006; Van Dam et al., 2006). The yield obtained from the experimental plots was converted to Kg ha^{-1}, and the simulated water balance components in mm were changed to volume basis (m^3 ha^{-1}). Then, the physical WP with respect to total applied water (I + R) for a specific growing season was calculated as:

$$WP_{I+R} = \frac{\text{Yield}\left(\text{Kg ha}^{-1}\right)}{\text{I} + \text{R}\left(\text{m}^3\text{ha}^{-1}\right)} \tag{24.9}$$

When considering only the usable form of water that is T, then the WP is given by:

$$WP_{T} = \frac{\text{Yield}\left(\text{Kg ha}^{-1}\right)}{\text{T}\left(\text{m}^3\text{ha}^{-1}\right)} \tag{24.10}$$

Distinguishing plant transpiration from soil evaporation (E) at field level is difficult. Therefore, the WP considering ET can be calculated as:

$$WP_{ET} = \frac{\text{Yield}\left(\text{Kg ha}^{-1}\right)}{\text{ET}\left(\text{m}^3\text{ha}^{-1}\right)} \tag{24.11}$$

If the percolated water is concerned, then the WP can be calculated as:

$$WP_{ETQ} = \frac{\text{Yield}\left(\text{Kg ha}^{-1}\right)}{\text{ETQ}\left(\text{m}^3\text{ha}^{-1}\right)} \tag{24.12}$$

The economic WP was also calculated using the cost of production and the selling price of the yield. Input cost data such as for seed, fertilizer, and pesticides was collected during the field experiment and through interviews with farmers. All fixed and variable costs specific to each crop type were included in the cost of production. Data on the selling prices of the yield were collected in the local markets. For each crop, the net benefit was calculated by deducting the total production cost from the total price at which the yield was sold. Then, the economic WP of each crop was calculated in the same manner as the physical WP, except the numerator (yield) was changed to USD ($).

24.3 Result and Discussion

24.3.1 Calibration of the SWAP Model

The result of model calibration with soil moisture indicated that the simulated soil moisture content is in agreement with observed values (Fig. 24.4). The observed and simulated soil moisture contents at the top depth (0–20 cm) were between 0.13

and 0.233 and 0.063 and 0.267 ($cm^3\ cm^{-3}$), respectively (Fig. 24.4a). Similarly, the observed and simulated soil moisture contents at the bottom depth (60–80 cm) were between 0.1 and 0.244 and 0.087 and 0.269 ($cm^3\ cm^{-3}$), respectively (Fig. 24.4d). The most sensitive parameters were empirical shape factors of α (cm^{-1}) and n (−), and they were adjusted based on the observed soil moisture contents. During the manual calibration of the model, optimizing the values of empirical shape factors using the OAT technique helps to increase the accuracy of simulated soil moisture. The optimized values of empirical shape factors of α (cm^{-1}) and n (−) were 0.057 and 1.57 at the top layer and 0.035 and 1.271 at the sublayer, respectively (Table 24.2). The same values were obtained when the optimization process was repeated with different initial values for α and n, demonstrating the uniqueness of the optimized parameter (Singh et al., 2006; Vazifedoust et al., 2008). The other parameters, such as θ_{res}, θ_{sat}, and λ, were less sensitive, and their influence on simulated output was insignificant. Initial values for less sensitive parameters were set using models and literature of similar soil physical properties. The initial value for saturated hydraulic conductivity K_{sat} was selected from the 'Soil Water Characteristics' model for a similar soil textural class. The RMSE between observed and simulated soil moisture data ranges from 0.031 to 0.041 (Table 24.3). Additionally, it was discovered that the average value of the NSE indicator was within the range of good performance.

The most important adjusted crop parameters are given in Table 24.4. Simulations of LAI, dry matter, and yield data for both researcher and farmer plots closely followed the field-measured data. The LAI was measured three times in the growing season for each crop. The model simulated the LAI accurately during the initial and development stages. However, relatively significant variations were observed between measured and simulated values at later growth stages. Low accuracy at a later growth stage may be caused by the fact that the leaf area was measured manually at the field level. The simulated yield (dry weight of living storage organ) of wheat

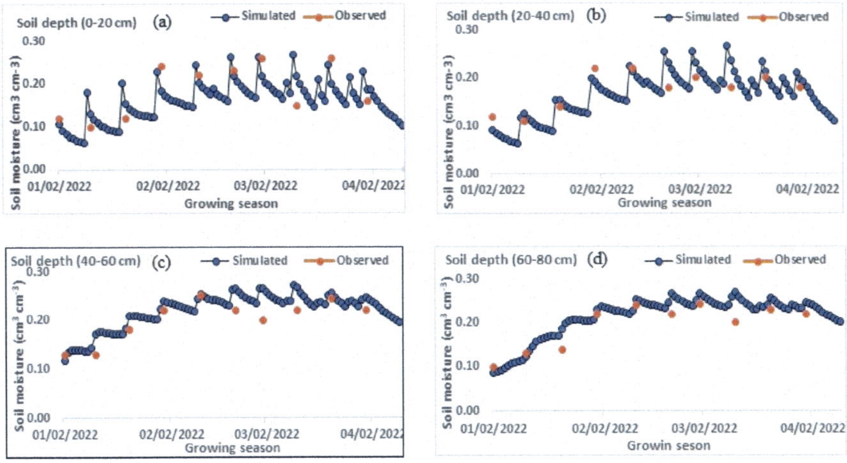

Fig. 24.4 Observed and simulated soil moisture content

Table 24.2 Soil hydraulic property parameters for the experimental location

Parameter	Top layer	Sub layer
Soil layer (cm)	0–40	> 40
Saturated water content, θ_{sat} (cm^3 cm^{-3})	31	30
Residual water content, θ_{res} (cm^3 cm^{-3})	0.01	0.01
Saturated hydraulic conductivity $K_{sa}t$ (cm d^{-1})	30.45	28.40
Soil texture	Sandy clay loam	Sandy clay loam
Shape parameter, α (cm^{-1})	0.0157	0.0135
Shape parameter, n	1.570	1.271
Shape parameter, λ	−1.6	−1.00

Table 24.3 RMSE and NSE of calibration with soil moisture contents

Soil depth (cm)	N	RMSE	NSE
0–20	10	0.041	0.52
20–40	10	0.031	0.40
40–60	10	0.035	0.48
60–80	10	0.033	0.62

Note N is number of observations

and onion was reasonably close to the observed yield; however, the simulated yield of tomato was quite smaller than the observed values. Dry matter production and dry weight of living storage organs were most sensitive to LAI and green vegetation cover (albedo) changes. This is because LAI determines how much light is intercepted and indirectly influences the dry matter accumulation in the plant organs (Vazifedoust et al., 2008). Dry matter production is also moderately sensitive to Minimum canopy resistance, Temperature sum from emergence to anthesis (TSUMEA), and Light use efficiency for real leaf. Generally, the results show that the model simulates the soil moisture content and crop growth practically well; however, future use will need further improvements.

Table 24.4 Main adjusted crop growth parameters in the SWAP-WOFOST detailed module

Parameter	Wheat	Onion	Tomato
Temperature sum from emergence to anthesis, TSUMEA (°C)	1400	1650	1750
Temperature sum from anthesis to maturity, TSUMAM (°C)	850	850	1000
Leaf area index at emergence (m^2/m^2)	0.01	0.01	0.04
Minimum canopy resistance, r_{crop} (S m^{-1})	70	70	70
Green vegetation cover (albedo)	0.19	0.22	0.22
Light use efficiency (kg ha^{-1} h^{-1}/J m^2 s^{-1})	0.45	0.45	1.0

24.3.2 Water Balance

Calculating reliable WP indicators requires an accurate simulation of the water balance component. In this study the water balance components were simulated using the calibrated model with field experimental data. The simulated water balance components are presented in Table 24.5. The rainfall contribution during the simulation period was low. The simulated irrigation depth was slightly superior to the applied depth in all plots. However, it was directly proportional to the applied depth on researcher and farmer plots and accurately represented the field condition. The simulated ET was 318.5 and 320.9 mm for wheat, 439.7 and 436.3 mm for onion, and 483 and 483.3 mm for tomato at the researcher and farmer plots, respectively. The simulated ET of wheat agreed with the mean ET obtained by (Mebrie et al., 2023) in northern Ethiopia. The simulated ET of onion was slightly higher than the ET that has been reported by (Bossie et al., 2009) in the central rift valley of Ethiopia. On the other hand, the simulated ET of tomato was less than the result of (Dirirsa et al., 2017) in the Central Rift Valley of Ethiopia. The ET variation is possibly due to climatic differences across locations in the country.

For all crop types, the simulated percolated depth of farmer plots was greater than that of researcher plots. As shown in Table 24.5, the percolated depth is 210.9 and 308.5 mm for wheat, 167.1 and 252.6 mm for onion, and 169.8 and 253.9 mm for tomato at the researcher and farmer plots, respectively. The farmer plots in all crop types were always over-irrigated throughout the growing season. Despite the higher irrigation depth at farmer plots, the simulated T and ET in the growing season were closely the same with researcher plots for the same crop type. This implies that there was a substantial amount of unusable water in farmer plots. Van Dam et al. (2006) reported a comparable finding in Sirsa district, India. Over-irrigation increases the percolation amount (Vazifedoust et al., 2008), which is unusable water. A fraction of the over-irrigated water could either evaporate or be stored in the soil. There was no significant difference in soil moisture storage between researcher and farmer plots; however, the storage of tomato plots was greater than that of wheat and onion in both researcher and farmer plots.

Table 24.5 SWAP simulated water balance at the researcher and farmers field plot

Water balance components	Researcher plot			Farmers field		
	Wheat	Onion	Tomato	Wheat	Onion	Tomato
Rainfall (mm)	55.9	55.9	55.9	55.9	55.9	55.9
Irrigation (mm)	524	606	693	627	685	780
Transpiration (mm)	223.3	364.3	437.1	227.5	361	437.5
Evaporation (mm)	95.2	75.4	45.9	93.4	75.3	45.8
Evapotranspiration (mm)	318.5	439.7	483	320.9	436.3	483.3
Percolation (mm)	210.9	167.1	169.8	308.5	252.6	253.9
Soil water storage (mm)	30.42	35.2	71.2	33.2	32.6	73.5

24.3.3 Water Productivity

Analysis of WP with different water balance components is beneficial for saving irrigation water (Singh et al., 2006). The physical and economic WP in both researcher and farmer plots was calculated using the simulated water balance components and field-measured yield data. Results indicated that the physical WP of all researcher plots was greater than that of farmer plots in all water balance components (Table 24.6). This was due to the fact that farmer plots received higher water depths than researcher plots for the same crop type. Farmers irrigated their fields, disregarding the crop's water requirement throughout its growth stage. Eshete et al. (2020) have reported excessive irrigation water application in the Ethiopian irrigation system, particularly in vegetable crops. The crop water requirement is based on climatic factors and the metabolic activities of the crop (crop factor) (Allen et al., 1998). Water used for irrigation in excess is non-beneficial, and it could percolate, evaporate, or be stored in the soil. Over-irrigation can reduce the WP in three ways: first, it increases the unusable form of water (evaporation, percolation, etc.) since crops cannot use water beyond their demand. Second, over-irrigation causes the leaching of essential soil nutrients, which are beneficial to increasing crop yield (Barton & Colmer, 2006). Third, it reduces root zone aeration (Cui et al., 2020), which affects crop development and yield production.

As shown in Table 24.6, the physical WP of wheat was less than that of onion and tomato in all water balance components. In both researcher and farmer plots, the physical WP of tomato was greater than that of onion and wheat in all water balance components. The physical WP_{ET} of wheat was 0.17 kg m^{-3} at the researcher plot and 0.11 kg m^{-3} at the farmer plot, which is much less than the results reported internationally by (Xue & Ren, 2017) and (Padhiary et al., 2020) and locally by (Meskelu et al., 2017). The reason for the low physical WP of wheat was a crop disease that severely affected the yield during the maturity stage. Eshete et al. (2020) stated that crop disease affects the water use efficiency of the Ethiopian irrigation system. This implies that comprehensive disease-control systems are necessary for improving the WP of crops. The physical WP_{ET} of onion was 1.56 kg m^{-3} at the researcher plot and 1.5 kg m^{-3} at the farmer plot. The result falls within the range of 1.36 and 1.65 kgm^{-3}, which has been noted by (Kifle et al., 2017) in northern Ethiopia. On the other hand, the physical WP_{ET} of tomato was 2.27 kg m^{-3} at

Table 24.6 Physical WP of wheat, onion and tomato at researcher and farmers plots

Physical WP	Researcher plot			Farmers plot		
	Wheat	Onion	Tomato	Wheat	Onion	Tomato
WP_{I+R} (kg m^{-3})	0.09	1.04	1.46	0.05	0.89	1.25
WP_T (kg m^{-3})	0.24	1.89	2.50	0.15	1.82	2.38
WP_{ET} (kg m^{-3})	0.17	1.56	2.27	0.11	1.50	2.16
WP_{ETQ} (kg m^{-3})	0.10	1.13	1.68	0.06	0.95	1.42

the researcher plot and 2.16 kg m^{-3} at the farmer plot. Gebru et al. (2018) reported 4.2 kg m^{-3} in northern Ethiopia. The reasons for spatial variation of crop WP_{ET} might be due to differences in climatic factors, irrigation water management practices, and crop disease outbreaks.

The physical WP_T of wheat, onion, and tomato were 0.24, 1.89, and 2.50 kg m^{-3} at researcher plots and 0.15, 1.82, and 2.38 kg m^{-3} at farmer plots, respectively. Variations in physical WP_T between different crops can be observed due to variances in the chemical composition and harvest index of crops (Vazifedoust et al., 2008). The percolated water reduced the physical WP_{ET} to WP_{ETQ} in both researcher and farmer plots (Table 24.6). In the study area, there is no shallow groundwater, and the source of water for irrigation is canal water. Therefore, the percolated water will not be recycled, and it can be considered non-beneficial. Generally, to improve physical WP, either yield must be increased while maintaining a constant water level or yield must be kept constant while using less water (Hamdy et al., 2003).

The economic WP was also calculated using the net benefits gained from each plot and the simulated water balance components. The economic WP_T of wheat, onion, and tomato were 0.10, 1.04, and 1.42 \$ m^{-3} at researcher plots and 0.03, 1.00, and 1.34 \$ m^{-3} at farmer plots, respectively (Table 24.7). For wheat, onion, and tomato, the economic WP_{ET} were 0.07, 0.86, and 1.28 \$ m^{-3} at the researcher plots and 0.02, 0.82, and 1.21 \$ m^{-3} at the farmer plots, respectively. Economic WP decreased in all plots by taking into account the percolated water. The result indicated that the economic WP of all researcher plots was greater than that of farmer plots in all water balance components. This is directly associated with water consumption level and production quantity. Yield volume differences between researcher and farmer plots have brought considerable variation in economic WP. On the other hand, wheat had a lower economic WP than onion and tomato, despite having a higher selling price per kg. In this study, tomatoes had a higher economic WP than wheat and onion. This was because tomato had a higher production volume, and the selling price per kg was also close to wheat during the data collection time for this study. In order to improve the economic WP, it is therefore clear that the economic viability of crops must be taken into account.

Table 24.7 Economic WP of wheat, onion and tomato at researcher and farmers plots

Economic WP	Researcher plot			Farmers plot		
	Wheat	Onion	Tomato	Wheat	Onion	Tomato
WP_{I+R} (\$ m^{-3})	0.04	0.57	0.83	0.01	0.48	0.70
WP_T (\$ m^{-3})	0.10	1.04	1.42	0.03	1.00	1.34
WP_{ET} (\$ m^{-3})	0.07	0.86	1.28	0.02	0.82	1.21
WP_{ETQ} (\$ m^{-3})	0.04	0.63	0.95	0.01	0.52	0.80

24.4 Conclusion

In this study, SWAP, the agrohydrological model, was used to determine the WP of the main irrigated crops in the Furfuro irrigation scheme. The SWAP agrohydrological model can be applied to simulate water balance components such as irrigation, evapotranspiration, transpiration, and percolation, as well as crop growth parameters such as dry matter and LAI. The model was manually calibrated using field-observed soil moisture and crop growth parameters. Since the observed and simulated values had good agreement, we confidently simulated the water balance components and, subsequently, calculated the WP of the main crops in the study area. A substantial amount of irrigation water is lost through percolation in the scheme due to over-irrigation in the farmer's field. Application of irrigation water based on crop water demand can enhance the WP, and over-irrigation and crop disease impacted the physical and economic WP of the crops in the study area. The determination of WP in terms of water balance components provides information about how much water is lost from irrigation. This can be used to save irrigation water and enhance WP. On the other hand, tomato demonstrated higher physical and economic WP in the study area compared with wheat and onion, and wheat achieved the lowest physical and economic WP. Generally, irrigation water application based on crop water demand, crop selection, and crop disease control systems can be used to enhance the WP.

Acknowledgements We wish to thank the Network of Excellence on Land Governance in Africa (NELGA)/ German Academic Exchange Service (DAAD). We wish to thank the National Meteorological Agency of Ethiopia (NMAE) for providing the climatic data of the study area. We are also grateful to the Worabe Agricultural Research Center for providing a vehicle to collect field data.

References

Abrar, H., Kura, A. L., Dube, E. E., & Beyene, D. L. (2023). AHP based analysis of groundwater potential in the western escarpment of the Ethiopian rift valley. *Geology, Ecology, and Landscapes*.

Allen, R. G., Pereira, L. S., Dirk, R., & Martin, S. (1998). *Crop evapotranspiration*. FAO irrigation and drainage paper No. 56.

Barton, L., & Colmer, T. D. (2006). Irrigation and fertiliser strategies for minimising nitrogen leaching from turfgrass. *Agricultural Water Management, 80*(1–3), 160–175.

Belay, A., Recha, J. W., Woldeamanuel, T., & Morton, J. F. (2017). Smallholder farmers' adaptation to climate change and determinants of their adaptation decisions in the Central Rift Valley of Ethiopia. *Agriculture and Food Security, 6*(1).

Bossie, M., Tilahun, K., & Hordofa, T. (2009). Crop coefficient and evaptranspiration of onion at Awash Melkassa, Central Rift Valley of Ethiopia. *Irrigation and Drainage Systems, 23*(1), 1–10.

Chakraborty, D., Nagarajan, S., Aggarwal, P., Gupta, V. K., Tomar, R. K., Garg, R. N., Sahoo, R. N., Sarkar, A., Chopra, U. K., Sarma, K. S. S., & Kalra, N. (2008). Effect of mulching on soil and plant water status, and the growth and yield of wheat (Triticum aestivum L.) in a semi-arid environment. *Agricultural Water Management, 95*(12), 1323–1334.

Cui, B. J., Niu, W. Q., Du, Y. D., & Zhang, Q. (2020). Response of yield and nitrogen use efficiency to aerated irrigation and N application rate in greenhouse cucumber. *Scientia Horticulturae, 265*.

Dirirsa, G., Hordofa, T., & Bekele, D. (2017). Crop coefficient and water requirement of tomato (Melka shola variety) at Melkassa, Central Rift Valley of Ethiopia. *Academic Research Journal of Agricultural Science and Research, 5*(September), 336–340.

Droogers, P., & Bastiaanssen, W. (2002). Irrigation performance using hydrological and remote sensing modeling. *Journal of Irrigation and Drainage Engineering, 128*(1), 11–18.

Eshete, D. G., Sinshaw, B. G., & Legese, K. G. (2020). Critical review on improving irrigation water use efficiency: Advances, challenges, and opportunities in the Ethiopia context. *Water-Energy Nexus, 3*, 143–154.

Gebru, A. A., Araya, A., Habtu, S., Wolde-Georgis, T., Teka, D., & Martorano, L. G. (2018). Evaluating water productivity of tomato, pepper and Swiss chard under clay pot and furrow irrigation technologies in semi-arid areas of northern Ethiopia. *International Journal of Water, 12*(1), 54–65.

Girma, M. M., Awulachew, S. B., & IWMI. (2007). *Irrigation practicies in Ethiopia charactersitics of selected irrigation schemes*. Working Paper 124.

Hamdy, A., Ragab, R., & Scarascia-Mugnozza, E. (2003). Coping with water scarcity: Water saving and increasing water productivity. *Irrigation and Drainage, 52*(1), 3–20.

Jiang, Y., Xu, X., Huang, Q., Huo, Z., & Huang, G. (2015). Assessment of irrigation performance and water productivity in irrigated areas of the middle Heihe river basin using a distributed agro-hydrological model. *Agricultural Water Management, 147*, 67–81.

Kang, S., Gu, B., Du, T., & Zhang, J. (2003). Crop coefficient and ratio of transpiration to evapotranspiration of winter wheat and maize in a semi-humid region. *Agricultural Water Management, 59*(3), 239–254.

Kassie, B. T., Rötter, R. P., Hengsdijk, H., Asseng, S., Van Ittersum, M. K., Kahiluoto, H., & Van Keulen, H. (2014). Climate variability and change in the Central Rift Valley of Ethiopia: Challenges for rainfed crop production. *Journal of Agricultural Science, 152*(1), 58–74.

Kaur, B., Sidhu, R. S., & Vatta, K. (2010). Optimal crop plans for sustainable water use in Punjab. *Agricultural Economics Research Review, 23*, 273–284.

Kifle, M., Gebremicael, T. G., Girmay, A., & Gebremedihin, T. (2017). Effect of surge flow and alternate irrigation on the irrigation efficiency and water productivity of onion in the semi-arid areas of North Ethiopia. *Agricultural Water Management, 187*, 69–76.

Kijne, J., Barker, R., & Molden, D. (Eds.) (2003). *Water productivity in agriculture: Limits and opportunities for improvement: Comprehensive assessment of water management in agriculture*, Series No. 1. CABI press.

Kroes, J., van Dam, J., Bartholomeus, R. P., Groenendijk, P., Heinen, M., Hendriks, R., Mulder, H., Supit, I., & van Walsum, P. E. (2017). *SWAP version 4; Theory description and user manual Wageningen, Wageningen environmetal research report*, 2780 (p. 244).

Ma, Y., Feng, S., & Song, X. (2015). Evaluation of optimal irrigation scheduling and groundwater recharge at representative sites in the North China Plain with SWAP model and field experiments. *Computers and Electronics in Agriculture, 116*, 125–136.

Mebrie, D. W., Assefa, T. T., Yimam, A. Y., & Belay, S. A. (2023). A remote sensing approach to estimate variable crop coefficient and evapotranspiration for improved water productivity in the Ethiopian highlands. *Applied Water Science, 13*(8).

Meskelu, E., Woldemichael, A., & Hordofa, T. (2017). Effect of moisture stress on yield and water use efficiency of irrigated wheat (Triticum aestivum L.) at Melkassa, Ethiopia. *Academy Research Journal Agriculture Science Research, 5*(2), 90–97.

Molden, D., Oweis, T., Steduto, P., Kijne, J., Hanjra, M., Bindraban, P., Bouman, B. A. M., Cook, S., Erenstein, O., Farnahi, H., Hachum, A., Mahoo, H., Nangia, V., Penden, D., Sikka, A., Silva, P., Turral, H., Upadhyaya, A., & Zwart, S. (2007). Pathways for increasing agricultural water productivity. In D. J. Molden, (Ed.), *Water for food, water for life: The comprehensive assessment*. Earthscan, London, and IWMI, Sri Lanka.

Molden, D., Oweis, T., Steduto, P., Bindraban, P., Hanjra, M. A., & Kijne, J. (2010). Improving agricultural water productivity: Between optimism and caution. *Agricultural Water Management, 97*(4), 528–535.

Molden, D., & Sakthivadivel, R. (1999). Water accounting to assess use and productivity of water. *International Journal of Water Resources Development, 15*(1–2), 55–71.

Mualem, Y. (1976). A new model for predicting the hydraulic conductivity of unsaturated porous media. *Water Resources Research, 12*(3), 513–522.

Muluneh, A. (2020). Impact of climate change on soil water balance, maize production, and potential adaptation measures in the Rift Valley drylands of Ethiopia. *Journal of Arid Environments, 179.*

Nagaz, K., Masmoudi, M. M., & Ben Mechlia, N. (2012). Yield response of drip-irrigated onion under full and deficit irrigation with saline water in arid regions of Tunisia. *ISRN Agronomy, 2012*, 1–8.

Orke, Y. A., & Li, M. H. (2022). Impact of climate change on hydrometeorology and droughts in the bilate watershed, Ethiopia. *Water (Switzerland), 14*(5), 729.

Padhiary, J., Swain, J. B., & Patra, K. C. (2020). Optimized irrigation scheduling using SWAT for improved crop water productivity. *Irrigation and Drainage, 69*(3), 387–397.

Qureshi, A. S., Ahmad, W., & Ahmad, A. F. A. (2013). Optimum groundwater table depth and irrigation schedules for controlling soil salinity in central Iraq. *Irrigation and Drainage, 62*(4), 414–424.

Ronco, P., Zennaro, F., Torresan, S., Critto, A., Santini, M., Trabucco, A., Zollo, A. L., Galluccio, G., & Marcomini, A. (2017). A risk assessment framework for irrigated agriculture under climate change. *Advances in Water Resources, 110*, 562–578.

Schneider, A. D., & Howell, T. A. (2001). Scheduling deficit wheat irrigation with data from an evapotranspiration network. *Transactions of the American Society of Agricultural Engineers, 44*(6), 1617–1623.

Singh, R., Van Dam, J. C., & Feddes, R. A. (2006). Water productivity analysis of irrigated crops in Sirsa district, India. *Agricultural Water Management, 82*(3), 253–278.

Spitters, C. J. T., Van Keulen, H., & Van Kraalingen, D. W. G. (1989). A simple and universal crop growth simulator: SUCROS'87. In R. Rabbinge, S. A. Ward, & H. H. Van Laar, (Eds.), *Simulation and systems management in crop protection, Simulation Monographs, Pudoc, Wageningen, The Netherlands* (pp. 147–181).

Supit, I., Hooijer, A. A., & van Diepen, C. A. (Eds.) (1994). *System description of the WOFOST 6.0 crop simulation model implemented in CGMS Vol. 1: Theory and algorithms* (vol. 146, pp. 1, 181). EUR Publication 15956, Agricultural Series, Luxembourg.

Van Dam, J. C., Singh, R., Bessembinder, J. J. E., Leffelaar, P. A., Bastiaanssen, W. G. M., Jhorar, R. K., Kroes, J. G., & Droogers, P. (2006). Assessing options to increase water productivity in irrigated river basins using remote sensing and modelling tools. *International Journal of Water Resources Development, 22*(1), 115–133.

van Genuchten, M. T. (1980). A closed-form equation for predicting the hydraulic conductivity of unsaturated soils. *Soil Science Society of America Journal, 44*(5), 892–898.

Vazifedoust, M., van Dam, J. C., Feddes, R. A., & Feizi, M. (2008). Increasing water productivity of irrigated crops under limited water supply at field scale. *Agricultural Water Management, 95*(2), 89–102.

Verma, A. K., Gupta, S. K., & Isaac, R. K. (2012). Long-term cyclic irrigation in subsurface drained lands: Simulation studies with SWAP. *Journal of Agricultural Science, 5*(1).

Xu, X., Sun, C., Qu, Z., Huang, Q., Ramos, T. B., & Huang, G. (2015). Groundwater recharge and capillary rise in irrigated areas of the upper Yellow River basin assessed by an agro-hydrological model. *Irrigation and Drainage, 64*(5), 587–599.

Xue, J., & Ren, L. (2017). Assessing water productivity in the Hetao irrigation district in inner mongolia by an agro-hydrological model. *Irrigation Science, 35*(4), 357–382.

Zhao, H., Xiong, Y. C., Li, F. M., Wang, R. Y., Qiang, S. C., Yao, T. F., & Mo, F. (2012). Plastic film mulch for half growing-season maximized WUE and yield of potato via moisture-temperature improvement in a semi-arid agroecosystem. *Agricultural Water Management, 104*, 68–78.

Chapter 25
Analysis of Future Demand and Resources in Abu Dhabi: A VAR Model Approach

Mahmoud E. Ahmed, Salem Issa, and Mohamed A. Hamouda

Abstract This study conducted a detailed analysis of water resource management in Abu Dhabi, projecting the region's water demand and supply amidst rapid urbanization and demographic shifts. This research incorporated historical data on population dynamics, water consumption, temperature fluctuations, and groundwater availability to forecast future scenarios in Abu Dhabi. In the study area, Abu Dhabi Distribution Company (ADDC) currently serves 341,162 registered customers, with anticipated significant growth in the industrial and commercial sectors over the next decade. A novel framework for water demand forecasting was utilized, employing historical trends and a refined Vector Autoregressive (VAR) model, which predicts demand and supply patterns in the study area until 2060. The research objectives were to forecast water demand and evaluate the sustainability of water resources under current consumption trends and demographic growth. By integrating historical data on population dynamics, water consumption, temperature variations, and groundwater availability, we developed a comprehensive model that allows for robust scenario planning. The model predicts that by 2050, the estimated water usage will be 875.1 Million Imperial Gallons per Day (MIGD), while the amount of groundwater available will be 474.5 km^3. By 2060, water consumption is expected to rise to 994.9 MIGD, with available groundwater resources declining to 416.5 km^3. Integrating demographic trends with modeling techniques allows decision-makers to

M. E. Ahmed (✉) · M. A. Hamouda
Civil and Environmental Engineering Department, College of Engineering, United Arab Emirates University, Al Ain P.O. Box 15551, UAE
e-mail: 700042847@uaeu.ac.ae

M. A. Hamouda
e-mail: M.hamouda@uaeu.ac.ae

S. Issa
Department of Geosciences, College of Science, United Arab Emirates University, P.O. Box 15551, Al Ain, UAE
e-mail: Salem.essa@uaeu.ac.ae

M. A. Hamouda
National Water and Energy Center, United Arab Emirates University, Al Ain P.O. Box 15551, UAE

A. Sefelnasr et al. (eds.), *Water Resources Management and Sustainability*, Water Science and Technology Library 114, https://doi.org/10.1007/978-3-031-80520-2_25

comprehend water demand patterns more comprehensively, make informed decisions on water resource allocation, and formulate effective water management strategies. These findings transcend Abu Dhabi, offering a valuable framework for arid regions globally facing analogous water resource management challenges and promoting data-driver approaches for regions with limited water resources.

Keywords Arid region · Climate · GIS · Geospatial modeling · Water management

25.1 Introduction

Water sustainability is a critical global concern, particularly in arid areas where water scarcity is an ongoing issue. These geographical areas encounter distinct difficulties as a result of their constrained water resources and elevated rates of evaporation (Momejian et al., 2019). Such regions are typified by the United Arab Emirates (UAE) Abu Dhabi region, which has low humidity, arid terrain, and only 78 mm of annual rainfall (Evolution of the Rainfall Regime in the United Arab Emirates—ScienceDirect, n.d.). Despite the prevalent challenge of water scarcity, the country and the region demonstrate a notably elevated level of water consumption, positioning them among the highest globally, with 650 L per capita daily (Elkaftangui & Mohamed, 2019). The UAE predominantly relies on groundwater reserves and alternative sources, particularly desalination, for its freshwater resources. The predominant use of groundwater is primarily for agricultural purposes, accounting for approximately 80% of total consumption (*Abu Dhabi Distribution Company (ADDC)*, n.d.; Qureshi, 2020).

Based on the continuation of prevailing behavioral consumption patterns and water management strategies, it is anticipated that Abu Dhabi will encounter challenges in fulfilling its projected water requirements by 2060. The claim is grounded on the assumption that several factors will impose substantial strain on the area's water resources, potentially resulting in severe water scarcity. Also, Abu Dhabi region is distinguished by its low-lying, arid topography and imposing sand dunes, which present a unique set of obstacles. The region's water resources face continuous strain due to an annual evaporation rate of 2–3 m and average temperatures ranging from 12 °C to 44 °C (Chowdhury et al., 2016; Darwish et al., 2021).

Despite existing studies on water demand in the UAE, this research introduces a VAR model that incorporates both short-term fluctuations and long-term trends in climate variables, population dynamics, and water consumption behaviors, offering a more integrated approach to understanding the interdependencies of these factors. By integrating the historical data, the model offers a comprehensive and nuanced perspective on water demand dynamics in the arid region. The approach enhances predictive accuracy by adjusting to policy and environmental conditions shifts, filling a critical gap in the continuous adaptation of water resource management strategies under changing climatic scenarios. This model's ability to incorporate real-time data updates and adapt its forecasts accordingly represents a significant advancement compared to other forecasting models. The results of this study are positioned to offer significant knowledge for policymakers, urban planners, and water resource managers in the UAE, providing a solid basis for making well-informed decisions. As water scarcity continues to be a global concern, the findings of this research pave the way for sustainable water management strategies in arid environments, with implications for similar regions. Specifically, the study focused on the following research questions:

RQ1—Will the region satisfy its water demand in 2050 and 2060?
RQ2—What are the proposed strategies to enhance water resilience in the region?
RQ3—What are the research gaps in the field of VAR modeling?

25.2 Background

Vector autoregression (VAR) is a statistical model that examines the interdependencies among various time series variables (Han et al., 2023). It is a popular tool for forecasting time series data, as it can capture complex systems' dynamic and interdependent nature (Garcia-Vega et al., 2020). Within the framework of water demand forecasting, VAR models can be used to examine the link between many influencing factors, such as population and weather, and water demand. By understanding how these factors impact water demand, VAR models can generate accurate forecasts of future demand, which is crucial for water resource management and planning (Zhao et al., 2022). A notable advantage of VAR models is their capacity to simultaneously handle several time series variables (Azhar et al., 2021). Furthermore, VAR models can be adapted to incorporate exogenous variables, which affect the system but are not directly included in the model's equations.

A study using VAR in arid contexts has demonstrated significant capability in detecting land-use variation with dam water levels, underscoring its robustness in capturing the impacts of land-use changes on water levels in arid (Ouma et al., 2022). However, while the VAR model excelled in correlating land-use changes with dam water levels, it encountered limitations in adequately capturing the influence of climate factors. Specifically, the stochastic VAR model could not establish a significant correlation between rainfall and temperature with the dam water levels without integrating four specific climate indices. This finding highlights the challenges in modeling the complex interdependencies between climate variables and water levels using standard VAR approaches, indicating a need for enhanced modeling techniques that can more effectively integrate multiple climate indices (Ouma et al., 2022). This study exemplifies the strengths and limitations of using VAR models to understand the multifaceted influences on water resources in arid environments.

Another study focused on causality analysis of groundwater dynamics in a semi-arid basin using a VAR model and investigated the causal relationships between observed groundwater levels (Mangiarotti et al., 2012). The analysis identified the predominant role of rainfall in the overall model structure, with varying contributions across different regions within the basin. Additionally, the study uncovered causal relationships between different parts of the basin, highlighting the importance of understanding the dynamics of groundwater interactions (Mangiarotti et al., 2012). The findings also emphasized the efficiency of the VAR model in detecting causal links within the network of groundwater measurements, particularly concerning the time scale for identifying these links (Mangiarotti et al., 2012). The study successfully derived an optimal model structure by employing a structural VAR model and applying model restrictions based on available knowledge. This process allowed for the elimination of spurious accumulations and ensured the adequacy of the VAR model in the specific context. Overall, the study's results underscored the potential of VAR modeling as a valuable tool for understanding causal relationships and dynamics in hydrological systems, particularly in arid regions.

A third study on rainfall and run-off modeling in an arid region confirmed the effectiveness of the VAR model in predicting rainfall and run-off as a joint time series (AL-Hadrami & Bahrawi, 2018). The First Order VAR model was found to be the most suitable for short-term forecasting of monthly rainfall and run-off levels (AL-Hadrami & Bahrawi, 2018). The study confirms that each time series in the existing measurement stations is stationary, supporting the applicability of the VAR model in this context. The results highlight the potential of the VAR model as an effective tool for rainfall and run-off modeling in arid and semiarid zones, contributing to the development of water resource systems in these challenging environments (AL-Hadrami & Bahrawi, 2018). The practical utility of the VAR model provides a valuable tool for practical solutions in water resource assessment, particularly in regions with limited data availability for storm run-off (AL-Hadrami & Bahrawi, 2018). The study suggests that the adoption of VAR modeling in hydrological studies has shown promise in arid contexts.

The VAR model has been extensively validated in environmental and hydrological studies due to its ability to handle complex interactions among multiple variables. Studies demonstrated that VAR models could explain over 80% of the variance in multi-variable systems involving economic indicators and environmental factors, highlighting its suitability for scenarios with intertwined data streams (Cao et al., 2023; Pandya et al., 2023). Moreover, VAR models are particularly advantageous for interpreting the impacts of policy changes or natural disturbances across different sectors of an ecosystem or economy.

In summary, the VAR model is widely recognized for its robustness in analyzing and forecasting systems where several time series variables interact. This model is particularly beneficial in environmental and hydrological studies, where it is essential to understand how variables influence each other over time. The VAR model's versatility in incorporating multiple time series data empowers researchers to capture the dynamic interdependencies among these factors. This comprehensive view is crucial for accurate forecasting and effective policy-making in resource management, instilling confidence in its applicability.

25.3 Methodology

The results provided in this study are derived from a VAR model constructed using historical data from 2003 to 2023. Moreover, the research facilitated the determination of the study region's capacity to meet its anticipated water requirements for the years 2050 and 2060.

25.3.1 Data Collection

ADDC, Meteoblue, and World Population Review served as the primary data sources for the analysis spanning from 2003 to 2023. In the study area, ADDC directly interacts with water and power consumers through the planning, management, ownership, and operation of network distribution assets (Abu Dhabi Distribution Company (ADDC), n.d.). The company guarantees the provision of water and electricity to over 600,000 service points via an innovative, resilient, efficient, and sustainable distribution network (Abu Dhabi Distribution Company (ADDC), n.d.). ADDC provided the historical data for the groundwater availability and water consumption from (5-Year Planning Statement 2023–2027 (Water).Pdf, n.d.; Spatial and Temporal Changes of Groundwater Storage in the Quaternary Aquifer, UAE, n.d.). ADDC has a strong reputation for providing reliable and accurate data, as governmental entities regulate it and operate under strict guidelines and standards. Additionally, ADDC's role as a critical player in the water and power sector in Abu Dhabi further enhances the credibility of the information provided. During the study period, there were some instances of missing values in the groundwater availability data where interpolation was utilized to address this. Additionally, the historical data relating to the temperature was acquired from Meteoblue (Climate Change Abu Dhabi, n.d.). Meteoblue is known for using advanced meteorological models, satellite observations, and ground-based data to provide accurate and up-to-date information, further strengthening its credibility. Lastly, the demographic information was acquired from the World Population Review (Abu Dhabi Population, 2023, n.d.). The World Population Review is a well-known source that provides accurate and up-to-date data on population demographics. The website is regularly cited by academic institutions, government agencies, and media organizations, which further validates its credibility. Additionally, World Population Review uses reliable sources, such as official government reports and international organizations, to gather their data, ensuring its accuracy and reliability.

In the data collection phase, the approach involved aggregating historical data from the mentioned reliable sources. Groundwater availability and water consumption data were sourced directly from ADDC, which utilizes advanced telemetry and sensor networks to monitor real-time water usage and resource levels. Additionally, temperature and demographic data were obtained from Meteoblue and the

World Population Review, respectively, which employ satellite data and population modeling techniques to provide updated and accurate information.

The accuracy of the three data sources makes them dependable and offers a solid foundation for examining how the different factors influence the consumption of water in Abu Dhabi region. The synergy of these reliable sources enhances the overall robustness of the study, instilling confidence in the findings and subsequent implications.

25.3.2 Data Preprocessing

A meticulous preprocessing phase was undertaken to ensure the integrity and accuracy of the data used in the VAR model. In the data preprocessing phase, a series of rigorous methods were employed to ensure our data's integrity and accuracy. This included using statistical outlier detection to identify and remedy anomalous data points, which could skew analysis results. Moreover, missing data points were addressed through interpolation methods, specifically using time series analysis techniques such as moving averages and spline interpolation to estimate missing values based on available data trends. This not only preserved the continuity of our datasets but also maintained their temporal integrity, which is crucial for the accuracy of time series modeling in a VAR framework.

25.3.3 Data Analysis

This paper considered evaluating the identified factors by identifying the vertical patterns, studying the correlations between the factors, and investigating the environmental implications of the factors. The growing population magnifies the need for water, worsening the depletion of groundwater and the increase of wastewater and urban runoff, leading to the deterioration of water quality. Simultaneously, increasing temperatures enhance evaporation rates and modify precipitation patterns, resulting in more intense droughts or flooding incidents that pose challenges to water management and ecological equilibrium. Excessive water usage, particularly in the agricultural and industrial sectors, results in unsustainable extraction methods, depleting groundwater levels. Moreover, the declining availability of groundwater leads to the loss of habitats, sinking of land, and higher levels of salt in soil and water, which substantially threaten agricultural output and natural ecosystems. Following the aforementioned, the results of the model were analyzed. Figure 25.1 illustrates the framework for the study.

The study investigated several factors influencing future water demand, intending to establish a solid basis for predicting water demand in 2050 and 2060. Throughout the analysis, population dynamics became apparent as a significant driver, exhibiting clear trends. The population's increase or decrease directly impacts the overall water

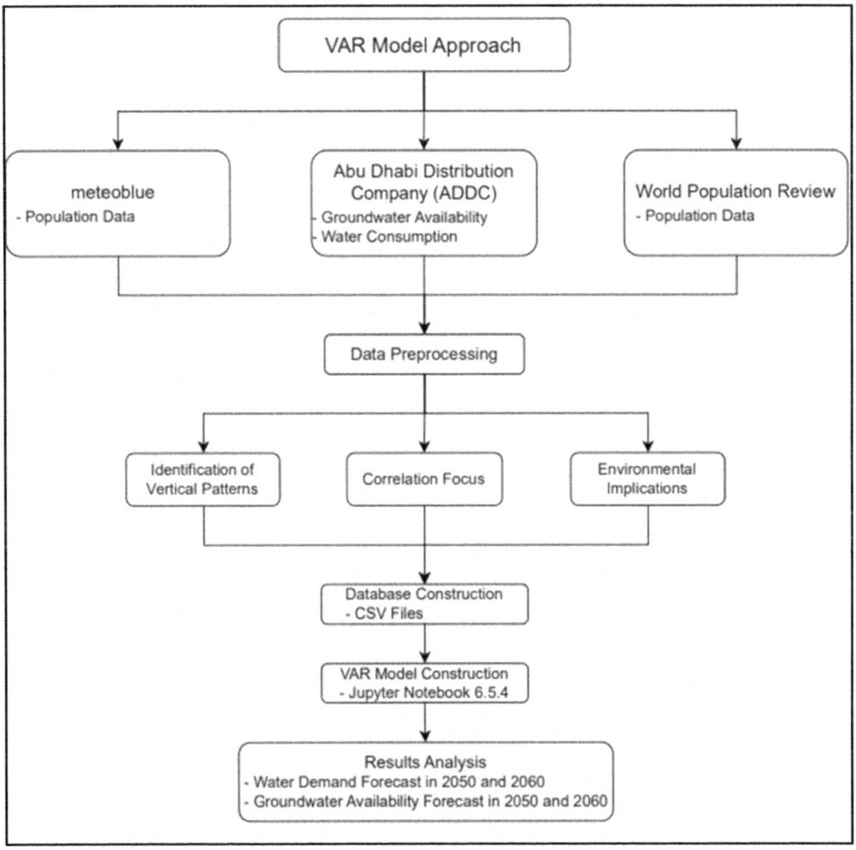

Fig. 25.1 The framework of the study

consumption patterns, making it a crucial factor in predicting future demand (He et al., 2021a, 2021b). Through the careful examination of population dynamics, the study not only assesses the present condition but also predicts future changes in demographics, allowing for a more precise estimation of water requirements in the upcoming decades.

Temperature fluctuations were another critical factor being examined, indicating the influence of climate change on water resources. Temperature variations have wide-ranging effects on the patterns of precipitation, rates of evaporation, and the overall availability of water (He et al., 2021a, 2021b). Given the unprecedented changes in global temperatures, it is crucial to accurately comprehend these variations to predict changes in regional water demand. The study's emphasis on temperature trends enhances our understanding of the complex relationship between climate dynamics and water resource management.

Another essential component of the analysis is the patterns of water consumption. Precise comprehension of these details is crucial for developing efficient water management plans, particularly when combined with the projected shifts in population and climate.

The availability of groundwater is crucial in linking all the factors mentioned above to the study's primary goal, which is to predict water demand for the years 2050 and 2060. Groundwater availability, being a finite and vital resource, is directly impacted by population growth, temperature variations, and water consumption patterns. The study examines the complex dynamics of groundwater resources, analyzing how they react to the identified patterns and changes. This approach enables a comprehensive evaluation of the sustainability of groundwater reserves in meeting future demands.

The comprehensive examination of population dynamics, temperature fluctuations, water usage, and groundwater accessibility between 2003 and 2023 forms the foundation for predicting future water demands in the upcoming years.

25.4 Results and Discussion

25.4.1 Identification of Vertical Patterns

Through a comprehensive analysis spanning from 2003 to 2023, we discerned prominent vertical patterns in critical variables, notably population dynamics, temperature variations, and water usage. The time series plots in the study were utilized to visually represent and analyze the longitudinal evolution of critical variables, including population dynamics, temperature variations, and water consumption rates. This approach allows us to observe and quantify the changes and trends that occur over the analyzed period, providing insights into how these factors behave and interact over time. Such analysis is crucial for understanding the temporal dynamics and dependencies within the system, which are essential for accurate forecasting and effective resource management. Figure 25.2 illustrates the studied factors' behavioral patterns concerning the study period.

Population dynamics data from 2003 to 2023 reveals a steadily increasing vertical pattern. Population growth has been steady from 590,288 in 2003 to 1,566,999 in 2023. The absence of population data peaks or fluctuations further proves this. Demographic growth, which affects water demand, is behind this steady rise. This stability is crucial for water demand forecasting, allowing for a more reliable projection of future needs. While growth is steady, some years may have slightly higher rates, requiring further analysis.

The water consumption data exhibit a recognizable vertical pattern with time. The values vary between 287.23 and 517.92 Imperial gallons per day (MIGD), indicating increased water consumption patterns. Upon initial examination, it is evident that there is an increasing trend from 2003 to approximately 2013, with the highest point being 536.59 MIGD. Following this peak, there is a subsequent decline. The initial increase and subsequent decrease in water consumption suggest possible changes in regional water demand patterns. The previous upward trend may correspond with

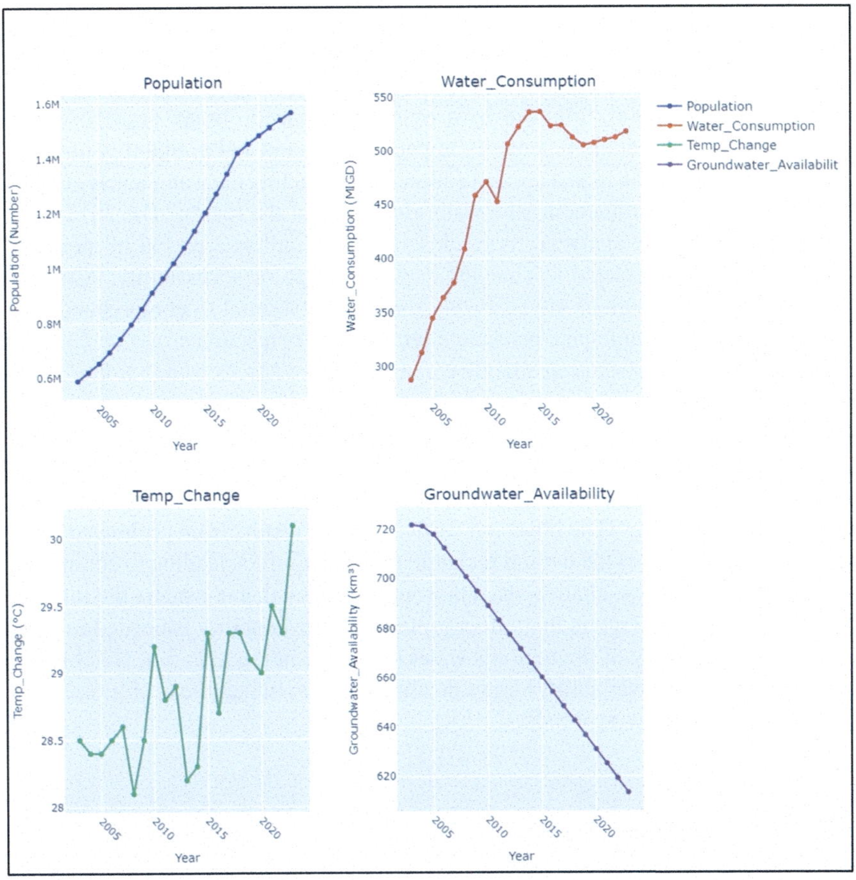

Fig. 25.2 Features over the years

the growth of the population and the expansion of industrial activities, increasing water demand. On the other hand, the subsequent decrease could indicate effective water conservation efforts, alterations in industrial procedures, or changes in population patterns. The data indicates a possible highest point in 2013, highlighting the importance of that specific year concerning increased water consumption. Additional examination of the factors surrounding this peak, such as simultaneous surges in population, expansions in industry, or climatic occurrences, would offer valuable context for comprehending the observed pattern.

The annual mean temperature data reveals a noticeable vertical pattern characterized by slight fluctuations. The temperatures fluctuated between 28.1 °C and 30.1 °C during the analyzed period, indicating a slight variation in the local climate. The general pattern indicates a relatively consistent environment, with no significant deviations from the average annual value. The consistent temperature observed here aligns with the overall context of climate stability, which is essential for comprehending its influence on water resources and usage patterns. Although the temperature data does not exhibit any distinct peaks or abrupt fluctuations, it is crucial to acknowledge the gradual rises observed in specific years. An incremental increase can be observed between 2003 and 2006, followed by a more pronounced climax in 2012. These moderate fluctuations could potentially impact the demand and availability of water. Higher temperatures can increase evaporation rates, affecting the water levels in reservoirs and potentially influencing the water dynamics of the region. The 2012 peak, specifically, should be analyzed more closely, as it signifies a period of increased warmth that may be related to changes in water usage or accessibility. Although the temperature fluctuations are relatively mild, it is essential to comprehend their subtleties in order to anticipate possible effects on water resources.

The data on groundwater availability, ranging from 721.75 to 613.6 km^3, indicates a significant decrease in the amount of groundwater resources available. The decreasing availability is causing significant concerns regarding the long-term water sustainability in the region. The progressive decline in groundwater availability corresponds to the broader understanding that excessive extraction or continuous demand depletes these resources. The observed vertical pattern results from prolonged water extraction that surpasses the natural recharge rates of aquifers, leading to cumulative impact. To address these challenges, a thorough strategy that considers the intricate interaction of various factors that affect groundwater availability is needed to guarantee the sustainability of the region's water resources in the long run. Notably, the data on groundwater availability shows no abrupt peaks; rather, the decline seems to follow a steady and steady trend.

25.4.2 *Correlation Focus*

The correlation analysis uncovers notable relationships among critical variables in Abu Dhabi's water ecosystem. Figure 25.3 shows the correlation focus between the factors considered in this study.

The correlation analysis uncovers notable relationships among key variables in Abu Dhabi's water ecosystem. The population exhibits a strong positive correlation with water consumption (0.844) and a positive link with temperature change (0.747). Remarkably, population growth is inversely correlated with groundwater availability (− 0.997). Water consumption demonstrates a moderate positive correlation with temperature change (0.523) but negatively correlates with groundwater availability (− 0.836). Temperature change also significantly negatively correlates with groundwater availability (− 0.760). These insights provide a foundation for the Vector Autoregressive (VAR) model, aiming to forecast water dynamics in Abu Dhabi until 2060 by considering the interplay of these vital factors.

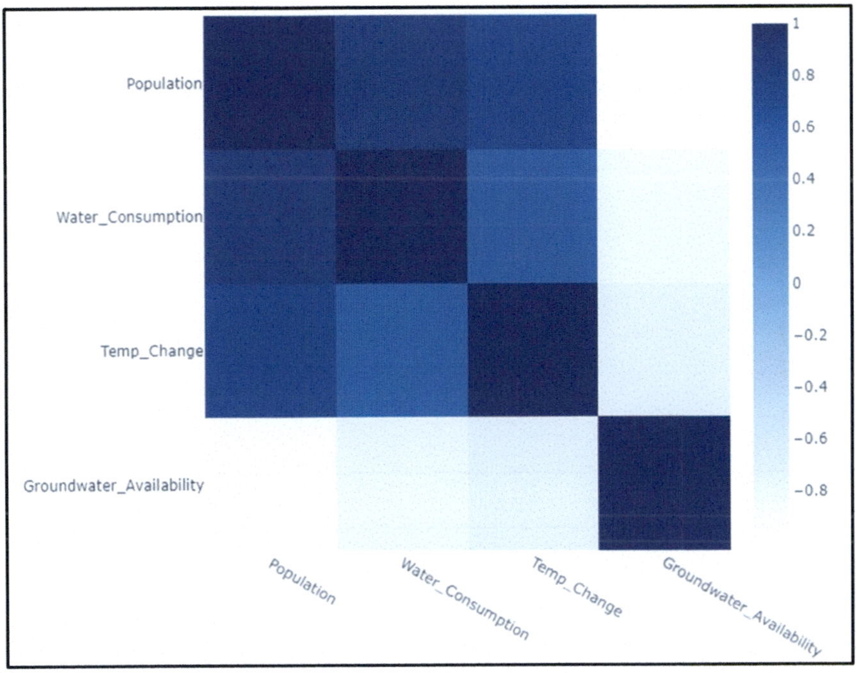

Fig. 25.3 Correlation focus

25.4.3 VAR Model Forecasts

The correlations and patterns identified played a pivotal role in shaping the VAR model's foundation. These critical insights, coupled with a comprehensive understanding of the interplay between population, temperature, and water use, provided a robust framework for forecasting. The VAR model was set up to incorporate population and temperature changes and slack qualities for groundwater supply and water use. Quantifiable standards and model diagnostics do not wipe out the inactivity.

The model utilized data from 2003 to 2023, and testing data was utilized to assess how well it was introduced. The model's strength was evaluated using a range of genuine measurements and assessments. The VAR model played a pivotal role in anticipating changes in water usage, effectively capturing the nuances influenced by climate variations and population growth. Also, the VAR model demonstrated remarkable accuracy in predicting shifts in water availability, accounting for the impacts of temperature fluctuations and population dynamics. The findings of the model are illustrated in Fig. 25.4.

The VAR model provided projections for the region's water sustainability, specifically forecasting water consumption and the availability of groundwater resources for the years 2050 and 2060. The visual representation illustrates an upward trajectory in consumption juxtaposed with a declining trend in availability. The model predicts

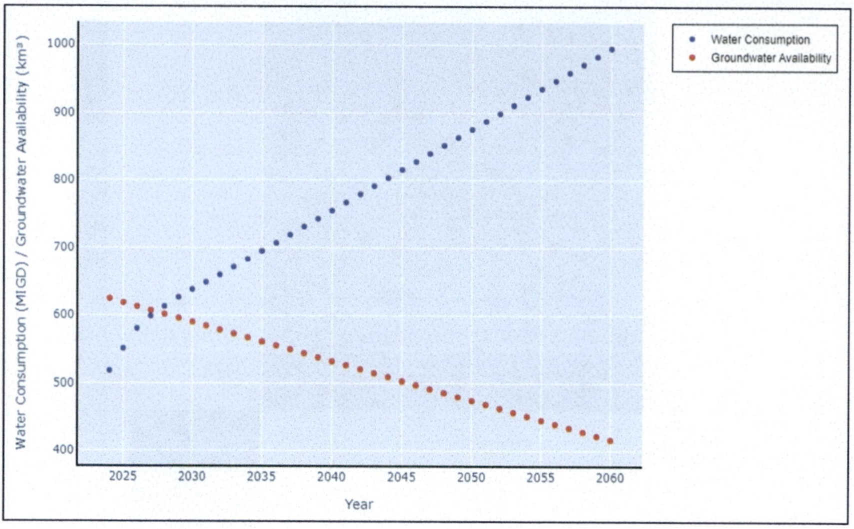

Fig. 25.4 Forecasted water consumption (2024–2060)

that in 2050 the estimated water usage will be 875.1 MIGD, while the amount of groundwater available will be 474.5 km^3. By 2060, the VAR model predicts that water consumption will rise to 994.9 MIGD, while the available groundwater resources are expected to decline to 416.5 km^3. According to the forecast, the available groundwater resources are expected to meet the water consumption by 2060. Nevertheless, the declining accessibility of groundwater gives rise to concerns and underscores the necessity for implementing sustainable water management strategies to maintain a stable balance between water availability and consumption.

25.4.3.1 OLS Regression

The Ordinary Least Squares (OLS) regression analyzed the model's factors more thoroughly. The results provided significant insights into the model's ability to predict using population, temperature change, and groundwater availability. The analysis suggests a moderate degree of effectiveness in predicting future water usage. The R-squared value of the model, which is 0.738, indicates that about 73.8% of the variation in water consumption can be explained by the selected indicator variables. The adjusted R-squared, which incorporates the number of predictors, is 0.691. The F-measurement of 15.94, along with a remarkably low p-value of 3.45e-05, strengthens the statistical significance of the model. Upon closer examination of the individual coefficients, it is observed that only the coefficient for groundwater availability is considered statistically significant. The *p*-value for temperature change is 0.941, while the p-value for population is 0.475. This implies that although the overall model is perceptive, the particular impacts of temperature change and population may have constraints in explaining variations in water consumption. This comprehensive analysis indicates that although the overall model offers valuable insights, it is essential to proceed with caution when interpreting the separate impacts of temperature change and population on variations in water consumption.

25.5 Discussions, Limitations and Research Gaps

25.5.1 Environmental Implications

Although the VAR model provided valuable predictions for water consumption and groundwater availability, adopting a more comprehensive and nuanced approach when analyzing forecasts is crucial, especially considering broader environmental implications.

The continuous population growth highlights the population dynamics and its environmental implications, which exert additional pressure on ecosystems, natural habitats, and biodiversity (Onyango et al., 2021). Urbanization and infrastructure development to support a growing population can result in fragmentation and loss of habitats, contributing to the deterioration of local ecosystems (Onyango et al., 2021). Consequently, this can worsen the exhaustion of resources and present obstacles to maintaining environmental sustainability.

Additionally, the rates at which water is consumed also directly impact ecosystems and aquatic habitats, especially as demand increases (Ayaz et al., 2022). Overly excessive water withdrawal can result in the exhaustion of surface water reservoirs, which has a detrimental impact on aquatic ecosystems and the plant and animal life that rely on them (Ayaz et al., 2022). In addition, increased water usage can potentially lead to elevated pollution levels. This occurs when runoff from urban and industrial regions carries contaminants into water bodies, negatively affecting water quality and the diversity of aquatic life (Yan et al., 2023).

Furthermore, fluctuations in temperature present environmental difficulties, which can have consequences for ecosystems, biodiversity, and the overall equilibrium of the environment (Roy et al., 2023). Temperature pattern changes can modify species' behavior and distribution, impacting their survival and disturbing intricate ecological relationships. Moreover, heightened temperatures can enhance the occurrence and intensity of weather phenomena, such as prolonged periods of dryness and extreme heat, thereby exerting a series of impacts on ecosystems and presenting difficulties for adaptation.

Lastly, the decreasing availability of groundwater has significant environmental implications. It impacts the stability of nearby ecosystems and increases the likelihood of land subsidence and saltwater intrusion (Chen et al., 2022). Diminished groundwater levels can result in the desiccation of wetlands and the depletion of vital habitats for diverse species. Furthermore, the decrease in groundwater availability may heighten the competition for water resources, thereby worsening environmental degradation.

25.5.2 Strategies Enhancing Water Resilience

In the past few years, various governmental entities, including ADDC, have implemented several strategies to enhance the efficiency and sustainability of the water supply system, minimize water losses, and encourage water conservation (5-Year Planning Statement 2023–2027 (Water).Pdf, n.d.; Ayaz et al., 2022; Mohamed et al., 2020).

One of the main strategies outlined involves restoring and renovating current pumping stations and water tanks. This will prolong the lifespan of assets and ensure their optimal functionality. The current process involves conducting a condition assessment of the water distribution network to identify pipes that require rehabilitation or replacement in the most cost-effective manner.

Another crucial strategy suggested is adopting a water balancing and accounting system, which will aid in evaluating the quantities of non-revenue water and effectively controlling water losses in the potable water distribution system. This entails the installation of interface meters, district meters, and sector meters to efficiently perform water balancing for primary feeder lines, pressure ring mains, and most of the distribution systems.

Additional demand-side management strategies included awareness-raising campaigns, programs to modify behavior, initiatives to optimize irrigation practices, and efforts to enhance industrial efficiency.

25.5.3 Limitations

While this study has provided insights, it is vital to recognize specific limitations that could affect the interpretation and applicability of the results. First, the precision of the water forecast is primarily dependent on the assumptions and parameters integrated into the VAR model. Possible biases may arise in the projections due to uncertainties or inaccuracies in the input data or model assumptions.

Furthermore, the study's emphasis on particular factors, such as population dynamics, temperature fluctuations, water consumption rates, and groundwater availability, may overlook other pertinent variables that could impact water demand and supply dynamics. The intricate nature of environmental systems poses a challenge in fully encompassing all the factors that contribute to them, which could potentially restrict the model's predictive capability.

Furthermore, the study is based on the premise that historical trends will persist without considering potential unforeseen technological breakthroughs or changes in policy. This constraint could affect the model's capacity to predict sudden changes or events that might profoundly influence the course of water demand and availability.

25.5.4 VAR Modeling Research Gaps

Although VAR modeling has proven to be a valuable tool for predicting water consumption and demand, several research areas must be addressed and considered in future studies. First, there is a limit to how many external factors and exogenous variables can be incorporated into traditional VAR models. Environmental and socio-economic factors, regulatory modifications, and the emergence of new technologies can substantially influence water consumption trends. By incorporating these external influences into the modeling framework, generating more reliable and flexible forecasts is possible, thereby overcoming the current constraints of primarily focusing on endogenous variables.

In addition, VAR models assume a linear correlation between variables, disregarding the possibility of non-linear relationships and complex interactions within the water system. Therefore, to accurately represent the system, it is essential to consider non-linearities in water demand responses to factors such as population growth and temperature changes.

Another aspect of the VAR model that needs to be addressed is its temporal dimension. Existing models frequently depend on yearly or monthly data, possibly disregarding shorter-term fluctuations or seasonal variations that can influence water usage. By examining data with higher frequencies and improving the level of detail in the model's time intervals, it is possible to improve its ability to capture more subtle trends and patterns.

Moreover, the presence of uncertainties related to climate change and its influence on temperature fluctuations presents a difficulty for VAR modeling. Integrating climate change projections and their potential impact on water demand could enhance the model's capacity to predict future scenarios amidst evolving environmental conditions.

To improve the reliability of water demand forecasts, future research should prioritize including external factors, address non-linear relationships, investigate smaller time intervals, and account for the uncertainties related to climate change.

25.6 Conclusions

In conclusion, this study applies an enhanced VAR model to predict water demand and supply in Abu Dhabi until 2050 and 2060, offering a data-based basis for strategic water management planning. By conducting a thorough analysis, noteworthy trends and connections among essential factors such as population changes, temperature variations, water usage rates, and groundwater availability were discovered. In accordance with the model, there will be a significant growth in water demand. It estimates that water consumption will increase from 875.1 MIGD in 2050 to 994.9 MIGD by 2060. The amount of available groundwater is anticipated to drastically decrease from 474.5 km^3 in 2050 to 416.5 km^3 in 2060, posing a significant challenge going forward.

The findings underscore the immediate need to implement sustainable water management strategies to tackle the projected water supply and demand deficits. The study recommends proactive measures, such as infrastructure maintenance and rehabilitation, as crucial for minimizing water losses. In addition, it suggests implementing demand-side management initiatives with the goal of decreasing individual water usage and establishing a thorough water balancing and accounting system to handle non-revenue water and enhance distribution network capacity efficiently.

This study provides a quantitative analysis of water resources and predicts future demands. It is a crucial tool for policymakers and urban planners in Abu Dhabi and offers a valuable framework for other arid regions with similar challenges. The numerical projections highlight the complex connections between population growth, climate fluctuations, and water resource management. These projections guide developing strategies that promote the region's long-term sustainability and resilience of water resources.

References

Abu Dhabi Distribution Company (ADDC). (n.d.). Retrieved December 6, 2023, from https://www.addc.ae/en-US/home/Pages/AboutUs.aspx

Abu Dhabi Population 2023. (n.d.). Retrieved December 6, 2023, from https://worldpopulationreview.com/world-cities/abu-dhabi-population

AL-Hadrami, A. S., & Bahrawi, J. (2018). Rainfall and runoff modeling as a joint time series case study: Southwest of Saudi Arabia. *International Journal of Civil Engineering and Technology, 9*(11), 1952–1969.

Ayaz, M., Namazi, M. A., ud Din, M. A., Ershath, M. I. M., Mansour, A., & Aggoune, el-H. M. (2022). Sustainable seawater desalination: Current status, environmental implications and future expectations. *Desalination, 540*, 116022. https://doi.org/10.1016/j.desal.2022.116022

Azhar, M. A. R., Adi Nugroho, H., & Wibirama, S. (2021). The Study of multivariable autoregression methods to forecast infectious diseases. In *2021 IEEE 5th international conference on information technology, information systems and electrical engineering (ICITISEE)* (pp. 83–88). https://doi.org/10.1109/ICITISEE53823.2021.9655881

Cao, L., Yuan, X., Tian, F., Xu, H., & Su, Z. (2023). Forecasting of water consumption by integrating spatial and temporal characteristics of short-term water use in cities. *Physics and Chemistry of the Earth, Parts A/B/C, 130*, 103390. https://doi.org/10.1016/j.pce.2023.103390

Chen, G., Xu, B., Bidorn, B., & Burnett, W. C. (2022). Effects of groundwater extraction and river regulation on coastal freshwater resources. In E. R. Urban Jr. & V. Ittekkot (Eds.), *Blue economy: An ocean science perspective* (pp. 123–152). Springer Nature. https://doi.org/10.1007/978-981-19-5065-0_5

Chowdhury, R., Mohamed, M. M. A., & Murad, A. (2016). Variability of extreme hydro-climate parameters in the north-eastern region of United Arab Emirates. *Procedia Engineering, 154*, 639–644. https://doi.org/10.1016/j.proeng.2016.07.563

Climate Change Abu Dhabi. (n.d.). Meteoblue. Retrieved December 6, 2023, from https://www.meteoblue.com/en/climate-change/abu-dhabi_united-arab-emirates_292968

Darwish, S. M., Saibi, H., Gabr, A., & Mohamed, M. M. (2021). Microgravity monitoring of groundwater dynamics in a shallow aquifer in Al-Ain (Abu Dhabi Emirate, United Arab Emirates): An opportunity for understanding aquifer hydrodynamics in arid regions. *Environmental Earth Sciences, 80*(17), 559. https://doi.org/10.1007/s12665-021-09854-6

Elkaftangui, M., & Mohamed, B. E. (2019). Rethinking the old residential sector in the UAE: A methodology for sustainable retrofitting. *International Journal of Services Technology and Management, 25*(1), 4–17. https://doi.org/10.1504/IJSTM.2019.096600

Evolution of the rainfall regime in the United Arab Emirates—ScienceDirect. (n.d.). Retrieved December 5, 2023, from https://www.sciencedirect.com/science/article/pii/S0022169414003060

Garcia-Vega, S., Zeng, X.-J., & Keane, J. (2020). Stock returns prediction using kernel adaptive filtering within a stock market interdependence approach. *Expert Systems with Applications, 160*, 113668. https://doi.org/10.1016/j.eswa.2020.113668

Han, X., Zhang, L., Wu, Y., & Yuan, S. (2023). *Algorithmic recourse for anomaly detection in multivariate time series*. (arXiv:2309.16896). arXiv. https://doi.org/10.48550/arXiv.2309.16896

He, C., Liu, Z., Wu, J., Pan, X., Fang, Z., Li, J., & Bryan, B. A. (2021). Future global urban water scarcity and potential solutions. *Nature Communications, 12*(1), Article 1. https://doi.org/10.1038/s41467-021-25026-3

He, G., Geng, C., Zhai, J., Zhao, Y., Wang, Q., Jiang, S., Zhu, Y., & Wang, L. (2021b). Impact of food consumption patterns change on agricultural water requirements: An urban-rural comparison in China. *Agricultural Water Management, 243*, 106504. https://doi.org/10.1016/j.agwat.2020.106504

Mangiarotti, S., Sekhar, M., Berthon, L., Javeed, Y., & Mazzega, P. (2012). Causality analysis of groundwater dynamics based on a vector autoregressive model in the semi-arid basin of Gundal (South India). *Journal of Applied Geophysics, 83*, 1–10. https://doi.org/10.1016/j.jappgeo.2012.04.003

Mohamed, M. M., El-Shorbagy, W., Kizhisseri, M. I., Chowdhury, R., & McDonald, A. (2020). Evaluation of policy scenarios for water resources planning and management in an arid region. *Journal of Hydrology: Regional Studies, 32*, 100758. https://doi.org/10.1016/j.ejrh.2020.100758

Momejian, N., Abou Najm, M., Alameddine, I., & El-Fadel, M. (2019). Groundwater vulnerability modeling to assess seawater intrusion: A methodological comparison with geospatial interpolation. *Water Resources Management, 33*(3), 1039–1052. https://doi.org/10.1007/s11269-018-2165-4

Onyango, D. O., Ikporukpo, C. O., Taiwo, J. O., & Opiyo, S. B. (2021). Monitoring the extent and impacts of watershed urban development in the Lake Victoria Basin, Kenya, using a combination of population dynamics, remote sensing and GIS techniques. *Environmental & Socio-Economic Studies, 9*(2), 11–25. https://doi.org/10.2478/environ-2021-0007

Ouma, Y. O., Moalafhi, D. B., Anderson, G., Nkwae, B., Odirile, P., Parida, B. P., & Qi, J. (2022). Dam water level prediction using vector autoregression, random forest regression and MLP-ANN models based on land-use and climate factors. Sustainability (Switzerland), *14*(22). https://doi.org/10.3390/su142214934

Pandya, U., Mudaliar, A., & Gaikwad, A. (2023). Forecasting of banana crop productivity using geospatial approach: A case study of Anand District. In ECWS-7 2023 (p. 38). https://doi.org/10.3390/ECWS-7-14248

Qureshi, A. S. (2020). Challenges and prospects of using treated wastewater to manage water scarcity crises in the gulf cooperation council (GCC) countries. Water, *12*(7), Article 7. https://doi.org/10.3390/w12071971

Roy, S., Debnath, P., Vulevic, A., & Mitra, S. (2023). Incorporating climate change resilience in india's railway infrastructure: challenges and potential. *Mechatron. Intell. Transp. Syst. 2*, 102–116. https://doi.org/10.56578/mits020205

Spatial and Temporal Changes of Groundwater Storage in the Quaternary Aquifer, UAE. (n.d.). Retrieved December 6, 2023, from https://www.mdpi.com/2073-4441/13/6/864

Yan, H., Lin, Y., Chen, Q., Zhang, J., He, S., Feng, T., Wang, Z., Chen, C., & Ding, J. (2023). A review of the eco-environmental impacts of the south-to-north water diversion: implications for Interbasin water transfers. *Engineering*. https://doi.org/10.1016/j.eng.2023.05.012

5-Year Planning Statement 2023–2027 (Water).pdf. (n.d.). Retrieved December 6, 2023, from https://www.addc.ae/content/Publications/5-Year%20Planning%20Statement%202023-2027%20%20(Water).pdf

Zhao, M., Wang, Y., Liu, S., Zhong, P., Liu, H., & Li, R. (2022). Correlation assessment of NDVI and land use dynamics with water resources for the southern margin of Mu Us Sandy Land. *China. Environmental Science and Pollution Research, 29*(12), 17049–17061. https://doi.org/10.1007/s11356-021-16757-3

Chapter 26
The Challenges of Water Management in the Canary Islands (Spain)

Juan C. Santamarta, Noelia Cruz-Pérez, Joselin S. Rodrígez-Alcántara, Jesica Rodríguez-Martín, Miguel Ángel Marazuela, and Alejandro García-Gil

Abstract Islands, due to their limited territory, are also limited in the resources available to them for the development of regular life. Spain can be included here, with an archipelago that is itself an outermost European region—the Canary Islands. The Canary Islands have unique characteristics due to richness in groundwater and the use of desalination to support water demand, which rises primarily due to agriculture and the tourism sector. However, the Canary Islands face significant environmental challenges, particularly in the management and preservation of their water resources. One of the most pressing issues is the contamination of coastal aquifers, primarily due to the intrusion of seawater. This problem is exacerbated by inadequate wastewater treatment facilities and the escalating reliance on desalination processes. The latter, while essential, demands a high energy consumption, posing sustainability and environmental concerns, especially in the context of the global push towards reducing carbon footprints. Therefore, this article analyses the current status of water resources in the Canary Islands, as well as future challenges that will

---

J. C. Santamarta (✉) · N. Cruz-Pérez · J. S. Rodrígez-Alcántara
Departamento de Ingeniería Agraria y del Medio Natural, Universidad de La Laguna (ULL), 38200 Tenerife, Spain
e-mail: jcsanta@ull.es

N. Cruz-Pérez
e-mail: ncruzper@ull.edu.es

J. S. Rodrígez-Alcántara
e-mail: jrodralc@ull.edu.es

J. Rodríguez-Martín
Departamento Técnicas y Proyectos en Ingeniería y Arquitectura, Universidad de La Laguna (ULL), 38200 Tenerife, Spain
e-mail: jrodrima@ull.edu.es

M. Á. Marazuela · A. García-Gil
Geological Survey of Spain (IGME-CSIC), C/ Ríos Rosas 23, 28003 Madrid, Spain
e-mail: ma.marazuela@igme.es

A. García-Gil
e-mail: a.garcia@igme.es

A. Sefelnasr et al. (eds.), *Water Resources Management and Sustainability*, Water Science and Technology Library 114, https://doi.org/10.1007/978-3-031-80520-2_26

arise in a scenario of climate change, where temperatures are expected to increase in Spain.

Keywords Archipelago · Climate change · Desalination · Groundwater · Water management

26.1 Introduction

The use, management and protection of water in a territory must be planned. Historically in Spain, water resources have been managed from the paradigm of increasing water supply (Jodar-Abellan et al., 2019). Increased agricultural extensions combined with massive urbanisation, especially in coastal areas, where there was also a shortage of water resources, increased the demand for water, which was very difficult to satisfy (MITECO, 2019). Water was considered another factor of production. When it was not possible to obtain water naturally, industrial production of fresh water from seawater was used (Santamarta et al., 2022a, 2022b, 2022c). In the worst cases, water was desalinated from abandoned wells, mainly coastal wells that became brackish, thus enhancing the effect of seawater intrusion (Llamas et al., 2015).

In recent years, there has been a paradigm shift in water management and planning. The approach is demand driven, which has meant imposing rational criteria to adjust demand to supply (Cruz-Pérez, et al., 2023a, 2023b). The environmental importance of water and the need for its protection is also highlighted (OCDE, 2015). Finally, a criterion of cost recovery in water use is established. This planning should not be static, but dynamic, and therefore needs to be reviewed periodically; in the case of hydrological plans, the deadline is six years (Escalante et al., 2020). The Water Framework Directive (WFD) is one of the major advances in water resources management for Europe. It has four pillars: management and planning at the scale of the river basin district, economic rationality, achieving good status in all water bodies, and transparency and social participation. Three planning cycles for the WFD have been developed: First planning cycle (2009–2015), Second planning cycle (2015–2021) and Third planning cycle (2021–2027).

The Canary Islands have seven Hydrographic Demarcations formed by intra-community river basin districts (Art. 5 BIS, of Law 10/2010, modification of Law 12/1990), which are the responsibility of the water administrations of the autonomous community. In the Canary Islands, documentation for the second cycle was submitted late, although work is currently underway on the third planning cycle. In this regard, it is worth noting that seven river basin districts exist, one for each island, except for La Graciosa (included in Lanzarote), organized through the Island Water Councils (Santamarta, 2011). Essentially, the hydrological planning process has the following phases: initial documents (environmental and economic characterization, registering of protected areas, identification of pressures); outline of important issues; establishment of objectives; hydrological plan of the demarcation and programme of measures; monitoring and evaluation. Public participation is integrated into the sections of the initial documents, in the outline of important issues, in the hydrological plan and in the programme of measures (CIATF, 2018).

When discussing water management, it is important to bear in mind that water management is also about energy management, especially in those facilities that require high levels of energy to produce water, such as desalination facilities (Santamarta et al., 2022a, 2022b, 2022c). Furthermore, considering that in the Canary Islands most of the energy produced in the archipelago (approximately 80%) comes from the burning of fossil fuels, and only less than 20% comes from renewable energies (Gobierno de Canarias, 2022), it is essential to highlight the importance of reducing greenhouse gas emissions associated with the production of drinking water in the archipelago (Nevado et al., 2016). To combat the impacts of climate change, a profound transformation of the energy and production model is required to eliminate dependence on fossil fuels. Prevention and adaptation to the transformations that have already begun are also necessary. The fight against the effects of this phenomenon is necessarily a cross-cutting policy, given that all areas of society and the economy have a bearing on the emissions indicated and will be affected by their impacts (Fagerberg et al., 2018).

With this in mind and considering the measures adopted by the government of the Canary Islands in its future Canary Islands Law on Climate Change and Energy Transition (BOC, 2023), several key aspects regarding water management in the Canary Islands are brought to the forefront:

(a) Climate change brings with it forecasts such as changes in rainfall patterns and higher temperatures (Millán, 2014). This will inevitably lead to changes in the reserves of water resources, so that operations such as artificial recharge of aquifers must be considered (Nijhawan et al., 2013; Valhondo et al., 2020).
(b) The water footprint of crops is high (Ledgard et al., 2020), and there are numerous studies that seek to minimize the use of potable water for crop irrigation and to increase the use of treated water for this purpose (Gu et al., 2016; Zhi et al., 2015).
(c) In all the Canary Islands (except for the island of La Palma) there are seawater desalination plants for tourist, residential and agricultural uses (González-Morales & Ramón-Ojeda, 2019). Numerous studies have been carried out to ascertain the energy demand of these facilities and to explore energy reductions in this activity (Ruiz-Rosa et al., 2016; Sadhwani et al., 2008).
(d) An increasing number of studies are interested in the natural purification of wastewater for crop irrigation (Gallego-Ayala & Gómez-Limón, 2009; Molleda et al., 2008; Verbyla et al., 2015).

The areas of water consumption in the Canary Islands are usually located in coastal areas. In the case of groundwater obtained by water galleries, these are located at high altitudes (300–2.000 m above sea level). In most cases, these water resources reach the areas of consumption by gravity, which is very positive from the point of view of energy efficiency (Santamarta & Lario-Bascones, 2015). In the case of wells, boreholes and seawater desalination plants, water must be pumped (Cruz-Pérez et al., 2022). This generates a strong water-energy nexus, which implies significant energy consumption linked to water resource management (Cruz-Pérez & Santamarta, 2021). It is in the production of renewable energy where we can alleviate the pressure of this activity on the environment (Schallenberg-Rodríguez et al., 2014), as well as in the correct management of the brine generated in the reverse osmosis process (Leon et al., 2021a, 2021b). Generally, the brine is discharged back into the sea, but it is necessary to ensure adequate dilution of the brine in the marine environment to prevent this salt concentrate from causing environmental damage in the area where it is discharged (Kress et al., 2020).

The main problems related to the quality of groundwater resources can be summarized as follows: contamination due to marine intrusion by chlorides, fundamentally in coastal water extraction areas (Gössling, 2001); contamination due to poor agricultural practices (nitrates) (Santamarta et al., 2022a, 2022b, 2022c); residual volcanic contamination (bicarbonates, sodium and fluoride) (Marrero et al., 2008); and emerging pollutants and microplastics (Gasco Cavero et al., 2023). Anthropic effects due to pumping and overexploitation of the aquifers, related to water quality, include marine intrusion, which produces salinization of coastal aquifers (Custodio et al., 2016). Marine intrusion is progressively evolving in most coastal aquifers and is the main aquifer management problem in these areas (Herrera & Custodio, 2003).

Within the Canary Islands archipelago, a distinction must be made between the eastern and western islands, since the water model varies depending on whether it is one group or the other. The eastern islands (Gran Canaria, Fuerteventura and Lanzarote), due to their proximity to the African continent and their orography (characterized by not having a high relief), have been pioneers in seawater desalination, establishing the first plants of this type in 1964 and 1974, in Lanzarote and Fuerteventura respectively (González-Morales & Ramón-Ojeda, 2019). The western islands (La Gomera, La Palma, El Hierro and Tenerife) are rich in groundwater, exploiting the aquifer mainly through water galleries and wells. The main difference between these two methods is that water galleries do not require a source of energy for their operation, as they have an inclination that favours the extraction of water, whereas wells do require energy to pump water from a lower level to a higher level on the island (Cruz-Pérez et al., 2023a, 2023b). The production of desalinated water has grown significantly in the Canary Islands from the 1970s to the present, from an annual production of < 50 hm^3 (1973) to almost 250 hm^3 in 2013 (Gómez-Gotor et al., 2018).

Therefore, a quantitative review has been conducted of the main sources of water in the Canary Islands—groundwater, surface water, desalinated water and reused water. An analysis was made of the main challenges of each of these water sources and the aspects that need to be improved in order to save energy and water resources, an extremely important factor that is even more relevant in the island context.

26.2 Results and Discussion

26.2.1 Groundwater and Surface Water Bodies

At the beginning of the twenty-first century, the Canary Islands are facing important and vital challenges for the survival of the archipelago with respect to water. Different studies show an irregular rainfall pattern and a general tendency for rainfall to decrease (Tejedor et al., 2017). The role of forests in the islands will be fundamental in this century, as elements that regulate water resources and favour the infiltration of water into the island aquifers and reduce erosion by conserving soils (Martín-Fernández & Martínez-Núñez, 2011).

The aquifers, drained by more than 1700 water galleries and 3000 wells and boreholes, are diminishing. Although seawater desalination is compensating for this reduction, the demands and pressures on subway resources are still high (Santamarta, 2014). It is necessary to order and control the use of subway resources, avoiding their overexploitation (Cruz-Fuentes et al., 2014). Sanitation and water treatment are pending issues in the Canary Islands, although this is being addressed with the construction of new infrastructures, given the threat of sanctions from Europe (Melián et al., 2010).

Surface water resources in the Canary Islands are scarce, accounting for no more than 10% of the overall resources available on the islands, and they are used mainly for agriculture (Santamarta et al., 2021). The islands with the largest number of dams are Gran Canaria and La Gomera, with 69 and 39, respectively (Gonzálvez, 2019).

Groundwater abstractions must be compensated for by natural recharge of the aquifer, which is constituted by a percentage of the water that infiltrates into the subsoil during rainfall events. Water infiltration is affected by soil characteristics (becoming runoff in impermeable areas such as asphalt surfaces, and may infiltrate in green areas covered by vegetation), the existence of vegetation cover (Neris et al., 2013), precipitation typology, etc. However, climate change is having a direct impact on the amount of precipitation expected in the coming years, which leads to a scenario where artificial recharge of aquifers is considered (Hernández Ríos et al., 2023).

26.2.2 Desalination

The history of desalination in the Canary Islands goes back more than 50 years, with the region being considered a pioneer in the installation of seawater desalination plants for human consumption (Sadhwani & Veza, 2008). At present, the production of desalinated water in the Canary Islands amounts to 600,000 m^3/d, between private and public plants (Arenas Urrea et al., 2019). The total value of desalinated water production in Spain is 5 million m^3/d, making Spain one of the largest producers of this type of water in the world (Khemlani et al., 2021).

It is important to note here that many hotels in the Canary Islands have their own desalination plants. Due to their proximity to the coast or difficult access, it is more viable to build their own plant than to access the public supply of the city where they are located (Cruz-Pérez et al., 2021). This means that the presence of desalination plants is growing in the Canary Islands, and there is some concern about their energy consumption on the islands, where the model is still far from being 100% renewable (Escamilla-Fraile et al., 2023). Due to the interest in the water-energy binomial on the islands, the aim has been to increase the sustainability of desalination plants through the use of renewable energies associated with them (Leon et al., 2021a, 2021b).

Furthermore, it is important to note that, although brine has traditionally been considered a source of coastal contamination in the Canary Islands, an increasing number of uses are associated with this by-product (Singh et al., 2022), which will allow, if we increase renewable energies in the Canary Islands, desalination to become a much more sustainable option than it currently is.

26.2.3 *Reused Water*

According to the Canary Islands Institute of Statistics, in 2018, of the total volume of water reused, the percentages were as follows: 66.8% for agriculture and 33.2% for gardens and leisure sports areas (ISTAC, 2021). It was not used for street washing or for industry. In general, the implementation of reused water in the hydraulic cycle of the Canary Islands archipelago is far from optimal, and this flow is not being used as it should be (Ruiz-Rosa et al., 2016), considering that agriculture in the Canary Islands is an important water demander, along with landscaped areas related to the tertiary sector (Cruz-Pérez et al., 2023a, 2023b).

26.3 Conclusions

The islands are particularly vulnerable to climate change, mainly due to rising sea levels and the expected decrease in rainfall and increase in temperatures. The increased use of sea water through desalination entails high energy consumption, which must be offset by an increase in renewable energies to embark on the ecological transition process in these archipelagos. One of the main aspects that should be further developed in water management on islands is wastewater treatment, which should be utilized in tourism and agricultural uses. This would eliminate much of the current pressure on aquifers. In conclusion, much work remains to be done and it is essential that public institutions commit themselves to the goal of sustainability, since water resources are an essential element in the sovereignty and livability of the islands.

Acknowledgements This research was partially supported by the European Union's Horizon 2020 Research and Innovation Program under grant agreement 101037424, project ARSINOE (Climate-resilient regions through systemic solutions and innovations).

References

Arenas Urrea, S., Díaz Reyes, F., Peñate suárez, B., de la Fuente Bencomo, J. A. (2019). Technical review, evaluation and efficiency of energy recovery devices installed in the Canary Islands desalination plants. *Desalination, 450*, 54–63. https://doi.org/10.1016/j.desal.2018.07.013

BOC. (2023). *Ley 6/2022, de 27 de diciembre, de cambio climático y transición energética de Canarias* (p. 64). Gobierno de Canarias. https://www.boe.es/eli/es-cn/l/2022/12/27/6

de Canarias, G. (2022). *Anuario Energético de Canarias 2021* (p. 392). Consejería de Transición Ecológica, Lucha contra el Cambio Climático y Planificación Territorial.

CIATF. (2018). *Plan Hidrológico de Tenerife. Ciclo de Planificación Hidrológica 2015–2021* (p. 960). Consejo Insular de Aguas de Tenerife.

Cruz-Fuentes, T., Heredia, J., Cabrera, M. C., & Custodio, E. (2014). Fonctionnement d'un petit aquifère volcano-sédimentaire bénéficiant de l'excédent d'eaux d'irrigation: La Aldea, Grande Canarie, Iles Canaries (Espagne). *Hydrogeology Journal, 22*(4), 865–882. https://doi.org/10.1007/s10040-013-1094-9

Cruz-Pérez, N., Rodríguez-Martín, J., Acosta Martín, J. F., García, C., Ruiz-Rosa, I., & Santamarta, J. C. (2021). Improvements in hotel water consumption: Case study of a five-star hotel (Canary Islands, Spain). *Urban Water Journal, 19*(1), 32–39. https://doi.org/10.1080/1573062X.2021.1949480

Cruz-Pérez, N., & Santamarta, J. C. (2021). *La Huella Ecológica del Agua en las Islas Canarias*. Universidad de La Laguna. https://doi.org/10.25145/b.HuellaEcoCanarias.2021

Cruz-Pérez, N., Santamarta, J. C., Gamallo-Paz, I., Rodríguez-Martín, J., García-Gil, A. (2022). A comparison between carbon footprint of water production facilities in the Canary Islands: Groundwater resources vs. seawater desalination. *Sustainable Water Resources Management, 8*(121). https://doi.org/10.1007/s40899-022-00706-0

Cruz-Pérez, N., Santamarta, J. C., & Álvarez-Acosta, C. (2023b). Water footprint of representative agricultural crops on volcanic islands: The case of the Canary Islands. *Renewable Agriculture and Food Systems, 38*. https://doi.org/10.1017/S1742170523000303

Cruz-Pérez, N., Rodríguez-Martín, J., García, C., Rodríguez-Lozano, P., Tirado, D., García-Gil, A., & Santamarta, J. C. (2023a). A carbon footprint: the full water cycle in the Balearic Islands. *Sustainable Water Resources Management, 9*(4). https://doi.org/10.1007/s40899-023-00901-7

Custodio, E., Andreu-Rodes, J. M., Aragón, R., Estrela, T., Ferrer, J., García-Aróstegui, J. L., Manzano, M., Rodríguez-Hernández, L., Sahuquillo, A., & del Villar, A. (2016). Groundwater intensive use and mining in south-eastern peninsular Spain: Hydrogeological, economic and social aspects. *Science of the Total Environment, 559*, 302–316. https://doi.org/10.1016/j.scitotenv.2016.02.107

Escalante, E. F., Casas, J. D. H., Medeiros, A. M. V., & Sauto, J. S. S. (2020). Regulations and guidelines on water quality requirements for managed aquifer recharge: International comparison. *Acque Sotterranee - Italian Journal of Groundwater, 9*(2), 7–22. https://doi.org/10.7343/as-2020-462

Escamilla-Fraile, S., Ramos-Real, F. J., Calero-García, F. J., & González-Díaz, B. (2023). A review of the energy policy and energy transition objectives for 2040 in the Canary Islands (Spain). *Energies, 16*(3). https://doi.org/10.3390/en16031321

Fagerberg, J., Laestadius, S., & Martin, B. R. (2018). The triple challenge for Europe: The economy, climate change, and governance. *Innovation, Economic Development and Policy: Selected Essays, 59*(3), 384–410. https://doi.org/10.1080/05775132.2016.1171668

Gallego-Ayala, J., & Gómez-Limón, J. A. (2009). Analysis of policy instruments for control of nitrate pollution in irrigated agriculture in Castilla y León, Spain. *Spanish Journal of Agricultural Research, 7*(1), 24. https://doi.org/10.5424/sjar/2009071-395

Gasco Cavero, S., García-Gil, A., Cruz-Pérez, N., Martín Rodríguez, L. F., Laspidou, C., Contreras-Llin, A., Quintana, G., Díaz-Cruz, S., & Santamarta, J. C. (2023). First emerging pollutants profile in groundwater of the volcanic active island of El Hierro (Canary Islands). *Science of the Total Environment, 872*(February), 162204. https://doi.org/10.1016/j.scitotenv.2023.162204

Gómez-Gotor, A., Del Río-Gamero, B., Prieto Prado, I., Casañas, A. (2018). The history of desalination in the Canary Islands. *Desalination, 428*, 86–107. https://doi.org/10.1016/j.desal.2017.10.051

González-Morales, A., & Ramón-Ojeda, A. Á. (2019). La desalación de agua de mar en las Canarias Orientales: los casos de Lanzarote y Fuerteventura. *Agua y Territorio, 13*, 15–26. https://doi.org/10.17561/at.13.3722

Gonzálvez, J. J. G. (2019). *Construcción de grandes presas en Canarias : la singularidad de Gran Canaria*. Agua: Reflexiones Para Una Gestión Eficaz, November, 37.

Gössling, S. (2001). The consequences of tourism for sustainable water use on a tropical island: Zanzibar, Tanzania. *Journal of Environmental Management, 61*(2), 179–191. https://doi.org/10.1006/jema.2000.0403

Gu, Y., Dong, Y. N., Wang, H., Keller, A., Xu, J., Chiramba, T., & Li, F. (2016). Quantification of the water, energy and carbon footprints of wastewater treatment plants in China considering a water-energy nexus perspective. *Ecological Indicators, 60*, 402–409. https://doi.org/10.1016/j.ecolind.2015.07.012

Hernández Ríos, I., Cruz-Pérez, N., Chirivella-Guerra, J. I., García-Gil, A., Rodríguez-Alcántara, J. S., Rodríguez-Martín, J., Marazuela, M. A., Santamarta, J. C. (2023). Proposed recharge of island aquifer by deep wells with regenerated water in Gran Canaria (Spain). *Groundwater for Sustainable Development*, *22*, 100959. https://doi.org/10.1016/j.gsd.2023.100959

Herrera, C., Custodio, E. (2003). Hipótesis sobre el origen de la salinidad de las aguas subterráneas en la isla de Fuerteventura, Archipiélago de Canarias, España. *Boletín Geológico y Minero 114*(4), 433–452

ISTAC. (2021). *Indicadores sobre suministro y tratamiento de agua por años* (p. 30). Gobierno de Canarias.

Jodar-Abellan, A., Fernández-Aracil, P., & Melgarejo-Moreno, J. (2019). Assessing water shortage through a balance model among transfers, groundwater, desalination, wastewater reuse, and water demands (SE Spain). *Water, 11*(5). https://doi.org/10.3390/w11051009

Khemlani, N., Delgado, Q., Soriano, J., Peñate, B., Minatta, A., & Basani, M. (2021). *Caso de Canarias: Ingenio e innovación en la cadena de valor del agua* (p. 77). Banco Interamericano de Desarrollo. IDB-TN-2130

Kress, N., Gertner, Y., & Shoham-frider, E. (2020). Seawater quality at the brine discharge site from two mega size seawater reverse osmosis desalination plants in Israel (Eastern Mediterranean). *Water Research, 171*, 115402. https://doi.org/10.1016/j.watres.2019.115402

Ledgard, S. F., Falconer, S. J., Abercrombie, R., Philip, G., & Hill, J. P. (2020). Temporal, spatial, and management variability in the carbon footprint of New Zealand milk. *Journal of Dairy Science, 103*(1), 1031–1046. https://doi.org/10.3168/jds.2019-17182

Leon, F., Ramos, A., Vaswani, J., Mendieta, C., & Brito, S. (2021b). Climate change mitigation strategy through membranes replacement and determination methodology of carbon footprint in reverse osmosis RO desalination plants for islands and isolated territories. *Water, 13*(3), 293. https://doi.org/10.3390/w13030293

Leon, F., Ramos, A., & Perez-Baez, S. O. (2021). Optimization of energy efficiency, operation costs, carbon footprint and ecological footprint with reverse osmosis membranes in seawater desalination plants. *Membranes, 11*(10). https://doi.org/10.3390/membranes11100781

Llamas, M. R., Custodio, E., de la Hera, A., & Fornés, J. M. (2015). Groundwater in Spain: Increasing role, evolution, present and future. *Environmental Earth Sciences, 73*(6), 2567–2578. https://doi.org/10.1007/s12665-014-4004-0

Marrero, R., López, D. L., Hernández, P. A., & Pérez, N. M. (2008). Carbon dioxide discharged through the Las Cañadas aquifer, Tenerife, Canary Islands. *Pure and Applied Geophysics, 165*(1), 147–172. https://doi.org/10.1007/s00024-007-0287-3

Martín-Fernández, L., & Martínez-Núñez, M. (2011). An empirical approach to estimate soil erosion risk in Spain. *Science of the Total Environment, 409*(17), 3114–3123. https://doi.org/10.1016/j.scitotenv.2011.05.010

Melián, J. A. H., Martín-Rodríguez, A. J., Araña, J., Díaz, O. G., & Henríquez, J. J. G. (2010). Hybrid constructed wetlands for wastewater treatment and reuse in the Canary Islands. *Ecological Engineering, 36*(7), 891–899. https://doi.org/10.1016/j.ecoleng.2010.03.009

Millán, M. M. (2014). Extreme hydrometeorological events and climate change predictions in Europe. *Journal of Hydrology, 518*(PB), 206–224. https://doi.org/10.1016/j.jhydrol.2013.12.041

MITECO. (2019). *Libro Verde de la gobernanza del agua en España* (p. 43). Ministerio para la Transición Ecológica. http://www.librogobernanzagua.es/DOCS/Libro-verde-de-la-gobernanza-del-agua.pdf

Molleda, P., Blanco, I., Ansola, G., & de Luis, E. (2008). Removal of wastewater pathogen indicators in a constructed wetland in Leon, Spain. *Ecological Engineering, 33*(3–4), 252–257. https://doi.org/10.1016/j.ecoleng.2008.05.001

Neris, J., Tejedor, M., Rodríguez, M., Fuentes, J., & Jiménez, C. (2013). Effect of forest floor characteristics on water repellency, infiltration, runoff and soil loss in Andisols of Tenerife (Canary Islands, Spain). *CATENA, 108*, 50–57. https://doi.org/10.1016/j.catena.2012.04.011

Nevado, S., Gomis, A., Mena, E., & Castro, M. (2016). Cálculo de la huella de carbono del ciclo urbano del agua. *La Experiencia De Emuasa. Tecnoaqua, 19*, 58–65.

Nijhawan, A., Labhasetwar, P., Jain, P., & Rahate, M. (2013). Public consultation on artificial aquifer recharge using treated municipal wastewater. *Resources, Conservation and Recycling, 70*, 20–24. https://doi.org/10.1016/j.resconrec.2012.11.001

OCDE. (2015). *Principios de Gobernanza del agua de la OCDE* (p. 24). http://www.oecd.org/water/regional/

Ruiz-Rosa, I., García-Rodríguez, F. J., & Mendoza-Jiménez, J. (2016). Development and application of a cost management model for wastewater treatment and reuse processes. *Journal of Cleaner Production, 113*, 299–310. https://doi.org/10.1016/j.jclepro.2015.12.044

Sadhwani, J., & Jaime; Veza, J. M. (2008). Desalination and energy consumption in Canary Islands. *Desalination, 221*, 143–150. https://doi.org/10.1016/j.desal.2007.02.051

Santamarta, J. C. (2014). Hidrología de las islas volcánicas; singularidades y contribución de la ingeniería forestal. *Montes, 116*, 26–31.

Santamarta, J. C., & Lario-Bascones, R. J. (2015). Improving training in the hydrogeology of Volcanic Islands by visiting the water galleries of the Canary Islands (Spain). *Procedia - Social and Behavioral Sciences, 191*, 1317–1322. https://doi.org/10.1016/j.sbspro.2015.04.521

Santamarta, J. C., Calvo, I., Rodríguez-Martín, J., & Cruz-Pérez, N. (2022b). Water status in the Canary Islands related to energy requirements. *Energy Efficiency, 15*(13), 12. https://doi.org/10.1007/s12053-021-10016-7

Santamarta, J. C., Machín, N., & Cruz-Pérez, N. (2022c). Irrigation efficiency in banana crops in the Canary Islands. *The Open Agriculture Journal, 17*(1), 1–7. https://doi.org/10.2174/18743315-v16-e221226-2022-49

Santamarta, Juan C., García, C., Rodríguez-Lozano, P., Rodríguez-Martín, J., Cruz-Pérez, N. (2021). Water footprint in the water cycle of the Canary Islands. In U. P. de Catalunya (Ed.), *II International conference on water and sustainability*. https://doi.org/10.3926/icws2021

Santamarta, J. C., Rodríguez-Martín, J., Poncela, R., Fontes, Jose C., Lobo de Pina, Antonio, Cruz-Pérez, N. (2022). Integrated water resource management in the Macaronesia. *International Review of Civil Engineering (I.RE.C.E.), 13*(4), 8. https://doi.org/10.15866/irece.v13i4.21523

Santamarta, J. C. (2011). El agua en Canarias: Historia, ciencia y tecnología. In Julio Antonio Yanes Mesa (Ed.), *Boletín de la Real Sociedad Económica de Amigos del País de Tenerife* (Vol. 1, pp. 348–377). Litografía Á. Romero, S.L.

Schallenberg-Rodríguez, J., Veza, J. M., Blanco-Marigorta, A. (2014). Energy efficiency and desalination in the Canary Islands. *Renewable and Sustainables Energy Review, 40*, 741–748. https://doi.org/10.1016/j.rser.2014.07.213

Singh, I., Hay, R., & Celik, K. (2022). Recovery and direct carbonation of brucite from desalination reject brine for use as a construction material. *Cement and Concrete Research, 152*, 106673. https://doi.org/10.1016/j.cemconres.2021.106673

Tejedor, E., Saz, M. A., Esper, J., Cuadrat, J. M., & de Luis, M. (2017). Summer drought reconstruction in northeastern Spain inferred from a tree ring latewood network since 1734. *Geophysical Research Letters, 44*(16), 8492–8500. https://doi.org/10.1002/2017GL074748

Valhondo, C., Martínez-Landa, L., Carrera, J., Díaz-Cruz, S. M., Amalfitano, S., & Levantesi, C. (2020). Six artificial recharge pilot replicates to gain insight into water quality enhancement processes. *Chemosphere, 240*. https://doi.org/10.1016/j.chemosphere.2019.124826

Verbyla, M. E., Cairns, M. R., Gonzalez, P. A., Whiteford, L. M., & Mihelcic, J. R. (2015). Emerging challenges for pathogen control and resource recovery in natural wastewater treatment systems. *Wiley Interdisciplinary Reviews: Water, 2*(6), 701–714. https://doi.org/10.1002/wat2.1101

Zhi, Y., Yang, Z., Yin, X., Hamilton, P. B., & Zhang, L. (2015). Using gray water footprint to verify economic sectors' consumption of assimilative capacity in a river basin: Model and a case study in the Haihe River Basin, China. *Journal of Cleaner Production, 92*, 267–273. https://doi.org/10.1016/j.jclepro.2014.12.058

Chapter 27
The Effect of Orifice Size and Retention Time of Local Tanks on Water Distribution Network

Syed Rizvi and Rabee Rustum

Abstract Private tanks are one of the most important components in the water distribution systems as there are a significant interaction between reservoirs and the water distribution networks because they affect the qualitative and quantitative aspects of the water. However, local private tanks are usually ignored during the design stage in favor of simplified network analysis. It is a common practice based on water distribution codes of many countries that hardly specify any range of values for the physical attributes of the private tanks. In fact, neglecting parameters such as inlet orifice size could result in tank failure if not designed properly. This study's purpose is to determine the optimum values of different tank parameters by conducting time and volume-based reliability analysis so that the pressure in the system does not lead to high heat loss. Two sample networks were taken from previous literature. The simulation of local tanks was conducted using the software 'WDnetXL' while the reliability analysis was conducted separately using MS EXCEL. The results showed that each local tank's orifice size and volume greatly influences its reliability. The general trend shows that the increasing the orifice sizes and volume capacity increases the value for both reliability indicators. In fact, it was further observed that changing the orifice diameter even slightly could easily help in achieving the required reliability than changing the volume capacity. This was indicated when the required reliability value of 1 was achieved by just increasing the orifice diameter from 3 to 5 cm. However, in the case of the capacity of the private tank, higher capacity often only results in providing additional volume. In a longer simulation run, the network would lose its reliability as water is consumed by the user. The results also discuss the impact of orifice sizes and volume capacity on the pump's carbon footprint. An increase of 5 cm of the orifice size of a single tank could result in an increase in the 10% of the overall carbon footprints. Moreover, the result indicated that the lowest orifice and volume size(s) to achieve the required reliability of unity is the optimum dimensions of the tank. Any increase on the optimum value of orifice size and retention time

S. Rizvi (✉) · R. Rustum
EGIS, Heriot-Watt University, Dubai 38103, UAE
e-mail: sar32@hw.ac.uk

R. Rustum
e-mail: R.rustum@hw.ac.uk

A. Sefelnasr et al. (eds.), *Water Resources Management and Sustainability*, Water Science and Technology Library 114, https://doi.org/10.1007/978-3-031-80520-2_27

would result in an increase in the pump's carbon footprint and drop in pressure. In fact, a drop of 10 m was observed when the orifice size(s) were increased by 5 cm of their optimum value. Hence, the incorporation of private tanks in the network model can be very vital as it brings changes to the resulting parameters of the network.

Keywords Water supply · Local tanks · Water distribution network · Orifice size · Retention time

27.1 Introduction

In the current scenario illustrated in Fig. 27.1, models developed for WDN design and operation generally assume the customer demand as fixed outflows concentrated at nodes without considering the nodal pressure, Demand Driven Analysis (DDA) techniques. This leads, in some locations, to negative pressure in the model, which is rectified by the designers using either increasing the pump head to increase the pressure in the system or increasing pipe diameters to decrease head loss (De Marchis et al., 2016; Mohamed et al., 2019).

However, both these options could yield high capital and operational costs for the network. Moreover, the local tanks cannot be modelled using DDA as the flow to the nodes (local tanks in most cases) is not based on consumer demand profiles. Instead, the flow to the tanks depends upon the function of orifice size, available storage, orifice level, available pressure in the network and coefficient of discharge, which are modelled using pressure-driven analysis (PDA). This is because the node of the network is connected to the orifice of the local tank and not directly to the consumers,

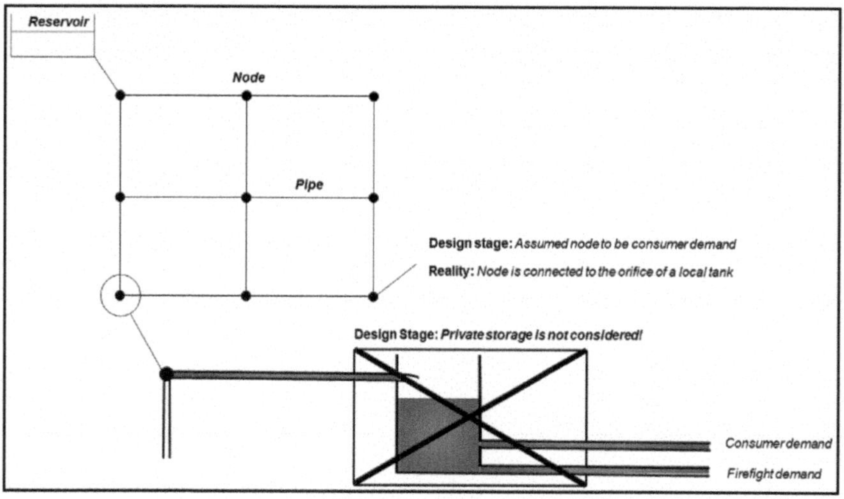

Fig. 27.1 Illustration of local tanks connection using DDA approach

as assumed during the design phase of the network. Hence, the current approach doesn't allow for assessing the reliability of the local tanks on the performance of the network with different constraints (orifice size, orifice level and volume of the local tanks), which leads to insufficient or overdesign sizing (Cobacho et al., 2008; Criminisi et al., 2009; De Marchis et al., 2010, 2018). In fact, the code of practice for many countries does not specify the range for these constraints. For example, having a very small inlet orifice of the tank could restrict the incoming flow into the tank, leading to tank failure if it cannot satisfy the required storage.

On the contrary, a large orifice size could increase the inflow towards the tank, which can fill the tank quickly and increase the head loss in the network due to the high flow rate in pipes and hence high velocity. This could eventually reduce the pressure in adjacent junctions and further spike the pumps' power, which would ultimately increase the network's carbon footprint. Similarly, having a large volume size tank could reduce the water quality as these tanks are not equipped with artificial mixing and heavily rely on the fill and empty process of the tank. Particularly, large volume sizes are more common as the customer would want to store more water in case of network failure or insufficient supply. Ultimately, these examples are small indications of the current research gap associated with the exclusion of private tanks in the network analysis.

To comply with this issue, the Behavior Simulation (BS) technique (McMahon et al., 2006) is one of the most valuable methods that can determine the reliability of the tanks in terms of time-based reliability (Rt) and volume-based reliability (Rv). Rv is the ratio of the total amount of water available to the customer and the desired amount demanded; Rt is the ratio between the total number of occasions in which the full demand was met to the number of periods in the simulation period (Adeloye et al., 2016).

The reservoir's capacity-demand function is affected by the flow to the tank, which is affected by the pressure in the network and the inlet's orifice size. This research aims to study the reliability and impacts of the local tank under PDA on the network.

27.2 Previous Literature

Throughout history, modelling water distribution networks has aimed to increase efficiency while reducing their total cost. Furthermore, the accuracy of the WDN computational models has been another constraint which is still being improved to bridge the gap between the design and actual operation of WDN. The gap created between the model and the real network can be associated with multiple factors like mathematical limitations, lack of topographical data, governmental regulation, user interference and mechanical interruption (Galaitsi et al., 2016). While these factors can be uncontrollable and unpredictable in some cases, there have been many studies conducted to reduce these challenges (Batchabani & Fuamba, 2014; Ciaponi & Creaco, 2018; Giustolisi et al., 2014; Vamvakeridou-Lyroudia et al., 2007). However, many issues have not been fully solved.

Particularly, mathematical limitations have been the focus of many scholars. The limitations occur mainly due to ignorance and approximation of certain parameters or components in favour of making the complex process much simpler to save time for the designers and modellers (Giustolisi, 2010). An example of such approximation is the procedure of Lumped Demand, where half of the total demand in a pipe is placed at each end. Hence, the pipe doesn't represent the original representation of the network in favour of reducing calculation constraints by disregarding the actual position of the demand. However, this can compromise the accuracy of the model and can generate recurring errors in the WDN model. One such limitation is that the networks are designed without private tanks (storage that is not hydraulically connected to the network). Hence, it is not easily simulated using currently available software(s) such as EPANET and its demand-driven analysis (DDA) family.

Integrating local tanks in WDN, the inflow at the orifice inlet should be calculated based on the local tank's pressure, orifice size, orifice level, and capacity of the tanks. Based on this philosophy, the filling/emptying process of these tanks could be simulated. This implies that the level of water in private tanks needs to be considered, which would govern the closure and opening of the inlet valve (orifice) (Giustolisi et al., 2011). This means that the valve will close if the local tank is full and vice versa. Then, a further model was developed that used a volume-controlled orifice in local tanks for their simulation tool 'WDnetXL' (Giustolisi et al., 2014). In this case, the relation between the tank's volume and the opening of the orifice was taken to be linear. However, another suggested that the opening of the orifice should behave non-linearly with the changing tank level or volume based on their experimental setup. Nevertheless, this behavior only occurs for a brief period during the opening and closing of the inlet float valves of the local tank; therefore, a linear relation can be assumed to compensate for the mathematical complication between the orifice inlet and available water in the tank (De Marchis et al., 2010).

27.3 Methodology

27.3.1 Sample Network

In this study, the data from a 2 WDNs are collected from the literature from different sources with every single attribute, including the nodal data, base water demand, pipe characteristics, valve settings, property of water sources, additional storage(s) data, pump curve(s), type(s) and operational patterns. From these attributes, some of the missing information for private tanks, such as the volume and orifice size(s), are initially assumed based on the given standards by Dubai Electricity and Water Authority (DEWA) and general practices from the local developers in UAE. In total, 3 sample networks are built to show the difference in the reliability analysis of private tanks in each case. The initial assumed orifice size was taken as 2 cm (minimum specification), and the volume size was assumed to have a retention time of 6 h.

The initial level of the tank was set to empty to account for the worst-case scenario. Furthermore, the minimum pressure to achieve the required demand varied between 10 to 15 m based on the practices from different countries and was taken as 12.5 m in this case.

Case 1: WDN in which the nodal junctions are supplied from the reservoir by three pumps. These pumps have similar characteristics and act as parallel type pumps (Suraksha, 2012).

Case 2: WDN that resembles the schematics of Anytown network. The layout of the networks shows that the junctional nodes are supplied by three sources (Two of them supply water with gravity and one of them uses pumped supply) to accommodate the high elevated nodes in the network (Devi Prasad and Tanyimboh, 2009) (Fig. 27.3).

27.3.2 Analysis and Working of the Simulation

Two main equations are used to model the dynamic behavior of the network: Mass balance and Energy balance. The energy balance equation represents the pressure difference between two connected nodes where the transported water loses energy mainly due to friction within the internal pipe walls. The mass balance equation conserves the flow at every node. The approach of PDA in 'WDnetXL' would be modified to integrate local tanks using the work from previous literature (Giustolisi et al., 2014) and reliability indicators from Eqs. 27.1–27.4 (Adeloye et al., 2016).

$$R_t = 1 - \frac{\sum_{t=1}^{N} f_t}{N} \tag{27.1}$$

$$R_v = 1 - \frac{\sum_{t=1}^{N} f_t(d_s - O_t)}{\sum_{t=1}^{N} d_s} \tag{27.2}$$

$$f_t = \begin{cases} 1, \ ifO_t < d_s \\ 0, \ otherwise \end{cases} \tag{27.3}$$

$$O_t = \begin{cases} I_t + S_{t-1}, \ if \ S_{t-1} + I_t \Delta t \leq \Delta t d_s \\ d_s, \qquad if \ d_s \Delta t \leq S_{t-1} + I_t \Delta t \leq K_a \quad WhereS_{t-1} isinitialstorage \end{cases} \tag{27.4}$$

where N is the total number of periods, ds is the desired consumer demand, and O_t is the actual outflow that depends on the case-by-case inequalities.

27.4 Result and Analysis

27.4.1 *Case 1: Demand is Supplied from a Single Source with the Three Pumps*

Figure 27.2 shows the schematic of the network that connects the node from a single source using three pumps. In this setup, the pumps keep running all day long together with the initial level of the tank set to empty. This case study's purpose is to find the optimum orifice and volume size for local tanks by using reliability analysis, pressure patterns and carbon footprint production.

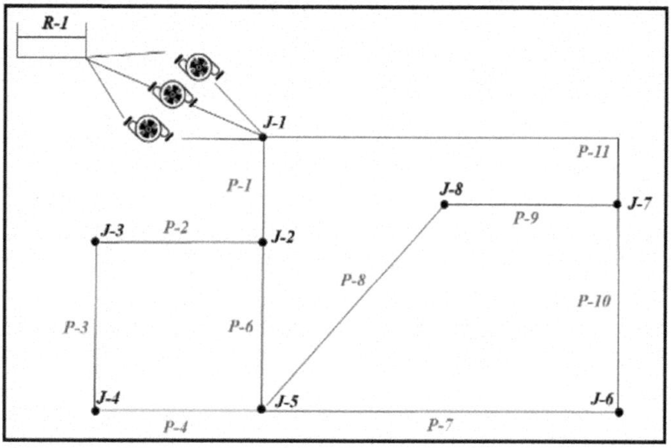

Fig. 27.2 Schematic of case 1

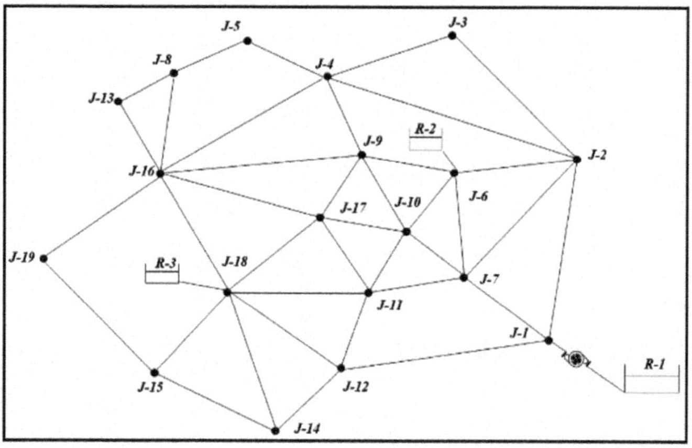

Fig. 27.3 Schematic of case 2

Table 27.1 shows the required orifice and volume size to achieve the required reliability of 1. Each local tank is tested with different orifice and volume sizes. It is observed that increasing the orifice diameter increases the reliability value while changing the retention time has little to no effect. It is further observed that the associated orifice size needs to be higher for higher base demand for the case of tanks 6 and 8. From these reliability values, the optimum value of the orifice and volume should be taken as the minimum to reach the required reliability. For example, tank number 6 should be assigned the orifice size of 4 cm and volume size equivalent to 6 h retention time. Any increase in the orifice and volume size, after that, would only increase the head loss and carbon footprint due to higher velocity in the network.

Hence, to further validate the facts that the orifice and volume size are optimum, Fig. 27.4 shows the designated flow path along with the hydraulic gradient line (HGL) in Fig. 27.5 for different orifice sizes and retention times. For tanks 2, 5 and 7, the minimum orifice size to achieve reliability was 2 cm. Hence, when increased to 5 cm, the pressure drop is approximately 30 m for retention time of 24 h in tank 7. As seen previously, increasing the orifice size tends to reduce the pressure at each nodal junction when tanks are filling. The hydraulic gradient shows that the head loss between the nodal junction is significant with each increased orifice size. In fact, an increased orifice with a high-capacity size has resulted in an even higher head loss. The same trend goes with the amount of carbon footprint produced, as indicated in Fig. 27.6. It again shows that the smallest orifice and volume size should be selected

Table 27.1 Reliability values for different orifice and volume size

Orifice sizes	Ret time	Tank 2		Tank 3		Tank 4		Tank 5		Tank 6		Tank 7		Tank 8	
		Rt	Rv	Rt	Rv	Rt	Rv	Rt	Rv	Rt	Rv	Rt	Rv	Rt	Rv
2 cm	6	1	1	1	1	1	1	1	1	0.32	0.46	1	1	0.25	0.42
	12	1	1	1	1	1	1	1	1	0.32	0.46	1	1	0.25	0.42
	18	1	1	1	1	1	1	1	1	0.32	0.46	1	1	0.25	0.42
	24	1	1	1	1	1	1	1	1	0.32	0.46	1	1	0.25	0.42
3 cm	6	1	1	1	1	1	1	1	1	0.78	0.96	1	1	0.41	0.75
	12	1	1	1	1	1	1	1	1	0.78	0.96	1	1	0.41	0.75
	18	1	1	1	1	1	1	1	1	0.78	0.96	1	1	0.41	0.75
	24	1	1	1	1	1	1	1	1	0.78	0.96	1	1	0.41	0.75
4 cm	6	1	1	1	1	1	1	1	1	1	1	1	1	0.75	0.92
	12	1	1	1	1	1	1	1	1	1	1	1	1	0.75	0.92
	18	1	1	1	1	1	1	1	1	1	1	1	1	0.75	0.92
	24	1	1	1	1	1	1	1	1	1	1	1	1	0.75	0.92
5 cm	6	1	1	1	1	1	1	1	1	1	1	1	1	1	1
	12	1	1	1	1	1	1	1	1	1	1	1	1	1	1
	18	1	1	1	1	1	1	1	1	1	1	1	1	1	1
	24	1	1	1	1	1	1	1	1	1	1	1	1	1	1

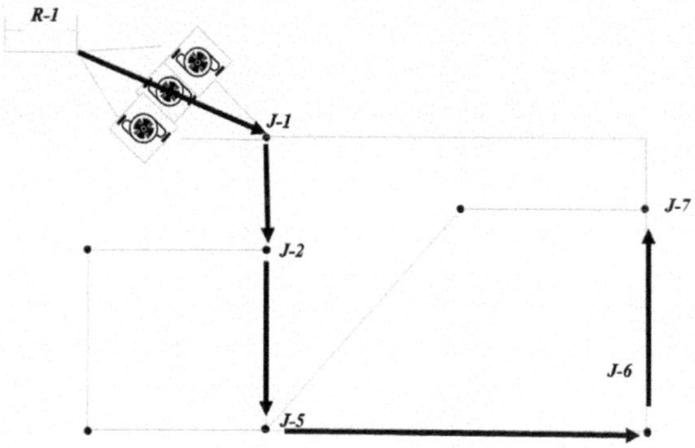

Fig. 27.4 Designated flow path

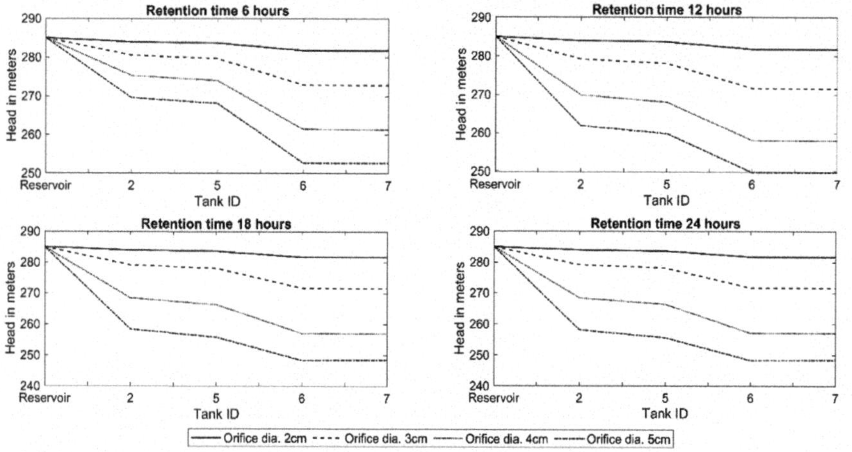

Fig. 27.5 HGL for the flow path with different orifice and volume sizes

to reach the required reliability, otherwise it could produce a higher carbon footprint and could reduce the sustainability of the network. Notably, increasing the retention time has a larger impact on the produced carbon footprint as the pump runs to fill the local tanks, which acquires more power.

27.4.2 Case 2: Anytown Network

Similar to the results in case 2, Table 27.2 provides the reliability value for different orifice sizes. The retention times are excluded in this table as they do not impact the reliability value in this case. As seen in the two cases before, the minimum orifice size to achieve the required reliability needs to be assigned as optimum; otherwise, it affects the energy and sustainability of the network. However, the idea here is to

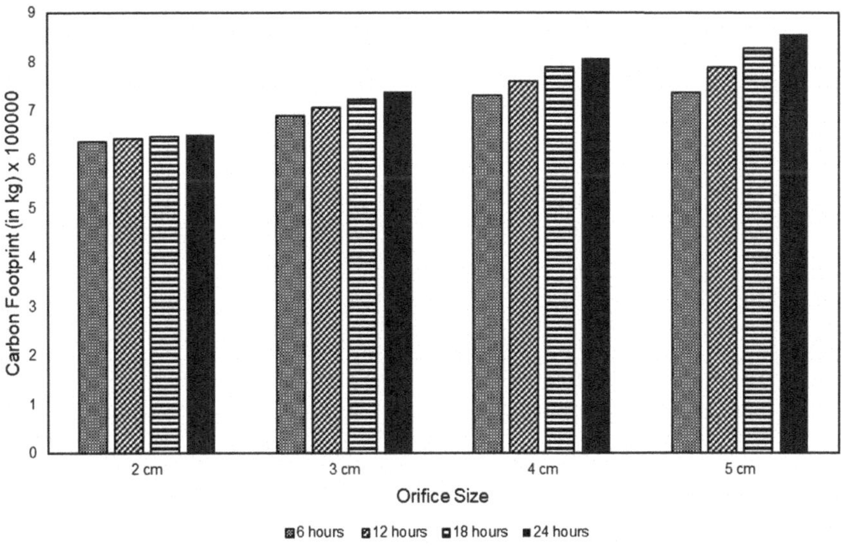

Fig. 27.6 Carbon footprint comparison with different orifice and volume size for case 1

show the effect of a single private tank on the pressure and carbon footprint. Hence, assigned flow paths, A and B, were analyzed.

For flow path A (Fig. 27.7), two simulations occurred, one with increasing the orifice size of tank 4 only and the other in which the orifice was increased for tank 1 only. The comparison was drawn with the optimum diameter for tank 1 and tank 4, 5 cm and 4 cm respectively.

All the other tanks are set to their optimum orifice and volume sizes based on the reliability analysis discussed above (see Table 27.2). When using a high orifice diameter for a single tank, the pressure at the other nodes could drop due to the high flow rate and increase the water head losses. To conserve the mass and energy balance equations, the pressure in the adjacent nodes will also decrease. This will reduce the actual inflow of tanks connected at the nodes and can affect their storage value. In Fig. 27.8, the impacts of increased orifice size of a singular tank (1 and 4) on the storage of adjacent tanks are shown using HGL. The effect of changing the orifice size on node 4 has little to no effect on HGL as it is far away from the source. However, for this flow path, the major impact is observed when the orifice size of tank 1 is increased to 10 cm; a pressure drops of around 10 m is observed from its optimum size. This has affected the adjacent nodes as highlighted from the HGL, which means that the pressure has been reduced on the adjacent tanks. However, the impact on other nodes is not significant as the network has multiple sources (further studied in flow path B). In the worst case (with a single source), the filling time for the other nodes could be delayed due to high head loss and can result in tank failure if tanks or orifice sizes are not integrated into the simulation.

Table 27.2 Reliability values for different orifice sizes (constant volume size)

	Orifice size					
	3 cm		4 cm		5 cm	
Tank ID	R(t)	R(v)	R(t)	R(v)	R(t)	R(v)
Tank 1	0.41	0.72	0.71	0.96	1	1
Tank 2	0.56	0.81	1	1	1	1
Tank 3	0.51	0.78	1	1	1	1
Tank 4	0.51	0.78	1	1	1	1
Tank 6	0.39	0.66	0.68	0.91	1	1
Tank 7	0.39	0.66	0.68	0.91	1	1
Tank 9	0.35	0.64	0.66	0.91	1	1
Tank 11	0.28	0.55	0.59	0.85	1	1
Tank 12	0.35	0.64	0.66	0.91	1	1
Tank 14	0.51	0.78	1	1	1	1
Tank 15	0.51	0.78	1	1	1	1
Tank 16	0.51	0.78	1	1	1	1
Tank 17	0.31	0.59	0.62	0.88	1	1
Tank 19	0.51	0.78	1	1	1	1

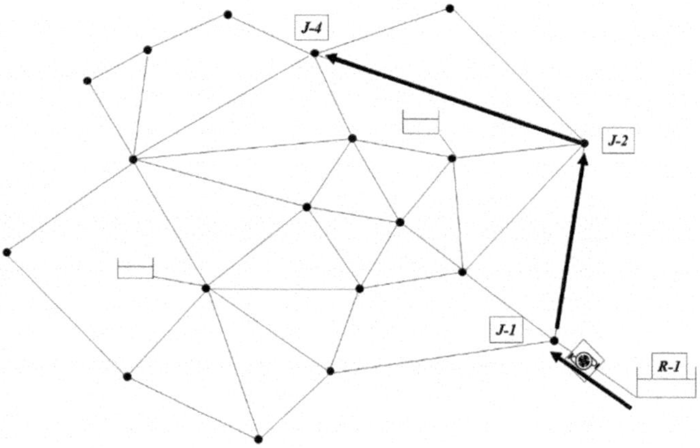

Fig. 27.7 Designated flow path A

A similar analysis in the different flow path (B), Fig. 27.9, was conducted with changing orifice size for tank 1 only.

Figure 27.10 results indicate that increasing the orifice size does not greatly impact the adjacent two nodes. This is because the nodal junctions 12 and 14 are closer to another reservoir in the network. Hence, these nodes are also receiving flow from a different source. In this case, the pressure drop is not significant with the adjacent

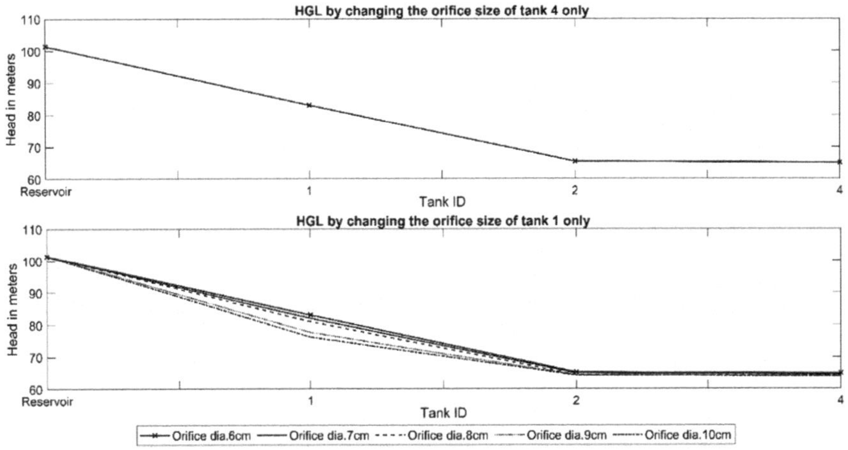

Fig. 27.8 Effect on the HGL when changing the orifice size of either tank 1 or 4 only in the designated flow path A

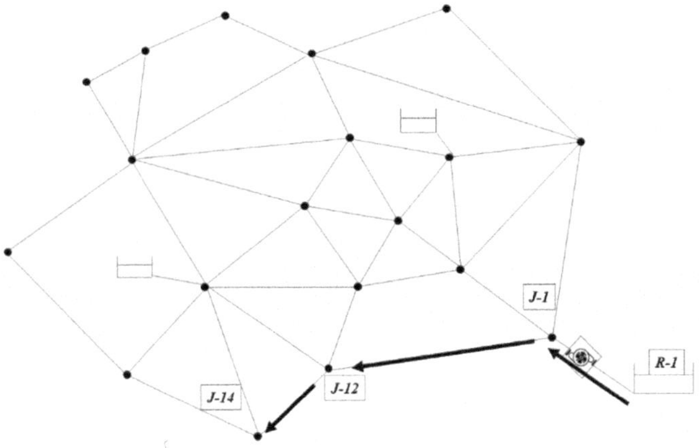

Fig. 27.9 Designated flow path B

nodal junction. In terms of carbon footprint (Fig. 27.11), the increased orifice size for a single tank, particularly tank 1 only, has further increased the overall power consumption of the pumps by 10%. This also highlights the fact that the pumps are losing their efficiency as more power is supplied to achieve the required flow rate due to larger orifices of a single tank. However, changing the orifice size for tank 4 only did not affect the overall production of carbon footprint. This is because tank 4 is fed by two sources, with the other source supplying the junction by gravity (instead of a pump). This reduces the pressure drop at the node when increasing the orifice size. Other than that, the volume capacity for tank 4 is relatively smaller (as the base demand is lower); hence the retention time of 6 h is not similar to that of tank 1.

Therefore, tank 4 gets filled quickly (with any increase in orifice size), and the flow rate becomes constant due to the level of the tank being close to full.

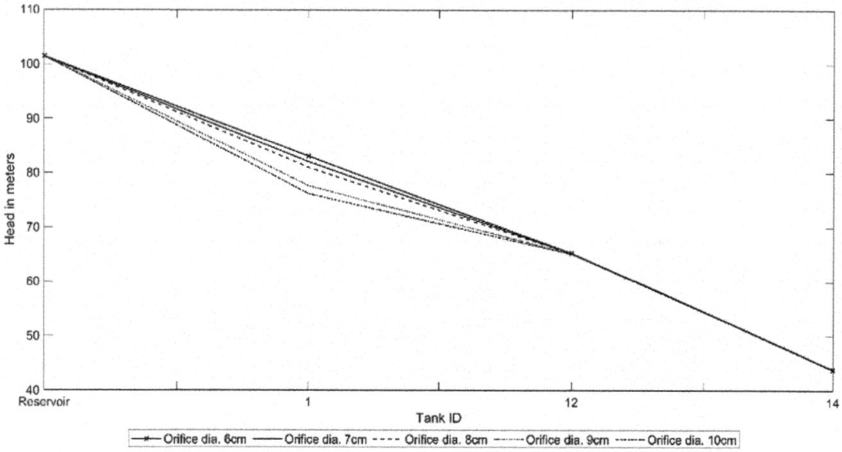

Fig. 27.10 HGL when changing the orifice size for tank 1 only for the flow path B

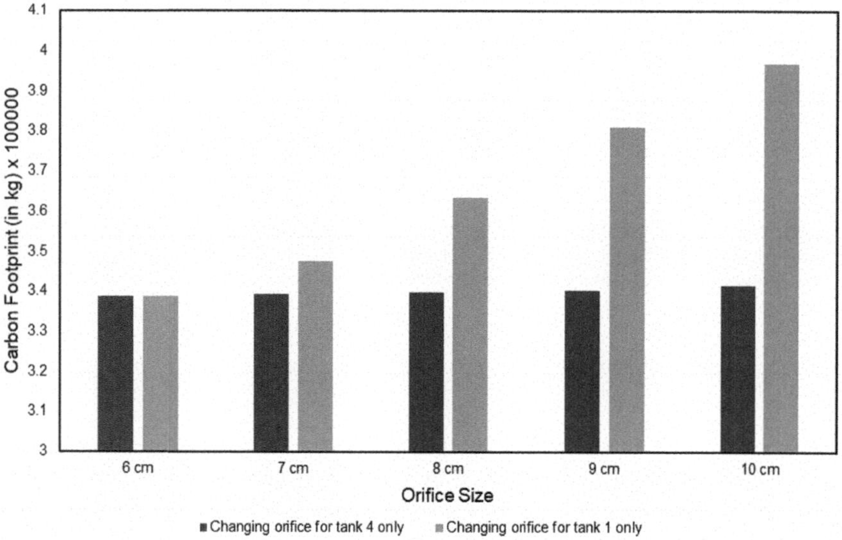

Fig. 27.11 Carbon footprint comparison with different orifice and volume size for case 2

27.5 Conclusion

The strategy to find the optimum volume and orifice size would allow the engineers to improve the current design of private tanks. As most of the associated code of practice recommends a 24-h retention time for local tanks is essential, this study shows that, in fact, the orifice diameter is a much more significant parameter to be optimized. Moreover, associating relatively high volume is not feasible as tank volume with 6 h retention time is observed to have enough capacity to satisfy the demand. The results have shown the importance of reliability indicators as they can assess the performance of the private tank by combining different factors that are affecting it. It is also found that there is a significant difference between the two indicators (time-based and volume-based). The difference in their value is due to the weightage provided by them. Nonetheless, each of them needs to reach the required reliability of 1 to satisfy the demand.

References

Adeloye, A. J., Soundharajan, B., & Mohammed, S. (2016). Harmonisation of reliability performance indices for planning and operational evaluation of water supply reservoirs. *Water Resources Management, 31*(3), 1013–1029. https://doi.org/10.1007/s11269-016-1561-x

Batchabani, E., & Fuamba, M. (2014). Optimal tank design in water distribution networks: review of literature and perspectives. *Journal of Water Resources Planning and Management, 140*(2), 136–145. https://doi.org/10.1061/(asce)wr.1943-5452.0000256

Ciaponi, C., & Creaco, E. (2018). Comparison of pressure-driven formulations for WDN simulation. *Water, 10*(4), 523. https://doi.org/10.3390/w10040523

Cobacho, R., Arregui, F., Cabrera, E., & Cabrera, E. (2008). Private water storage tanks: Evaluating their inefficiencies. *Water Practice and Technology, 3*(1). https://doi.org/10.2166/wpt.2008.025

Criminisi, A., Fontanazza, C. M., Freni, G., & Loggia, G. L. (2009). Evaluation of the apparent losses caused by water meter under-registration in intermittent water supply. *Water Science and Technology, 60*(9), 2373–2382. https://doi.org/10.2166/wst.2009.423

De Marchis, M., Fontanazza, C. M., Freni, G., Loggia, G. L., Napoli, E., & Notaro, V. (2010). Analysis of the impact of intermittent distribution by modelling the network-filling process. *Journal of Hydroinformatics, 13*(3), 358–373. https://doi.org/10.2166/hydro.2010.026

De Marchis, M., Freni, G., & Milici, B. (2016). Experimental evidence of the discharge law in private tanks connected to water distribution networks. *Procedia Engineering, 154*, 115–122. https://doi.org/10.1016/j.proeng.2016.07.428

De Marchis, M., Freni, G., & Milici, B. (2018). Experimental analysis of pressure-discharge relationship in a private water supply tank. *Journal of Hydroinformatics*. https://doi.org/10.2166/hydro.2018.135

Devi Prasad, T., & Tanyimboh, T. T. (2009). *Entropy based design of '"Anytown"' water distribution network*. https://doi.org/10.1061/41024(340)39

Galaitsi, S., Russell, R., Bishara, A., Durant, J., Bogle, J., & Huber-Lee, A. (2016). Intermittent domestic water supply: A critical review and analysis of causal-consequential pathways. *Water, 8*(7), 274. https://doi.org/10.3390/w8070274

Giustolisi, O. (2010). Considering actual pipe connections in water distribution network analysis. *Journal of Hydraulic Engineering, 136*(11), 889–900. https://doi.org/10.1061/(asce)hy.1943-7900.0000266

Giustolisi, O., Berardi, L., & Laucelli, D. B. (2011). Generalizing WDN simulation models to variable tank levels. *Journal of Hydroinformatics, 14*(3), 562–573. https://doi.org/10.2166/hydro.2011.224

Giustolisi, O., Berardi, L., & Laucelli, D. (2014). Modeling local water storages delivering customer demands in WDN models. *Journal of Hydraulic Engineering, 140*(1), 89–104. https://doi.org/10.1061/(asce)hy.1943-7900.0000812

McMahon, T. A., Adeloye, A. J., & Zhou, S.-L. (2006). Understanding performance measures of reservoirs. *Journal of Hydrology, 324*(1–4), 359–382. https://doi.org/10.1016/j.jhydrol.2005.09.030

Mohamed, H. I., El-dardir, M. S., & Abdel-Hafez, E. A. (2019). Effect of roof tanks filling on the hydraulic performance of water supply pipes network. JES. *Journal of Engineering Sciences, 47*(3), 280–295. https://doi.org/10.21608/jesaun.2019.109855

Suraksha,. (2012). Integration model of ISO 14001 with lean principles. *American Journal of Applied Sciences, 9*(12), 1974–1978. https://doi.org/10.3844/ajassp.2012.1974.1978

Vamvakeridou-Lyroudia, L., Savic, D., & Bragi Walters, G. (2007). *Tank simulation for the optimization of water distribution networks. 133*(6), 625–636. https://doi.org/10.1061/(asce)0733-9429(2007)133:6(625)

Chapter 28
Environmental Valuation for Appraising the Economic Level of Leakage in Arid Environments: A Case of Qassim, Saudi Arabia

Husnain Haider

Abstract With incredibly high-water stress conditions, water conservation is an uphill task for the municipalities in the Kingdom of Saudi Arabia (KSA). Most of these municipalities rely on periodic active leakage control programs for their water distribution systems (WDS) to meet the ministerial target of allowable water losses of up to 10%. This target does not appear to be ambitious for countries in arid regions where the impacts of droughts limit water resources, and population growth and urbanization overexploit them. In addition to the overuse of water, consumers are least bothered about the possible leakage on their side due to low water rates. For intermittent water supplies in arid regions, a framework based on the environmental valuation of water loss was proposed to establish the economic level of leakage. The process initiates with an estimation of real losses and their comparison with the unavoidable real losses. The framework propagates the idea of reducing the real losses up to the unavoidable real losses with an infrastructure leakage index of '1'. Implanting the framework on a WDS in the Qassim region reveals that conventional strategies, including active and passive leakage control, pressure management, and management of consumer-side issues, can reduce the Infrastructure Leakage Index (ILI) from 20 to 5.26. Recognizing and implementing the environmental cost of water loss in arid environments using an environmental valuation technique (i.e., based on evoking consumer preferences about the expected degradation of existing water resources) can further reduce the ILI with a nominal increase in water prices.

Keywords Sustainability · Economic level of leakage · Environmental valuation · Infrastructure leakage index · Intermittent water supply · Leakage management

H. Haider (✉)
Department of Civil Engineering, College of Engineering, Qassim University, Buraydah, Qassim, Saudi Arabia
e-mail: h.chaudhry@qu.edu.sa

A. Sefelnasr et al. (eds.), *Water Resources Management and Sustainability*, Water Science and Technology Library 114, https://doi.org/10.1007/978-3-031-80520-2_28

List of Abbreviations

ALC	Active leakage control
CARL	Current annual real losses
CVM	Contingent value method
ELL	Economic level of leakage
ILI	Infrastructure leakage index
KSA	Kingdom of Saudi Arabia
NWR	Nonrevenue water
OFWAT	Office of water services
PLC	Passive leakage control
RRR	Rehabilitation, renewal, and replacement
UARL	Unavoidable annual real losses
UFW	Unaccounted-for water
WDS	Water distribution systems
WHO	World health organization
WTA	Willingness to accept
WTP	Willingness to pay

28.1 Introduction

Most countries in the Arabian Gulf region, including the Kingdom of Saudi Arabia (KSA), are located in the arid environmental region of the globe with extremely high-water stress conditions (World Resources Institute, 2018). In addition, increasing population and industrial growth rates have led to the overuse of limited freshwater resources in this region (Ahmed et al., 2015; Al-Salamah et al., 2011). Unlike public and private water suppliers in developed countries, water utilities in most of these countries are motivated to provide safe and ample water to their consumers at reasonable prices (Haider et al., 2019). Besides, these utilities endeavor to minimize water losses in their water distribution systems (WDS) to conserve their limited water resources (Water Scarcity Solution, 2018). Water utilities in arid regions are facing severe water scarcity issues and consider intermittent water supply a rational solution for the long-term sustainability of their water resources. Disappointingly, intermittent supply has not been recognized as a successful case in several studies (Erickson, 2016).

Several aspects need to be considered while planning the WDS with intermittent supplies. Firstly, under the existing water scarcity condition, it is essential to appreciate the actual 'environmental value of water' instead of depending on the conventional 'cost of water production and distribution' (Ashton & Hope, 2001; Haider et al., 2019). Secondly, intermittent supplies have resulted in the deterioration of linear assets (i.e., water mains) and consequently increased water losses, system inefficiency, and the cost of rehabilitation, renewal, and replacement (RRR)

(Agathokleousa & Christodouloub, 2016). Finally, non-continuous operations deteriorate water quality due to biofilm development and intrusion of contaminated water from surroundings (Hashwa & Tokajian, 2004). The present study focuses on the first aspect.

Leakage in distribution networks' pipes is a primary component of nonrevenue Water (NRW). In KSA, unaccounted-for water (UFW) of up to 40% has been reported in large cities (Global Water Market, 2017). Municipalities are trying to bring the UFW as low as a ministerial target, i.e., 10% (Hashwa & Tokajian, 2004). Due to the absence of fresh water, the production cost of desalinated (i.e., thermal or reverse osmosis) saline (or brackish) water in the country is very high (Haider, 2017). Considering the importance and cost of desalinated water, keeping the UFW as low as possible is paramount for water utilities in arid regions.

Input volumes to a water supply system either used by the consumers, i.e., authorized consumption, or lost due to leakage in the system components, i.e., losses. Excluding the 'billed authorized consumption,' all the remaining supply, including 'unbilled authorized consumption', and water losses, is counted as NRW. Water losses are apportioned into 'apparent losses' and 'real losses'. Apparent losses (e.g., theft, billing errors, meter inaccuracies, etc.) can vary from 1 to 9% (Lambert, 2002). If the meter inaccuracies, the possibility of theft, and unbilled authorized consumption are neglected in KSA (due to the installation of new meters and replacing old meters with smart meters in the recent past, trust in stringent regulations against theft, and low water rates) primarily real losses are the actual amount of UFW. Real losses may be generated from leakage through pipes, valves, service connections, and consumer-side issues. These assumptions hold for most residential urban neighborhoods in KSA, mainly newly developed ones.

Current Annual Real Losses (CARL) are generally assessed for a one year. Unavoidable Annual Real Losses (UARL) are the minimum losses that always occur at a given operating pressure without any structural or economic constraint. The Infrastructure Leakage Index (ILI) is the ratio between CARL and UARL and a helpful water loss indicator in several countries (WLRandA and ILMSS, 2018; Haider et al., 2014). ILI is '1' if CARL is reduced to UARL, which is usually not economically viable. Therefore, the Office of Water Services (OFWAT) defined the Economic Level of Leakage (ELL) as *"the level at which further reduction in leakage becomes more expensive than producing water from another source."* Logically, ELL is attained when the economic cost of leakage control actions equals the economic value of conserved water.

Figure 28.1 presents the conventional relationship between CARL, UARL, ILI, and ELL. The left arrow corresponds to the Passive Leakage Control (PLC), which includes reducing the time of unplanned interruptions through efficient repairs in case of emergency leaks. The right arrow relates to identifying and repairing unreported leaks in the pipe network through planned Active Leakage Control (ALC). Pressure management, i.e., the top arrow, maintains the operating pressure to provide the desired residual head while minimizing leakage throughout the network. The bottom arrow reduces the losses through risk-based RRR strategies (Francisque et al., 2017).

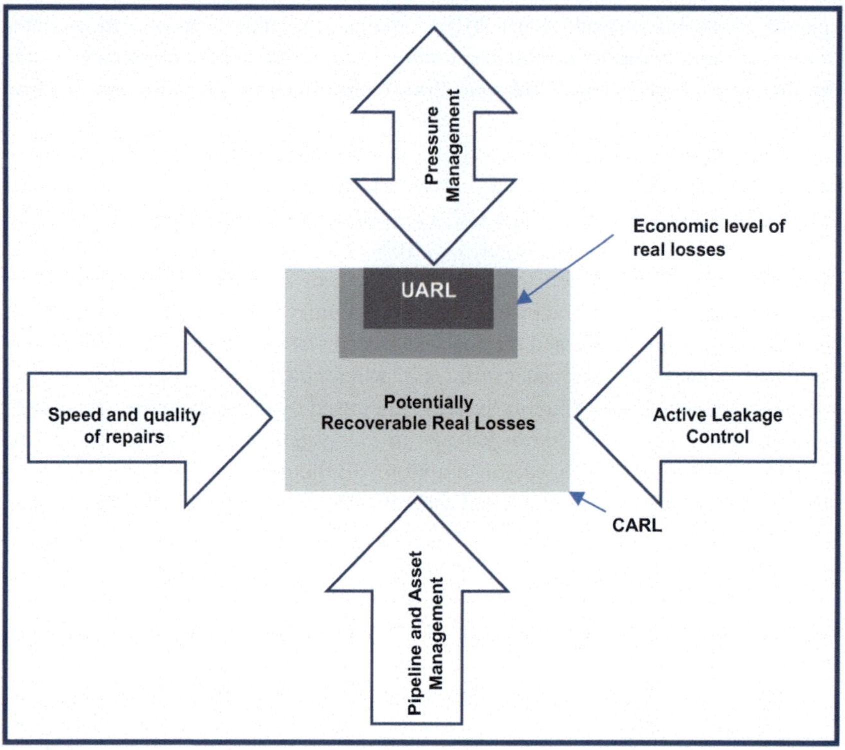

Fig. 28.1 Relationship between CARL, UARL, and ELL (Lambert & McKenzie, 2002)

All the water loss control strategies in Fig. 28.1 have been effectively applied within the distribution network up to the service connection to attain the ELL. For instance, privately owned systems in the UK and Wales reduced the leakage by up to 35% in less than a decade (i.e., between 1992 and 2000) through effective control measures (OFWAT, 2000). However, the definition of ELL stated above should be rationally applied to arid environments. Considering the existing water scarcity situation in the Gulf region and Saudi Arabia, ELL needs to be established based on the environmental valuation of limited available water instead of relying on the conventional way of comparing the 'cost of water lost' with the 'cost of leakage management. Hence, setting goals such as < 10% (i.e., ministerial target) for acceptable water losses must be re-evaluated by giving anticipated importance to every drop of water for the long-term sustainability of existing resources.

Water supply is intermittent in most parts of the KSA. Consequently, consumers are bound to store water in underground tanks, which might be leaking and old. Due to the relatively low water prices, customers may not report any leakage in their in-house plumbing system to the water supplier. Water suppliers generally apply the control actions in Fig. 28.1 up to the service connection (Islam & Babel, 2013). Investigating the root causes of issues on the consumer side would undoubtedly be helpful.

However, with additional human and financial resources, enhancing the effectiveness of Active Leakage Control (ALC) can further reduce the CARL. Appropriate environmental valuation methods can assess the actual value of water loss and determine how much the consumers can contribute to the cost of these extra control measures.

This paper proposes a framework to establish the ELL using environmental valuation techniques and conventional leakage controls for WDS in arid regions. After defining the main inputs required and outputs of the framework in the methodology section, valuable methods used to calculate the ELL are briefly outlined. The concept of environmental valuation for water loss management is also introduced. Finally, the main findings of the framework implementation in a case study of the Qassim Region in KSA are summarized. For details on data collection and analysis of various control actions, interested readers are referred to Haider et al., (2019). Consumer side issues and environmental valuation are given primary attention in the present study for water loss management.

28.2 Methods

28.2.1 Environmental Valuation Framework

While establishing the ELL for countries in the Gulf region like the KSA, some critical considerations are intermittent water supply, decreasing groundwater table, lack of freshwater resources, the exceptionally high environmental value of limited available water, and low water rates. Figure 28.2 illustrates the proposed methodological framework for establishing the *ELL* for such considerations. The framework begins by evaluating water losses before any improvement action to evaluate the existing scenario by taking the difference between the measured supply and the metered consumption. This step evaluates CARL through night-time flow analysis. The next step calculates UARL by utilizing data on water infrastructure, such as main lengths, number and average length of service connections, and average operating pressure. Subsequently, the utility can assess the percentage of water loss to compare it with the ministerial target. If the water loss is less than the ministerial target, the framework recommends checking the system for ILI or 'go for improvements'. ELL is established if the ILI is equal to or less than '1'; otherwise, the system needs water loss management. If economically viable, it is recommended for arid environments to bring ILI near to a value of '1'. Municipalities can attempt to achieve this over-ambitious goal by implementing improvement actions step-by-step until the ELL is established. In this study, the impacts of ALC, pressure management, Passive Leakage Control (PLC), management of consumer side issues, and improving the efficacy of ALC based on environmental valuation of water loss have been investigated. In this regard, scenario analysis accompanied by cost–benefit analysis is conducted on a WDS in Buraydah, Qassim, KSA, to evaluate the pragmatism of the proposed framework.

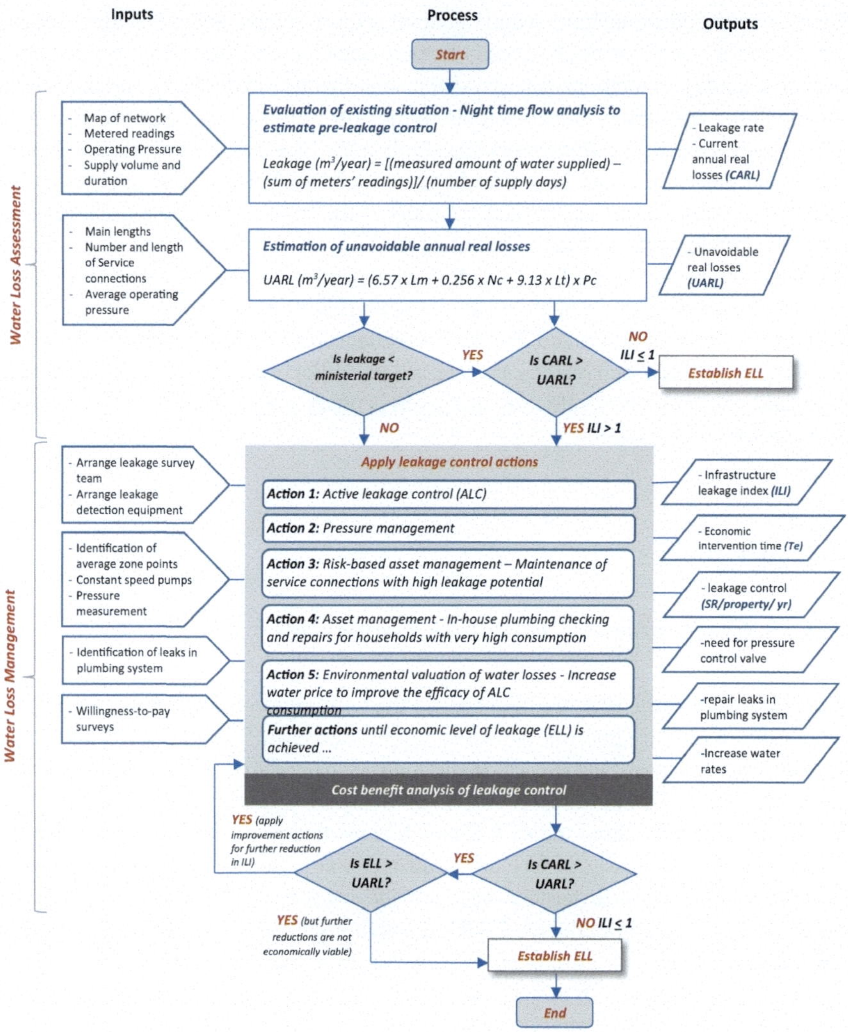

Fig. 28.2 Methodological framework to establish the ELL for WDS in arid environments (modified after Haider et al., 2019)

28.2.2 Study Area

The study area, shown in Fig. 28.3, is a small WDS serving around 1500 households in the city of Buraydah, located in the Qassim region of the KSA. The city accommodates almost 40% of the province's population. The primary source of water supply in Qassim is the Saq Aquifer, which has constantly been decreasing while serving the growing population, agriculture, and industrialization in and around the city (Haider et al., 2017). The responsible municipality has initiated a planned ALC program once every three years to meet the regional water loss benchmark of 8%. Infrastructure and consumer data were obtained from the municipality to implement the proposed framework (Fig. 28.2), e.g., ALC program information, intermittent supply hours, average operating pressure, number of customers, etc.

The total length of water mains is around 39,000 m. The distribution network consists of ductile iron (DI), plasticized polyvinyl chloride (UPVC), and high-density polyethylene pipe (HDPE) pipes of 300 mm in size. Intermittent water supply through direct pumping for three days/week is the current practice in the study area. Consumers store this water in underground storage tanks to ensure a 24/7 supply within the house.

28.2.3 Data Collection and Water Loss Indicators

The following equation estimated the *UARL* (European Union, 2015):

$$UARL\left(m^3/year\right) = (6.57 \times Lm + 0.256 \times Nc + 9.13 \times Lt) \times Pc \tag{28.1}$$

where Lm = lengths of mains (km), Lt = length of service connections between the municipality line and the consumer meter (km), Nc = number of service connections, and Pc = average operating pressure (m).

For intermittent water supply, i.e., some days in a week time, a CARL can be estimated through a planned survey by monitoring the amount of supplied water as:

$$\begin{aligned} CARL\left(m^3/day\right) = {} & \left[\left(amount\ of\ water\ supplied\ during\ the\ supply\ period\ in\ m^3\right)\right. \\ & \left.-\left(amount\ of\ water\ consumed\ during\ the\ supply\ period\ in\ m^3\right)\right] \\ & /(duration\ of\ supply\ period\ in\ days) \end{aligned} \tag{28.2}$$

Equation (28.3) calculated ILI.

$$ILI = \frac{CARL}{UARL} \tag{28.3}$$

The detailed methodology adopted for estimating average zone pressure and night flow analysis for intermittent supply can be seen in ILMSS (2013) and Van Zyl et al. (2017).

28.2.4 Environmental Valuation of Water Losses

The literature on assessing natural environmental resources' economic (monetary) value is limited (Ojeda et al., 2008). UKWIR/WRc (1994) states that the ELL is

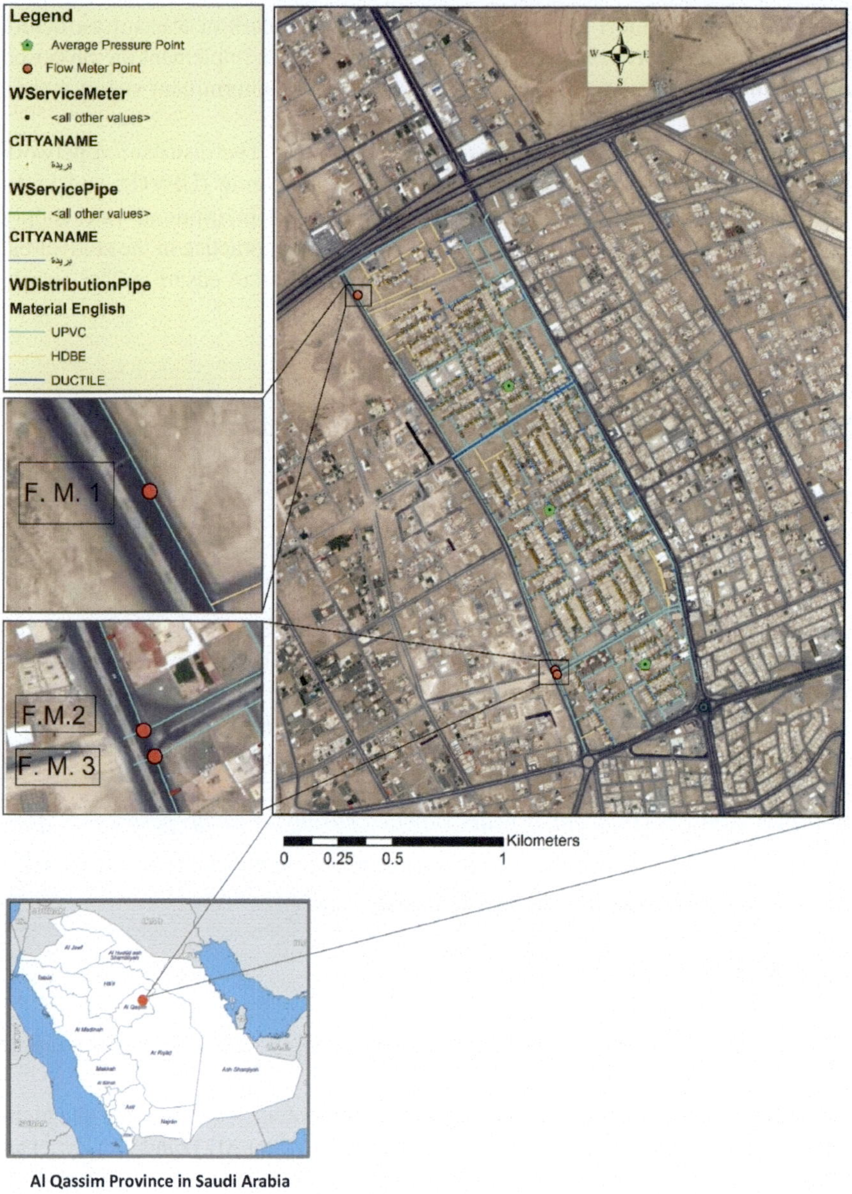

Fig. 28.3 High-resolution satellite imagery of the study area in Buraydah, Qassim Province, showing flow and pressure moinitoring points *Source* Haider et al., 2019

achieved when *"the financial cost of 1 m^3 leakage reduction becomes equal to the financial value of 1 m^3 of water conserved."* After the report was published in 1994, most of the attention was paid to evaluating the effect of leakage control on the water supply demand and balance. Later, in 1999, UKWIR recommended that social (i.e., arises from the activities borne by the society) and environmental (i.e., a particular sub-group of social costs such as pollution) costs and associated benefits should be included in the establishment of ELL (UKWIR, 1999).

For a specific scenario of arid environments, it is inevitable to incorporate the economic valuation of the environment in the ELL assessment process. Environmental valuation (i.e., economic valuation of environmental impacts) endeavors to evoke consumer preferences regarding the expected degradation of existing water resources and translates these preferences' monetary value. To evaluate the economic value of an environmental resource (i.e., treated ground or surface water), the values of 'use of the environmental resource' and 'non-use of that resource related to its discontinuation due to over-exploitation need to be distinguished. The concept is based on a primary assumption that the worth of an environmental resource based on human judgment is acceptable. Ashton and Hope (2001) reviewed different environmental cost and benefit analysis methodologies. Table 28.1 summarizes these methods; for details, the readers refer to Ojeda et al. (2008).

Table 28.1 shows that the Contingent value method (CVM) is the most commonly used method for the environmental valuation of water resources. After obtaining the basic information about the respondent, i.e., gender, age, education level, existing average monthly bill, number of family members, etc., it is crucial to inform the consumers in arid environments about the importance of limited availability of existing water resources, impacts of climate change, population increase, etc. Subsequently, surveys from individuals evaluate the following main (not limited to) questions:

- *Has he/she ever heard about water scarcity in the country?*
- *Is he/she and other family members of the household trying to conserve water in their routine activities?*
- *What type of water conservation practices are they using at the household level?*
- *How much is the consumer's willingness to pay (WTP) for conserving the water lost due to leakage?*
- *How much is the consumer's willingness to accept (WTA) for compromising this precious natural resource?*
- *How much can he/she pay in addition to his/her existing water rate?*

An increase in water rates can be offered as bid levels as percentage increases in existing water prices, e.g., 1, 2, 5%, etc. The dichotomous choice method can be used to find a suitable increase in water rates. The data can be arranged in the form of a four-column table. An example is shown in Table 28.2.

The results of the CVM survey provide the basis for municipality managers to increase water rates rationally. The mean WTP from the survey results can be estimated using the following equation (Ojeda et al., 2008):

Table 28.1 Summary of environmental valuation methods (*Source* Ashton & Hope, 2001)

No.	Method	Description	Application
1	Contingent value method (CVM)	• Willingness-to-pay (WTP) surveys are conducted to assess: • How much an individual can pay to maintain the status quo? • How much can an individual pay for a positive change in the existing environment? • How much would an individual be willing to accept (WTA) for discontinuing the benefits of the environment? • How much can an individual experience the increase in environmental dis-benefit due to the discontinuation of this specific resource? • The best way of using CVM is not yet clear. Strategies used in selecting sample populations, question types, and the survey method can significantly affect the outcomes of the method • Regardless of survey cost, in-person interviews are more reliable than postal questionnaires and telephonic interviews	• Economic valuation of environmental uses with sustained flows in Yaqui River (Ojeda et al., 2008) • Economic of watershed management for affected communities in Nicaragua (Johnson & Baltodano, 2004) • Estimation of WTP from neighborhoods to protect undeveloped land with natural views, open spaces, and wildlife habitat in Colorado (Breffle et al., 1998) • Loss estimation of the value of water due to pesticide contamination in Vietnam (Phuong & Gopalakrishnan, 2010)
2	Travel cost method (TCM)	• In TCM, users' preferences are evaluated in terms of how much they spend on visiting that particular environmental site • The method can only asses the 'use' value of the resource	• To evaluate the benefits of water quality improvements in rivers (Smith & Desvousges, 1985)
3	Hedonic price method (HPM)	• The method uses the market prices of the goods or services, which can be affected by the environmental impacts of concern. e.g., the impact of noise levels on the price of a house	• To estimate the demand for characteristics of a house (Palmquist, 1984)
4	Replacement cost method	• This method estimates the cost required for re-establishing an affected environmental attribute	• For managing water resource development impacts on indigenous customary economies in Australia (Jackson et al., 2014)

(continued)

Table 28.1 (continued)

No.	Method	Description	Application
5	Effect-on-production method	• The method assesses the economic loss of business due to the deterioration of an environmental attribute	–
6	Avertive expenditure method	• In this method, the financial cost of active mitigation actions required to prevent the deterioration is assessed, and the environmental cost is assessed	• Case studies from Kenya to value the environment (Emerton, 1996)
6	Dose–response method	• The method assesses the value of impacts on the health of humans, plants, and animals due to increasing pollution levels	(WHO, 2005)
7	Use of expert opinion	• This method was found particularly useful for the decision-making of controversial projects • The method may affect the general public's confidence due to inherent subjectivity	• Has been used to establish ELL (Yorkshire Water Services Limited, 1997)
8	Benefit transfer method (BTM)	• This method transforms the findings of the environmental valuation study of the study site to the 'policy site' • The difference between the two sites may lead to erroneous results	(Phuong and Gopalakrishnan, 2010)

Table 28.2 Example of Dichotomous choice results (modified after Ojeda et al., 2008)

Bid level in Saudi Riyals' SR' (X_i)	Number of respondents per given bid level	Number of respondents agreed to the bid level (Y_i)	Number of respondents found not agree with the bid level (Z_i)
1%	10	10	0
2%	9	7	2
3%	11	8	3
…	…	…	…
Total	N	Y	Z

$$WTP = \frac{1}{N}\sum_{i=1}^{N} X_i Y_i \tag{28.4}$$

Where N = the total number of respondents, Xi = bid level, and Yi = the number of respondents found to agree to the bid level with a '*Yes*' response, details of the method can be seen elsewhere (Carson, 2000; Ojeda et al., 2008).

In any leakage control program, a situation comes when the economic returns start diminishing against the human and financial resources spent. The return rate is high when identifying and repairing initial leaks. Subsequent reductions of hidden leaks are generally difficult to achieve economically (Ashton & Hope, 2001). As per AwwaRF (2007) report, 50% of recoverable real losses are reduced per leakage detection survey, and an additional 50% of losses (out of remaining losses) can be reduced through a subsequent survey. In this way, leakage detection costs exponentially increase with the increasing efficiency of the leakage survey. For example, if 100,000SR is required to reduce 30% losses to 15%, the same amount will reduce the losses from 15 to 7.5%, and so on. In addition, as per a leakage detection study conducted in New Mexico (Hardeman, 2008), the cost of a leakage detection survey can vary from one WDS to another.

In arid environments, the additional revenues generated from increased water rates, based on CVM surveys, can be utilized to improve the effectiveness of the ALC program, for instance, pressing the ministerial allowable losses of 10% to 5% or even more.

28.3 Results and Discussion

In 2016, the municipality performed a planned ALC program for leakage detection using acoustic techniques and subsequent water mains and service connection repairs. Flowmeters monitored the flows during the survey period; the survey results were obtained from the municipality. According to the findings of this survey, the water loss was reduced to 666.08 m^3/weak (i.e., less than the regional target of 8%).

The present study carried out a three-week survey from 17 March to 2 April 2017. Figure 28.3 shows the positions of flow and pressure measuring points. The average daily estimated flows were found to be 24.4 m^3/h at F.M.1, 55.2 m^3/h at F.M.2, and 52.8 m^3/h at F.M.3. Highest water demand was observed during the daytime, while the minimum at the nighttime. On the contrary, the pressure was maximum during the nighttime throughout the supply period. The third (i.e., the last) day of supply is the most critical supply period when all the in-house storage tanks are full, and a very high pressure of up to 3 bars can undoubtedly lead to high water losses through pipes and service connections.

Therefore, minimum night flow analyses were conducted between 2:00 AM and 4:00 AM on the last supply day to develop the correlation between leakage and operating pressure. Details of night flow analysis can be seen in Haider et al. (2019).

A total water loss of 716.18 m^3/week was observed in the study area, which agreed with the findings of the survey (i.e., 666.8 m^3/week) conducted by the municipality. As per these results, the *CARL* was calculated as 34,642 m^3/year after completion of the ALC program. Using equation [1], the estimated value of *UARL* was 5115 m^3/year, resulting in an *ILI* value of around 7. According to the framework proposed in Fig. 28.2 to establish ELL in arid environments, this significant gap after ALC (which met the regional target of 8%) suggests additional improvement measures, such as pressure management and additional asset management of consumer side issues like cracks in the underground storage and out-of-service float valves.

Different scenarios were developed for the area under study to evaluate the existing situation, conventional improvement actions (i.e., ALC, pressure management, and PLC), and additional management measures to resolve consumer-side issues. Table 28.3 presents these scenarios, their annual cost, water losses, and the corresponding ILI values achieved due to their implementation.

As part of the present study, the study area (see the boundary of three zones in Fig. 28.3) was disconnected from the remaining WDS. Supplied water was monitored at three flow meters, and the operating pressures were measured at the average zone points. According to the information obtained from the municipality, water losses were higher than 8% before the implementation of the ALC program (i.e., ***S1: No Control***) with an ILI value of around 20 (refer to Table 28.3).

Subsequently, the municipality hired a contractor for the planned ALC (once in 3 years), i.e., ***S2: Control 1***. The contractor was supposed to perform the following major tasks:

(i) identification of defective meters,
(ii) identification of leaks in mains and service connections,
(iii) perform RRR of leaking mains, and
(iv) perform RRR of service connections.

The above tasks either continued to achieve the regional benchmark (8%), or the efforts continued to exhaust the given amount for ALC. Sometimes, the main leaks are either fewer or easy to identify; in this case, the contract performs a complete leakage survey with all the mandatory tasks stated above, and the water losses may be reduced to less than the benchmark value.

An ILI was reduced from 20 to 7.28 after implementing the ALC, and the CARL was decreased to 2.78%. These results were encouraging and showed the effectiveness of the ALC program. As the subsequent ALC program is scheduled every three years, a regular PLC program is desired to minimize the water losses, which certainly increase with time. Therefore, the cost of Control 1 in Table 28.3 includes the cost of PLC. Instead of relying on arbitrarily selecting the frequency (which is the current practice) of an ALC program, for details on the method to calculate the economic frequency and budget of a planned leakage control program, see Haider et al. (2019).

The value of ILI after ALC is much higher than '1', and the remaining water losses need to be reduced. As mentioned above, significant water losses occurred during the last period of intermittent supply due to high pressure; effective pressure management can minimize the water losses in the mains and service connections. The

Table 28.3 Developed scenarios for establishing *ELL* (*Source* Haider et al., 2019)

No.	Description	Annual cost (SR/year)	CARL, after implementing the controls (m^3/year)	UARL (m^3/year)	ILI
S*1	***No Control:*** Water losses before leakage control	0	102,310	5115	20
S2	***Control 1:*** Water losses after conventional active leakage control (ALC) and passive leakage control (PLC) in case of main breaks in the study area	70,000	27,820	5115	7.28
S3	***Control 2:*** Water losses after Pressure management	16,667	22,412	4049	5.53
S4	***Control 3:*** Water losses after managing (checking and repairing) consumer side issues of top 5% (75 service connections) high water use households	3725	21,700	4049	5.35
S5	***Control 4:*** Water losses after managing consumer side issues of more than 5% (i.e., top 10%) of households with higher water consumption	3725	21,334	4049	5.26

(continued)

Table 28.3 (continued)

No.	Description	Annual cost (SR/year)	CARL, after implementing the controls (m^3/year)	UARL (m^3/year)	ILI
S6	***Control 5:*** Water losses after managing consumer side issues of an additional 40% (i.e., top 50%) of households with higher consumption	89,250	20,182	4049	4.98
S7	***Control 6:*** A hypothetical analysis to evaluate the impact of increasing the effectiveness of ALC based on a given increase in water rate	Variable	Variable	–	Variable

* S: Scenario

study revealed that reducing the existing operating pressure by 10m can reduce the CARL to 431m^3/week, i.e., ILI = 5.53. Hence, pressure management (***S3: Control 2***) can further reduce 40% of losses without affecting the quality of service. Installing a pressure control value is not economically viable when water production costs are low (Creaco & Walski, 2017). However, such control values would be feasible for WDS in arid environments where the treatment cost is high and the limited water's environmental value is innumerable.

Sorting the water consumption data of 1497 service connections in descending order revealed that 5% of the households were responsible for 27.33% (i.e., 55.5 m^3/connection) of the total weekly supply to the service area. Based on these findings, we investigated the conventional plumbing system used in the study area. Consumers store water in underground tanks and then lift the water to the rooftop for in-house supply. Figure 28.4, illustrating a typical cross-section of the in-house plumbing scenario, shows that the underground tanks' cover, bottom slab, and side walls are made of reinforced concrete up to the float valve of the underground storage tank. While the side walls above the float valve are constructed with brick masonry, the possibility of cracks in this part of the tank is higher. Consequently, the damaged float valves may increase the amount of leakage through possible cracks compared to the lower part of the tank with concrete construction.

Identifying and resolving consumer side issues (i.e., ***S4: Control 3***) for these households would undoubtedly be helpful; however, the possibilities of large

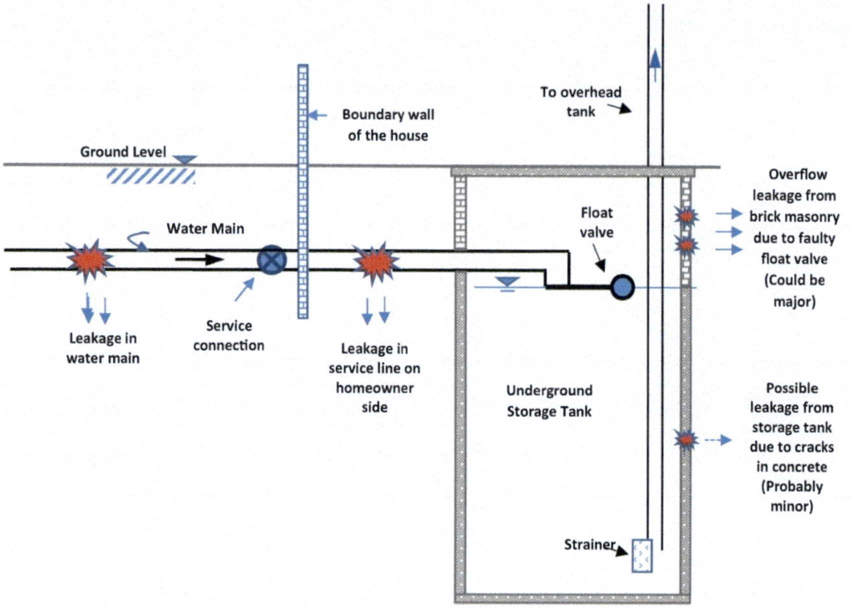

Fig. 28.4 Vignette showing a typical house connection with an underground storage tank in the study area and possible locations of leakage (*Source* Haider et al., 2019)

family members and swimming pools should be considered. Heading towards ELL, resolving consumer-side issues for the top 5% of households with the highest water consumption can reduce the *ILI* to 5.36. Such a minor drop in *ILI* can conserve more than 700 m^3/year, which is highly valuable for a small residential neighborhood.

Further analysis of water consumption for each service connection revealed that 10% of households consumed about 41% of the total supply. Identifying and managing these issues in the households with the highest water consumption (***S5: Control 4***) can further drop the *ILI* to 5.26 with more than 1050 m^3 water savings per annum. Although ILI can be reduced to 4.98 by checking an additional 40% of households (***S6: Control 5***), this action needs careful cost–benefit analysis (refer to Table 28.3).

Figure 28.5 illustrates different components of water loss analysis, including ELL, UARL, and the cost of water losses. As per the total cost curve, significant water losses were reduced through conventional measures, i.e., ALC, pressure management, and PLC. While the ELL was achieved after implementing Control 4, the cost exponentially increased with further efforts.

Water loss up to this point was primarily managed through conventional measures. Figure 28.6 demonstrates a significant gap between the UARL and ELL. Enhancing the effectiveness of ALC through subsequent surveys to identify the remaining leaks,

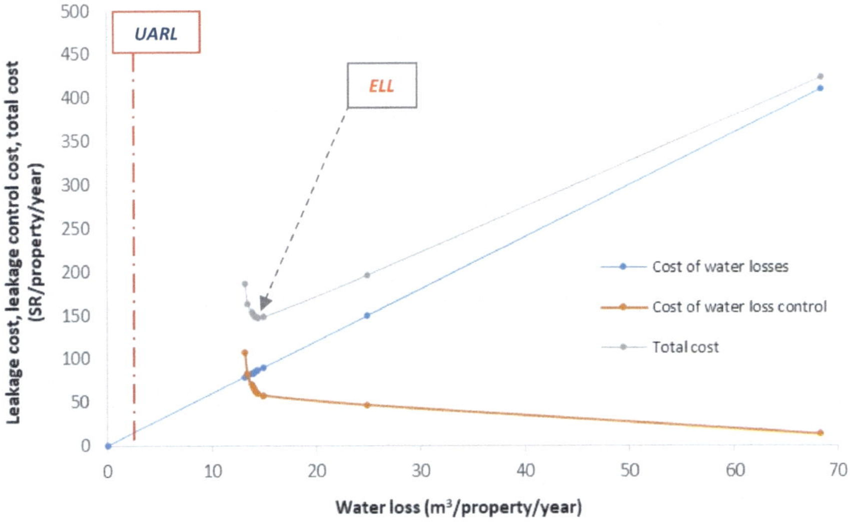

Fig. 28.5 Economic level of leakage (ELL) analysis for the study area

which are challenging to locate by the survey teams, needs additional human financial resources. Thus, the relationship between water loss reduction and the associated costs is not linear (Hardeman, 2008). Figure 28.6 shows a non-linear behavior between the rate of change of cost and the reduction in unit water loss. The present study developed Fig. 28.6 using the knowledge obtained from literature, municipality personnel, and experts from practice and academia.

The next step, while appraising the impact of environmental valuation on ELL, is arranging the additional financial resources. Generally, these resources can be generated through a rational increase in water rates. Although existing water rates in the study areas are substantially low (i.e., 1.6 US$/m^3) compared to developed countries, evaluating the WTP after a significant price increase in the recent past is essential. Under the existing sociopolitical conditions of KSA, WTP surveys are practically challenging to conduct for real-time environmental valuation using CVM, i.e., Eq. (28.4). Therefore, a hypothetical analysis evaluated the impact of the increase in water price on ELL. The results presented in Fig. 28.7 show that sufficient finances can be raised for enhanced ALC with a slight price increase of 2% in existing water rates. The gap between the ELL and UARL can be further reduced with a 4% increase in water rates. Considering the existing subsidized water pricing system in KSA, such a minor rise can come up with considerable environmental benefits for the long-term sustainability of limited water resources in the country.

aquifers using the DRASTIC model and GIS techniques. *Environmental Earth Sciences, 74*(2), 1303–1318. https://doi.org/10.1007/s12665-015-4120-5

Al-Salamah, I. S., Ghazaw, Y. M., & Ghumman, A. R. (2011). Groundwater modeling of Saq aquifer Buraydah Al Qassim for better water management strategies. *Environmental Monitoring and Assessment, 173*(1–4), 851–860. https://doi.org/10.1007/s10661-010-1428-1

American Water Works Association Research Foundation (AwwaRF). (2007). *Evaluating Water Loss and Planning Loss Reduction Strategies*, Report 91163.

Ashton, C. H., & Hope, V. S. (2001). Environmental valuation and the economic level of leakage. *Urban Water, 3*, 261–270. https://doi.org/10.1016/S1462-0758(01)00046-2

Breffle, W. S., Morey, E. R., & Lodder, T. S. (1998). Using contingent valuation to estimate a neighbourhood's willingness to pay to preserve undeveloped Urban land. *Urban Studies, 35*(4), 715–727. https://doi.org/10.1080/0042098984718

Carson, R. T. (2000). Contingent valuation: A user's guide. *Environmental Science & Technology, 34*(2000), 1413–1418. https://doi.org/10.1021/es990728j

Creaco, E., & Walski, T. (2017). Economic analysis of pressure control for leakage and pipe burst reduction. *Journal of Water Resources Planning and Management, 143*(12), 04017074. https://doi.org/10.1061/(ASCE)WR.1943-5452.000084

Emerton, L. (1996). *Valuing the environment: Case studies from Kenya, African Wildlife Foundation* (pp. 12–13). Singal Press Ltd.

Erickson, J. J. (2016). *The effects of intermittent drinking water supply in Arraiján, Panama* [Dissertation]. University of California, Berkeley.

European Union (EU). (2015). *EU Reference document good practices on leakage management WFD CIS WG PoM, main report, office for official publications of the european communities, Luxembourg*.

Francisque, A., Tesfamariam, S., Kabir, G., Haider, H., Reeder, A., & Sadiq, R. (2017). Life cycle cost analysis for water mains renewal of small to Medium sized utilities: A decision support tool. *Uran Water Journal, 14*(5), 493–501. https://doi.org/10.1080/1573062X.2016.1223321

Global Water Market. (2017). *Meeting the world's water and wastewater needs until 2020*. Volume 4: Middle East and Africa. Saudi Arabia, p 1379–1385.

Haider, H. (2017). Performance assessment framework for groundwater treatment plants in arid environments: A case of Buraydah, Saudi Arabia. *Environmental Monitoring and Assessment, 189*, 544. https://doi.org/10.1007/s10661-017-6271-1

Haider, H., Al-Salamah, I. S., & Ghumman, A. R. (2017). Development of groundwater quality index using fuzzy-based multicriteria analysis for Buraydah, Qassim, Saudi Arabia. *Arabian Journal of Science and Engineering, 49*(9), 4033–4051. https://doi.org/10.1007/s13369-017-2578-z

Haider, H., Al-Salamah, I. S., Ghazaw, Y. M., Abdel-Maguid, R. H., Shafiquzzaman, M., & Ghumman, A. R. (2019). Framework to establish economic level of leakage for water distribution systems in arid environments. *Journal of Water Resources Planning and Management, 145*(2), 05018017. https://doi.org/10.1061/(ASCE)WR.1943-5452.000102

Haider, H., Sadiq, R., & Tesfamariam, S. (2014). Performance indicators for small and medium sized water supply systems: A review. *Environmental Reviews, 22*(1), 1–40. https://doi.org/10.1139/er-2013-0013

Hardeman, S., (2008), *A Cost-Benefit Analysis of Leak Detection and the Potential of Real Water Savings for New Mexico Water Systems*. Water Resources Professional Project Reports, UNM Digital Repository, University of New Mexico.

Hashwa, F., & Tokajian, S. (2004). *Intermittent water supply and domestic water quality in the middle east, Chapter 2, water in the middle east and north Africa* (pp. 157–166). Springer.

ILMSS. (2013). Guidelines relating to the assessment and calculation of average pressure in water distribution systems and zones. In *International Leakage Management Support Services, ILMSS Ltd*. UK.

Islam, M. S., & Babel, M. S. (2013). Economic analysis of leakage in the bangkok water distribution system. *Journal of Water Resources Planning and Management, 139*, 209–216. https://doi.org/10.1061/(ASCE)WR.1943-5452.0000235

Jackson, S., Finn, M., & Scheepers, K. (2014). The use of replasscement cost method to assess and manage the impacts of water resource development on Australian indigenous customary economies. *Journal of Environmental Management, 135*(15), 100–109. https://doi.org/10.1016/j.jenvman.2014.01.018

Kirchoff, S., Colby, B. G., & LaFrance, J. T. (1997). Evaluating the performance of benefit transfer: An empirical inquiry. *Journal of Environmental Economics and Management, 33*, 75–93. https://doi.org/10.1006/jeem.1996.0981

Lambert, A. O. (2002). International report: Water losses management and techniques. *Water Supply, 2*(4), 1–20. https://doi.org/10.2166/ws.2002.0115

Lambert, A. O., & McKenzie, R. D. (2002). Practical experience in using the infrastructure leakage index. In *Proceedings of the IWA Conference—Leakage Management: A Practical Approach, Lemesos, Cyprus*.

Lambert, A. O., Brown, T. G., Takizawa, M., & Weimer, D. (1999). A review of performance indicators for real losses from water supply systems, AQUA—water Infrastructure. *Ecosystem, and Society, 48*(6), 227–237. https://doi.org/10.2166/aqua.1999.0025

OFWAT. (2000). *1999–2000 report on leakage and water efficiency*. Office of Water Services, Birmingham. https://www.nao.org.uk/wp-content/uploads/2000/12/9900971.pdf. Accessed December 26, 2022.

Ojeda, M. I., Mayer, A. S., & Solomon, B. D. (2008). Economic valuation of environmental services sustained by water flows in the Yaqui River Delta. *Ecological Economics, 65*(2008), 155–166. https://doi.org/10.1016/j.ecolecon.2007.06.006

Palmquist, R. B. (1984). Estimating the demand for the characteristics of housing. *The Review of Economics and Statistics, 66*(3), 394–404. https://doi.org/10.2307/1924995

Phuong, D. M., & Gopalakrishnan, C. (2010). An application of the contingent valuation method to estimate the loss of value of water resources due to pesticide contamination: The case of the Mekong Delta, Vietnam. *International Journal of Water Resources Development, 19*(4), 617–633. https://doi.org/10.1080/0790062032000161409

Smith, V. K., & Desvousges, W. H. (1985). The generalized travel cost model and water quality benefits: A reconsideration. *Southern Economic Journal, 52*(2), 371–381. https://doi.org/10.2307/1059623

UKWIR. (1999). *The environmental and social value of leakage reduction, UKWIR Report 99/WM-08/24*. United Kingdom Water Industry Research Ltd.

UKWIR/WRc. (1994). *Managing leakage (Report C): Setting economic leakage targets*. UK Water Industry Research, Water Research Center, Water Services Association, Water Companies Association, UK.

Van Zyl, K., Lambert, A., & Collins, R. (2017). Realistic modelling an intrusion flows through leak openings in pipes. *Journal of Hydraulic Engineering ASCE, 143*(9), 04017030.

Water Scarcity Solutions. (2018) Water Resources Group, https://waterscarcitysolutions.org/ajax/results. Accessed 15 Nov 2022.

WHO (World Health Organisation). (2005). *Air quality guidelines: Global update 2005*. WHO.

WLRandA, and ILMSS. (2018). *Water leakage and pressure management*. http://www.leakssuite.com. Accessed May 15 2022.

World Resources Institute. (2018). *Water stress by country*. http://www.wri.org/resources/charts-graphs/water-stress-country. Accessed June 3 2022.

Yorkshire Water Services Ltd. (1997). *Establishing the economic level of leakage*. Yorkshire Water Services Ltd., Leeds. UK.

Chapter 29
How Can Water Resources Be Used Economically for Agro-irrigation in Water-Limited Regions? Example of Semi-arid Konya Ecology, Turkey

Bilal Acar, Yasemin Demir Yurteri, Nadia Riaz, Nurcan Yavuz, and Duran Yavuz

Abstract Konya basin is a great example of severe climate change, with recurring spells of agricultural drought. Climate change has had detrimental effects on the region, including decreased surface water levels and, in some cases, full lake desiccation. This research included a thorough assessment of the Konya Basin's current water and plant production assets, and suggestions for the judicious use of existing water resources in agriculture. Unregulated groundwater withdrawals for irrigation have resulted in significant groundwater level depletions in several parts of Konya Plain. The region's overuse of water resources is driven by two key factors: uncontrolled extension of irrigated land and increased cultivation of high water-demand crops such as sugar beet and maize. The first step in fostering sustainable water resource management is to coordinate crop patterns with present water supplies and to promote the use of water-efficient irrigation technologies such as sprinkler and drip systems. Additionally, within field circumstances, it is recommended to apply deficit irrigation for certain crops such as sugar beet, maize, sunflower, common bean, and potato.

Keywords Crop pattern · Irrigation systems · Water science · Water conservation · Water shortage

29.1 Introduction

Water and sanitation were formally recognized as basic human requirements by the United Nations (UN) in 2012 (De Albuquerque & Leo, 2014). Nonetheless, the current situation demonstrates that 786 million people lack access to adequate,

B. Acar (✉) · Y. D. Yurteri · N. Riaz · N. Yavuz · D. Yavuz
Department of Farm Building and Irrigation, Faculty of Agriculture, University of Selçuk, Konya P.O. Box 42130, Turkey
e-mail: biacar@selcuk.edu.tr

A. Sefelnasr et al. (eds.), *Water Resources Management and Sustainability*, Water Science and Technology Library 114, https://doi.org/10.1007/978-3-031-80520-2_29

uncontaminated water sources (Gupta et al., 2016). This situation is aggravated by a variety of reasons, including flooding, the effects of climate change, and water pollution, all of which contribute to a rising shortage of freshwater (Gupta et al., 2020). Protecting water resources depends on establishing a good water management system. The water management system covers a wide range of areas, from the protection and use of water resources to the pollution that occurs after use and the purification of water. The water management system proposed for developing countries where water shortages are most common today is likely to create more problems rather than solve them (Kilic, 2008). Figure 29.1 shows the enhancement of crop yield to the potential transformation of low flow hydrology.

Policy formation, strategic planning, quality preservation, investment, monitoring, regulatory permits, inspections, regulation enforcement, and coordination initiatives are all part of water management. These multimodal activities attempt to achieve the beneficial use of water resources while limiting their negative consequences and taking into consideration the demands of various ecosystems and sectors (Kirtorun & Karaer, 2018). Water management's primary goal is to ensure fairness in water accessibility, long-term stewardship of water resources, methodical development, equitable distribution, and efficient utilization (Elerman et al., 2022). Water use can be broken down into three major areas globally: agriculture (71%), industry (18%), and domestic usage (11%). These proportions, however, differ dramatically across developed and developing countries. Over 90% of water is utilized for agriculture in

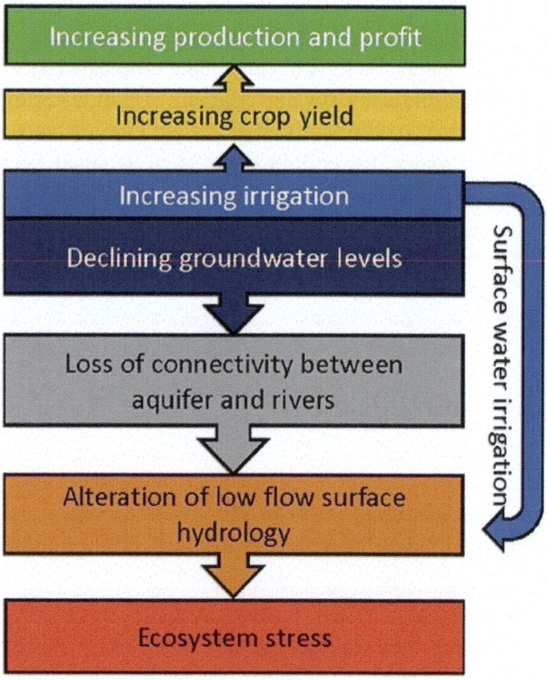

Fig. 29.1 This conceptual visualization evaluates the trade-offs of extending irrigation in order to boost crop productivity and profit while recognizing the possible ecological stress, particularly on low flow hydrology (Yasarer et al., 2020)

developing countries, whereas it is only about 16% in developed nations. The 73% of fresh water in Turkey is allocated to agriculture, 11% to industry, and 16% to home usage (Nayır & Somuncu, 2018).

Sustainable water resource management necessitates an integrated approach to water management that covers both immediate problems and prospective issues (Loucks, 2000). Current water sector issues may be ascribed to causes such as fast-growing populations, industrial growth, changing consumption patterns, growing urbanization, and the effects of climate change-induced drought (Kirtorun & Karaer, 2018; Saatsaz, 2020). With a growing global population and fluctuating climatic patterns, efficient water management is becoming increasingly important, particularly to fulfill the expanding need for food production. Sustainable irrigation technologies provide a way to improve water consumption efficiency while reducing waste and boosting agricultural production. Additionally, proper water management in agriculture is critical to protecting water supplies, lowering stress on freshwater ecosystems, and enhancing environmental sustainability (Saatsaz, 2020). Climatic challenges, especially severe drought, highlight the critical importance of optimizing the protection of current water resources and ensuring the efficient operation of existing structures (Tanışıklı & Çakmak, 2023). It is critical to note that drought has a reversible influence on groundwater, resulting in a dynamic interaction between the two (Parween et al., 2021). This study is focused on analyses of Konya Basin water resources regarding agricultural water management.

29.2 Methods

In this study, the current water and plant production conditions in the Konya basin were evaluated. In this context, the sustainability of groundwater resources has been analyzed. Data about water resources and the current planting rate was taken from the results of research conducted by the authors previously. Hence, with this objective, some common field crops are examined.

The irrigation programs of alfalfa (Fig. 29.2), sugar beet (Fig. 29.3), sunflower (Fig. 29.4), and dry bean plants, which are widely cultivated in the region, were examined. Data regarding the irrigation program of the field crops in question, which are widely cultivated, were obtained from farmers' statements as a result of farmer visits to the field. To assess the irrigation performance by farmers more appropriately, some mathematical expressions were required. In that regard, the seasonal amounts of irrigation water given to sugar beet, alfalfa, wheat, and barley plants irrigated by sprinkler irrigation under farmer conditions were calculated with the help of the Eq. (29.1) below:

$$I_s = Ip \times Ta \tag{29.1}$$

where I_s is the applied water with sprinkler system (mm), Ip is the sprinkler precipitation rate (mm/h) and Ta is the seasonal irrigation time (h). Similarly, the amount

Fig. 29.2 Alfalfa (Medicago sativa) crop grown in the field under sprinkler irrigation

Fig. 29.3 Sugar beet (Beta vulgaris) crop grown in the field under sprinkler irrigation

of seasonal irrigation water given to grain-corn (Figs. 29.5 and 29.6), sunflower, and dry bean plants irrigated with drip irrigation under farmer conditions was calculated with the help of the Eq. (29.2) below:

$$d_v = \frac{q}{ds \times ls} \quad (29.2)$$

Fig. 29.4 Oil sunflower (Helianthus annuus) crop is harvesting

where d_v, q, ds *and* ls is the dripper velocity (mm/h), dripper flow rate (L/h), dripper space (mm), and lateral space (m) respectively. As given in Eq. (29.3):

$$Id = dv \times Ta \tag{29.3}$$

where *Id* represents applied water with drip system (mm). The results obtained were compared with universal literature and the necessary steps to use water resources efficiently in the region were explained in detail.

29.3 Results and Discussion

29.3.1 Assessment of Water and Crop Potential in Konya Plain

Konya, Turkey's biggest province, occupies an extensive area of 40,838 km^2, accounting for 5.24% of the country's massive 780,043 km^2 landmass, including lakes. Agriculture thrives across 18,590,788 decares inside its limits. The province's elevation ranges from 570 to 1700 m above sea level. Annual precipitation varies

Fig. 29.5 Corn (Zea mays) crop-grown irrigation through single lateral design for two crop rows

Fig. 29.6 Ripened drip-irrigated corn on the cob

greatly over the region, ranging from 300 to 760 mm. Notably, the province's eastern and northern expanses, collectively known as the legendary 'Konya Plain,' receive inadequate rainfall, falling below 400 mm in several parts and even falling below the 300 mm threshold in certain local pockets, prompting the adoption of dry farming methods. Furthermore, according to statistics from the 4th and 18th Regional Directorates of the State Hydraulic Works (DSI), Konya province has an irrigated agricultural domain covering 609,299 hectares. The introduction of fruit-bearing and non-fruit-bearing tree farming, aided by irrigation expansion, progressively modifies the plain's steppe-like appearance (Anonymous, 2021). Table 29.1 shows the land utilization in Konya by comparison to Turkey. It can be observed that Konya province has approximately 9.5% of Turkey's field crop cultivation.

According to statistics from the 4th and 18th Regional Directorates of the State Hydraulic Works (DSI), Konya Plain presently benefits from irrigation across 609,299 hectares of land. A scenario expands in which all subterranean and surface irrigation networks operate as closed systems to supply the water requirements for crops planted across a spreading 1,859,079-hectare area. In this framework, in-field irrigation is carried out using precision drip/sprinkler systems that are suited to crop varieties and managed in accordance with plant needs while taking meteorological variables into account. This whole strategy demands 4.5 billion cubic meters of water every year. In contrast, the accessible subsurface and surface water resources in Konya province are 3.1 billion cubic meters. Even when agricultural science's finest practices are completely implemented, this sharp difference indicates a severe water shortfall. Groundwater's importance as an invisible supply of water within the world of freshwater resources highlights its special relevance (Yılmaz, 2021).

The fall in groundwater levels within the Konya plain has grown more significantly, particularly during the periods 1989–2019 (Fig. 29.7). The reasons for the groundwater level reductions are increasing the irrigation areas and huge land size of cultivation areas with high water consuming crops such as maize and sugar beet. Therefore, overwater extractions to meet the crop water requirement of those high-water consuming crops are present from the basin's groundwater resources. This has resulted gradual water level depletions of groundwater.

This depletion pattern is expected to worsen in coming years, with periods of less winter rainfall and, conversely, unexpected and strong spring rainfalls (Orhan,

Table 29.1 Land utilization in Konya and Turkey (Anonymous, 2021)

	Total cultivation land (ha)	Field crops (ha)	Fellow (ha)	Vegetable (ha)	Fruit plants (ha)	Ornamental plants (ha)
Konya	1,859,079	1,473,258	307,158	30,881	47,698	84
Turkey	23,136,584	15,614,972	3,173,252	779,246	3,563,707	5407
Konya/TR (%)	8.04	9.43	9.68	3.96	1.34	1.55

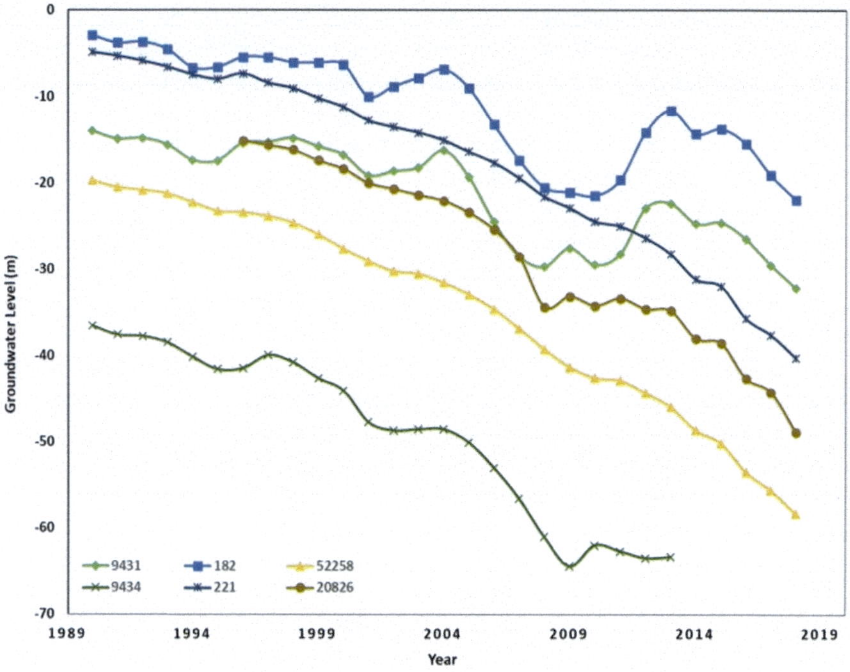

Fig. 29.7 Annual level changes in the six groundwater between 1990 and 2018 in Konya Basin (Orhan, 2021)

2021). These climatic conditions have resulted in a significant decline in groundwater levels, which have exceeded 20 m in certain regions. An average reduction of around 2.5 m was reported over the whole Konya closed basin. In contrast to 1960, groundwater levels dropped by over 50 m across the basin. Groundwater depths have dropped below 400 m in critically water-stressed agro-districts such as Çeltik, Yunak, Cihanbeyli, and Altnekin in the basin's northern borders, casting question on the sustainability of groundwater-dependent agriculture in these areas. The growth of water-intensive crops such as sugar beet, silage maize, and alfalfa is a major contributor magnifying the requirement for groundwater in the basin. While claiming for a total ban on these crops in the basin is unfeasible, there is space for change in terms of the amount to which they are grown. The basin has between 35,000 and 40,000 permitted wells and, frighteningly, three times as many unregistered wells and farmers are therefore compelled to utilize these as a water resource. The basin utilizes 4.5 billion cubic meters of water each year, with approximately 2 billion cubic meters squeezed from surface water and a similar amount drawn from subsurface reservoirs (https://www.karamanhabercisi.com/konya-ovasindaki-yer-alti-su-kaynaklari-her-gecen-gun-dusuyor-78816h.htm-access in September 18, 2023) (Fig. 29.8).

Fig. 29.8 Sinkholes in some parts of the Konya basin (Sabah, 2019, 2022)

The Konya closed basin is Turkey's greatest groundwater-consuming basin. Fortunately, the basin is rich in ground water supply but not used sustainably. Water consumption limitation can help to prevent sinkhole formation in the short and medium run. The influence of water on sinkhole formation can be considerably decreased by having control over groundwater consumption within the basin—a crucial contributing component. Sinkholes have grown fast over the region in recent years, owing mostly to extensive and uncontrolled water extraction, which has been exacerbated by recurrent drought. The excessive removal of water from several freshly constructed boreholes in the Konya plain, along with the resulting decline in groundwater levels, clearly stands out as a stimulant for sinkhole development (Ökten, 2012). The primary challenge in agricultural irrigation within the Konya plain is inefficient water resource management. Notably, there is an overreliance on subterranean water supplies for irrigation, owing especially to the region's substantial cultivation of water-intensive crops. Another important problem is the lack of adequate water application measures in the irrigation of crops (Acar et al., 2023a).

29.3.2 Importance of Pressurized Irrigation Systems for Conserving Water Resources

High and quality plant production can only be achieved by fully meeting the irrigation water needed by the plants in the plant root environment throughout the growing season. Decreases in yield and quality may be observed with increased water stress in the root zone. If water stress exceeds a certain level, plants may die. In order to prevent this situation, the missing water in the plant root area should be supplemented regularly through irrigation practices (Essien & Essien, 2014). Acar et al. (2022) emphasized the importance of pressurized irrigation methods in the economical use of water in agriculture, especially in arid and semi-arid areas where water resources are scarce. They underlined that those techniques' system components must be designated properly and managed well under field conditions to save a significant amount of irrigation water. Some researchers (Acar, 2019; Darko et al., 2017; Nassab et al., 2013) state that with well-designed and operated irrigation systems, irrigation water will be stored more homogeneously in the plant root zone with very little water loss. They added that the efficiency of plant fertilizer use will increase as a result of preventing the plant nutrients from being washed away under the plant root zone, and because of all this, there is no doubt that plant production and yield quality will increase.

The drip irrigation method is indeed a specially designed irrigation technique that can apply the desired amount of irrigation water. Thanks to this system, irrigation water is applied to the plant root environment at frequent intervals, regularly and slowly. In cases where the land is designed, laid out and managed properly, the effectiveness of water application from a drip irrigation system can be around 90–95% (Goyal, 2012). It was reported (Acar et al., 2023b) that in order to apply equal irrigation water to the plants at the desired rate with the drip irrigation system, the system must first be planned correctly and installed on the land with qualified workmanship. In addition, they recommended some physical, chemical and biological processes be done depending on the clogging situation to prevent clogging of the drippers.

Therefore, the drip irrigation system alone is not sufficient to achieve a high-water application performance. High water application efficiency can be achieved from the system with the quality of the system elements, correct hydraulic design of the system and healthy operation of the system (Luquet et al., 2005). Sprinkler and drip irrigation techniques can be recommended to apply irrigation water to the plants in the desired homogeneity (AR-GE, 2012). This basin is known for the wide-spread practices of sprinkler irrigation systems in Turkey and due to the agricultural drought in basin, the drip irrigation system is getting the popularity.

29.3.3 *Assessment of Irrigation Scheduling of Some Field Crops*

In research fields, a water source is a deep well with capacity of 130–140 m^3/h. Sugar beet, alfalfa, wheat and barley have been irrigated by sprinkler system. The sprinkler arrangement was (10 × 10) m. Sprinkler flow rates for sugar beet and alfalfa were 0.6 m^3/h, and for wheat and barley was 0.9 m^3/h. Grain maize, sunflower, dry bean have been irrigated by drip system. One lateral design for two crop rows, for example 90 cm for dry bean, and 140 cm for both the grain-maize and sunflower plants. The emitter space and lateral tube diameter, and emitter discharge were 30 cm, 22 mm and 1.6 L/h for all drip irrigation systems respectively. The irrigation interval was 7 days for those examined crops. Table 29.2 shows the irrigation practices with yields of research field crops.

The yields of sugar beet, alfalfa, grain corn, sunflower, dry beans, wheat and barley, which were the subject of the research, were 75–100 t/ha, 12–15 t/h, 16–17 t/ha, 3.5–4.0 t/ha, 4.0 t/ha, 7–8 t/ha and 8–9 t/ha, respectively. The results obtained

Table 29.2 Irrigation practices for common field crops in Konya—Kadınhanı Province

Crops	Irrigation number	Yield (t/ha)
Sugar beet	12	75–100
Grain-maize	8–9	16–17
Sunflower	7	3.5–4.0
Common bean	8	4.0
Alfalfa	8–10	12–15*
Wheat	2–3	7–8
Barley	2	8–9

* (one cut at two irrigation)

are satisfactory because farmers in the basin are quite familiar to all the agricultural activities including irrigation system utilization.

The seed yield of the Sandro hybrid sunflower variety, irrigated with furrow irrigation in clayey soil in Bursa conditions, was reported as 4,056 t/ha under full irrigation conditions. The amount of irrigation water applied under full irrigation conditions was found to be 355 mm (Göksoy et al., 2004). They also stressed that deficit irrigation should be avoided during the flowering period. The seed yield obtained from the current research is quite close to their findings. Yavuz (2021) determined the grain yield in dry beans irrigated with drip irrigation as 2.36 t/ha and the amount of irrigation water applied as 508 mm in full irrigation, where irrigation is done during the entire plant growing period in the clay loam soil in Konya. Although the amount of irrigation water applied in this research is approximately similar to Yavuz (2021), the grain yield obtained, 4.0 t/ha, is higher than the result of Yavuz, (2021). Although both studies were conducted in the same region, the high grain yield in the current study may be due to the bean seed characteristics used in both studies.

The alfalfa hay yield obtained from this research is similar to Li & Su (2017). The amount of irrigation water applied to alfalfa irrigated with border irrigation under clayey soil conditions in Davis, California was determined as 1423 mm as a two-year average (Hanson et al., 2007). The results from the current study are consistent with Hanson et al., (2007). Table 29.3 shows the irrigation practices performed by the farmers in Konya, Kadınhanı.

Farmers prefer 0.6 m^3/h small flow sprinkler heads for sugar beet and alfalfa irrigation. On the other hand, they prefer sprinkler heads with a flow rate of 0.9 m^3/h for wheat and barley irrigation. As a result of the application of these sprinkler flow rates in a (10 × 10) m sprinkler arrangement, the rainfall rates are 6 mm/h and 9 mm/h, respectively, and these values are close to or slightly lower than the soil water intake rate values. Therefore, these sprinklers do not cause ponding problems on the soil surface during irrigation. In addition, considering that the average water resource capacity in the region is 135 t/h, 225 sprinklers that give a flow rate of 0.6 m^3/h and 150 heads that give a flow rate of 0.9 m^3/h are used in irrigation. In this way, as the size of the irrigated area will increase, labor costs will also decrease significantly. In this research, the amounts of irrigation water applied to sugar beet, alfalfa, wheat,

Table 29.3 Irrigation scheduling under farmer's practices for common field crops in Konya—Kadınhanı Province

Irrigation system	Crops	Irrigation time (h)	Irrigation number	Applied water (mm)
Sprinkler	Sugar beet	12	12	864
	Alfalfa	15	10	900
	Wheat	10	3	270
	Barley	10	2	180
Drip	Grain-maize	15	8	757
	Sunflower	13	7	347
	Dry bean	12	8	569

barley, grain corn, sunflower and dry beans are 864 mm, 900 mm, 270 mm, 180 mm, 757 mm, 347 mm, and 569 mm, respectively. As seen clearly here, the plant to which the most water is applied is alfalfa, followed by sugar beet.

Sprinkler irrigation used in alfalfa farming in the Zanjan region of Iran. It was determined that the amount of irrigation water applied to alfalfa by sprinkler irrigation in the enterprises where alfalfa cultivation was carried out in the region in question varied between 984.9 mm and 2057.6 mm, and the average was 1425 mm. Hay yield of alfalfa as varying between 6.5 t/ha and 14.1 t/ha, with the average being 10.4 t/ha (Ojaghlou et al., 2023). The grain yield of barley under irrigation around Diyarbakır, located in the Southeastern Anatolia region, varies between 3880 and 4870 kg/ha (Karahan & Sabancı, 2010). Wang (2017), grain yield from winter wheat on silty loam soil in two different agricultural lands of Shandong, China, was 9560 kg/ha as an average of seven years (2008–2014). Therefore, China has successful potential in terms of productivity.

Therefore, in terms of efficient use of water, the plants examined are irrigated with sprinkler and drip irrigation, which shows that local farmers are conscious about irrigation water management. Considering the plant pattern, in terms of sustainability in agricultural irrigation, it can be strongly recommended to increase the planting areas of plants with low water consumption such as wheat, barley, dry beans, sunflower, chickpeas and lentils in the region. Especially for winter grains such as wheat and barley, as well as summer crops such as lentils and chickpeas, it is possible to obtain economic efficiency from these plants with one or two irrigations during critical periods when the plants are affected by water stress. In this context, it can be recommended that the state provide agricultural financial support to farmers who grow these plants to prevent excessive water withdrawal from underground water reserves for irrigation purposes.

In addition, irrigation energy costs constitute the highest production cost in agricultural areas where pressurized irrigation is carried out with groundwater in the region. As a matter of fact, in a study conducted in the Konya plain (Yavuz et al., 2016), it was determined that the cost of irrigation energy for the potato plant was approximately 44% of all production inputs applied during the growth periods of the product. It is obvious that excessive irrigation water not only reduces irrigation water use efficiency but also causes an increase in irrigation energy costs. In this context, farmers show planned behavior in terms of irrigation duration and number of irrigations. The hay yield of the Emperor Alfalfa variety, which is sprinkler-irrigated in the Gansu region in the northern part of China, was found to be 15.2 t/ha as a two-year average, depending on the irrigation doses (Li & Su, 2017).

Kaya et al. (2012) recommended that 6 irrigations were appropriate for sugar beet under loamy soil conditions in Erzurum and that 458.6 mm of net irrigation water was applied. In Bursa conditions, the amount of irrigation water applied to sugar beets irrigated with drip irrigation in clay soil was found to be 758 mm, plant water consumption was 863.45 mm and root production was 80.44 t/ha (Yetik & Candogan, 2022). Some studies can be found around the world regarding the amount of irrigation water applied seasonally to corn plants. For example, in the Pingtung region in the southern part of Taiwan, this value for grain corn is 375–555 mm (Greaves & Wang, 2017), in a study conducted in the Mardan region of Pakistan (Tariq & Khalid, 2009), this value for grain corn was 661.44 mm; and this value was determined as 486.8 mm in the Metekel region of Ethiopia (Tefera & Mitku, 2017). The amount of irrigation water applied to the alfalfa plant under Harran conditions was determined as 1270.3 mm (Nacar, 2021). The findings obtained from the current study are in general in conformity with the studies' results of authors analyzed above.

29.4 Recommendations

A thorough strategy to water conservation in agriculture is required particularly for water stressed regions like Konya basin. The primary solution deduced is that the crop patterns must be designated in accordance with current water resources to mitigate environmental problems. In addition, a deficit irrigation program should be implemented on field scales to conserve the water resources. Widespread adoption of pressurized irrigation system utilization is also another solution for water productivity. Agriculture water cost based on usage is critical for promoting appropriate use. Marginal water could be recommended for some special crops for minimizing water stress on the water resources.

29.5 Conclusions

Arid climatic conditions have a significant impact on crop selection, driving the cultivation of water-efficient crops as a strategy to reduce the water burden on the agricultural sector, which is a major consumer of water resources. To use the water resources sustainably particularly in water shortage regions, comprehensive irrigation planning at the agricultural basin level should be considered. This attempt aids in determining the most economically viable crops for each region of the country, hence reducing water and financial constraints. In our studied region, we found that saving up to 25% of irrigation water in the cultivation of crops such as sugar beet, corn, potato, dry beans, and sunflowers did not significantly reduce production or quality. In addition, to meet the agricultural sector's requirements for water, water transportation from external basins to our basins must be explored. Improving water transportation and application efficiency necessitates the use of mass pressure pipes over open channels, as well as the development of drip and sprinkler irrigation technologies. Drought-related research should go beyond just determining precipitation deficits.

29.6 Competing Interests

Authors have declared that no competing interests exist.

Acknowledgements We extend our heartfelt gratitude to the farmers of Konya-Kadınhanı—Kolukısa town for their invaluable contributions, including the provision of essential data, information, and pertinent documents integral to the study's context.

References

Acar, B., Direk, M., Yurteri, Y. D., Sayman, O., & Muradi, S. M. (2022). Field water management for saving water in water-starved environments such as konya plain Türkiye. *Green Reports, 3*(8), 48–50.

Acar, B. (2019). Deficit irrigation effect on water use efficiency of crops in arid and semi-arid regions. *International Journal of Agriculture and Economic Development, 7*(2), 18–25.

Acar, B., Uğurlu, N., Afacan, S., Gülen, E., Açık, N., Kökdere, E., & Taştan, Y. (2023a). Why is Correct AgriculturalWater Management Necessarily a Prerequisites in Water Shortage Regions?.In S. Haddout, A.M. Hoguane, & P.K. Lakshmi, *Springer Proceedings in Earth and Environmental Sciences* (pp. 164–170).

Acar, B., Afacan, S., & Açık, N. (2023b). Examining of Water Distribution Uniformity in Drip Irrigation Systems: A review of various recents Studies world wide. *International Scientific Journal Mechanization in Agriculture & Conserving of the Resources, 67*(1), 6–8.

AR-GE. (2012). *Bitkisel Üretim Çiftçi Rehberi. Konya Şeker.*

Darko, R. O., Shouqi, Y., Junping, L., Haofang, Y., & Xingye, Z. (2017). Overview of advances in improving uniformity and water use efficiency of sprinkler irrigation. *International Journal of Agricultural and Biological Engineering, 10*(2), 1–15.

De Albuquerque, C., & Leo, H. (2014). *Common violations of the human rights to water and sanitation, united nations human rights report. 2014.* Availabel online: https://www.ohchr.org/EN/Issues/WaterAndSanitation/SRWater/Pages/CommonHRViolations.aspx (Accessed on 30 September 2020).

Elerman, O., Yazıcı, T. G., & Gültekin, N. (2022). Su Yönetiminde Alternatif Su Kaynağı Sarnıçlar: Bodrum Yarımadasından Örnekler. *İdealkent, 14*(Özel Sayı), 510–535.

Essien, O. E., & Essien, M. (2014). Design and evaluation of gravity-fed perforated tube drip irrigation for dry season irrigation of *Citrus Sinensis. International Journal of Engineering Inventions, 3*(6), 28–35.

Göksoy, A., Demir, A., Turan, Z., & Dağüstü, N. (2004). Responses of sunflower (Helianthus annuus L.) to full and limited irrigation at different growth stages. *Field Crops Research, 87*(2–3), 167–178.

Goyal, M. R. (2012). *Management of drip/trickle or micro irrigation.* CRC Press.

Greaves, G. E., & Wang, Y. M. (2017). Yield response, water productivity, and seasonal water production functions for maize under deficit irrigation water management in southern Taiwan. *Plant Production Science, 20*(4), 353–365.

Gupta, A., Mishra, S., Bokde, N., & Kulat, K. (2016). Need of smart water systems in India. *International Journal of Applied Engineering Research, 11*(4), 2216–2223.

Gupta, A. D., Pandey, P., Feijóo, A., Yaseen, Z. M., & Bokde, N. D. (2020). Smart water technology for efficient water resource management: A review. *Energies, 13*(23), 6268.

Hanson, B., Putnam, D., & Snyder, R. (2007). Deficit irrigation of alfalfa as a strategy for providing water for water-short areas. *Agricultural Water Management, 93*(1–2), 73–80.

https://www.karamanhabercisi.com/konya-ovasindaki-yer-alti-su-kaynaklari-her-gecen-gun-dusuyor-78816h.htm-access in September 18, 2023.

Karahan, T., & Sabancı, C. O. (2010). Determination of yield and yield components of some barley (Hordeum vulgare L.) cultivars in Southeast Anatolia ecological conditions. *Derim.*

Kaya, S., Meral, R. & Demir, A. D. (2012). Şeker Pancarı Su İlişkileri. *1. Uluslararası Anadolu Şeker Pancarı Sempozyumu*, pp. 20–22 Eylül, 2012, Kayseri, 209–214 (In Turkish)

Kilic, S. (2008). Managing water in the processes of global climate change. *Istanbul University Journal of Faculty of Political Sciences-Siyasal Bilgiler Fakültesi Dergisi I, 39*, 161–186.

Kirtorun, E., & Karaer, F. (2018). Water management and sustainability of water. *Journal of Sustainable Engineering Practices and Technological Developments, 1*(2), 151–159.

Li, Y., & Su, D. (2017). Alfalfa water use and yield under different sprinkler irrigation regimes in north arid regions of China. *Sustainability, 9*(8), 1380.

Loucks, D. P. (2000). *Sustainable Water Resources Management. Water International, 25*(1), 3–10.

Luquet, D., Vidal, A., Smith, M., & Dauzat, J. (2005). 'More crop per drop': How to make it acceptable for farmers? *Agricultural Water Management, 76*(2), 108–119.

Nacar, M. S. (2021). *Determining the water consumption of alfalfa under harran plain condition by the use of weighing lysimeter and comparison with the predicting methods Harran University.* Graduate School of Natural and Applied Sciences.

Nassab, S. B., Beiranvand, Z. M., Maleki, A., & Panah, Z. (2013). Technical evaluation of sprinkler irrigation systems in khorramabad Iran. *World Rural Observations, 5*(3), 23–27.

Nayır, M.M., & Somuncu M. (2018). Agricultural water management in Ankara province. In *International Geography Symposium on the 30th Anniversary of Tucaum, 3–6 October 2018* (pp. 772–791), Ankara (In Turkish).

Ojaghlou, H., Ojaghlou, F., Jafari, M. M., Misaghi, F., Nazari, B., & Karami Dehkordi, E. (2023). Effect of irrigation management on water productivity indicators of alfalfa. *Water and Soil, 37*(2), 165–185.

Orhan, O. (2021). Monitoring of land subsidence due to excessive groundwater extraction using small baseline subset technique in Konya Turkey. *Environmental Monitoring and Assessment, 193*(4), 174.

Ökten, S. (2012). Important stations about water administration in Konya District, Suleyman Demirel University. *The Journal of Visionary, 3*(6).

Parween, F., Kumari, P., & Singh, A. (2021). Irrigation water pricing policies and water resources management. *Water Policy, 23*(1), 130–141.

Saatsaz, M. (2020). A historical investigation on water resources management in Iran. *Environment, Development and Sustainability, 22*, 1749–1785.

Sabah, D. (2019). *2 Giant sinkholes open up in Turkey's central Konya province.* Retrieved 10/24/2023 from https://www.dailysabah.com/turkey/2019/02/19/2-giant-sinkholes-open-up-in-turkeys-central-konya-province

Sabah, D. (2022). *Hundreds of sinkholes plague Türkiye's Konya Plain.* https://www.dailysabah.com/turkey/hundreds-of-sinkholes-plague-turkiyes-konya-plain/news

Tanışıklı, B., & Çakmak, B. (2023). Evaluation of water management activities at çorum water user association. *Türk Tarım-Gıda Bilim ve Teknoloji dergisi, 11*(5).

Tariq, J. A., & Khalid, U. (2009). Regulated deficit irrigation scheduling of maize crop. *Sarhad Journal of Agriculture, 25*(3), 441–450.

Tefera, A. H., & Mitku, D. T. (2017). Determination of optimum irrigation scheduling and water use efficiency for maize production in north-west ethiopia. *Environmental Science. Journal of natural science research,* 559722514.

Wang, D. (2017). Water use efficiency and optimal supplemental irrigation in a high yield wheat field. *Field Crops Research, 213,* 213–220.

Yasarer, L. M., Taylor, J. M., Rigby, J. R., & Locke, M. A. (2020). Trends in land use, irrigation, and streamflow alteration in the Mississippi River Alluvial Plain. *Frontiers in Environmental Science, 8,* 66.

Yavuz, D., Suheri, S., & Yavuz, N. (2016). Energy and water use for drip-irrigated potato in the middle a natolian region of Turkey. *Environmental Progress & Sustainable Energy, 35*(1), 212–220.

Yavuz, N. (2021). The response of dry bean to water stress at various growth cycles in a semi-arid region. *Selcuk Journal of Agriculture and Food Sciences, 35*(2), 91–100.

Yetik, A. K., & Candogan, B. N. (2022). Optimisation of irrigation strategy in sugar beet farming based on yield, quality and water productivity. *Plant, Soil and Environment, 68*(8), 358–365.

Yılmaz, M. (2021). Examining the legal and institutional perspective of groundwater management in Turkey. *Dicle University Journal of the Social Science Institute, 27,* 134–150.

Chapter 30
Reliable Short Term Water Demand Forecast Using Machine Learning—Part 1

Waleed Eldamaty, Mohammed Abdallah, and Khalid Al Zaabi

Abstract Water scarcity is one of the main challenges that Gulf Cooperation Council (GCC) countries have been facing for multiple decades. The challenge grows bigger as the region experiences rapid social and economic transformation, leading to an increase in water demands. Out of the various solutions that were implemented to face the challenge, demand and supply side management were heavily investigated. The balance between water supply and demand requires efficient water management system techniques, which are highly dependent on the use of accurate forecasting methodologies and tools. Accurate demand forecasting is a necessary input to many water processes, such as determining the required water reserve precisely as well as developing optimum operational plans for pumping stations and water production plants. There is no single global optimum method used to forecast water demand. It is more of a case-by-case approach depending on the network complexity, operational limitations, available data, forecast horizon, intuitiveness of the tool, and accepted percentage of deviation between actual and forecasted demand. The purpose of this paper is to achieve the global sustainability goal by addressing the gap that is currently present between water supply and demand. This is done by presenting an innovative and accurate methodology to forecast water demand using machine learning (ML) for the short-term. The results for a water utility in the United Arab Emirates (UAE) showed that the mean absolute percentage error (MAPE) between the actual and forecasted demand was reduced from 5.42% for the conventional forecasting method to 3.25% for the proposed ML forecasting method. Similarly, the root mean square error (RMSE) was reduced from 11.14 million imperial gallons per day (MIGD) for the conventional forecasting method to 7.6 (MIGD) for the proposed ML forecasting method. Additionally, the total difference per year between the actual and forecasted

W. Eldamaty (✉) · M. Abdallah · K. Al Zaabi
Emirates Water and Electricity Company, United Arab Emirates University, P.O. Box 22219, Abu Dhabi, UAE
e-mail: waleed.eldamaty@ewec.ae

M. Abdallah
e-mail: mohammed.abdulla@ewec.ae

K. Al Zaabi
e-mail: khalid.alzaabi@ewec.ae

A. Sefelnasr et al. (eds.), *Water Resources Management and Sustainability*, Water Science and Technology Library 114, https://doi.org/10.1007/978-3-031-80520-2_30

demand was reduced from 2683 million imperial gallons (MIG) for the conventional forecasting method to 898 (MIG) for the proposed ML forecasting method. This shows that by having an accurate demand forecast, the gap between actual and forecasted demand can be reduced, which will improve the overall efficiency and performance of the water management system.

Keywords Demand forecast · Machine learning · Water demand · Sustainability · Water scarcity · Water management

30.1 Introduction

Forecasting water demand is crucial for effective water management systems, allowing for the optimization of water supply and demand. Therefore, various approaches to water demand forecasting have been explored in recent times. The estimation of water consumption costs in Malaysia was conducted through the application of Holt-Winter's and auto-regressive integrated moving average (ARIMA) models. The outcomes revealed that the ARIMA (2,1,4) model displayed lower mean absolute percentage error (MAPE) and mean absolute deviation (MAD) values, as well as enhanced accuracy when compared to Holt-Winter's model (Razali et al., 2018).In Spain, a study was conducted on various double seasonal univariate time series models, including generalized autoregressive conditional heteroskedasticity (GARSH), ARIMA, and Holt-Winters. The findings revealed that combined forecasts yield higher accuracy for short-term demand forecasting. Conversely, when it comes to specific days of the week, individual forecasts from GARCH models and Holt-Winters exponential smoothing can improve forecast precision (Caiado, 2010). The investigation in Cyprus involved the use of multiple linear regression and three types of multilayer perceptron artificial neural networks (ANN) to analyze the short-term forecast for peak demand. Among these methods, the Levenberg–Marquardt ANN model proved to be the most accurate in predicting the forecast (Adamowski & Karapataki, 2010). Additional research was conducted to compare the effectiveness of combining ANN and discrete wavelet transforms (WA) with multiple linear regression (MLR), multiple nonlinear regression (MNLR), and ARIMA in Canada. The study revealed that the WA-ANN models yielded a more accurate demand forecast (Adamowski et al., 2012). Research was carried out to develop a model for forecasting water demand in six different areas of the Netherlands over the next 48 h, with 15-min time intervals. The study revealed that the mean absolute percentage error (MAPE) for the 24-h forecasts ranged from 1.44 to 5.12%, and for the 15-min forecasts from 3.35 to 10.44% (Bakker et al., 2013). In order to plan for short-term water grids, a set of methodologies was devised to formulate optimal operational plans (Ashbolt et al., 2014). The research pointed out that different objectives could influence the decision on which demand forecast technique to use for operational planning. These objectives encompass water security, operational expenses, energy utilization, environmental consequences, water quality, storage tank levels, and so forth. Data availability and complexity of the network have a big impact on the choice of the best demand forecast method. The performance of different water demand forecasting techniques was examined in a study by (Vijai & Bagavathi Sivakumar, 2018). Methods such as Gaussian process regression (GPR), least square support vector machine (LSSVM), random forest (RF), and multiple regression were analyzed for their effectiveness. The study highlighted the good performance of ANN models in short-term predictions. (De Souza Groppo et al., 2019) carried out a review focusing on the use of artificial intelligence in forecasting water demand. The study explored different techniques based on soft computing, as well as models developed through conventional statistical methods like linear regression or time-series analysis. It was noted in the paper that there is no single global optimal model that can be universally applied to all cases in all regions. Each region poses unique challenges and factors that impact water demand forecasting. The long short-term memory (LSTM) model is used to forecast the long-term water demands in UAE considering multiple factors including population growth, gross domestic product (GDP), consumer price index (CPI) and the mean temperature (Ahmed et al., 2020). Two hybrid models for water demand forecasting are utilized to study future water consumption in Spain (Pandey et al., 2021). In one method, difference pattern sequence forecasting (DPSF) is combined with ensemble empirical mode decomposition (EEMD), while in another method, EEMD is combined with DPSF and ARIMA. The (EEMD-DPSF) approach achieved higher prediction accuracy than other advanced methods. Short-term water demand forecast for smart water grids (SWG) are studied in Korea (Koo et al., 2021). The performance of the model was assessed by using the Pearson correlation coefficient, normalized root mean square error, residual, root mean square error, and Nash–Sutcliffe efficiency as indices.

The results showed that the ARIMA, RBF-ANN, and LSTM models are limited in terms of application for the management of a water supply system, considering only usage time and water consumption data. Hence, more factors need to be added to the model for optimum results. Deep learning techniques usually yield the best results. However, they have a high forecasting error at extreme points. (Salloom et al., 2021) studied this issue in China by inserting virtual data within the actual data to relieve the nonlinearity around them and developing a gated recurrent unit (GRU) model along with an unsupervised classification method (k-means) to create new features to improve forecasting accuracy. The results showed that extending the data set reduces errors by 30%, however it increases the training time. Manual feature extraction in deep learning is one of its main issues. In India, (Kavya et al., 2023) conducted a comparison between deep learning and machine learning methodologies. The results showed that for the univariate model and multivariate model, long-short term memory (LSTM) model showed the best results, with a mean absolute error of 0.11 m^3/hr. and 2.96 m^3/hr., respectively.

As can be seen from the literature review, different techniques and models were investigated to forecast water demand. However, it is clear that there is no single optimum technique that can be used for any region or any water network. Based on the conducted literature survey, the novelty of this work stems from the lack of investigation of short-term water demand forecasts in GCC countries and especially within the UAE. In addition, certain approaches cited in the literature may not be feasible in our specific scenario due to their incapacity to deliver real-time short-term demand forecasts, considering their complexity and lengthy model training time. Furthermore, the majority of short-term demand forecasting methods examined in the literature were trialed in areas with water supply and demand trends that varied from those seen in GCC countries. Moreover, the paper demonstrates the utilization of Microsoft's software Power BI, and its new machine learning features in training and optimizing the models. Hence, all of the aforementioned reasons highlight the importance and originality of this work. The aim of this paper is to bridge the existing gaps mentioned previously between water supply and demand, especially in the UAE, in order to fulfill the global sustainability goal by presenting a new reliable short term water demand forecast method based on machine learning.

30.2 Method

Many methods have been used in recent years to forecast the water demand, ranging from time series analysis, exponential smoothing, ARIMA to regression analysis, artificial neural networks, and hybrid methods. However, what is agreed upon is that there is no single global optimum method used to forecast water demand. It is more of a case-by-case approach depending on the network complexity, situation, operational limitations, available data, forecast horizon, degree of model complexity, etc. With that being said, machine learning models in particular have proved to have a leading edge over the other methods as they have the ability to deal with complex systems. Our focus in this paper will be on supervised machine learning, specifically regression modeling, as it has the ability to predict an output based on one or more inputs.

30.2.1 Regression Modeling

The primary objective of the regression model is to determine the correlation between a dependent variable (y) and independent variables. (x).

$$y = \alpha + \beta x + \varepsilon \tag{30.1}$$

The y-intercept ("Constant") is the mean value of the independent variable when all the dependent variables in the model are equal to zero (value of y when x = 0). (β) is the average change in the independent variable for a one unit increase in the dependent variable. The error (ε) is the difference between a predicted value and the observed actual value.

30.2.2 Regression Model Variables

The water demand forecast is influenced by a variety of factors. These factors can be divided into two main categories: short-term forecasts are affected by climatic conditions like temperature, relative humidity, and precipitation, while long-term forecasts are impacted by socio-economic factors such as population growth, GDP, and tariff rate. In this discussion, we will primarily focus on the first category, specifically daily temperature. Another factor the model will take into consideration is the day type and whether it's a weekend or a weekday. This will help the model capture the daily network behavior profile as it shifts.

30.2.3 Machine Learning Software

The regression model creation process in Microsoft Power BI software involves leveraging its automated machine learning (AutoML) feature. This powerful functionality enables users to train, validate, and deploy machine learning models directly within Power BI. By integrating with Azure Machine Learning, AutoML in Power BI automatically identifies the most relevant features, selects appropriate algorithms, and optimizes and validates the ML model.

30.2.4 Proposed Regression Model Architecture

As it can be seen in Fig. 30.1, the input dataset into the software consists of three variables: the historical daily water demand of a water utility in the UAE for 2017, its corresponding day name, and the hourly temperature for 2017 as collected from the National Center of Meteorology (NCM) in the UAE. The data is cleaned to address any anomalies or errors. Moreover, the data is processed by converting the hourly temperature to average daily temperature, and the day name is converted to the day type (weekend, weekday). Afterwards, the cleaned and processed data is fed to the machine learning model. Subsequently, the machine learning model splits the provided input data into training (80%) and testing (20%) datasets. To determine the line that best fits through the data, the model keeps iterating using different algorithms ranging from the light gradient boosting model (GBM) regressor, elastic net, random forest regressor, extra trees regressor to decision tree regressor, and stochastic gradient descent (SGD) regressor until it reaches the best model performance by validating it against the testing dataset. The optimum model is then used to forecast water demand for 2018 based on an unobserved dataset that comprises the processed average daily temperature of 2018 and its corresponding day type.

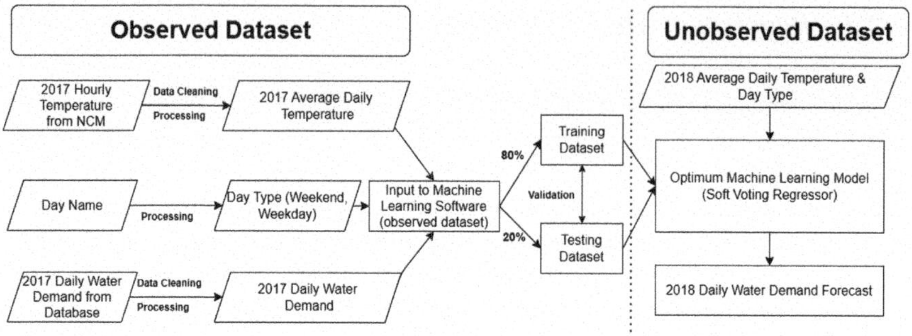

Fig. 30.1 Proposed regression model architecture

30.3 Results and Discussion

The results of the AutoML Modelling in Power BI are divided into 2 main sections: the model training details and model performance.

30.3.1 Regression Model Training on Observed Dataset

The training details displayed in Fig. 30.2 offer insights into the model's training procedure. It encompasses various aspects, such as the total number of sampled rows utilized for training (365), the number of rows on which the model was trained (292), the total number of iterations executed to discover the optimal model (25), the algorithm employed in each iteration, the model's performance at each iteration, and the ultimate model employed. In our case, the pre-fitted soft voting regressor algorithm proved to be the most effective in delivering desirable outcomes.

30.3.2 Voting Regressor

Ensemble learning relies on voting to merge predictions from different models into a single prediction. Hard voting and soft voting are two prevalent types of voting methods in ensemble techniques. Hard voting is a technique that entails aggregating the predictions from various base models and selecting the class with the highest number of votes as the final prediction.

Soft voting combines predictions from multiple regression models through averaging. The training data is employed to train the individual models. During the prediction phase, each model generates its own prediction, which is subsequently combined using averaging to yield the final ensemble prediction. In soft voting, one

Fig. 30.2 Regression model training details

can opt for simple averaging with equal weights for all models or weighted averaging with different weights assigned to each model's prediction. Weighted averaging enables us to prioritize the predictions of specific models that are likely to excel on certain subsets of the data. The inclusion of diverse models, ranging from traditional to tree-based models, is imperative. The models present diverse strengths and weaknesses, which can be efficiently utilized to reinforce each other when incorporated in a voting regressor. The voting regressor ensemble method improves the accuracy and robustness of final predictions. This results in more reliable outcomes compared to relying on individual models. In our case, as can be seen in Fig. 30.3, the optimum model used was the soft voting regressor with an ensemble of methods, including extra trees regressor, elastic net, and light GBM regressor.

30.3.3 Regression Model Performance on Observed Dataset

The data presented in Fig. 30.4 showcases the assessment of the model, detailing the overall R-Squared (model performance) and the total number of rows used for testing and validation (73). R-squared serves as a measure of how well the regression model aligns with the actual data. A higher R-squared value indicates a stronger fit of the model to the data. This value, known as the coefficient of determination, represents the portion of variance in water demand that the model can account for or forecast.

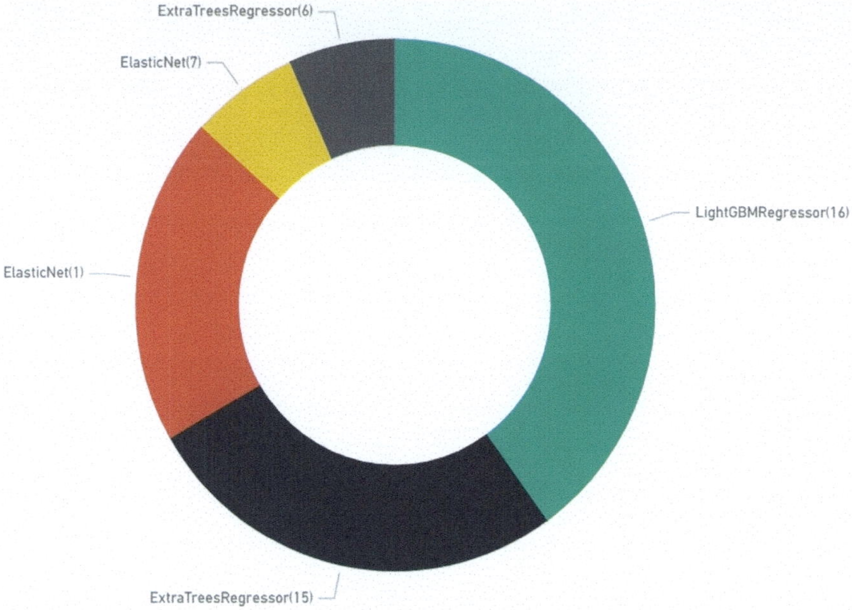

Fig. 30.3 Optimum model used—soft voting regressor

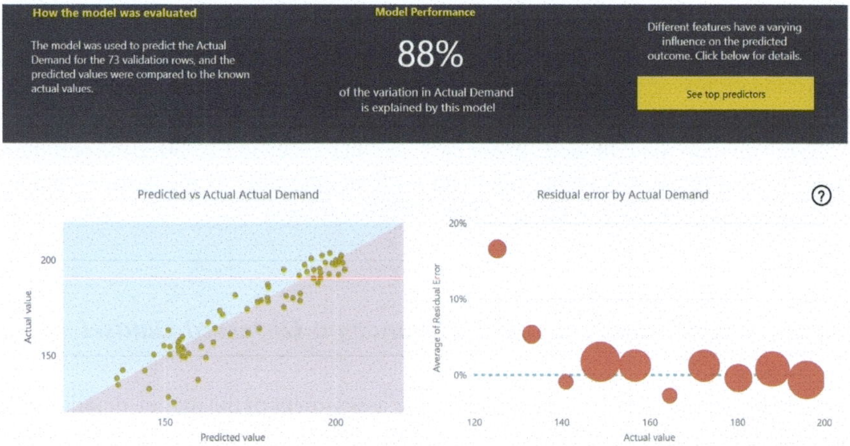

Fig. 30.4 AutoML model performance

30.3.3.1 Predicted vs Actual Demand

By comparing predicted versus actual demand, Fig. 30.4 showcases the predicted values derived from the model alongside the known actual values for the validation rows in the dataset. The model tends to underestimate values in the blue area, while overestimating values in the red area. The error in prediction is measured by the distance from the diagonal. Ideally, a proficient model would have the majority of test samples grouped near the diagonal.

30.3.3.2 Residual Error by Demand

In Fig. 30.4, the residual error illustrates how the average error percentage is distributed across different values in the sample data set. The vertical coordinate is the average residual error within the range, and the bubble's horizontal coordinate represents the mean of the actual value within the range. To illustrate, an average residual error of 3% suggests that the model will overestimate the value in that range by 3%. The frequency of the values in the validation dataset within that range is shown by the size of the bubble.

30.3.4 Evaluation of Regression Model

In order to compare the accuracy of predictions from multiple models, the following error metrics were used.

30.3.4.1 Mean Squared Error (MSE)

The mean squared error is the average of the squared distance between the actual and predicted values.

$$MSE = \frac{1}{n}\sum_{i=1}^{n}(y_i - p_i)^2 \tag{30.2}$$

The (y_i) is the $\left(i^{th}\right)$ observed value, (p_i) is the corresponding predicted value for (y_i), and (n) is the number of observations. The $(\sum)$ indicates that a summation is performed over all values of (i).

30.3.4.2 Mean Absolute Error (MAE)

The mean absolute error is the average of the absolute distance between actual and predicted values.

$$MAE = \frac{1}{n}\sum_{i=1}^{n}|y_i - p_i| \tag{30.3}$$

30.3.4.3 Mean Absolute Percentage Error (MAPE)

The mean absolute percentage error is the mean absolute error, converted to a percentage.

$$MAPE = \frac{1}{n}\sum_{i=1}^{n}\frac{|y_i - p_i|}{y_i} \tag{30.4}$$

30.3.4.4 Root Mean Square Error (RMSE)

The root mean square error is the measure of the average deviation of the estimates from the observed values.

$$RMSE = \sqrt{\frac{1}{n}\sum_{i=1}^{n}(y_i - p_i)^2} \tag{30.5}$$

30.3.5 Regression Model Performance on Unobserved Dataset

In order to evaluate the accuracy of the water demand forecast of the developed machine learning model, it is compared to the conventional demand forecast of a water utility in the UAE for the year 2018. As can be seen in Figs. 30.5 and 30.6, the ML demand forecast is much closer to the actual demand than the conventional demand forecast. The conventional demand forecast is consistently overestimating the amount of demand throughout the year, while the machine learning model is better at predicting the behavior of the network. This is also clearly highlighted in Fig. 30.7 and Table 30.1, where the MSE was reduced from 124 MIGD for the conventional demand forecast to 58 MIGD for the machine learning model, the MAE was reduced from 8.85 MIGD for the conventional demand forecast to 5.25 MIGD for the machine learning model. Additionally, the MAPE was reduced from 5.42% in the conventional demand forecast to 3.25% in the machine learning model, the RMSE was reduced from 11.14 MIGD to 7.6 MIGD. Moreover, the total yearly difference between actual and forecasted demand was reduced from 2683 to 899 MIG.

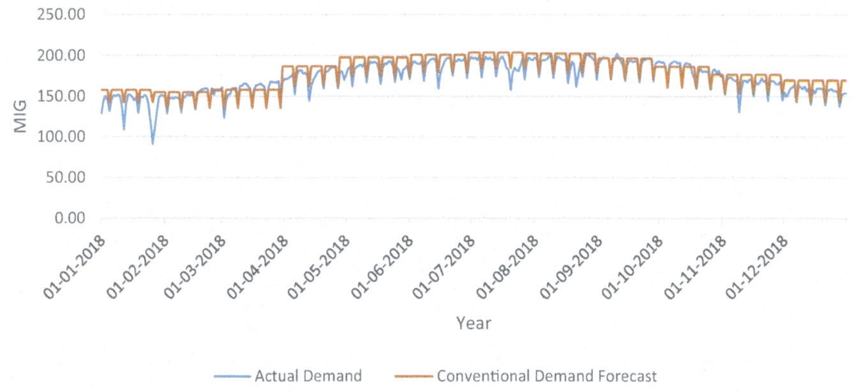

Fig. 30.5 Actual versus conventional demand forecast

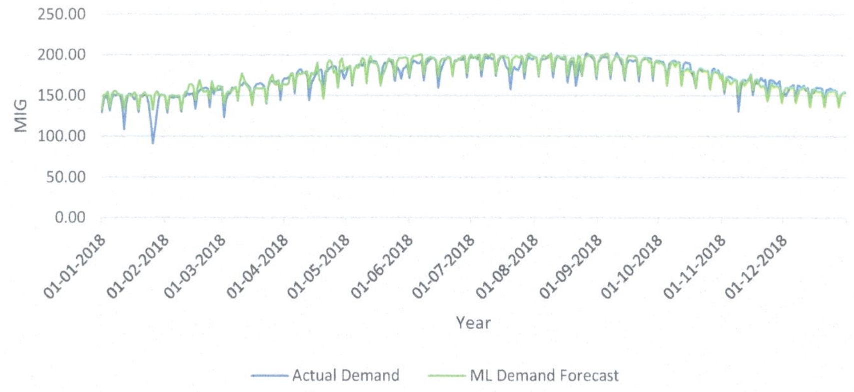

Fig. 30.6 Actual versus ML demand forecast

30.4 Conclusions

A machine learning water demand forecasting model was developed and tested by utilizing the AutoML feature in Microsoft Power BI. 2017 daily water demand for a water utility in the UAE was used as the input dataset along with the average daily temperature and day type. The optimum model yielded was used to forecast the water demand for 2018 based on unobserved data, which comprises the average daily temperature of 2018 and its corresponding day type. The results showed a big improvement in the forecasting accuracy compared to the conventional demand forecast, where the MAPE was reduced from 5.42% for the conventional demand forecast model to 3.25% for the machine learning demand forecast model. Additionally, the RMSE was reduced from 11.14 MIGD for the conventional demand forecast model to 7.6 MIGD for the machine learning demand forecast model. More importantly, the conventional demand forecast model overestimated the amount of

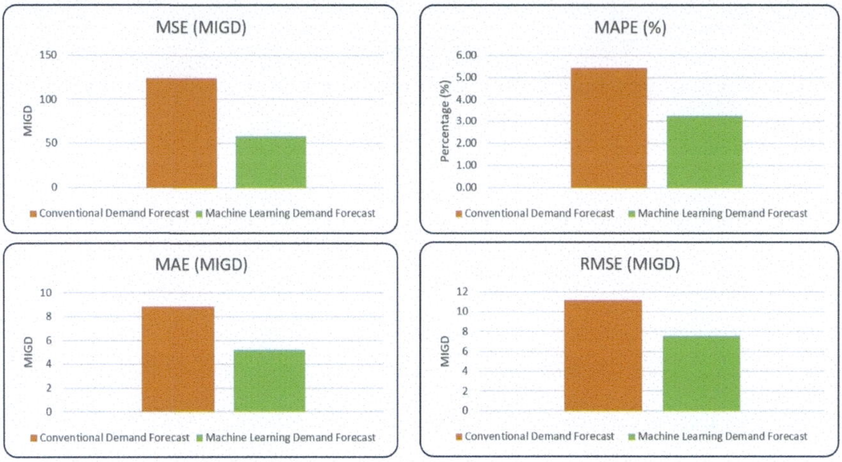

Fig. 30.7 Error metrics evaluation (Conventional vs. ML)

Table 30.1 Model performance evaluation

Error Metric	Conventional demand forecast	Machine learning demand forecast
MSE	124 MIGD	58 MIGD
MAE	8.85 MIGD	5.25 MIGD
MAPE	5.42%	3.25%
RMSE	11.14 MIGD	7.6 MIGD
Total Difference / Year	2683 MIG	898 MIG

water by 2683 MIG. This value was reduced to 898 MIG by using the machine learning demand forecast model, saving 1785 MIG in the process. Those extra saved million imperial gallons can now be utilized to adequately plan for water production and transmission outages, identify potential water supply shortages in advance, thus improving the mitigation procedures carried out by stakeholders. Additionally, it can be utilized to precisely determine the required water reserve as well as develop optimum operational plans for pumping stations and water production plants. This will reduce operational costs, improve the overall efficiency and performance of the water management system, and reduce the gap between actual and forecasted demand, helping to achieve the global water sustainability goal.

Acknowledgements We would like to acknowledge the National Center of Meteorology (NCM) in the UAE for providing the climatic data in a timely manner.

References

Adamowski, J., Fung Chan, H., Prasher, S. O., Ozga-Zielinski, B., & Sliusarieva, A. (2012). Comparison of multiple linear and nonlinear regression, autoregressive integrated moving average, artificial neural network, and wavelet artificial neural network methods for urban water demand forecasting in Montreal, Canada. *Water Resources Research, 48*(1). https://doi.org/10.1029/2010wr009945

Adamowski, J., & Karapataki, C. (2010). Comparison of multivariate regression and artificial neural networks for peak urban water-demand forecasting: Evaluation of different ANN learning algorithms. *Journal of Hydrologic Engineering, 15*(10), 729–743. https://doi.org/10.1061/(asce)he.1943-5584.0000245

Ahmed, V., Saad, A., Saleh, H., Saboor, S., Kasianov, N., & Alnaqbi, T. (2020). *Implementation of water demand forecasting model to aid sustainable water supply chain management in UAE.* https://doi.org/10.20944/preprints202011.0205.v1

Ashbolt, S., Maheepala, S., & Perera, B. J. (2014). A framework for short-term operational planning for water grids. *Water Resources Management, 28*(8), 2367–2380. https://doi.org/10.1007/s11269-014-0620-4

Bakker, M., Vreeburg, J., Van Schagen, K., & Rietveld, L. (2013). A fully adaptive forecasting model for short-term drinking water demand. *Environmental Modelling & Software, 48*, 141–151. https://doi.org/10.1016/j.envsoft.2013.06.012

Caiado, J. (2010). Performance of combined double seasonal univariate time series models for forecasting water demand. *Journal of Hydrologic Engineering, 15*(3), 215–222. https://doi.org/10.1061/(asce)he.1943-5584.0000182

De Souza Groppo, G., Costa, M. A., & Libânio, M. (2019). Predicting water demand: A review of the methods employed and future possibilities. *Water Supply, 19*(8), 2179–2198. https://doi.org/10.2166/ws.2019.122

Kavya, M., Mathew, A., Shekar, P. R., & Sarwesh, P. (2023). Short term water demand forecast modelling using artificial intelligence for smart water management. *Sustainable Cities and Society, 95*, 104610. https://doi.org/10.1016/j.scs.2023.104610

Koo, K. M., Han, K. H., Jun, K. S., Lee, G., Kim, J. S., & Yum, K. T. (2021). *Performance assessment for short-term water demand forecasting models at an end-use level in Korea.* https://doi.org/10.20944/preprints202104.0332.v1

Pandey, P., Bokde, N. D., Dongre, S., & Gupta, R. (2021). Hybrid models for water demand forecasting. *Journal of Water Resources Planning and Management, 147*(2). https://doi.org/10.1061/(asce)wr.1943-5452.0001331

Razali, S. N., Rusiman, M. S., Zawawi, N. I., & Arbin, N. (2018). Forecasting of water consumptions expenditure using Holt-winter's and ARIMA. *Journal of Physics: Conference Series, 995*, 012041. https://doi.org/10.1088/1742-6596/995/1/012041

Salloom, T., Kaynak, O., & He, W. (2021). A novel deep neural network architecture for real-time water demand forecasting. *Journal of Hydrology, 599*, 126353. https://doi.org/10.1016/j.jhydrol.2021.126353

Vijai, P., & Bagavathi Sivakumar, P. (2018). Performance comparison of techniques for water demand forecasting. *Procedia Computer Science, 143*, 258–266. https://doi.org/10.1016/j.procs.2018.10.394

Chapter 31
Artificial Groundwater Recharge in the Kingdom of Spain. Analysis of Suitability and Identification of Optimal Places to Propose Its Application in the Face of the Effect of Climate Change

José Manuel Murillo and Alberto Padilla

Abstract The resilience to drought and climate change -in 1251 Hydrogeological Enclosures (HE_S)- is analyzed in this paper, as well as the suitability that these can present in the face of a possible artificial recharge that mitigates the effect of an unfavorable situation, such as a prolonged drought or the climate change. A Hydrogeological Enclosure (HE_S) is the basic unit of work that has been used by the Spanish hydraulic administration to evaluate and quantify groundwater resources in the Kingdom of Spain. It has been defined as each of the different hydrogeological sectors that within the same body of groundwater (GB) drain into different bodies of surface water (SB). The methodology that has been used in this paper makes use of the Standardized Precipitation Index (SPI); since this index is a direct indicator of water scarcity due to drought, as well as of the daily climatic series of precipitation provided by the State Meteorological Agency of the Government of Spain (AEMET), and of the Post-process carried out by the same Organization of the European project ENSEMBLES. This project systematically and comprehensively integrates several climate prediction models (21) that provide more concrete and accurate climate change projections on a temporal and geographical scale. As for the calculation of the SPI index for a given period and in a particular locality, that has a rainfall record for a sufficiently long period of time, it is determined by adjusting that record to a theoretical probability distribution, which is then transformed into a normal distribution so that the average SPI is zero. For the evaluation of the suitability of proposing or carrying out artificial recharge of aquifers, the depletion coefficient, and the semi-emptying time of the Hydrogeological Enclosures ($HE_{S)}$ have been used, since the

J. M. Murillo (✉)
Freelance Hydrogeologist, General Lacy 58, 28045 Madrid, Spain
e-mail: jm.murillo955@gmail.com

A. Padilla
Aljibe Consultores, José Luis Pérez Pujadas 7, 18006 Granada, Spain
e-mail: a.padilla@aljibeconsultores.com

A. Sefelnasr et al. (eds.), *Water Resources Management and Sustainability*, Water Science and Technology Library 114, https://doi.org/10.1007/978-3-031-80520-2_31

faster the discharge of an aquifer, the worse the behavior they present against an artificial recharge operation. The depletion coefficient and the semi-emptying time have been obtained, whenever possible, from data of hydrographs, obtained by foronomic control, to which the Maillet function has been adjusted. Where it is possible to proceed in this way, Rorabaugh's expression or bibliographical data have been used. The results obtained are very encouraging, since 45% of the Hydrogeological Enclosures (HE_S) have a favorable aptitude a in the face to the realization or proposal of artificial recharge of aquifers. Large detrital depressions are the aquifers that offer the best prospects of artificial recharge, while the worst are provided by small karst aquifers. The effect of climate change will increase the need to propose the realization of more artificial recharge in large areas of the Spanish territory, where currently the simulations carried out do not detect deficit. The methodology designed and used can be described as easy to apply and quick to execute and interpret.

Keywords Artificial recharge · Climate change · Depletion coefficient · Drought · Managed recharge

31.1 Introduction

Artificial groundwater recharge (AGR) is a water management technique that is not exempt from some operational complexity, since it requires technological and infrastructure interventions, as well as other legal, administrative, regulatory, and economic ones. Todd (1959) defines AGR as the practice of increasing by unnatural means the amount of water entering an aquifer. However, there are many other definitions of AGR that only vary in minor nuances as can be seen from the bibliography collected in Signor et al. (1970). A more up-to-date definition could be the following: technique pursuing, after carrying out technical-economic-administrative feasibility studies, the planned and programmed introduction of water into an aquifer to increase the degree of guarantee and availability of water resources, as well as the protection and improvement of their qualitative aspects, environmental protection and sustainability of surface and groundwater. These last two aspects, which until recently had been marginal, are destined to acquire remarkable prominence in the immediate future.

Proposing an AGR operation requires that there is a real or potential demand for water that justifies it or an environmental problem that can be solved with the use of this technology, such as the overexploitation of aquifers, marine intrusion, maintenance of ecological flows or the protection and recovery of ecosystems dependent on groundwater. AGR requires two conditions totally necessary to be technically viable: (1) An aquifer with adequate hydrodynamics to allow the introduction, storage and recovery of recharged water in a satisfactory timely and manner to the uses and demands for which it has been designed. (2) That there is sufficient water in quality and quantity to be able to carry out the AGR operation.

The other topic discussed in this paper is the duality of groundwater resources and climate change, since climate change studies have focused mainly on the field of surface waters, with groundwater usually being relegated to the background (Barbieri et al., 2021). The importance that AGR can have in water planning and adaptation to climate change in the Kingdom of Spain is highlighted in the Regulation of Hydrological Planning (Royal Decree 1159/2021), specifically in its article 4-bis "Adaptation to climate change" where it is said that actions consistent with the principles of the Water Strategy for the Ecological Transition must be identified and evaluated in order to: increase the resilience of water bodies and help maintain their good status, increase security of supply and cope with extreme events.

31.2 Methodology

In the present work it has not been entered into assessing whether there will be water sources in the future that allow the programming of AGR, since what is analyzed in it is, on the one hand, the possible shortage of water due to a decrease in the variable precipitation (drought or climate change) that may affect the natural recharge to aquifers and, on the other, the favorable aptitude of the HE_S to store water in aquifers. or in the part of these that are part of those, for a sufficiently long time to allow an effective and sustainable use.

31.2.1 Evaluation of the Favorable Aptitude of the HEs in the Face to an AGR Operation

AGR is always effective in overexploited aquifers until the piezometric level rise caused by such an operation reaches the aquifer drainage level, or until it is ensured that the aquifer discharges a certain ecological flow into the surface hydrographic network. From that moment on, and as in aquifers that are not overexploited, AGR will only be effective if the circulation rate of the groundwater flow is slow enough so that the recharged water, or the hydrodynamic effect it causes on the water stored in the aquifer, does not cause drainage. partial or total, through the natural controls of the aquifer, before the AGR operation achieves the objective for which it was designed.

The greater or lesser ease with which an aquifer discharges water to the surface hydrographic network, or to the sea, is one of the variables that determine the suitability of a AGR proposal. In this regard, Custodio and Vilaró (1976) point out that an operation of artificial aquifer recharge requires, as a necessary condition to be able to propose it, that there are empty holes in the aquifer capable of being filled by the water to be recharged, as well as a time of permanence of this in the aquifer that allows its subsequent use, both in the quantity demanded and in the quality required for the use to which it is intended. The parameter that can evaluate and quantify this "suitability" is the depletion coefficient (α), but it is more intuitive and easier to handle what is called semi-emptying time $\left(t_{1/2} = \mathrm{Ln}\ 2/\alpha\right)$; which is the time it takes for an aquifer, during a period when it does not receive any water recharge, to halve the volume of water stored above its discharge level. From the expression $t_{1/2} = \mathrm{Ln}\ 2/\alpha$ follows that the higher the depletion coefficient (α) the faster the aquifer is discharged. On the contrary, the lower this variable, the longer the retention time of the water stored in the aquifer, so that high semi-emptying time indicate that aquifers present in principle good hydrodynamic conditions to carry out AGR operations, while low values determine the opposite.

The depletion coefficient is determined from the decomposition of the hydrographs to which the Maillet function is adjusted (Maillet, 1905). However, there are particularized solutions, such as those aquifers whose geometry and hydrogeological properties can resemble rectangular and homogeneous aquifers drained on one side by a completely penetrating river (Rorabaugh, 1963). In that case, depletion coefficient equation is:

$$\alpha = \pi^2 T/4SL^2 \tag{31.1}$$

where:

T is the transmissivity in the direction of flow (m^2/d)

S, the storage coefficient

and L (m) the distance from the river to the impermeable edge of the aquifer or the hydrogeological divide.

From the expression $\alpha = \pi^2 T/4SL^2$ follows that the AGR will present a greater probability of success the larger the aquifer and the lower its diffusivity, as well as the farther the recharge facility is from the rivers and springs that drain the aquifer. On the other hand, the diffusivity presents low values if the storage is high and the transmissivity is reduced, and high when the opposite occurs. If the storage coefficient remains constant, the diffusivity, provided that the permeability remains constant, will decrease if the saturated thickness decreases, since transmissivity is defined as the product of permeability times the saturated thickness. On the other hand, if the saturated thickness is one that remains constant, the diffusivity will increase if the permeability increases. Consequently, the same value of the depletion coefficient or the semi-emptying time can be the result of a casuistry derived from the different values that may take permeability, storage coefficient, size of the aquifer and saturated thickness. This amalgam of combinations makes it possible for aquifers with quite different hydrogeological characteristics to present the same potential for success or failure in the face to an AGR operation.

31.2.2 Assessment of Possible Water Scarcity Due to Reduced Rainfall (Drought or Climate Change) and Its Effect on HE_S

To assess the water deficit, which results from a prolonged drought or of the possible effect of climate change, it is necessary to analyze the behavior and evolution of both phenomena, both in terms of the most recent past and the most immediate future. In the case of the present investigation, both periods have been 70 years.

The State Meteorological Agency of the Government of Spain (AEMET) has provided the historical daily precipitation data (the most recent past). These are distributed in a grid of 5x5 km. The period to which these data correspond is January 1951 to December 2020. The rainfall that could occur under the effect of climate change has been obtained from the daily climatic series provided by the post-process carried out by the AEMET of the ENSEMBLES project (Van der Linden & Mitchell, 2009). The generated series consists of a first-time interval called "known", which covers from 1961 to 2000, and a second, designated as "forecasted", which extend from 2011 to 2100. Between the "historical" series and the generated ("known" and "forecasted") series, there is a period of overlap. This has been used to make corrections, using transfer functions. In the present paper, the forecasts of the CNRM-RM5.1-ARPEGE models (Dosio & Paruolo, 2011) have been used, which use simple transfer functions to estimate the climatic series at the reference points.

To compare the historical data with those generated ("known" and "forecasted") the coordinates of the centroids of the HES, in number of 1251 there are in the Spanish peninsular territory and the Balearic Islands, have been identified. Each of these centroids has been assigned the historical and generated series that is located closest to each of them. Figure 31.1 shows the HE_S with their centroids, the points where the historical series are located according to their greater proximity to the centroids of the HE_S, and the points where the generated series are located according to their greater proximity to the centroids of the HE_S. The risk of drought, and therefore the foreseeable effect of climate change, has been quantified by the Standardized Precipitation Index (SPI) (Mckee et al., 1993), as this index is a direct indicator of drought-induced water scarcity and therefore of the potential consequences of climate change. In this sense, those HE_S where the risk of water deficit, due to a reduction in the precipitation variable, is high, will be more conducive to receiving an additional supply of water through the application of the AGR technique.

In each HE_S, all historical periods of drought between 1951 and 2020 have been calculated for each of the characteristic climatic series of the HE_S. This calculation has also been made for the series forecasted for climate change between 1921 and 2090. Drought episodes have been characterized by their duration (D), intensity (I) and magnitude (M), the last two with positive values. Each episode j has been evaluated according to the expression: DRj = (2xDj+Ij+Mj).

The variable "duration" has doubled because it has been estimated that the persistence of drought has a more pronounced effect on groundwater resources since aquifers have a greater inertia to climatic episodes. The risk of drought (DR) of

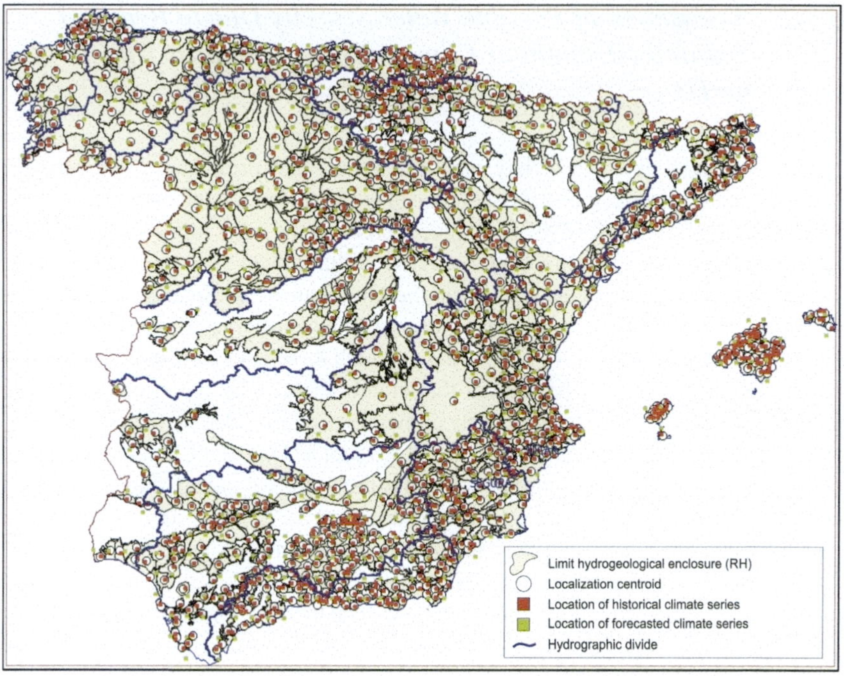

Fig. 31.1 Location of the centroids of the HE_S of mainland Spain and Balearic Islands, as well as of the points where the control stations of the historical and forecasted series

each of the series assigned to the HE_S, has been quantified as the sum of the different values of the DRj parameter corresponding to each drought episode that has been detected: DR=DRjΣ.

31.3 Results and Discussion

31.3.1 Suitability to the AGR of HE_s

The number of hydrogeological enclosures (HE_S) in which the depletion coefficient has been determined has been 1200. This amount represents 96% of those that exist in mainland Spain and the Balearic Islands.

The logarithmic distribution of the semi-emptying time value, obtained from the depletion coefficient of the 1200 HE for which data are available, is shown in Fig. 31.1. This distribution fits well to a logarithmic normal distribution of a mean of 2.6 and a standard deviation of 3.7. According with the results shown in Fig. 31.2-Left, a classification has been established (Table 31.1) to evaluate the aptitude, favorable, that presents the HE in the face of an AGR operation according to

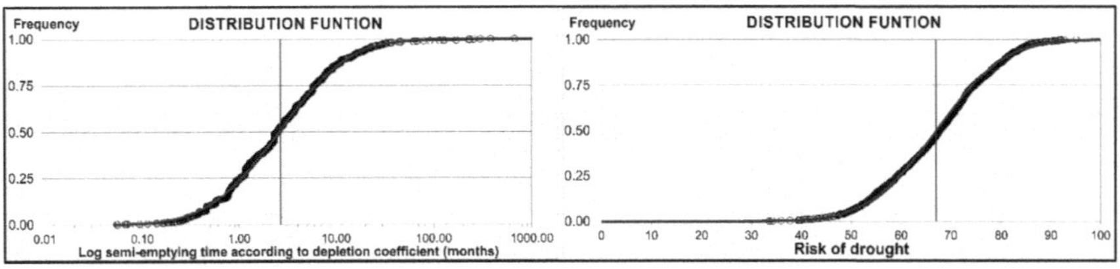

Fig. 31.2 The left graphic shows the distribution function of the semi-emptying time estimated at 1200 HE_S and in continuous line the adjusted log-normal distribution of mean 2.6 and standard deviation 3.7. The right graphic shows distribution function of the historical drought risk for 1251 HE_S. In continuous line the adjusted normal distribution of mean 67.2 and standard deviation 10.8

the value of its semi-emptying time. This classification is similar to proposed by Estrela, (2000) to assess the resilience of aquifers to drought. Figure 31.3 shows the spatial distribution of the HE according to its suitability for artificial recharge: very high red, high orange, medium green, and low yellow.

According to the results shown in Table 31.1 and in Fig. 31.3, the HE_S with a semi-emptying time equal to or greater than 6 months would be the most suitable to propose AGR. This proposition seems logical, since the beneficial effect that such an operation would produce could be diluted, without effective use, if the semi-emptying time were shorter (less than 3 months). It is not possible to rule out those HE_S with a semi-emptying time between 3 and 6 months, although the AGR operation would be more complicated. The consideration made in the previous paragraph is especially significant in aquifers that are preferably used in agriculture, since the period in which there is surplus water to perform the artificial recharge operation is usually -in the temperate countries of the northern hemisphere- between December and March, while the season of irrigation with water from the aquifer begins -generally- in mid spring. Therefore, water that is recharged in December could remain still stored in the aquifer in early April or May if this is moderately fast. In the southern hemisphere, the dates in which irrigation is conducted and the periods in which water is available for AGR are different, but the time intervals between them have a similar breadth.

Table 31.1 Suitability of HE_S for water storage by AGR according to its semi-depletion time or its depletion coefficient

Semi-emptying time (months)	Depletion coefficient (d^{-1})	Suitability AGR	Hydrogeological enclosures (HE) number	%
< 3	> 0.0077	Low	670	56
3–6	0.0077–0.0038	Middle	227	19
6–12	0.0038–0.0018	High	171	14
>12	0.0018	Very high	132	11
Total			1200	100

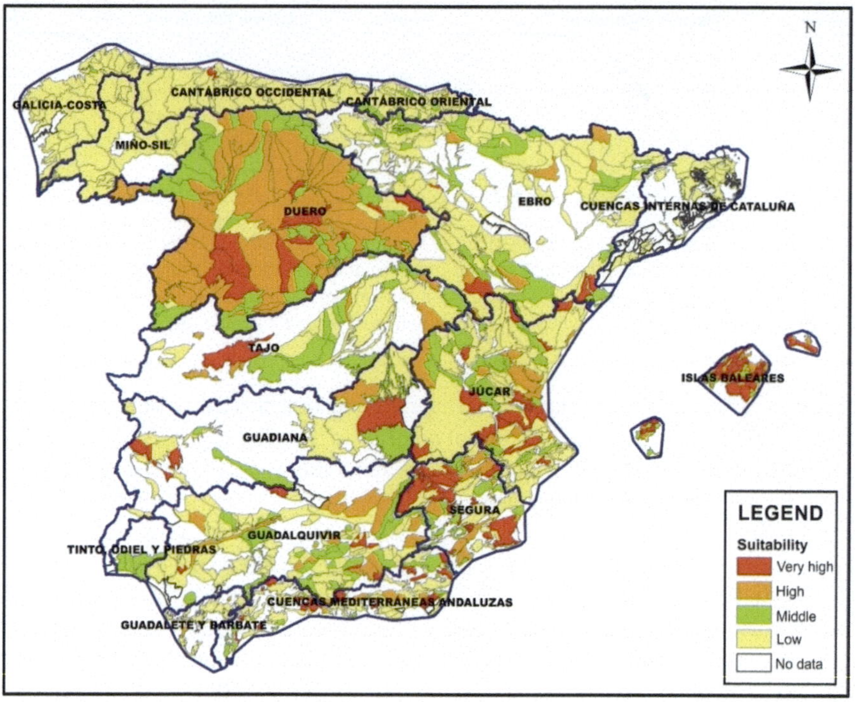

Fig. 31.3 Distribution of hydrogeological enclosures (HE_S) classified by their suitability to AGR according to semi-emptying time criterion

The same happens in areas of the planet where there are only dry and wet periods, although the spacing between the two is greater than in the previous cases.

Of the 1,200 HE_S in which depletion coefficient data are available, 303 (25%) have "high" or "very high" suitability in the face to the positive effect that AGR can have (Table 31.1). Figure 31.3 shows that the most suitable river basins to propose AGR actions are, both for their extension and for the number of HE_S that present suitability to AGR, those of, the Balearic Islands with 135 HE (74%), the Duero with 86 HE (60 %) and Segura with 44 HE (49%), while in the northern of Spain river basins (Miño-Sil, Galicia-Costa, Cantábrico Occidental y Oriental) there are hardly any HE_S suitable for proposing AGR actions (only two) because their emptying is very fast. If the analysis performed included those HE_S that present a semi-emptying time value of 3 to 6 months, it would be possible to propose AGR operations in approximately 45% of the HE_S identified in Table 31.1.

31.3.2 Water Scarcity Due to Reduced Rainfall (Drought-Climate Change) Effect on HE_S

The HEs have been classified by their drought risk (DR) into four categories according to an increasing risk (Table 31.2) that has been established from the distribution function of the historical precipitations period (Fig. 31.2). Figure 31.4 shows the spatial distribution of the HE_S classified according to historical and future drought risk. As for the future risk of drought (Fig. 31.4-right), most of the enclosures become at "very high" risk (note the red color that extends over most of the Spanish peninsular territory and the Balearic Islands in Fig. 31.4-right). In contrast to this general trend, emphasis should be placed on the eastern coastal strip where the future risk of drought is medium-low. Specifically, the HE_S with a very high risk of drought go from 305 to 1045, so their number increases by 738, which represents a percentage increase of 59% compared to the situation detected for the risk of historical drought. In global Figures, enclosures with a very high risk of future drought represent 83% of the total. The greatest increase in future drought irrigation compared to the historical one occurs in the Eastern Cantabrian River basin (4.14 times higher than the historical one) and in the south of the peninsula: Guadalquivir, Guadalete-Barbate and Andalusian Mediterranean basins with respectively 4.03; 3.14 and 2.97 times higher than historical. In the rest of the hydrographic basins, it is between 1.02 times higher than the historical one in the Júcar River basin, so it hardly suffers variation, and the 1.88 of the Guadiana River basin.

Figure 31.5 shows the relationship between expected (future) and historical drought risk. The latter graphical representation indicates the proportion by which drought risk is expected to vary in each HES. The largest increases are detected in the HE_S belonging to the hydrographic basins of the Eastern and Western Cantabrian with HE_S where values that triple those of the risk of historical drought are manifested. On the contrary, in the oriental coastal strip reductions are detected, evidently moderate, so the situation improves concerning the risk of historical drought.

Table 31.2 Distribution of HE_S by its risk of drought (DR)

Probability	Drought risk	Drought risk assessment	Hydrogeological enclosures (HE) number	
			Historical	Future
< 25%	<59	low	300	88
25%-50%	59-68	Middle	335	59
50%-75%	68-75	Hight	311	61
>75%	>75	Very hight	305	1043
Total			1251	1251

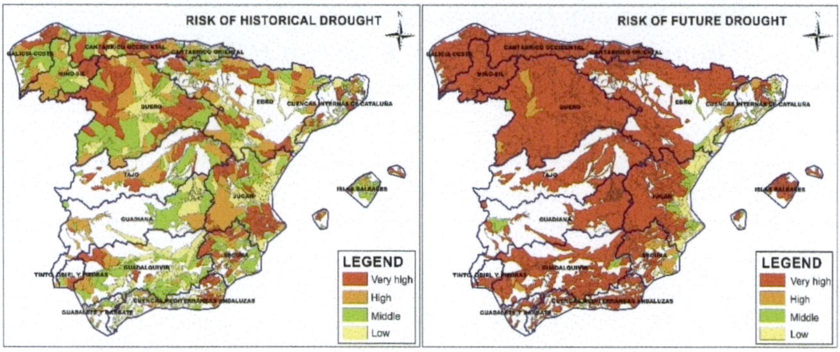

Fig. 31.4 Distribution of HEs classified by their risk of historical drought (left figure) and risk of future drought (right figure)

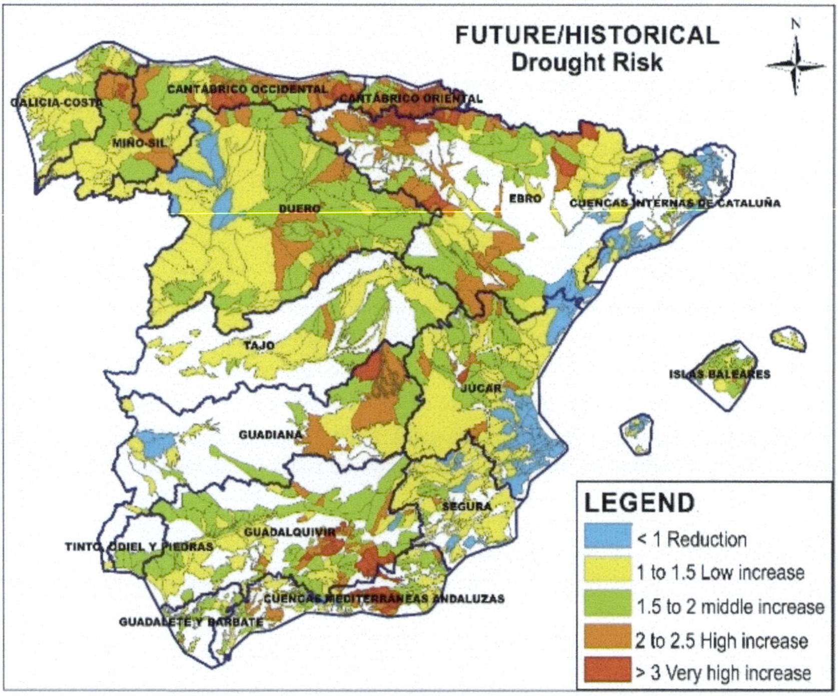

Fig. 31.5 Distribution of HEs classified by their variation in the risk of future drought with respect to the historical

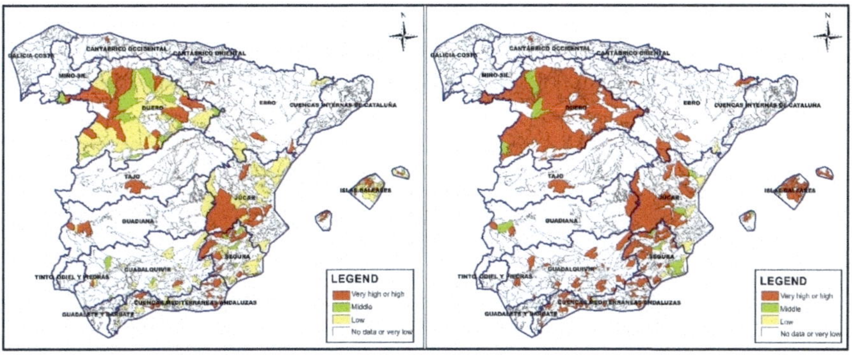

Fig. 31.6 Spatial distribution of HE_S where, according to the criterion of semi-emptying time, it is ideal to propose artificial recharge operations in case of historical drought (left figure) and future drought (right figure)

31.3.3 Joint Analysis of the Semi-emptying Time and of the Drought Risk

The uses of the coefficients of semi-emptying time or of depletion, as an evaluating criterion of the viability that an artificial recharge proposal can present, provides an assessment of the intrinsic potential that an aquifer presents against said possibility and, therefore, are independent of aspects such as the availability of water to carry it out or even the real need to do it. On the other hand, drought risk analysis, whether it responds to a historical period located in the past or to a hypothetical future situation, only provides information on the places -in this case aquifers or HE_S—where water may have been lacking or where in the future it may be scarce due to a lower rate of precipitation. That is, where there has been or is foreseeable that there is less natural recharge to aquifers and, therefore, AGR could or may be proposed.

Figure 31.6 shows the spatial distribution of HE_S where, according to the criterion of semi-emptying time, it is ideal to propose artificial recharge operations in case of historical drought (left Figure) and future drought (right Figure). In both situations the priority HE_S has been marked in red.

31.4 Conclusions

In mainland Spain and the Balearic Islands, 45% of the EH_S present, according to their semi-emptying time or depletion coefficient, a hydrodynamics suitable for AGR operations. On the other hand, can be predicted that the number of EH_S identified, which have historically been classified as having a "high" or "very high" drought risk, will increase significantly in the future because of climate change. Specifically, by 59 %, so AGW can be a palliative technique to this situation. However, the availability of water for its application at present is scarce, since the storage and consequent regulation of surface water resources are very high in Spain, so in areas where there are unmet demands, natural resources that can be stored not only superficially, but also underground, are very scarce. About the future situation, it should be noted that in a study such as the one presented, where a 70-year simulation scenario is proposed, it is plausible to consider any alternative that can provide water to hypothetical AGW facilities, since a reduction in desalination costs, using alternative energy sources, it is not disposable in the medium term.

References

Barbieri, M., Barberio, M.D., Banzato, F., Billi, A., Boschetti, T., Franchini, S., Gori, F. & Petitta, M. (2021). Climate change and its effect on groundwater quality. *Environment Geochemistry and Health*, pp. 1–12.

Custodio, E. & Vilaró, F. (1976). Métodos de recarga artificial: aspectos generales. In E. Custodio, & M.R. Llamas (Eds.), *Hidrología Subterránea* (pp. 1970–1991). Barcelona: Editorial Omega.

Dosio, A. & Paruolo, P. (2011). Bias correction of the ENSEMBLES high-resolution climate change projections for use by impact models: Evaluation on the present climate. *Journal of Geophysical Research-Atmospheres, 116*.

Estrela Monreal T. (2000). Estimación de un indicador de la resistencia de los acuíferos españoles a las sequías. Monográfico: Aguas subterráneas y sequías. *Revista de la Real Academia de Ciencias Exactas, Físicas y Naturales, 94*(2), pp. 173-181.

Van der Linden, P., & Mitchell, J. (Eds.) (2009): ENSEMBLES: Climate change and its impacts: Summary of research and results from the ENSEMBLES project (p. 164). Met Office Hadley Centre, FitzRoy Road, Exeter EX1 3PB, UK.

Maillet, E. (1905). *Essais d'hydraulique souterraine et fluviale* (p. 218). Paris: Librairie Scientifique A. Hermann.

McKee, T.B., Doesken, N.J., & Kleist, J. (1993). The relationship of drought frequency and duration to time scale. In *Proceedings of the Eighth Conference on Applied Climatology, Anaheim, California. American Meteorological Society* (pp 179–184), Boston.

Rorabaugh, M. I. (1963). Estimating changes in bank storages in groundwater contributions to streamflow. *International Association of Scientific Hydrology., 63*, 432–441.

Signor, D. C., Growitz, D. J. & Kam, W. (1970). *Annotated Bibliography on Artificial Recharge of Ground Water, 1955–1967* (p. 141). Geological Survey Water-Supply. Paper 1990 A sequel to Water-Supply Paper 1477.

Todd D. K. (1959). *Annotated Bibliography on Artificial Recharge of Ground Water Through 1954* (p. 115). Geological Survey Water-Supply Paper 1477.

Chapter 32
Exploring the Evolution of Inland and Coastal Water Bodies in the UAE Through Remote Sensing and GIS Analysis

Mubbashra Sultan, Salem Issa, and Nazmi Saleous

Abstract The United Arab Emirates (UAE) faces significant water scarcity due to its arid conditions and rising water demands driven by population growth and economic expansion. Despite relying heavily on desalinated seawater and limited groundwater, inland and coastal water bodies in the region play a crucial role in preserving biodiversity, cultural heritage, economic prosperity, groundwater replenishment, and climate regulation. The waters within the land borders of the UAE, regardless of their salinity, are critically limited yet hold great significance. Therefore, effective preservation and management of scarce water resources is imperative. This study conducts a detailed multi-temporal analysis of inland and coastal water bodies within the current land borders of the UAE. The analysis utilizes satellite imagery spanning from 1972 to 2021 and GIS techniques. It incorporates machine learning algorithms and supplementary datasets to enhance the classification process and improve the accuracy of classified images. The Random Forest classifier demonstrated higher accuracy in land use land cover (LULC) mapping within the study area compared to the CART and SVM classifiers. Consequently, the Random Forest classifier was selected for the mapping process in the study area. The total area occupied by inland surface water bodies, including those near the coastline (coastal water), has fluctuated from 365 km^2 in the early 1970s to approximately 154 km^2 in 2021, indicating a general decline in surface water bodies over the study period. During this study period, new artificial lakes, such as Al Qudra, Al Wathba, and Zakher, have emerged. However, coastal water areas have experienced a decline due to activities like land reclamation and the construction of artificial islands. The study highlights

M. Sultan (✉) · S. Issa
Geosciences Department, United Arab Emirates University, P.O. Box 15551, Al Ain, UAE
e-mail: 700039508@uaeu.ac.ae

S. Issa
e-mail: salem.essa@uaeu.ac.ae

N. Saleous
Department of Geography and Urban Sustainability, College of Humanities and Social Sciences, United Arab Emirates University, Al Ain, Abu Dhabi, UAE
e-mail: nazmi.saleous@uaeu.ac.ae

A. Sefelnasr et al. (eds.), *Water Resources Management and Sustainability*, Water Science and Technology Library 114, https://doi.org/10.1007/978-3-031-80520-2_32

that the creation of small lakes and the loss of coastal water are primarily driven by anthropogenic factors. This study sheds light on the spatiotemporal dynamics of inland and coastal water bodies within the existing land borders of the UAE, offering valuable insights to assist in the planning and management of the region's limited existing surface water bodies.

Keywords Arid regions · Change analysis · Geo-informatics · LULCC · Machine learning · Modeling

32.1 Introduction

Water, a fundamental life-sustaining element (Altman, 2002), is a valuable resource facing scarcity, especially in regions with low precipitation rates and high evaporation (Naorem et al., 2023). While surface water bodies traditionally include rivers, natural lakes, and streams, it's important to note that artificial water bodies such as recreational or treated water lakes, dams, and ponds have also become increasingly prevalent in recent years. These additions further underscore the importance of water resources in various forms. Inland water bodies, whether natural or artificial, hold immense significance for economic growth, groundwater replenishment, ecosystems, and climate control (Rinke et al., 2019). They are, however, vulnerable to diverse environmental, human-induced, and climatic influences, impacting their size, scope, quality, and functionality. Changes in land use, like urbanization, agriculture, and infrastructure expansion, can alter the hydrological cycle, diminishing their availability and affecting their purity (Cheng et al., 2022; ElHaj et al., 2023). Additionally, climate change can disturb precipitation patterns, temperatures, and evaporation rates, finally altering the balance and dynamics of these water bodies. Meanwhile, coastal water bodies including those located on the coast or near seas or oceans; cover a wide range of distinctive environments and provide services such as tourism, leisure activities, transportation, and fisheries (Hinrichsen, 2013). These coastal water bodies can be affected by a combination of marine and land-based elements, such as salinity, tide patterns, water flow, and land use. These factors make the administration of these systems very complex (Crain et al., 2009), making their continuous monitoring and assessment of high priority.

Water distribution on Earth is not equal, and regions like the United Arab Emirates (UAE) face water scarcity due to their arid nature and accelerating consumption (Shahin & Salem, 2015). The UAE has experienced significant transformations in recent decades, including population and urban growth (Aldogom et al., 2019; Dahy et al., 2021), leading to higher water demand. However, the UAE relies on limited groundwater, treated wastewater, or desalinated water to meet its needs, as natural rivers or lakes are absent (Ahmed, 2010). Inland water sources in the UAE are linked to valleys, where heavy mountain precipitation results in occasional flooding, forming temporary water bodies like ponds, lakes, and wetlands. Surface water runoff also comes from the Oman Mountains in the east and south of the country (Rizk &

Alsharhan, 2003). The UAE's inland water resources are highly restricted, posing a challenge to sustainable development. Despite its dry terrestrial environments, the UAE has a rich coastal ecosystem; although the ongoing coastal development is leading to degradation of these coastal environments (Burt, 2014). Various water bodies in the UAE encounter challenges; for instance, coastal waters are under pressure from erosion and land reclamation (Subraelu, P., et al., 2022), while inland freshwater bodies, both natural and artificial, are overexploited and experience losses due to reduced replenishment and increased evaporation (Mangoosh, 2004).

Given the critical issue of water scarcity in the UAE due to its arid nature and rising consumption, it is imperative to comprehensively assess and manage the limited freshwater resources and unique coastal water bodies. Understanding the dynamics of water bodies within the study area is essential for effective water resource management and sustainable development, particularly because of the continuous conversion of desert areas into urban centers and vegetated areas resulting in increasing pressure on those limited water resources over the years. While numerous studies focus on other land uses in the UAE, some focus on water resources of the country such as spatiotemporal assessment of groundwater level (Yilmaz, et al., 2020), water quality of distribution network of Sharjah (Mortula, et al., 2019), Dubai Creek water quality mapping (Ali, et al., 2016), artificial islands on coastal areas of the UAE (Subraelu, P., et al., 2022). However, none of the studies covers a long-term assessment of water resources over the entire country. This study aims at filling the gap by assessing and monitoring UAE water resources bodies using innovative techniques such as remote sensing and GIS over a 50-year period.

Along with coastal waters, the UAE has inland freshwaters, including natural resources replenished through flash floods and springs, as well as artificially created water bodies such as lakes from treated waste and seawater. Therefore, this study aims to comprehensively assess and map the status of both inland and coastal water bodies in the UAE, bridging the knowledge gap and facilitating informed decision-makers to address water scarcity challenges. Traditional methods like field surveys and measurements have limitations in terms of time, cost, and consistency, often lacking satisfactory historical and current information about inland water bodies. Remote sensing (RS) and geographic information system (GIS) offer potential solutions for mapping water bodies (Dahy et al., 2021; Wang & Xie, 2018). These technologies offer multiple advantages over traditional methods, enabling the capture of extensive, high-resolution spatial and temporal data, access to historical and current satellite imagery, integration of diverse data types, and the application of various analytical techniques and models. RS and GIS play a pivotal role in detecting and examining the spatiotemporal alterations of water bodies, identifying the underlying factors and main drivers for these changes (Zhou Y. et al., 2017).

This study aims to explore the transformation of freshwater and coastal water bodies in the UAE spanning the period 1972–2021, employing RS data and GIS techniques including machine learning classifiers and supplementary datasets. By exploring the evolution of freshwater and coastal water bodies in the UAE, the outcomes will be used to enhance our understanding of water resource dynamics of the UAE. The insights garnered from this study hold significance for policymakers, land managers, and researchers engaged in water resource planning and governance within the region.

The subsequent sections of this paper are structured as follows: Sect. 32.2 delineates the study area, data sources, and the methodology applied for the analysis. Section 32.3 presents and deliberates upon the results derived from the analysis. Lastly, Sect. 32.4 concludes the study and offers recommendations for future research endeavors.

32.2 Methods

32.2.1 Study Area

The study area encompasses the United Arab Emirates (UAE), which is located between the Arabian Gulf and the sea of Oman. The UAE has a coastline that stretches approximately 800 km. The country covers a land area of 83,600 km^2 (8,360,000 ha) and is situated between latitudes 24° 08´ N and 26° 04´ N, and longitudes 51° 35´ E and 56° 22´ E (Fig. 32.1).

The UAE experiences an arid climate, characterized by mild winters and hot and humid summers (Aldababseh et al., 2017). The average annual rainfall in the UAE ranges from 50 to 150 mm, with the eastern mountainous regions receiving slightly higher amounts (Tourenq, et al., 2011). The temperature in the UAE can reach as high as 50 °C during the summer months, while winter temperatures range from 10 °C to 25 °C (Aldababseh et al., 2017).

The country's landscape is dominated by mountains and deserts, with loose sand and gravel covering the barren land (Pain & Abdelfattah, 2015). The main natural resources of the UAE include petroleum, natural gas, and marine resources (Beydoun, 1986). However, the UAE faces major environmental challenges due to its

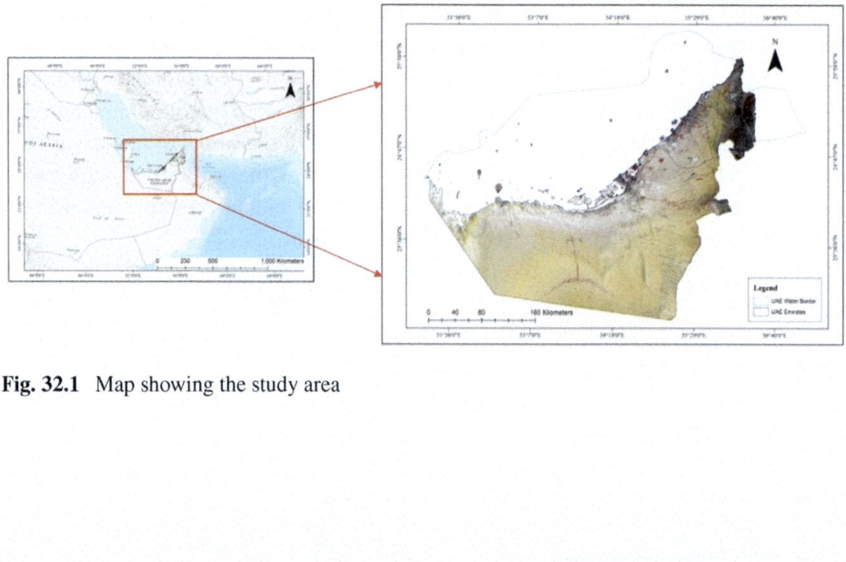

Fig. 32.1 Map showing the study area

Table 32.1 Landsat image information and spatial enhancement methods

Landsat image year	Sensor	Original resolution	Resolution enhancement method	Spatial resolution
1972	MSS	60 m	Resampled	15 m
1986	TM	30 m	Resampled	15 m
1992	TM	30 m	Resampled	15 m
1997	TM	30 m	Resampled	15 m
2002	ETM +	30 m	Pan-sharpening	15 m
2013	OLI	30 m	Pan-sharpening	15 m
2017	OLI	30 m	Pan-sharpening	15 m
2021	OLI	30 m	Pan-sharpening	15 m

limited natural freshwater resources, which necessitate heavy reliance on large-scale desalination plants (Shahin & Salem, 2015).

32.2.2 Data Collection and Spatial Enhancement

The study utilized multitemporal Landsat images spanning fifty years from 1972 to 2021, comprising a total of 8 years: 1972, 1986, 1992, 1997, 2002, 2013, 2017, and 2021. The images covering the study area and study period were downloaded from the USGS website. Landsat satellites typically offer a spatial resolution of 30 m, which can be enhanced for improved data quality. The dataset used and spatial enhancements applied are detailed in Table 32.1.

Spatial enhancement approaches were employed in the study to achieve a resolution of 15 m. As Landsat 7 and 8 have a panchromatic band with 15 m resolution hence these images (2001, 2013, 2017 & 2021) were spatially enhanced through pan-sharpening, a technique that combines the multispectral bands with the higher-resolution panchromatic band to achieve a resolution of 15 m. For older images (1972, 1986, 1992 & 1997) from Landsat 1 and 5, i.e. MSS and TM respectively, due to the absence of panchromatic bands pan sharpening wasn't possible, a resampling process was implemented to adjust the pixel size to 15 m. This ensured consistency in spatial resolution across all images.

32.2.3 Data Preparation

In regions with complex geomorphology, medium-resolution satellite images such as Landsat have limitations in producing accurate LULC classification results. To address this challenge, four indices—Normalized Difference Vegetation Index (NDVI), Normalized Difference Built-up Index (NDBI), Modified Normalized

Difference Water Index (MNDWI), and Desert Brightness Soil Index (DBSI)—along with a Digital Elevation Model (DEM) were utilized as ancillary data in the classification process. The formulations of these indices are provided below:

$$NDVI = \frac{NIR + Red}{NIR - Red} \tag{32.1}$$

$$NDBI = \frac{SWIR + NIR}{SWIR - NIR} \tag{32.2}$$

$$MNDWI = \frac{Green + SWIR}{Green - SWIR} \tag{32.3}$$

$$DBSI = \frac{SWIR + Red}{SWIR - Red} \tag{32.4}$$

where:

Green = Green band reflectance.

Red = Red band reflectance.

NIR = Near Infrared band reflectance.

SWIR = Shortwave Infrared band reflectance.

After preparing input layers stacking was done in this step along with the multispectral satellite images the 4 indices layers and DEM were used to create a raster of 9 layers.

32.2.4 LULC Classification

The next step was LULC classification, while there are numerous supervised and unsupervised classification algorithms being used for mapping LULC classes. It is noteworthy that the traditional parametric classifiers are not that accurate (Basheer et al., 2022). Hence researchers have turned to machine learning algorithms for LULC classification. Machine learning algorithms offer the advantage of automating the classification process by learning patterns and relationships from training data (Talukdar et al., 2020). This allows for more objective and consistent classification results, as well as the ability to handle large datasets and complex classification problems as in our case where we had used multispectral bands along with DEM and indices to improve classification accuracy. Various machine learning algorithms give varying degrees of accuracy depending on the dataset used and the nature of the study area. In addition to image quality and quantity, selection of the most appropriate machine learning classifier can be challenging, as the choice of classifier and its configuration and tuning can impact the outcomes of LULC mapping (Nichols et al., 2019).

In this study, three machine learning algorithms—Classification and Regression Trees (CART), Support Vector Machines (SVM), and Random Forest (RF) were initially tested, and their results compared for LULC classification in the UAE. These algorithms were chosen based on their popularity and proven effectiveness in remote sensing applications (Talukdar et al., 2020). After careful evaluation, the Random Forest algorithm was selected as the most suitable algorithm for this study due to its higher accuracy and easier implementation compared to the other two algorithms.

Training data plays a critical role in LULC classification, the training samples were used to extract various attributes such as spectral (mean of DN value), spatial (variance of DN value), topographic (elevation), and object aspect (compactness). These attributes were saved as shapefiles. Next, the RF algorithm was employed to train the machine intellect using the training data. This allowed for generalization and prepared the algorithm for the actual classification process. The spatial model used for classification was developed using the ERDAS Imagine Spatial Modeler tool. Finally, the classification was conducted to generate thematic maps consisting of four distinct classes i.e., urban, vegetation, water, and desert. Later we applied 2-class mask of the classified images and reclassified thematic images as: water class & others (desert, vegetation and urban class). The water class included coastal and freshwater sources within the land borders of the UAE.

32.2.5 Post Classification Processing

Although the classification images had a considerably high overall accuracy ranging between 88 and 98%, there were some discrepancies in the water class which is the class under consideration in this study. The hill shades and some vegetation areas were misclassified as water bodies. Hence, further cleaning was performed to improve the results. To this end, a filtering process was implemented. This involved applying a Majority Filter with a window size of (3 $\times$ 3) to remove isolated single pixels. Additionally, manual cleaning was performed using the Fill Tool in ERDAS Imagine correcting any misclassified pixels by replacing them with the appropriate values after verifying and validating with the original multispectral image as well as google earth historical images. After manual cleaning in ERDAS Imagine, data was exported to ArcMap 10.8.2 for creating final maps used in the study and area calculation to figure out the changes in area covered by water over the years.

32.3 Results and Discussion

This section offers insights into the temporal evolution of water-covered areas within the UAE's land borders throughout the study period, considering the dynamic environmental conditions and the human-induced impacts on water resources in the UAE. Figure 32.2 presents the maps from 1972 to 2021 showing the water bodies in blue color and rest of the LULC classes in light yellow color.

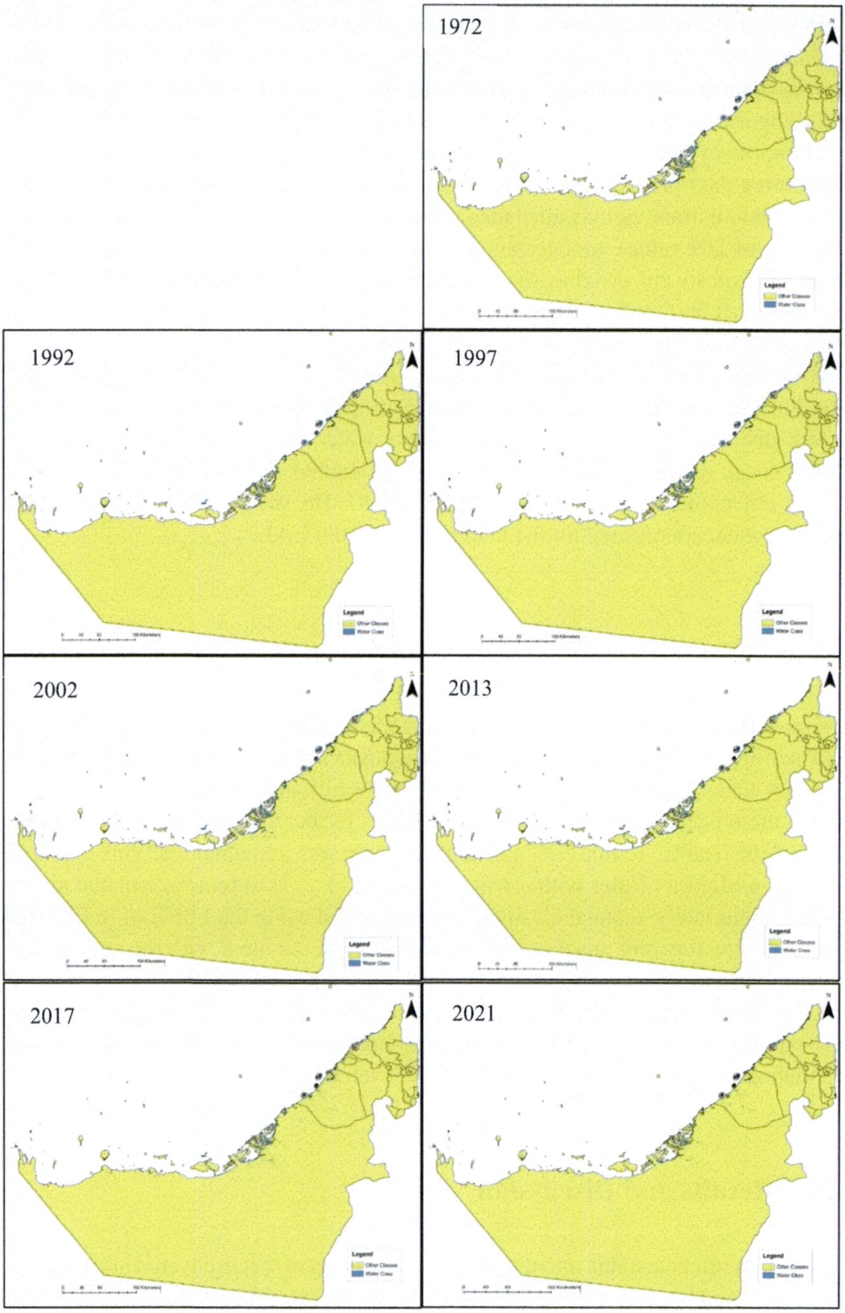

Fig. 32.2 Figures showing the water bodies in dark blue color while other classes in yellow color

Visual representation of maps shows that area covered by water is critically low and can hardly be viewed on maps. This observation was further confirmed by the quantitative assessment as the ratio of water as compared to the overall study area is very low.

Table 32.2 presents the changes in inland and coastal water areas within the current land borders of the UAE from 1972 to 2021. The percentage of water, as compared to the overall study area, varied over the years. In 1972, the percentage of inland and coastal water was 0.51%, which equates to an estimated area of 365.99 square kilometers. This percentage decreased to 0.33% in 1986, indicating a decrease in the presence of water. Notably, a significant decline occurred in the coastal areas of Abu Dhabi, particularly affecting Al Hudariyat Island, Umm Yifenah Island, and Samaliyah Island. Other islands in Abu Dhabi and coastal regions of Dubai and Umm Al Quwain also experienced declines, albeit to a lesser extent.

In 1992, the percentage further decreased to 0.28%, showing a continued decline in water coverage. Coastal water declined more than inland water during this period. Abu Musa Island, Sir Abu Nu'Ayr, and Abu Al Abyad Island experienced a significant reduction in coastal water area, due to natural and human-induced factors, mainly land reclamation and urbanization. In contrast, some small inland water bodies were seen in Abu Dhabi emirate, such as Al Wathba Lake, Al Fayah, and Al Houma, and other small water bodies in Sharjah. These inland water bodies are also manmade. However, in 1997, there was a slight increase in the percentage of inland water to 0.28%. The rise can be associated with the floods that occurred in 1995 and 1997 in the country (Yagoub & Yammahi, 2022). In 2002, the percentage declined slightly to reach 0.27%, indicating a relatively stable inland water area. While minor rise in inland water could be seen through some water reservoirs like Ham Dam is in Fujairah, Bih Dam is in Ras Al Khaimah, and Gulfa Dam is in Ajman (Al-Nuaimi & Murad, 2007).

The trend of declining water areas continued in 2013, with the percentage dropping to 0.24% and an area of 175.51 square kilometers. However, there was a slight increase in 2017, with the percentage rising to 0.33% and the area expanding to 234.89 square kilometers. The most recent data in 2021 shows a significant decrease

Table 32.2 Changes in Inland and coastal water areas in the UAE from 1972 to 2021

Year	Inland and coastal water (sq km)	Inland and coastal water (%)
1972	365.99	0.51
1986	238.54	0.33
1992	200.68	0.28
1997	201.96	0.28
2002	194.09	0.27
2013	175.51	0.24
2017	234.89	0.33
2021	153.64	0.21

in both the percentage (0.21%) and the corresponding area (153.64 square kilometers) of inland and coastal water in the UAE. Some minor changes in water area cover were observed during this period. As it has been observed that major variations in water are induced by changes in coastal areas that too result from the artificial islands created for recreational purposes and land reclamation projects. Most of the reclamation projects have been completed by this time (Subraelu et al., 2022).

The low percentage of inland water compared to the overall study area is expected in arid and dry regions like the UAE. Water bodies in the UAE are primarily confined to coastal areas and the minimal inland waters are influenced by seasonal variations and precipitation-induced temporary flooding or manmade artificial lakes. For its water needs, the UAE relies mainly on desalination and groundwater. However, the country also has some perennial and artificial surface water sources, such as wadis, springs, dams, and islands that contribute to its water security and diversity. This section shows the variations in inland and coastal waters within the country's land borders. While the overall presence of water in the study area is low, its significance cannot be denied.

To provide comprehensive understanding, we analyzed the change in the water covered area in the different emirates of the UAE. Table 32.3 provides information about the areas covered by water in the seven emirates for eight different years from 1972 to 2021. Each row represents a different year, and each column represents a specific emirate.

The analysis highlights Abu Dhabi's consistent dominance in water among all the emirates over the years. Conversely, Ajman exhibits the lowest water cover. Figure 32.3 shows the variations in the water cover of the Emirates over the years.

In Abu Dhabi the water area decreased significantly from 268.26 sq km in 1972 to 69.48 sq km in 2021, reflecting a consistent decline over time This decline in water area can be attributed to various factors, including land reclamation and urbanization, leading to a reduction of coastal water combined with a loss of natural vegetation and wetlands (Yagoub et al., 2006). Additionally, the creation of artificial inland water bodies for recreational purposes and storage of treated water was also observed during the period. In contrast, Sharjah consistently maintains a small water area, except for

Table 32.3 Showing area covered by water (Square km) in each Emirate

Year	Abu Dhabi	Sharjah	Dubai	Ajman	Umm Al Quwain	Ras Al Khaimah	Fujairah
1972	268.26	9.15	48.48	2.96	7.5	23.9	5.74
1986	161.31	6.86	41.13	0.52	3.1	17.48	8.14
1992	116.69	8.22	40.79	1.14	4.89	21	7.95
1997	127	6.92	41.46	0.63	1.98	15.73	8.24
2002	118.68	8.79	41.03	1.02	2.2	14.29	8.08
2013	89.43	9.83	17.69	1.4	5.3	15.98	5.88
2017	174.9	10.98	17.93	1.88	6.6	14.9	7.7
2021	69.48	9.45	11.58	1.47	5.7	14.42	6.54

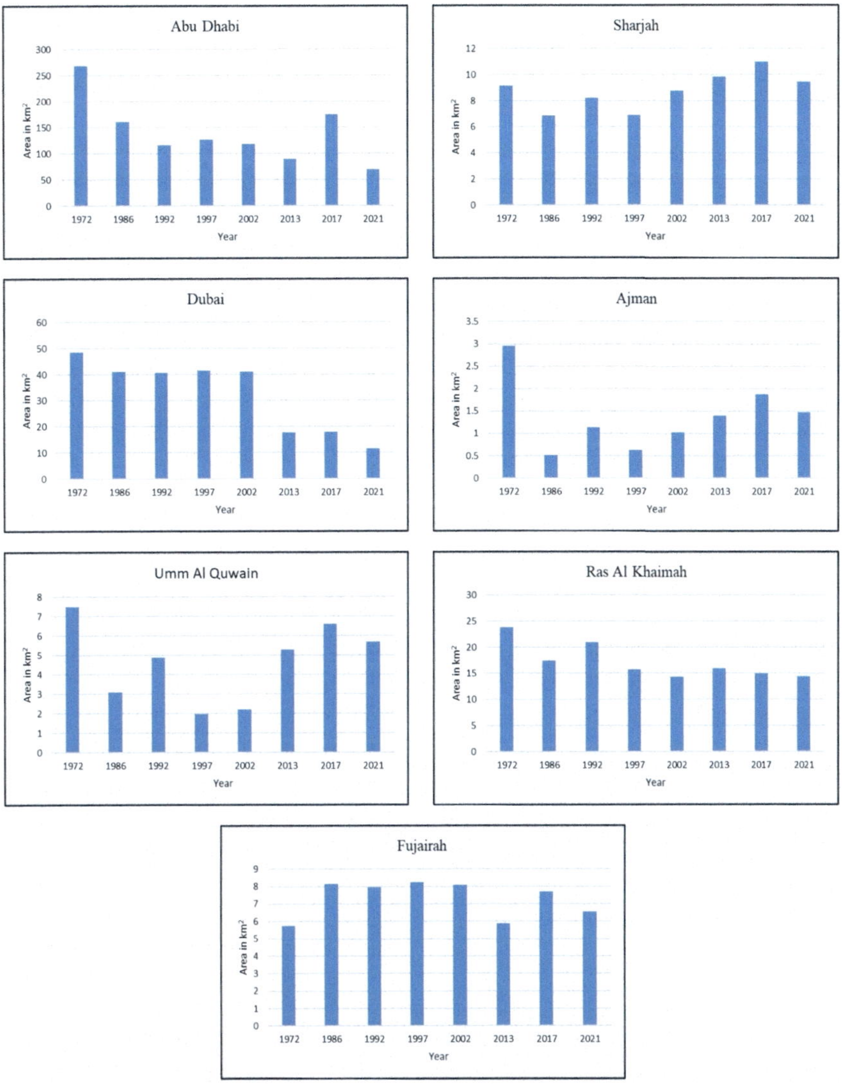

Fig. 32.3 The area coverage by water bodies in the Emirates of the UAE

1986 and 1997 when Ajman recorded the lowest value. Sharjah's water area ranges from 6.86 sq km in 1986 to 10.98 sq km in 2017 showing a slight increase over time.

Dubai ranks third in water area among the Emirates but has experienced a significant decline in recent years, fluctuating from 48.48 sq km in 1972 to 11.58 sq km in 2021. Ajman consistently maintains the second smallest water area among all emirates ranging from 0.52 sq km in 1986 to 2.96 sq km in 1972. There is no clear trend in the water area over time for Ajman. Umm Al Quwain's water area has moderately declined over time, ranging from 1.98 sq km in 1997 to 7.5 sq km in 1972. Ras Al Khaimah consistently holds the fourth largest water area among the emirates, while Fujairah shows slight fluctuations in water area over time, with no clear trend.

It is noteworthy that the decline in water area is not limited to any specific emirate, but is a trend observed in the entire UAE. These trends emphasize the importance of implementing sustainable practices and addressing the various factors that contribute to the decline and limited water coverage in the region.

The overall trend in water coverage throughout the study period indicates a decline, particularly in coastal waters within the land borders. Coastal areas are highly attractive for development activities, which can have both positive and negative impacts on the surrounding areas. However, these activities must adhere to defined guidelines to minimize negative consequences. Formulated guidelines for such activities can be found in Mangor, et al. (2008). Amidst the decline in coastal waters, there is a notable emergence of inland artificial lakes. These lakes are primarily created for recreational purposes or water storage. For example, Al Qudra lake in Dubai was created in 2013, serves as habitat for birds and contributes to biodiversity. This evolution of inland water bodies over the years, including the emergence of new lakes like Love Lake in 2021, highlights the changing landscape and the need for sustainable water management practices (Fig. 32.4).

The significance of artificial lakes cannot be denied; for example, Al Wathba Lake (Fig. 32.5), an artificial lake in the Al Wathba wetland reserve, holds great ecological significance as the first Ramsar site of UAE (Al Dhaheri & Saji, 2013). The area covered by lakes has varied over the years. Warsen Lake (Dubai), which was first observed in the 2002 map and had largest area covered during that year i.e., 0.34 km^2 while the area reduced to 0.17 km^2 in 2013, it remained same in 2017 while a minor decline was found in 2021. On the other hand, Al Sheba lakes complex, appearing

Fig. 32.4 The evolution of Al qudra lakes over the time

on maps in 2013 and covering an area of 0.22 km^2, maintained consistency until 2021, showcasing effective management over time. The inland water bodies exhibit fluctuating trends over the years. Figure 32.5, for example, shows the evolution of Al Wathaba Lake in Dubai and wetlands from 1972 to 2021, showing some water shrinking while others expanded during the study period. Evolution reflects human intervention in managing the natural wetland. While the original wetland area was diminishing the creation of new lakes has helped sustain the ecosystem. Artificial lakes have also been used for economic gains such as Hatta Lake created in 2006 which has been used for electricity generation since then.

Figure 32.6 shows the emergence of inland water bodies in Dubai, alongside the conversion of coastal areas for constructing new artificial islands. The significant

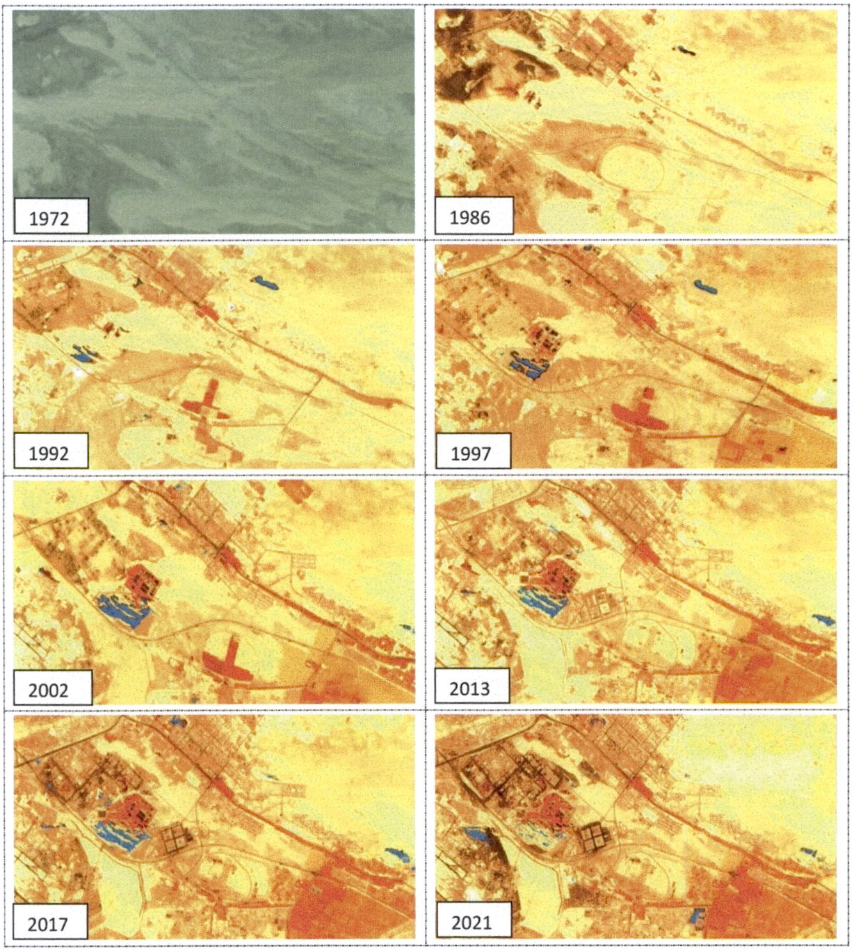

Fig. 32.5 The evolution of Al Wathaba Lake and wetlands over the years

decline in water areas can be attributed to the larger size of the reclaimed coastal areas converted into islands compared to the areas occupied by newly created artificial inland water bodies.

The establishment of artificial lakes demonstrates the UAE's commitment to enhancing its water security and environmental sustainability, while also boosting its tourism and recreation sectors. Additionally, these artificial lakes offer essential ecosystem services, including water purification, carbon sequestration, and habitat provision.

However, these artificial lakes cannot compensate for the loss of natural water bodies, particularly the coastal areas, which face threat from sea level rise, coastal

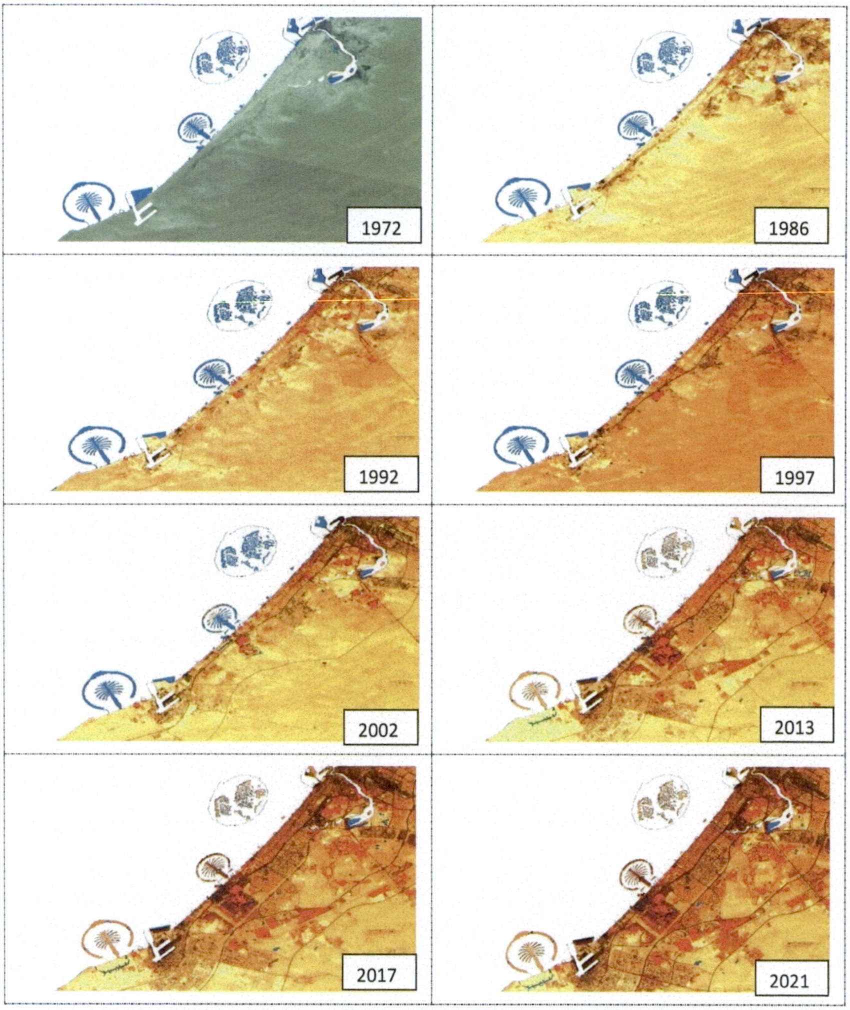

Fig. 32.6 The evolution of new islands and inland water bodies in Dubai

erosion, and land reclamation (Subraelu et al., 2022). Inland water bodies, primarily replenished through treated wastewater, desalinized water and occasional rainfall, are under threat due to anthropogenic activities, and high evapotranspiration rates. For instance, Zakher Lake in Al Ain, initially created from treated wastewater, its area was 0.62 km in 2013 reduced to 0.3 and 0.1 km in 2017 and 2021 respectively (Ghebreyesus et al., 2016). Figure 32.7 shows the different stages of evolution of Zakher Lake. This decline is attributed to higher evaporation and lower replenishment rate. This would have negative impacts on the economy, society, and ecology of the UAE, as the water sources are vital for trade, tourism, fisheries, and biodiversity and climate regulation. On the other hand, there are multiple factors responsible for the decline of coastal waters in the UAE, such as coastal development, which affects the natural landscape and hydrology of the coast and leads to habitat loss and degradation. Land reclamation also contributes to coastal area decline as it reduces the water depth and flow, increases the sedimentation and turbidity, and affects the marine life and ecosystems. To ensure the stability of artificially created lakes and protect the natural and cultural heritage of the UAE, it is important to adopt integrated water resource management strategies that balance the needs of development and conservation.

The coastal areas are experiencing decline due to conversion into artificial islands and other anthropogenic activities, with land reclamation and erosion being major contributing factors, along with urbanization. Conversely, artificial inland water bodies are being created and managed. It is believed that these newly created and artificially maintained water bodies are contributing to rising groundwater levels through recharge, providing habitats for bird species, contributing to ecosystem maintenance, and offering economic benefits and recreational opportunities. However, these water bodies may also pose threats such as the presence of infectious bacteria (Alsharhan

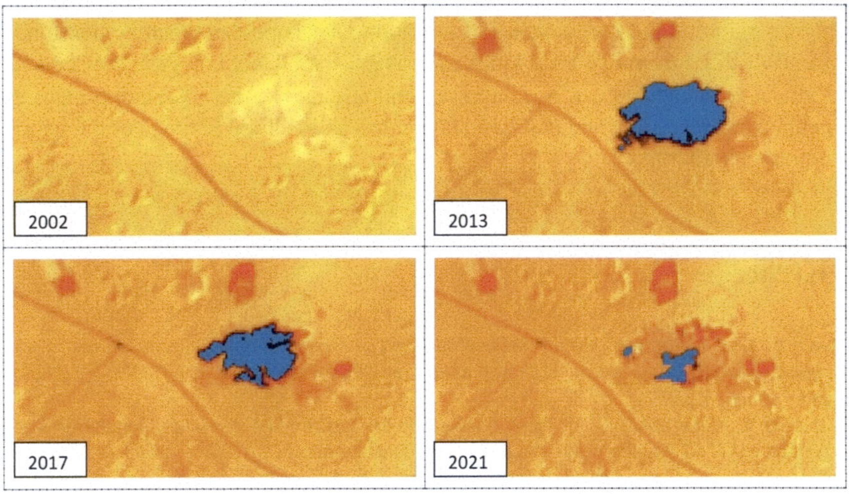

Fig. 32.7 The evolution of Zakher Lake over years

et al., 2020), depending on the level of treatment carried out for waste and sea water. Hence, there is a continued need to monitor the water bodies.

32.4 Conclusions

This study aims to track and analyze changes within both inland and coastal water bodies within the present land borders of the UAE spanning five decades, from 1972 to 2021, using Landsat imagery. Despite the proportion of water concerning the total study area remaining notably low, it holds significant importance in an arid region like the UAE. Its area percentage dynamics display variability, ranging from 0.51% in the 1970s to approximately 0.21% in recent years. The extent of water coverage across different emirates varies over time too, with Abu Dhabi boasting the largest water area and Sharjah exhibiting the smallest in most years. Inland water bodies are predominantly artificial in nature and are created either to store rainwater or through treated waste and sea water. These water bodies are influenced by seasonal changes and temporary flooding triggered by precipitation, rate of replenishment and evaporation and consumption rates. Man-made lakes like Al Qudra and Al Wathba and Zakher lakes, contribute positively to biodiversity and the economy. Yet, these artificial water bodies cannot substitute for the natural ones, particularly the vulnerable coastal zones encountering threats such as rising sea levels, erosion, and land reclamation. Remote sensing and GIS integrated with machine learning algorithms have played a crucial role in mapping water bodies and assessing changes in their areas over the years. The study identified major causes of decline in coastal water bodies, primarily due to anthropogenic activities such as urbanization, land reclamation, and the creation of artificial islands. Inland water bodies were also at risk due to higher evaporation rates as well as to the same anthropogenic factors mentioned above. By using innovative techniques such as geospatial technologies, both coastal and inland water bodies can be conserved and sustained. The UAE can protect its valuable water resources and maintain the natural and cultural heritage of the region for future generations.

References

Ahmed, A. (2010). An overview of conventional and non-conventional water resources in arid region: assessment and constrains of the United Arab Emirates (UAE). *Journal of Water Resource and Protection, 2010.*

Al Dhaheri, S., & Saji, A. (2013). Water quality and brine shrimp (Artemia sp.) population in Al Wathba Lake, Al Wathba wetland reserve, Abu Dhabi emirate, UAE. *International Journal of Biodiversity and Conservation, 5*(5), 281–288.

Aldababseh, A., & Temimi, M. (2017). Analysis of the long-term variability of poor visibility events in the UAE and the link with climate dynamics. *Atmosphere, 8*(12), 242.

Aldogom, D. et al. (2019, October). Multi temporal satellite images for growth detection and urban sprawl analysis; Dubai City, UAE. In *Remote Sensing Technologies and Applications in Urban Environments IV* (Vol. 11157, pp. 71–81). Spie.

Ali, T. A., et al. (2016). A GIS-based spatiotemporal study of the variability of water quality in the Dubai Creek, UAE. *Water Quality Research Journal of Canada, 51*(3), 219–232.

Al-Nuaimi, H. S., & Murad, A. A. (2007). The role of dams in securing the surface water in the northern and eastern parts of the United Arab Emirates (UAE). *IAHS Publications-Series of Proceedings and Reports, 315*, 206–214.

Alsharhan, S. et al. (2020). Water resources and water demands in the UAE. *Water Resources and Integrated Management of the United Arab Emirates*, 757–791.

Altman, N. (2002). *Sacred water: The spiritual source of life.* Paulist Press.

Basheer, S., et al. (2022). Comparison of land use land cover classifiers using different satellite imagery and machine learning techniques. *Remote Sensing, 14*(19), 4978.

Beydoun, Z. R. (1986). The petroleum resources of the middle east: A review. *Journal of Petroleum Geology, 9*(1), 5–27.

Burt, J. A. (2014). The environmental costs of coastal urbanization in the Arabian Gulf. *City, 18*(6), 760–770.

Cheng, C., et al. (2022). What is the relationship between land use and surface water quality? A review and prospects from remote sensing perspective. *Environmental Science and Pollution Research, 29*(38), 56887–56907.

Crain, C., et al. (2009). Understanding and managing human threats to the coastal marine environment. *Annals of the New York Academy of Sciences, 1162*(1), 39–62.

Dahy, B. et al. (2021). A review of land change modelling techniques using Remote sensing and GIS. In *42nd Asian Conference on Remote Sensing, ACRS 2021. Asian Association on Remote Sensing.*

ElHaj, K. et al. (2023). Modeling and prediction of groundwater level fluctuations using geoinformatics and artificial neural networks in Al Ain City, UAE. In *Water Resources Management and Sustainability: Solutions for Arid Regions* (pp. 205–217). Cham: Springer Nature Switzerland.

Ghebreyesus, D. et al. (2016, April). Monitoring water resources from space in an arid watershed of Al Ain city. In *EGU General Assembly Conference Abstracts* (pp. EPSC2016–5461).

Hinrichsen, D. (2013). *Coastal waters of the world: Trends, threats, and strategies.* Island Press.

Mangoosh, A. (2004). *Envisioning a Comprehensive Earth Information System for Improving Water Resource Assessment in the UAE* (Doctoral dissertation).

Mangor, K., et al. (2008). Shoreline management of the Dubai coast. *COPEDEC VII, 15*, 134–145.

Mortula, M. M., et al. (2019). Spatiotemporal assessment of water quality in the distribution network of City of Sharjah, UAE. *International Journal of Information and Decision Sciences, 11*(2), 116–128.

Naorem, A. et al. (2023). Soil constraints in an arid environment—challenges, prospects, and implications. *Agronomy, 13*(1), 220.

Nichols, A., et al. (2019). Machine learning: Applications of artificial intelligence to imaging and diagnosis. *Biophysical Reviews, 11*(1), 111–118.

Pain, C. F., & Abdelfattah, M. A. (2015). Landform evolution in the arid northern United Arab Emirates: Impacts of tectonics, sea level changes and climate. *CATENA, 134*, 14–29.

Rinke, K et al. (2019). Ecosystem services from inland waters and their aquatic ecosystems. *Atlas of ecosystem services: Drivers, risks, and societal responses*, 191–195.

Rizk, Z. S., & Alsharhan, A. S. (2003). Water resources in the United Arab Emirates. In *Developments in water science* (Vol. 50, pp. 245–264). Elsevier.

Shahin, S. M., & Salem, M. A. (2015). The challenges of water scarcity and the future of food security in the United Arab Emirates (UAE). *Natural Resources and Conservation, 3*(1), 1–6.

Subraelu, P., et al. (2022). Land in water: The study of land reclamation and artificial islands formation in the UAE coastal zone: a remote sensing and GIS perspective. *Land, 11*(11), 2024.

Talukdar, S., et al. (2020). Land-use land-cover classification by machine learning classifiers for satellite observations—A review. *Remote Sensing, 12*(7), 1135.

Tourenq, C., et al. (2011). Hydrogeology of Wadi Wurayah, United Arab Emirates, and its importance for biodiversity and local communities. *Hydrological Sciences Journal, 56*(8), 1407–1422.

Wang, X., & Xie, H. (2018). A review on applications of remote sensing and geographic information systems (GIS) in water resources and flood risk management. *Water, 10*(5), 608.

Yagoub, M. M., & Al Yammahi, A. A. (2022). Spatial distribution of natural hazards and their proximity to heritage sites: Case of the United Arab Emirates. *International Journal of Disaster Risk Reduction, 71*, 102827.

Yagoub, M. M., & Kolan, G. R. (2006). Monitoring coastal zone land use and land cover changes of Abu Dhabi using remote sensing. *Journal of the Indian Society of Remote Sensing, 34*, 57–68.

Yilmaz, A. G., et al. (2020). Spatio-temporal trend analysis of groundwater levels in Sharjah, UAE. *International Journal of Environmental Science and Development, 11*(1), 9–14.

Zhou, Y., et al. (2017). Open surface water mapping algorithms: A comparison of water-related spectral indices and sensors. *Water, 9*(4), 256.

Chapter 33
Solar-Based Atmospheric Water Absorber: Design and Performance Analysis

Anwur Alenezi, Abouelyazed Kuliab, and Yousef Alabaiadly

Abstract Solar system for harvesting water from moisture in the atmospheric air, referred to as a solar still unit. The solar still unit is a simple but cost-effective method that works as an absorption-based Atmospheric Water Absorber (AWA) system, taking advantage of the decent climatic conditions in areas with high temperatures and humidity. The study followed an experimental approach, denoting a cost-effective method for harvesting water from humid air using solar energy and an absorption process. The design of the experiment integrates absorption at night and evaporation–condensation during the day, using lithium bromide (Li-Br) as the absorbent because of its favorable properties. The device used in the experiment features an evaporation basin, glass cover, circulating pump, and absorbent solution. Different instruments were used to record measurements of ambient temperature, Li-Br solution temperature, and humidity. The performance of the device at different climatic conditions was assessed using energy balance equations focusing specifically on the rates of absorption and desorption. The operational procedures of the device involved absorbent circulation and strategic placement of the glass cover to optimize the condensation of water vapor. Overall, the design of the experiment presents a simple but low-cost method to water production from the atmosphere, with possible environmental and health benefits. The experiments were conducted from 20/0/2022 to 20/7/2022. Relative humidity reached 69%, with the wind speed being recorded as 5.5 m/s and ambient temperature as 38°C. The absorbent mass flux was fixed at 3 kg/hr m^2. Water production increased until midday, influenced by the effect of solar radiation on the

A. Alenezi (✉)
Water Research Centre, Kuwait Institute for Scientific Research, 13109 Safat, Kuwait
e-mail: anwenezi@kisr.edu.kw

A. Kuliab
New and Renewable Energy Authority, Ministry of Electricity and Renewable Energy, Cairo, Egypt
e-mail: Abouelyazed.koluib@nrea.gov.eg

Y. Alabaiadly
Studies and Research Department, Ministry of Electricity and Water, Ministries Zone, 12010 South Surra, Kuwait
e-mail: yaalobidly@mew.gov.kw

A. Sefelnasr et al. (eds.), *Water Resources Management and Sustainability*, Water Science and Technology Library 114, https://doi.org/10.1007/978-3-031-80520-2_33

basin glass cover. There was also an increase in water production with rising temperature because of enhanced water vapor evaporation. Li-Br concentration had an effect on water production, peaking at 40%. The absorbent mass flux had an impact on the rate of water production, with the maximum yield being recorded as 3 kg/hr m^2. The device yielded 1.5–2 L of water daily, highlighting its potential for water harvesting under different climatic conditions. The study findings are particularly relevant to Kuwait, a country that features unique climatic conditions. Traditional sources of water can be scarce in the country, and the process of harvesting fresh water from the atmosphere can be a viable solution. The device proposed could be effective in Kuwait, a country with high temperature and humidity.

Keywords Absorption · Humid air · Lithium bromide · Solar energy · Water absorber

33.1 Introduction

Humans and other living things depend heavily on the availability of clean drinking water. However, water resources are becoming increasingly scarce on a global scale, leading to socio-hydrological impacts on water distribution processes (Kharraz et al., 2012; Morante-Carballo et al., 2022). The increasing lack of freshwater, driven by a growing demand in water and change in climate, poses a significant risk to the global economy. Urgent action is required to transition to sustainable water use, with some of the initiatives likely revolving around educating consumers, running initiatives in the private sector, following regulations set by the government, and implementing targeted investments. Recently, the World Economic Forum ranked water scarcity as one of the three most critical systemic risks globally (The Global Water Crisis, 2023). This assessment was based on a comprehensive review on the perception of risk among business, academia, governments, civil society, and global institutions. The effects of freshwater scarcity are evident in reducing groundwater tables, declining river flows, lakes becoming smaller, and waters that end up heavily polluted. Additionally, there is a notable increase in the costs of supply and treatment, intermittent water supplies, and conflicts over water resources. Several drivers contribute to future water scarcity, including population and economic growth, higher demands for animal products and biofuels, and the impact of climate change (Hoekstra, 2014).

The atmosphere holds an approximate 14,000 km^3 of water vapor, making it a potentially sustainable source of water (Hamed, 2000). In select regions worldwide, dew water can complement the drinking water supply, benefiting plants and animals living in dry or semiarid environments. However, a recent report indicates that by 2030, water demand in certain developing regions may surpass supply by 50% (Farahani, 2020). Water serves a crucial responsibility in the worldwide economy, often being used as a solvent for different chemical substances and providing helpful in facilitating cooling and transportation in different industries. Agriculture consumes approximately 70% of the freshwater available. Over the past few decades, access to safe drinking water has improved in all parts of the world. Notably, there exists a clear correlation between access to safe water and gross domestic product (GDP) per capita. Nevertheless, some observers estimate that by 2030, over half of the world's population will face water-related vulnerability (Water Resources Group, 2009).

A method has been discovered to harvest water from the air without the need for electricity or fuel supply, along with the required apparatus (Zimmerman, 2003). The apparatus includes a rotatable, hygroscopic moisture-absorbing rotor with at least one region allowing air to pass through its thickness, a regeneration passage that circulates air for regeneration with openings at opposing surfaces of a regeneration region (part of the moisture-absorbing rotor) where the absorbed moisture is regenerated, and a drain hole for harvesting condensed water from the regeneration passage (Zimmerman, 2003).

Additionally, devices capable of generating drinking water using sunlight are currently under development. One such device requires approximately 310 Watts-hours to produce a liter of water. It operates by releasing water when the sun heats graphite (Patel, 2017).

Porous materials are known for their efficiency as desiccants, but they must be carefully designed to prevent water molecules from becoming trapped in their tunnels. Additionally, the exit pathways should be smooth to facilitate effective water removal. A metal–organic framework (MOF) can effectively gather water molecules by forming clusters or layers. The process of desorption can be influenced by the hydrophilicity of the functional groups within the MOF (Kim et al., 2018). Fathieh et al. (2018) designed and assessed a solar panel-powered desorption and water collection method. Their system achieved a water yield of 0.7–1 L per kilogram of MOF in the Mojave Desert in California.

In May 2022, another set of researchers introduced a super hydroscopic film composed of a plant-based polysaccharide and lithium chloride (Guo et al., 2022). They claimed that this innovative material can be incorporated into a device operating at a rate of 14–24 cycles per day in arid regions, producing an impressive water yield ranging from 6 to 13 L. The process involves heating the film for approximately 30 min to release bound water. Once exposed to air, the film readily absorbs moisture (Guo et al., 2022).

The objective of the current study is to design a straightforward solar system for harvesting water from moisture in the atmospheric air. The system does not require any cooling surface. The technique relies on utilizing an absorbent water material in conjunction with a solar heating device. This device is referred to as a solar still unit. Both theoretical and experimental approaches are employed in this study. The test rig is specifically designed and manufactured in Kuwait. To absorb humidity from the air, Lithium–Bromide is used as the absorbent solution. Subsequently, the unit undergoes testing during both nighttime and daytime conditions. The study also investigates various factors that can influence the system's performance.

33.2 Methods

There are different techniques that can be used to harvest water from humid air. The first technique involves lowering the ambient air's temperature to below the dew point. The second technique involves collecting water vapor by heating the adsorbent and then liquefying it. The acquisition of fresh water from the atmosphere seems to suit climatic conditions with high temperature and humidity. A case in example is the Gulf countries, most of which have such climatic conditions. In these environments, the ambient atmosphere usually has water in the form of vapor in quantities ranging between 9.6 and 55.3 g per m^3 of dry air at saturation temperatures between 10 and 48 °C.

33.2.1 Experimental Design

To design a cost-effective unit for producing fresh water from the atmosphere using solar energy and the absorption method, various climatic variables like insolation, wind velocity, temperature, and humidity must be considered, along with absorbent characteristics. The system incorporates two processes: absorption at night and evaporation–condensation during the day in the same unit. Lithium bromide (Li-Br) is chosen as the suitable absorbent due to its satisfying properties. Policies shaping water resource management may provide subsidies to support the adoption of Atmospheric Water Absorber (AWA) systems, leading to environmental benefits and improved health outcomes. This paper explored a simple and low-cost approach to expand water production from the atmosphere.

The key features of the device are shown in Fig. 33.1 and include an evaporation basin, glass cover, circulating pump, absorbent solution (lithium bromide), tanks, control valves, and collector tubes. The evaporation device is designed as a triangle solar still unit. The unit was assembled from aluminum sheet metal and wood. Wood is used for the outer surface of the solar still device body; however, the sheet metal was used to manufacture the evaporation basin. Fiberglass material is used as insulation between the metal and wood surface. The absorbent tank was made from aluminum sheet metal too. The dimensions of the box are 75 × 60 cm and its total area amounts to 0.45 m^2. The box includes a flat, blackened, and slanted surface that has a layer of single glazing with a 45 cm air gap. The bottom part of the unit features fiberglass insulation with an average thickness of 6 inches. Towards the outer surface of the inner basin, a black material coating was included to increase the absorptivity of the device to solar radiation.

Shown in Fig. 33.2 is a picture of the Atmospheric Water Absorber AWA unit used in this study. The unit has a tilt angle of 45° and a width of 55 cm. It is equipped with a 4 mm thick clear single glass cover, which serves as the condensation surface during solar heating and helps distribute the Li-Br solution to absorb water vapor from the atmosphere at night. Additionally, the glass cover facilitates water vapor

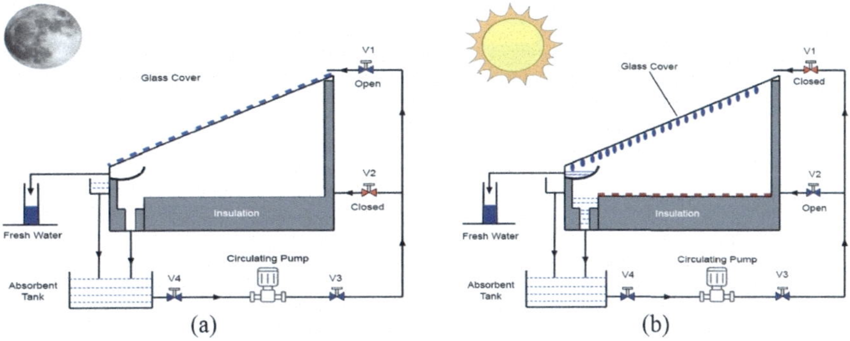

Fig. 33.1 (a) AWA nighttime (b) AWA daytime

condensation during daylight hours when heated. To distribute the Li-Br solution across the glass cover surface, a small circulating pump with a flow rate of 5 L per minute was utilized.

Various parameters were measured during the study, including ambient temperature, basin Li-Br solution temperature, and humidity. The ambient and basin Li-Br solution temperatures were recorded using a digital thermometer with an accuracy of $\pm 0.2°$. To measure the flow rates of Li-Br solution, a pulse turbine digital flow meter with an accuracy of $\pm$ 0.002 L per minute was employed. Humidity measurements were taken using a hygrometer, while relative humidity was determined using a sling psychrometer (see Fig. 33.3).

This study analyzes the unit's performance by using energy balance equations to compute the absorption rate at night and desorption rate during the day, in one-hour intervals. Changes in climatic conditions affect the unit's performance, impacting desiccant temperature, water absorption, and desorption rates. Data from Kuwait's climate was used in the calculations. The analysis assumes steady-state conditions,

Fig. 33.2 Photo of AWA unit

Fig. 33.3 Sling psychrometer with the wet-bulb thermometer located on top

equal areas for the glass cover and absorber surface, and negligible back and edge losses. The energy balance equation for a unit absorber area during daytime operation is given as follows (Habeebullah, 2009):

33.2.2 *Operation Procedure*

The system functions as follows: During night-time operation, the concentrated absorbent flows over the glass cover, absorbing moisture from the ambient air. During the day, the solar energy on the absorber surface heated the weak absorbent, resulting in a rise of water vapor and its eventual condensation on the cool glass cover. This causes valve V2 to close and V1 to open and absorb the water vapor. In low-humidity climates, a stronger solution (e.g., 55% lithium bromide) may be used. Day-time operation involves opening V2, allowing solar radiation to heat the absorbent, releasing the water vapor absorbed at night, and condensing it on the glass cover. The system's night-time performance depends on mass transfer potential, while day-time performance relies on the difference in water vapor pressure between the absorbent and glass cover. To minimize overheating and maintain optimal performance, the glass cover is placed in a strategic position at a distance from the absorber surface. Such an arrangement helps reduce the temperature of the glass cover during the day when solar radiation is absorbed. During night-time operation, an alternative method can be used where the glass cover is removed. This allows the system to take up the moisture from the ambient air more efficiently, as there is no barrier between the absorbent and the environment.

33.3 Results and Discussion

A series of experiments were conducted between 20/6/2022 and 20/7/2022. The data was collected and filtrated. Figures 33.4, 33.5, 33.6 and 33.7 show a sample of measured data on 25/6/2022. The relative humidity on this day reached 69%, along with a high wind of 5.5 m/s and a mean ambient temperature of 38°C. The mass flux of absorbent was fixed as 3 kg/hr m^2.

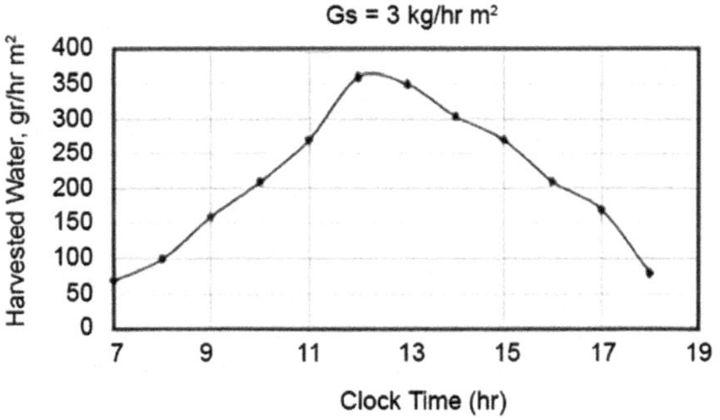

Fig. 33.4 AWA unit performance during daytime

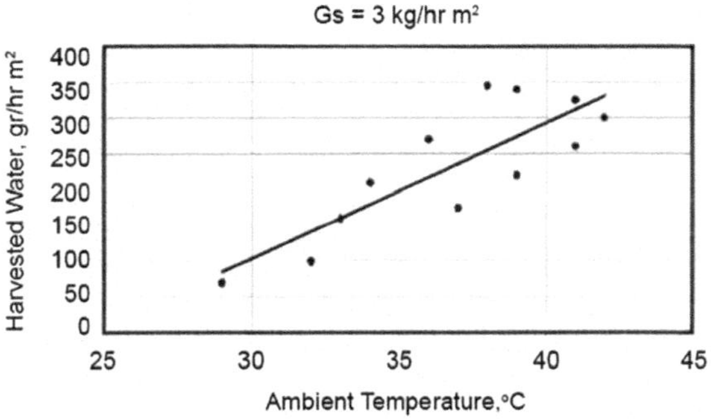

Fig. 33.5 Harvested water rate with ambient temperature

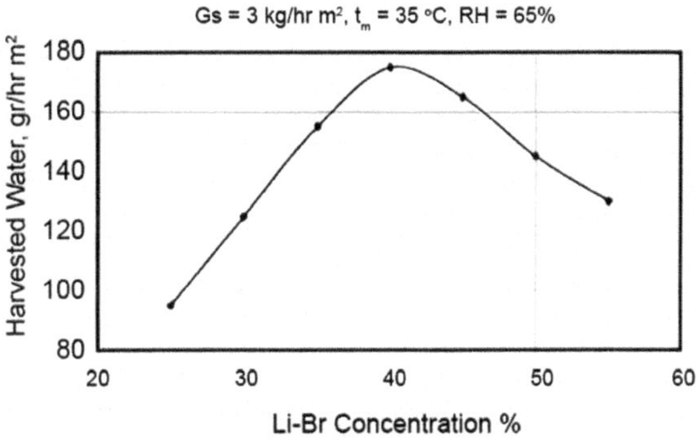

Fig. 33.6 Harvested water rate with Li-Br concentration

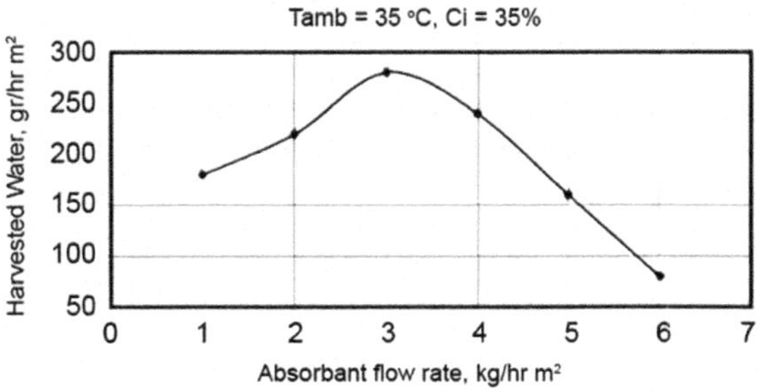

Fig. 33.7 Harvested water rate with absorbent flow rate

As depicted in Fig. 33.4, the water production gradually increases until midday and then decreases at the same rate. This pattern is influenced by the solar radiation's effect on the basin glass cover. The peak solar intensity occurs around midday (approximately 1 PM), leading to higher condensation rates during this period.

As the ambient temperature increased, the rate of water harvested increased too due to the increase in the rate of water vapor evaporation in the basin. An increase in the ambient temperature led to an increase in the Li-Br solution temperature which increased the rate of heat transfer through the unit and the rate of water evaporation. Figure 33.6 illustrates the effect of Li-Br concentration on the rate of freshwater production. It demonstrates that at a concentration of 40%, the rate of water production is at its maximum. Beyond this concentration value, the water produced decreases. The mean temperature during the experiment was "$t_m = 35\ °C$", influencing the rate of water harvesting in the system. As the ambient temperature increases, it raises the solution and basin temperatures, boosting the mean temperature difference (t_m), eventually increasing heat transfer efficiency and boosting the rate of water evaporation.

Figure 33.7 demonstrates the impact of absorbent mass flux (Gs) on the water production rate from the AWA unit. The maximum water production occurred at a mass flux of 3 kg/hr m^2 with a fixed concentration of 35%. A range of factors influence the operation and performance of the AWA unit, such as humidity ratio in the atmosphere, solar glass cover angle, absorbent flow during daytime, ambient temperature, available solar energy, absorbent concentration, and system design including the type of glass, insulation, tanks, and wind velocity on-site. Through a series of experiments, the proposed AWA unit in this study yielded 1.5–2 L of water per day, based on climate data during the operational interval.

33.4 Conclusions

This study investigated a simple but cost-effective method for harvesting water from the atmosphere, concentrating on the design and performance of an Atmospheric Water Absorber (AWA) system. The experiments conducted in this study were done under Kuwait's summer weather conditions, revealing crucial insights into the dynamics of water production. The findings revealed that higher absorbent solution flow rates enhanced the rate of water absorption and reduced water desorption during daytime operation. The rate of water production was influenced by factors such as ambient temperature, absorbent concentration, and relative humidity. The findings also highlighted desiccant-based systems as suitable solutions for harvesting passive water, making use of solar energy as the source of heat. According to measured data, the average water output was approximately 0.24 L/hr during solar hours under certain ambient conditions. These findings highlight the potential of AWA systems for addressing the challenges of waster scarcity in different locations.

To ensure comprehensive evaluation, it is advisable to re-test the unit throughout all months of the year, especially during periods of remarkably high relative humidity and ambient temperatures. Optimizing the angle of the glass cover is essential to enhance performance. Continued research and development will contribute to harnessing atmospheric water absorber as a sustainable solution for water scarcity challenges. Further research directions include looking into desiccant cooling effects, optimizing the angle of the glass cover, and studying the absorbent concentration to enhance the efficiency and scalability of the system. Economic studies also need to be carried out to examine the feasibility and viability of AWA systems on a wider scale.

References

Farahani, M. H. D. A. (Ed.). (2020). *Desalination: Challenges and opportunities*. IntechOpen. https://www.intechopen.com/books/desalination-challenges-and-opportunities

Fathieh, F., Kalmutzki, M. J., Kapustin, E. A., Waller, P. J., Yang, J., & Yaghi, O. M. (2018). Practical water production from desert air. *Science Advances*, *4*(6), eaat3198. https://doi.org/10.1126/sciadv.aat3198

Guo, Y., Guan, W., Lei, C., Lu, H., Shi, W., & Yu, G. (2022). Scalable super hygroscopic polymer films for sustainable moisture harvesting in arid environments. *Nature Communications, 13*(1), 2761. https://doi.org/10.1038/s41467-022-30505-2

Habeebullah, B. A. (2009). Potential use of evaporator coils for water extraction in hot and humid areas. *Desalination, 237*(1), 330–345. https://doi.org/10.1016/j.desal.2008.01.025

Hamed, A. M. (2000). Absorption–regeneration cycle for production of water from air-theoretical approach. *Renewable Energy, 19*(4), 625–635. https://doi.org/10.1016/S0960-1481(99)00068-3

Hoekstra, A. Y. (2014). Water scarcity challenges to business. *Nature Climate Change*, *4*(5), Article 5. https://doi.org/10.1038/nclimate2214

Kharraz, J. E., El-Sadek, A., Ghaffour, N., & Mino, E. (2012). Water scarcity and drought in WANA countries. *Procedia Engineering, 33*, 14–29. https://doi.org/10.1016/j.proeng.2012.01.1172

Kim, H., Rao, S. R., Kapustin, E. A., Zhao, L., Yang, S., Yaghi, O. M., & Wang, E. N. (2018). Adsorption-based atmospheric water harvesting device for arid climates. *Nature Communications*, *9*(1), Article 1. https://doi.org/10.1038/s41467-018-03162-7

Morante-Carballo, F., Montalván-Burbano, N., Quiñonez-Barzola, X., Jaya-Montalvo, M., & Carrión-Mero, P. (2022). What do we know about water scarcity in semi-arid zones? A global analysis and research trends. *Water*, *14*(17), Article 17. https://doi.org/10.3390/w14172685

Patel, P. (2017, April 13). *Solar-Powered Device Pulls Water Out of Thin (and Pretty Dry) Air*. IEEE Spectrum. https://spectrum.ieee.org/solarpowered-device-pulls-water-out-of-thin-and-pretty-dry-air

The Global Water Crisis. (2023). Water.Org. https://water.org/our-impact/water-crisis/global-water-crisis/

Water Resources Group. (2009). *Charting our water future—Economic frameworks to inform decision-making*. https://www.2030wrg.org/wp-content/uploads/2014/07/Charting-Our-Water-Future-Final.pdf

Zimmerman, A. (2003). *Method and apparatus for extracting water from atmospheric air* (China Patent CN1103843C). https://patents.google.com/patent/CN1103843C/en

Chapter 34
Bridging the Waters of Time: Integrating Traditional Water Management in Semi-arid Regions—Case Study of Jaipur, India

Kanika Bimrah and Izuru Saizen

Abstract Water, an indispensable element of life, has been a fundamental resource for the sustainability and development of human civilization since its inception. Throughout history, societies have settled around water sources, developed advanced irrigation systems, and devised strategies to preserve and manage these resources. However, arid and semi-arid regions worldwide present a unique paradox of water challenges and scarce environments rich in technology and innovation. One such paradigm is exemplified by Jaipur's traditional water management system, which is a testament to human ingenuity during times of adversity. Jaipur, the capital city of Rajasthan, lies in the semi-arid desert region of India. This geographical location has a unique combination of weather conditions with extremely hot summers and uneven and scant rainfall throughout the year. This study aims to explore the significance of traditional water management systems in Jaipur and examines how these structures are used for water-utilitarian solutions in semi-arid regions. The water management system of Jaipur was designed to efficiently accommodate the water needs of its residents through rainwater harvesting. These structures were designed to capture rainwater and store it for dry periods, reflecting a deep understanding of local hydrology and exhibiting a harmonious integration of engineering with functionality. Nevertheless, the challenges of the modern era, such as population growth, urbanization, climate change, and increased water demand, have prompted the need for innovative solutions beyond traditional methods. Hence, a comprehensive research methodology with a multidisciplinary research design was adopted to achieve this objective. It involved an architectural analysis that examined the structural layout, focusing on features such as stepwells, underground channels, and storage reservoirs, and hydrological assessments that involved studying the hydrology of the

K. Bimrah (✉) · I. Saizen
Laboratory of Regional Planning, Graduate School of Global Environmental Studies, Kyoto University, Kyoto, Japan
e-mail: kanika.bimrah.24e@st.kyoto-u.ac.jp

I. Saizen
e-mail: saizen.izuru.4n@kyoto-u.ac.jp

A. Sefelnasr et al. (eds.), *Water Resources Management and Sustainability*, Water Science and Technology Library 114, https://doi.org/10.1007/978-3-031-80520-2_34

region and catchment areas of the forts, along with rainwater harvesting and groundwater recharge techniques. By employing this comprehensive research methodology, this study highlights the adaptation of these structures to modern water needs and their contribution to the community's water security and examines how these water management techniques are used worldwide as innovative technologies to transform water management. The relevance of this study extends beyond historical curiosity as it provides valuable insights for present-day water resource management in arid and semi-arid regions that deal with water scarcity exacerbated by climate change. By analyzing the sustainability of these traditional water management systems, this study offers potential solutions for adapting and modernizing these techniques to the current challenges. This bridges the gap between historical wisdom and modern innovation, underlining the importance of incorporating traditional knowledge into sustainable water management strategies. By examining the traditional water management systems in Jaipur, this study sheds light on the enduring wisdom of the past and its applicability in addressing existing water challenges. These findings have implications for policymakers, urban planners, and researchers working toward sustainable water management solutions in arid and semi-arid regions worldwide.

Keywords Traditional water management system · Integrated water resource management · Semi-arid region · Rainwater harvesting · Water catchment area

34.1 Introduction

Since the beginning of human civilization, water, an essential component of life, has been a vital resource for sustainability and advancement. Humans started to establish permanent settlements approximately 10,000 years ago after adopting an agrarian lifestyle and began establishing sociocultural settlements (Vuorinen et al., 2007). All great ancient settlements, such as the Egyptian, Mesopotamian, Roman, and Indus Valley civilizations, were established at places where water resources were readily available and thrived due to their sustainable water management practices (Hosseiny et al., 2021). As water was the foundation for ancient civilizations, humans had comprehensive knowledge of the hydrological cycles and the characteristics and patterns of its diverse components. The traditional knowledge by ancestors regarding water resources was so effective that it played a vital role in ensuring their survival for centuries (Kumar Singh et al., 2020). Nonetheless, certain areas, notably arid and semi-arid regions, face additional challenges owing to the uneven spatial and temporal distribution of water resources (Cosgrove & Loucks, 2015). Remarkably, in certain instances, traditional water management systems implemented centuries ago in semi-arid and arid regions have persistently operated to this day, effectively meeting the water needs of the local population.

An exemplary illustration of enduring traditional water management systems exists within the forts of Jaipur city, located in the semi-arid region of India. The walled city of Jaipur was the first planned city in India and is now a UNESCO World

Heritage Site (Hooja, 2022). It is situated in the foothills of the Aravalli range, surrounded by hillocks on the northern and eastern sides, and plains on the western and southern sides. The summers in Jaipur are very hot, with temperature rising up to 47 °C in May while winters are extremely cold. The region is characterized by relatively low and erratic precipitation, with the monsoon season ranging from July to September (Tyagi et al., 2011). Jaipur has witnessed sporadic but intense rainfall with an average rainfall of 633 mm/year over the last five decades (India Meteorological Department, 2021). Despite being situated in a desert region characterized by fluctuating rainfall patterns, the city's water management structures were ingeniously designed to capture and store rainwater effectively. The city, surrounded by the Aravalli Range in the southwest, took advantage of the various catchment areas that were available for direct storage in response to geophysical conditions.

This study aims to explore the significance of traditional water management systems in Jaipur and examines how these structures are used for water-utilitarian solutions in semi-arid regions. The primary focus of this research was to investigate the water management systems within Jaipur's forts, which were constructed in the sixteenth century. The three major forts lie in the proximity of Jaipur: Amber Fort, Jaigarh Fort, and Nahargarh Fort, as illustrated in Fig. 34.1. This study utilized a holistic methodology that included structural and hydrological evaluations conducted through a primary survey. It highlights the commonalities between these water management systems by closely examining case studies. This study attempts to identify the commonalities among these historical forts through a detailed comparative analysis and how the lessons learned from these case studies can be practically applied in modern situations.

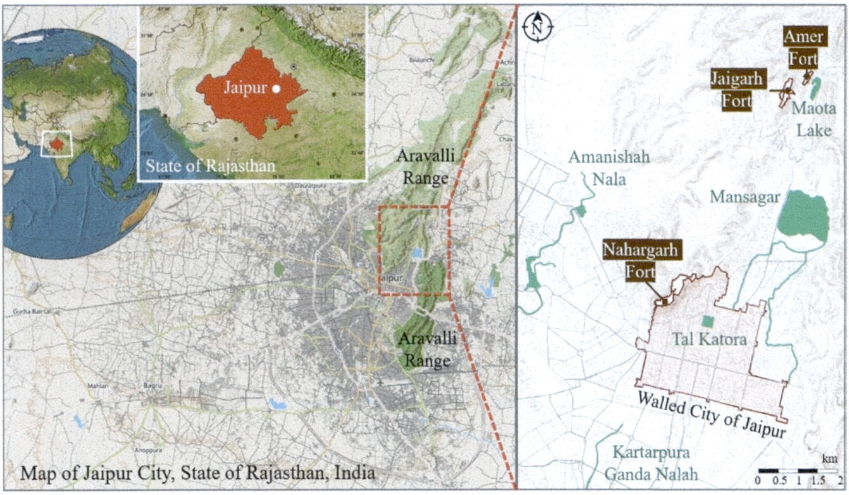

Fig. 34.1 Location map of case study area: Jaipur, India

34.2 Method

A comprehensive research methodology inspired by the case study method defined by COSMOS Corporation (Yin, 1983) was developed for a multidisciplinary research design. This methodology is structured into a three-step framework comprising (i) define & design, (ii) prepare, collect & analyze, and (iii) analyze & conclude, as shown in Fig. 34.2. The cases selected for this study were water management systems in three major forts in Jaipur. Case studies were conducted through a primary survey involving site visits from April 2023 to May 2023. To ensure the reliability of data collected from secondary sources, collaboration was conducted with information centers, insights from tourist guides, and consultations with officials from the State Tourism Authority. This study entails an architectural analysis that scrutinizes the structural layout, emphasizing features, such as step wells, underground channels, and storage reservoirs. Additionally, hydrological assessments explored the hydrology of the region and the catchment areas of the forts, including rainwater harvesting and groundwater recharge techniques. Through the application of this comprehensive research methodology, this study highlights the adaptation of these structures to modern water needs and their substantial contributions to the community's water security.

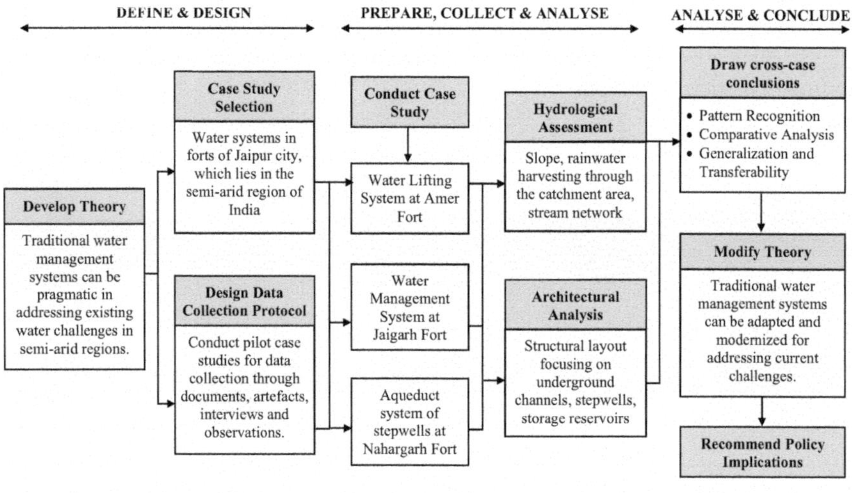

Fig. 34.2 Research methodology

34.3 Results and Discussion

Raja Man Singh I constructed the Amber Fort in 1592, close to the Amber settlement which was established in 967 (Sinha & Rajora, 2014). In the early eighteenth century, around 1726, Jaigarh Fort was constructed by Maharaja Sawai Jai Singh II to serve as a military stronghold and to safeguard Amber Fort and the city of Amber. However, as Amber City is surrounded by steep terrain, it could no longer expand and faced challenges in accommodating its growing population. Furthermore, the area around Amber faced water security issues, and there was a need for a location with a reliable and sustainable water supply (Alain Borie, 2020). As a result of these considerations, Maharaja Sawai Jai Singh II founded Jaipur in 1727, and it became the new capital of the Kachwaha kingdom. It was the first planned city in India and eventually earned the name 'Pink City' because of the predominant color of its buildings. After the foundation of Jaipur City in 1734, the construction of the Nahargarh Fort began as part of the defense system through fortified walls along the Aravalli Hills (World Heritage Centre, 2013). This section provides a detailed discussion of the case study analyses conducted for each of these forts, shedding light on the specific features, functions, and implications of their water management systems.

34.4 Amber Fort

Amber Fort has two primary sources of water: *Maota lake* and the rainwater collection system of the fort area (Shrestha, 2019). The rainwater falling on the roofs of the fort is collected in 'Mansingh Nahal ka Tanka' which has a capacity to hold approximately 300,000 L of fresh water. Monsoon runoff from the hill ranges surrounding Amber was collected in *Maota lake* through an extensive water catchment, collection, and distribution system. The community's water intelligence, or traditional knowledge of the water system, is remarkable in all aspects. Water was transported from the hills to vast macro-catchments located in the valleys through canals and aqueducts. Within the palace interior, water collected from the lakes was distributed to the tankas (underground tanks) using pulleys, draught animals, and manpower (Sinha & Rajora, 2014). The water system at Amber Fort can be classified into four stages, as discussed in detail below.

34.4.1 Stage 1: Initial Lifting from Primary Source

Water was drawn from the *Maota lake* along the eastern facade of the *Kesar Kyari* by draught animals through a series of pulleys and leathern bags. The water drawn through this process was gathered in two storage tanks constructed on the terraces overlooking the garden. Subsequently, a sectioned clay pipeline approximately 125

(a) Stage 1: Intial Lifting From Water Source

(b) Stage 2: Vertical Lifting

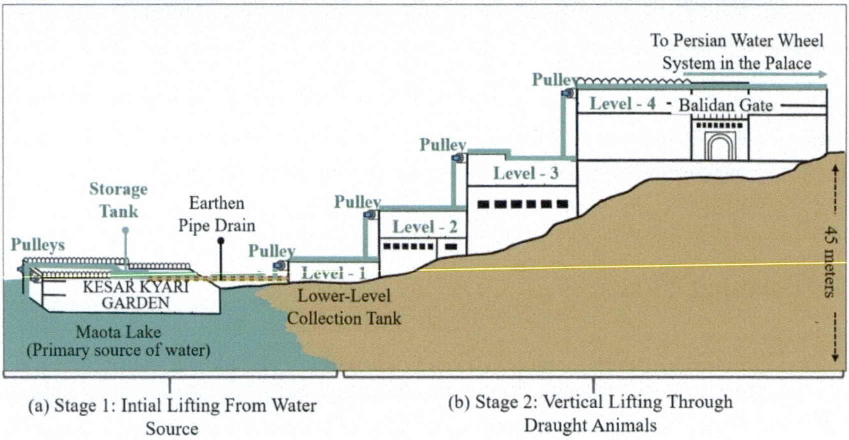

Fig. 34.3 Water lifting system at Amber Fort

m in length directed the water to another storage tank situated at the base of the second stage, as shown in Fig. 34.3a (Munshi et al., 2017). It acted as a settlement tank for the water to be aerated and desilted.

34.4.2 Stage 2: Vertical Lifting Through Draught Animals

The second stage comprises a four-story lifting system situated in the southeastern part of the palace. Due to the steep incline near the water source at Amber Fort, a four-tiered pulley system was implemented to improve the stability of the structure and enhance its structural integrity (Fig. 34.3b). Each of the four structures had its own pulley and rope arrangements, operated by draught animals. Each structure featured its own intake cum-storage tank at the base. Subsequently, water from the lowermost storage tank was transported to the higher tank using a leather bag suspended over

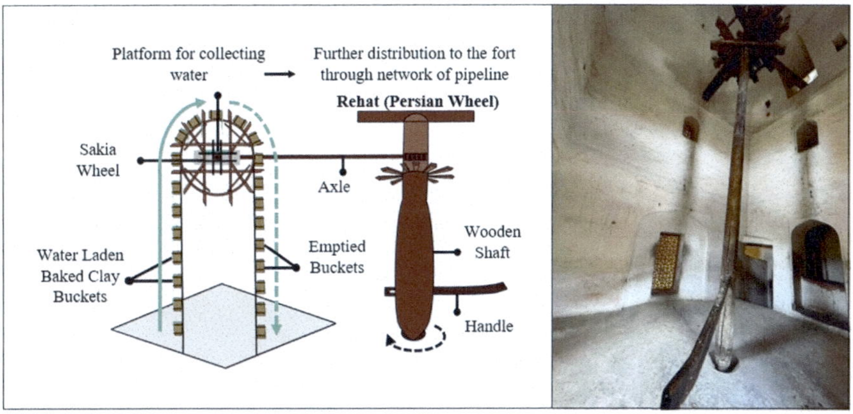

Fig. 34.4 Water lifting through rehat (Persian Wheel)

a pulley with a rope. Water was taken up from the first storage tank situated at the lowest level to the last one situated on the first floor of *Balidan Gate*. The structures are between 10 and 13 m in height, achieving a lift of approximately 45 m using this system. Each of these structures can store approximately 45,000 L of water (Suman, 2019).

34.4.3 Stage 3: Water Lifting Through Rehat *(Persian Wheel)*

The final phase of the lift involved the use of a Persian water wheel, *Rehat* or Sakia. The *Rehat* was manually operated and has been used in Persia since as far back as 500 BC. It consists of a long wooden shaft rotating along its axis, providing power to the drum axle. The axle had a rope attached to several water-laden baked-clay buckets (Fig. 34.4). The axle is attached to the Sakia wheel, and as it rotates, the rope and its attached buckets move elliptically down and up through the water in the tank. During this process, the buckets were filled with water and subsequently carried upward, pouring the water into a collection tank at the top of the mechanism, which has a capacity of 100,000 L. Water was circulated throughout the palace via a network of earthen pipelines.

34.4.4 Stage 4: Water Distribution System Through Pipelines

The collected water is subsequently distributed throughout the main palace through an intricate network of earthen pipelines. This fort demonstrates a sophisticated

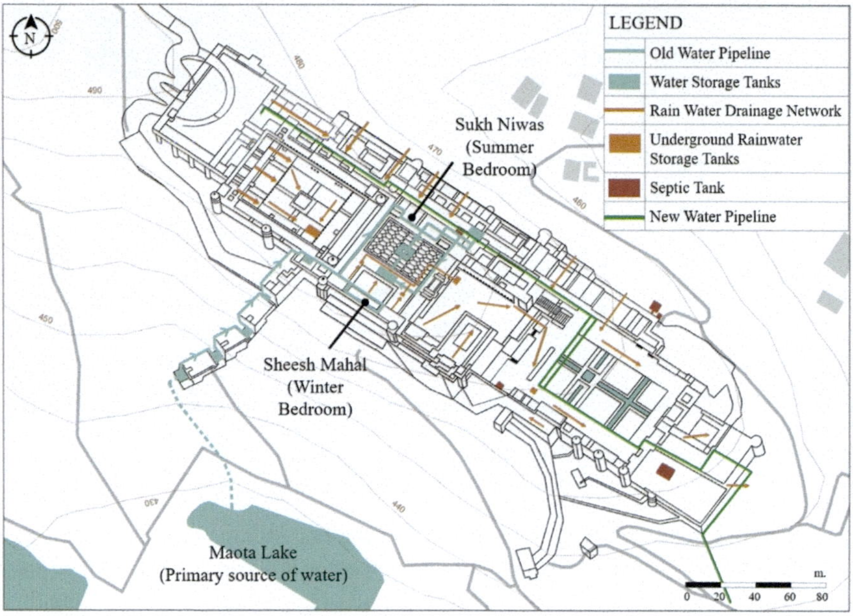

Fig. 34.5 Water distribution system at Amber Fort

rainwater collection system that ensures a sustainable water supply. The terraces of all buildings within the fort have a slight slope for water to flow into the drains, where it was collected. The collected water was efficiently stored in strategically located water tanks in various sections of the fort, as shown in Fig. 34.5. One of the smallest underground water tanks in the fort is the Jaleb Chowk Tank, which has a capacity of 50,000 L. Moreover, thoughtful provisions have been made for septic tanks within the fort to manage wastewater effectively. Notably, architectural foresight extends to the rulers' living quarters. Separate bedrooms were designated for the ruler's use during the winter and summer seasons and were strategically aligned based on the direction of the sun. This thoughtful arrangement aims to minimize the evaporation of water stored in the water tanks of the fort, showcasing a nuanced consideration for environmental sustainability.

34.5 Jaigarh Fort

Jaigarh Fort is a magnificent historical site on the hilltop of the Aravalli Hills at an altitude of 600 m above the MSL. This is a state-of-the-art rainwater collection technology. Ingenious water conservation structures collect and store rainwater in large tanks, demonstrating foresight for sustainable water management (Singh, 2019). The hydraulic system of the fort is an integrated water engineering and conservation system that can be divided into three parts based on the functions being performed by each part separately: (i) water channels outside the fort, (ii) water channels inside the fort, and (iii) water reservoir and storage tanks.

34.5.1 Water Channels Outside the Fort

Water channels outside the fort form a network of aqueducts encircling three hills with various contours. They were constructed in such a manner that the rainwater falling on the hills is not allowed to drain down, but is collected in these channels to be directed towards the fort owing to gravity flow, as shown in Fig. 34.6a. The course of the channels starts from a hillock, situated at a distance of 3.5 km from the southern ramparts of the Jaigarh fort, which is aligned in north-west and south-east direction. (Sharma & Rezavi, 1993) The total catchment area of the aqueducts is 0.13 km^2 and has the capacity to collect 42.57 million liters of water at an average, which might vary based on the annual rainfall received. The hillock is surrounded by water ducts, forming an elliptical circle, around a kilometer in length. Here, the water is directed through a channel towards the west to an aqueduct, which carries the water from one hillock to another, avoiding an otherwise sudden dip between the hills. The point from which the aqueduct commences is constructed as a rectangular platform that is meant to accommodate the person responsible for regulating the flow of water. At this junction, the aqueduct is 3 m wide and gradually tapers to 1.33 m only, which is the general width of the channels. After 330 m, the channel bifurcates into two ducts, forming a second loop around the hillocks. These two ducts rejoin each other 500 m apart and immediately separate to form an irregular third loop. At all points where the western and eastern channels join, a triangular shape is provided to allow easy flow of water without causing any erosion or damage to the embankment or the floors of the ducts. Along this route, the channels change direction at sharp angles to ensure the sedimentation of runoff debris. Immediately before entering the fort, the eastern duct zigzags into seven sharp bends. According to hydraulic engineering and the characteristics of the assimilated impurities, such angular directional hindrances are meant to eliminate floatable and carried unwanted materials.

34.5.2 Water Channels Inside the Fort

The rainwater carried through the channels entered the iron grill (lattice) rectangular opening of the impurity settling tunnel outside the fort, as shown in Fig. 34.6b. Runoff from the initial rainfall is invariably allowed to enter the sluice gate of the unfinished open reservoir, commonly known as *Nathawaton ka Tanka*, to ensure water clarification. Although it is an unfinished reservoir, its terraced embankments are covered with dressing stones (Singh, 2019). The reservoir is 106.68 m long, 60.96 m wide and 4.6 m deep, and was presumably meant to cater to the needs of

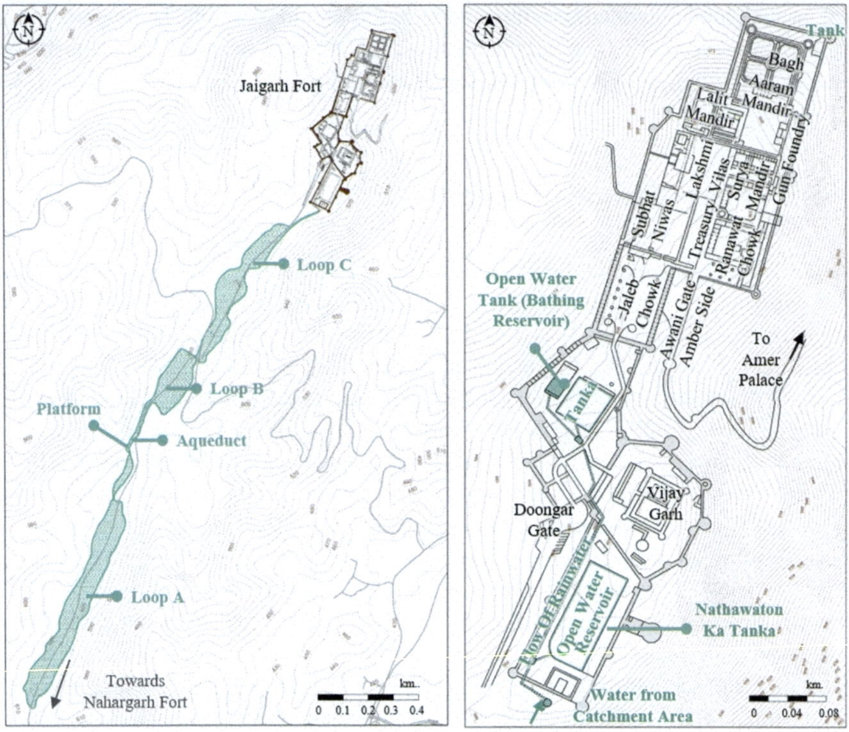

(a) Water Channel outside the fort (b) Water Channel inside the fort

Fig. 34.6 Water management system at Jaigarh Fort

the animals inside the fort. Just after the initial two to three washes, the collected rainwater was transferred to the central water storage network through the sluice gate and the 125 m-long covered subterranean channel up to the second gate (from *Doongar Darwaza*) of the fort. Water moves in the open channel to join the deep rectangular cistern, which acts as an interceptor for grit, silt, and dirt. The distribution cistern has two sluice openings and allows the clarified water to enter the largest water reservoir, popularly known as *Tanka*, which is completely covered. When *tanka* is either filled or in the event the water is diverted onwards through the second sluice, water passes through an open drain to the open reservoir. This rectangular open tank is popularly known as a bathing reservoir, with a volumetric capacity of more than 1.2 million liters.

34.5.3 Water Reservoir and Storage Tank

The largest tank, *tanka*, had a ceiling that rested on 81 arched tanks and had a capacity of 22.7 million liters. On top of the ceiling and in its middle, a square platform with an opening was constructed to facilitate the drawing of water (Fig. 34.7). The windows provided at regular intervals on the south and west sidewalls of the reservoir (above the ground) in the *tanka* acted as ventilators and provided some amount of sunlight to enter the tanks. Towards the north of the reservoir, a flight of the staircase descends from both the east and west (a walkway), which enables water to be drawn (Misra, 2001). The ingenuity of constructing underground and covered water tanks had its own advantages, such as minimum water loss through evaporation, the fact that stored water could be kept unpolluted for a long time, and the fact that water thus preserved remained cool owing to the thermostat effect of the air trapped inside the tank covered with a roof.

This ingenious technique of captivating rainwater on the hills and then directing it to the fort enabled Jaigarh Fort to withstand a long siege. This technique also solved the problem of carrying water from lakes situated far below the hillocks.

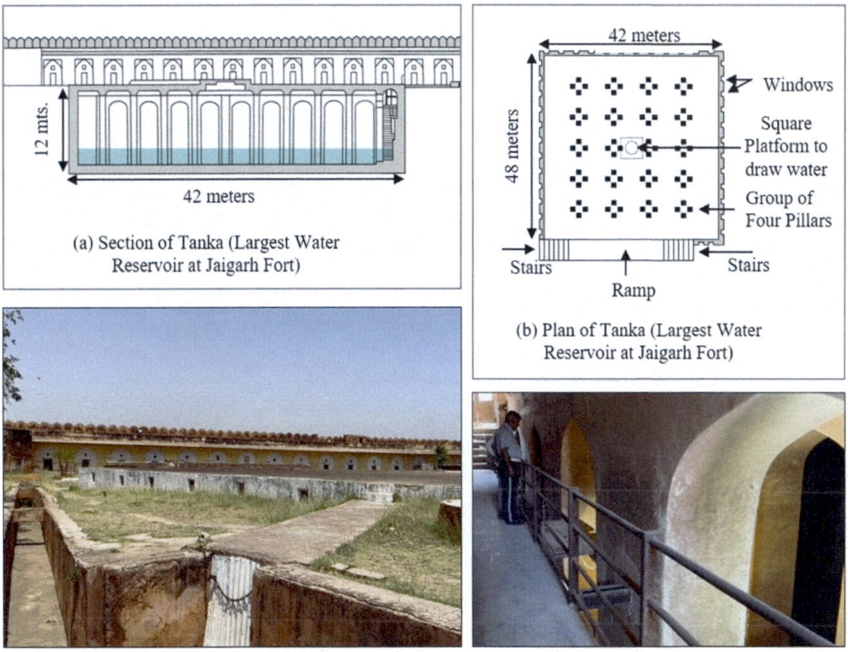

(a) Section of Tanka (Largest Water Reservoir at Jaigarh Fort)

(b) Plan of Tanka (Largest Water Reservoir at Jaigarh Fort)

(c) Photograph showing the exterior of covered Tanka and channels diverging into it

(d) Photograph showing the interior part of Tanka

Fig. 34.7 Largest water reservoir (*Tanka*) at Jaigarh Fort

34.6 Nahargarh Fort

Nahargarh Fort was built in 1743 by Maharaja Sawai Jai Singh II, shortly after the founding of the walled city, as a defense fort on Cheel ka Teela (Hill of Eagles), overlooking the city of Jaipur. It functioned primarily as a residential royal fort with a labyrinth of a courtyard and chambers for nine queens. The water management system at the Nahargarh Fort can be divided into three major components: (i) the catchment network outside the fort, (ii) the outer stepwell, and (iii) the inner stepwell, each of which is detailed below.

34.6.1 Catchment Network Outside the Fort

Similar to Jaigarh Fort, the catchment area of Nahargarh Fort comprises a network of five closed catchments connected through small canals and aqueducts laid outside and within the fort, as shown in Fig. 34.8. Rainwater from higher altitudes on hills is carried through masonry canals to stepwells at lower altitudes with the assistance of gravitational flows (Gangane & Gupta, 2022). The catchment beds and canals were designed considering the gradient and undulating surfaces. The total catchment area is 0.097 km^2 and has the capacity to capture 31.68 million liters of water on average, which varies according to the yearly rainfall. Runoff from these beds was collected and directed through a long drain that entered the fort and bifurcated at a distribution point. The canal is divided into two branches at the drain divergence point: one proceeds to the outer stepwell, and the other proceeds to the stepwell inside the fort. This intersection features a ground-level slit and a cistern, where a stone slab could be inserted to obstruct and switch the flow of water between the outer and inner step wells.

34.6.2 Outer Stepwell

The flow of water within the canals is regulated by the gradient of the hill before it enters the outer stepwell. It is an aesthetically structured step well with a nearly triangular shape that shrinks downwards, as illustrated in Fig. 34.9a. A unique spiral structure was installed near a stepwell for water purification. An elaborate swirling mechanism was designed to ensure that the flow of water could be controlled and that excess water flow would not cause any damage to the structure. It also acts as a filtration mechanism to separate particles such as leaves and stones, which may enter the stepwell along with rainwater. Water from the stepwell was primarily used for cleaning purposes. The unusual asymmetrical step also characterized a separate pathway that was wide enough for buffaloes and goats, and the doorway was high

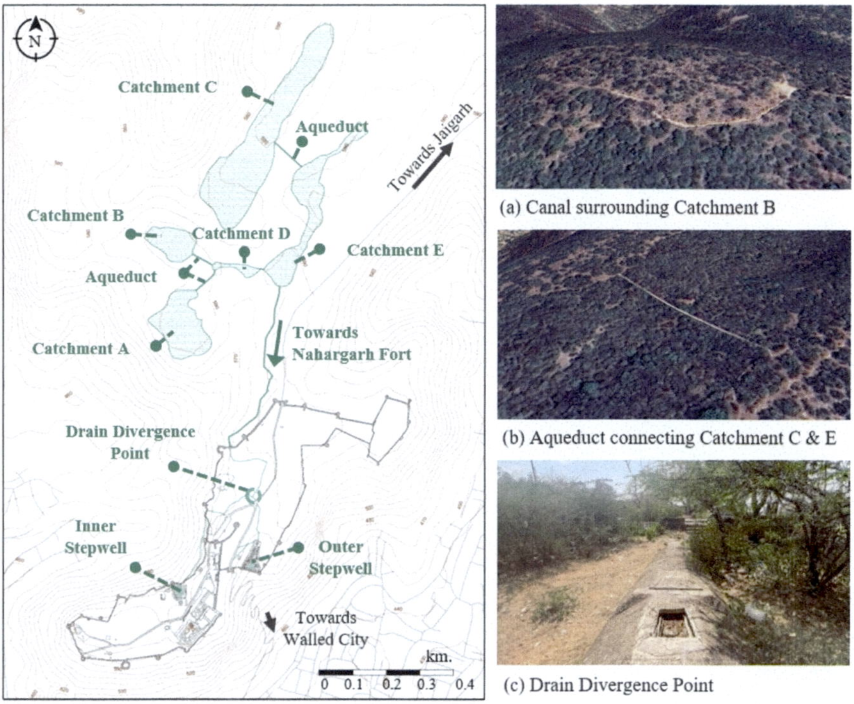

(a) Canal surrounding Catchment B

(b) Aqueduct connecting Catchment C & E

(c) Drain Divergence Point

Fig. 34.8 Catchment network outside Nahargarh Fort

enough for a man on a horse (Livingston & Beach, 2002). Furthermore, outlet holes were designed in the walls of the step well to drain excess water.

34.6.3 Inner Stepwell

This medieval step well (Fig. 34.9b) was built to collect and conserve rainwater flowing through bifurcated underground channels laid out on the hills of Nahargarh. The water passes through small filtration chambers where all the impurities are removed, and clean water flows towards the step well inside the fort along the gradient of the hill (Mirajkar & Agarwal, 2020). Rainwater harvested from within the fort was also collected and stored in the stepwell through water inlet points designed on the wall of the stepwell. The square rainwater harvesting tank known as '*tanka*,' located next to the stepwell is utilized for collected rainwater coming down from the terrace of Shri Mandhvendra Bhawan (within the fort) and the surrounding hills. Water from the inner stepwells and *tanka* was used by the residents of the royal family.

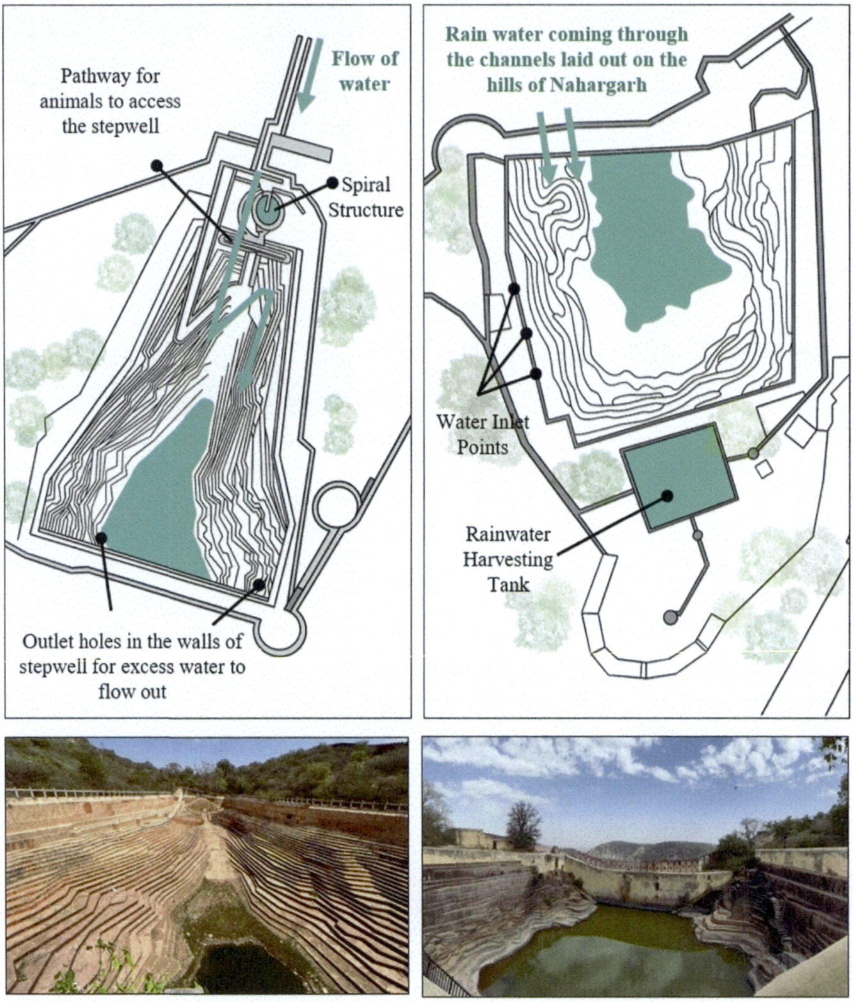

Fig. 34.9 Stepwells at Nahargarh Fort

34.7 Cross-Case Inferences

34.7.1 Pattern Recognition

After completing the structural and hydrological assessments of the case studies outlined in the research methodology, striking similarities become apparent. Across all three forts, water management systems share a common thread, characterized by intricate networks of aqueducts, stepwells, tanks, and reservoirs seamlessly integrated into the fort's architectural layout. Notably, these features were strategically positioned to capitalize on the natural topography, facilitating the capture and storage of rainwater. The systems exhibit a shared design principle by leveraging the gravitational forces inherent in the landscape. Elaborate channels and aqueducts play a pivotal role in directing rainwater runoff to strategically positioned storage facilities. This collective approach ensured a reliable and self-sufficient water supply throughout the year.

34.7.2 Comparative Analysis

Although the water management systems of these forts exhibit notable similarities in design principles, each fort possesses distinct and unique characteristics. For instance, Jaigarh Fort distinguished itself from an underground reservoir by adding a distinctive feature to its water infrastructure. In contrast, Amber Fort showcases a notable water-lifting system, setting it apart in terms of engineering ingenuity. Additionally, the Nahargarh Fort featured unique stepwell attributes, further emphasizing the individuality of its water management approach. The stages of the water system for each forest are compared in Table 34.1.

Table 34.1 Comparative analysis of water management systems in forts of Jaipur

Parameter	Amber Fort	Jaigarh Fort	Nahargarh Fort
Source of water	Surface water and rainwater collection system	Rainwater from catchment area	Rainwater from catchment area
Water collection system	Water lifting system through pulleys and Persian wheel	Network of aqueducts encircling three hills	Network of five closed catchments connected through canals and aqueducts
Water collection capacity	Water is collected in *Maota lake**	42.57 million liters	31.68 million liters
Water Storage System	Storage tanks integrated with collection system	Underground Water Reservoir	Stepwells
Water storage capacity	630,000 L	23.9 million liters	Undetermined**
Water filtration mechanism	Settlement tank	• Irregular shape of aqueduct • Iron grill (lattice) rectangular opening • Lime mortar	• Spiral structure at the stepwell • Iron grill to filter out particles such as leaves, stones etc

* The capacity of *Moata lake* could not be determined

** Accurate volume of the stepwells could not be calculated due to their irregular shape

34.7.3 *Generalization and Transferability*

The underlying concept of water management planning for these forts is straightforward. The planners devised a system that was sufficiently robust to capture every raindrop within the vicinity of the fort, coupled with an effective means to store this water for year-round usage without contamination. This uncomplicated yet effective approach provides valuable insights that can be readily applied to enhance water management systems. The transferability of this idea is evident when similar case studies worldwide are considered. For instance, the design principle underlying Jaigarh Fort's underground water reservoir resonates with other parts of the world, albeit on a larger scale. Locations such as the underground water reservoir in Budapest (Comrade, 2022), Bracebridge Heath in the UK (Anglian Waters, 2020), and the underground water reservoir in Japan (Japan Experience, 2014) mirror the simplicity and efficacy of the concept witnessed at Jaigarh Fort. These international examples underscore the universal applicability and adaptability of the water management principles implemented within these forts.

34.8 Conclusions

Geographical disparities in water availability across the globe, especially in arid and semi-arid regions, highlight the importance of developing practical approaches to meet the water needs of local populations. These areas are vulnerable to water stress and often struggle to manage scarce water resources. The challenge lies in developing solutions that not only address the immediate water needs of communities in semi-arid regions, but also cater to long-term resilience and sustainability, considering the changing climatic conditions and growing populations.

Strategically positioned in India's semiarid region, the forts of Jaipur provide an exemplary illustration of the effective management of scarce water resources. This study explores the timeless knowledge of the past and how it can be applied to contemporary water issues by examining the city's traditional water management systems. Each fort in Jaipur had been equipped with a special water management system suited to its own needs through efficient rainwater harvesting and groundwater recharge techniques. This demonstrates the flexibility of traditional approaches in addressing the problem of water scarcity.

Comparable water management strategies have been adopted worldwide, as seen in the forts of Jaipur, highlighting the replicability and transferability of these approaches. This demonstrates their effectiveness in resolving water management issues in various climatic and geographic settings. Although traditional methods of managing water have proven effective, there is still a gap between conventional wisdom and sustainable methods of water management that must be acknowledged. The need to engage the community and pass on traditional knowledge to younger generations has become increasingly evident over time to close this gap and incorporate traditional knowledge into sustainable water management strategies.

The significance of this study goes beyond historical curiosity because it offers insightful information for managing water resources in arid and semi-arid areas, where water scarcity is worsened by climate change. These conclusions are useful for researchers, decision makers, and urban planners seeking sustainable water management plans, especially in arid and semi-arid regions. The knowledge gained from Jaipur's forts can be a useful tool for creating proactive situation-specific strategies to address water-related issues in these areas. Water management practices worldwide can become more resilient and sustainable if these insights are incorporated into planning and policymaking.

Acknowledgements We would like to express our sincere gratitude to JSPS KAKENHI for their substantial grant (Grant Number 21KK0014), which played a vital role in successfully conducting the field visit and was crucial in accomplishing this research.

References

Alain Borie, F. C. R. P. (2020). *Jaipur: A planned city of Rajasthan* (First). Altrim Publishers.

Anglian Waters. (2020). Anglian water completed inspection of 100 year old Underground Reservoir. *Water Industry Journal.*

Comrade K. (2022, May 3). *Exploring Budapest's Dystopian Underground Water Reservoir.* https://comradekiev.com/blog/budapest-dystopian-underground-water-reservoir/

Cosgrove, W. J., & Loucks, D. P. (2015). Water management: Current and future challenges and research directions. In *Water Resources Research* (Vol. 51, Issue 6, pp. 4823–4839). Blackwell Publishing Ltd. https://doi.org/10.1002/2014WR016869

Gangane A.S., & Gupta A.R. (2022). *Review Study of Step Wells. 6*(2), 90–92. http://tiikmpublishing.com/data/conferences/doi/ace/26731029.2020.1107.pdf

Hooja, R. (2022). The city of Jaipur and its traditional water systems. *Jaipur Unchartered.* https://jaipuruncharted.in/the-city-of-jaipur-and-its-traditional-water-systems/

Hosseiny, S. H., Bozorg-Haddad, O., & Bocchiola, D. (2021). Water, culture, civilization, and history. *Economical, Political, and Social Issues in Water Resources*, 189–216. https://doi.org/10.1016/B978-0-323-90567-1.00010-3

India Meteorological Department. (2021). *Monsoon Report 2021 Rajasthan.*

Japan experience. (2014, November 10). *Kasukabe Underground Flood Protection Tank.* https://www.japan-experience.com/all-about-japan/saitama/attractions-excursions/kasukabe-underground-flood-protection-tank

Kumar Singh, P., Dey, P., Kumar Jain, S., & Mujumdar, P. P. (2020). Hydrology and water resources management in ancient India. *Hydrology and Earth System Sciences, 24*(10), 4691–4707. https://doi.org/10.5194/hess-24-4691-2020

Livingston M., & Beach M. (2002). *Steps to Water: The Ancient Stepwells of India.* Princeton Architectural Press

Mirajkar, D., & Agarwal, N. (2020). *Unfolding Aqueduct System of Stepwell at Nahargarh, Jaipur*, pp. 104–112. https://doi.org/10.17501/26731029.2020.1107

Misra, S. K. (2001). *Jaigarh water management the world's unparallel scientific and engineering marvel.* The Jaigarh Public Charitable Trust Amber.

Munshi, S., Bapat, S., Vaishnav, S., Ghosh, S., & Pratap Singh Sekhawat, Y. (2017). *Water Lifting System at Amber Palace: Learning from Traditional Practices.*

Sharma R., & Rezavi S.A.N. (1993). Aspects of Hydraulic Engineering in Medieval Rajasthan: A case study of water system in Jaigarh Fort. *Art and Culture.*
Shrestha H. R. (2019). Ancient practice of rainwater harvesting in the Forts of Jaipur, India. *National Conference on Rainwater Harvesting on Livelihood, Community and Ecosystem Resilience.*
Singh, V. K. (2019). Waterworks at Jaigarh Fort: An archaeological study of hydraulic technology. *Indian Journal of Archaeology, 4*(1). www.ijarch.org
Sinha, A., & Rajora, N. (2014). Gaze and the picturesque landscape of Amber, India. *Studies in the History of Gardens and Designed Landscapes, 34*(4), 309–322. https://doi.org/10.1080/14601176.2013.874305
Suman. (2019). *Water Storage System at Amber Fort by Heritage Water Walks.* https://nomadicshoes.com/things-to-do/water-storage-system-at-amber-fort-by-heritage-water-walks/
Tyagi, A., Singh, O. P., Singh, S. S., & Kumar, S. (2011). *Climate of Jaipur.*
Vuorinen, H. S., Juuti, P. S., & Katko, T. S. (2007). History of water and health from ancient civilizations to modern times. In *Water Science and Technology: Water Supply* (Vol. 7, Issue 1, pp. 49–57). https://doi.org/10.2166/ws.2007.006
World Heritage Centre. (2013). *Hill Forts of Rajasthan (India).*
Yin, R. K. (1983, September). *The Case Study Method*. COSMOS Corporation.

The manufacturer's authorised representative in the EU is Springer Nature Customer Service Centre GmbH, Europaplatz 3, 69115 Heidelberg, Germany. If you have any concerns regarding our products, please contact ProductSafety@springernature.com

Printed and bound by CPI Group (UK) Ltd, Croydon, CR0 4YY
01/10/2026
02225958-0017